Linear and Nonlinear Crystal Optics

This book provides both theoretical foundations and practical insights into linear and nonlinear crystal optics. It bridges the fundamental principles of light-crystal interactions with advanced applications in optics.

Linear and Nonlinear Crystal Optics offers a thorough exploration of electric susceptibility—the cornerstone of optical properties through electric constitutive relations and Maxwell's equations. Key features include comprehensive tensor representations, detailed angular distributions of optical parameters, complete mathematical demonstrations, and original research insights developed over decades. The book covers essential topics including phenomenological introduction to light-matter coupling, mathematical background in orientation and thermodynamic symmetries, electric susceptibility and field tensors, polarization calculations, linear optical effects linked to the real and imaginary parts of the refractive index, and both second-order and third-order interactions. Each chapter concludes with a summary and references, providing readers with clear learning pathways and resources for further exploration.

This book is ideal for advanced undergraduate students, graduate students, and researchers in physics, optical engineering, materials science, and crystallography. It serves as an essential reference for professionals working in fields requiring deep understanding of crystal optics, including photonics, laser technology, telecommunications, and materials characterization.

Optical Sciences and Applications of Light

Founding Editor: Brian J. Thompson, University of Rochester Rochester, New York

Biochemical Applications of Nonlinear Optical Spectroscopy
Edited by Vladislav Yakovlev

Optical Methods of Measurement
Wholefield Techniques
Second Edition
Rajpal Sirohi

Fundamentals and Basic Optical Instruments
Daniel Malacara Hernández

Advanced Optical Instruments and Techniques
Daniel Malacara Hernández

Entropy and Information Optics
Connecting Information and Time
Second Edition
Francis T.S. Yu

Handbook of Optical Engineering
Second Edition, Two Volume Set
Daniel Malacara Hernández

Optical Materials and Applications
Moriaki Wakaki

Photonic Signal Processing
Techniques and Applications
Le Nguyen Binh

Practical Applications of Microresonators in Optics and Photonics
Edited by Andrey B. Matsko

Near-Earth Laser Communications
Second Edition
Edited by Hamid Hemmati

Nonimaging Optics: Solar and Illumination System Methods, Design, and Performance
Roland Winston, Lun Jiang, Vladimir Oliker

Introduction to Optical Metrology
Rajpal Sirohi

Linear and Nonlinear Crystal Optics
Phenomenology, Symmetries and Angular Distributions
Benoît Boulanger, Simon Joly, Patricia Segonds, and Yannick Petit

For more information about this series, please visit: https://www.routledge.com/Optical-Sciences-and-Applications-of-Light/book-series/CRCOPTSCIAPP

Linear and Nonlinear Crystal Optics

Phenomenology, Symmetries, and Angular Distributions

Benoît Boulanger,
Simon Joly, Patricia Segonds, and Yannick Petit

CRC Press
Taylor & Francis Group
Boca Raton London New York

CRC Press is an imprint of the
Taylor & Francis Group, an **informa** business

Designed cover image: Simon Joly

First edition published 2026
by CRC Press
2385 NW Executive Center Drive, Suite 320, Boca Raton FL 33431

and by CRC Press
4 Park Square, Milton Park, Abingdon, Oxon, OX14 4RN

CRC Press is an imprint of Taylor & Francis Group, LLC

ISBN: 978-1-498-77456-7 (hbk)
ISBN: 978-1-041-16496-8 (pbk)
ISBN: 978-0-429-09173-5 (ebk)

DOI: 10.1201/9780429091735

Typeset in Times
by Apex CoVantage, LLC

Contents

Foreword

The concept of symmetry plays a key role in the field of nonlinear optics. The symmetry properties on a material system determine what sort of nonlinear optical properties can occur in that material. The best-known example of such behavior is probably the result that second-order nonlinear optical processes, such as second-harmonic generation, cannot occur in a material that possesses inversion symmetry. The proof of this result is straightforward and is presented in many textbooks on nonlinear optics. However, the connection between symmetry and the allowed form of the nonlinear optical response holds more generally. This idea can be quantified by determining the allowed form of the nonlinear optical susceptibility tensor for each of the crystal classes, that is, the allowed forms of a crystal structure as determined by the requirement that atoms are arranged in accordance with the periodicity of the crystal structure. As shown in this book, there are precisely 32 different ways in which a crystal can be so arranged. Each of these crystal classes permits a particular form of the susceptibility tensor for each order of nonlinearity, that is, for $\chi^{(N)}$ where N is the order of the response. The allowed forms of $\chi^{(N)}$ for N = 1, 2, and 3 are well known and are presented in standard textbooks. However, the mathematical derivation of these relations is rarely provided. A significant learning opportunity is thereby lost. It is often not enough to know that a certain physical result is correct; one also wants to know *why* it is correct.

The monograph of Boulanger and coauthors, Joly, Segonds and Petit, fills this need. It provides a satisfying mathematical development of how the symmetry properties of a material system determine the allowed form of the nonlinear susceptibility tensor for each order N of nonlinearity. The mathematical discipline known as group theory provides this relationship. Following this line of thinking, but considering all possible polarization configurations of the optical field for the three optical classes (*i.e.* isotropic, uniaxial and biaxial), this book describes the symmetry of the field tensors $F^{(2)}$ and $F^{(3)}$, which are formed from the tensor product of unit electric fields of interacting waves, the contraction product of $\chi^{(N)}$ by $F^{(N)}$ partly governing the magnitude of nonlinear energy conversion efficiency. Another issue covered here is the general tensor approach to quasi-phase-matching, which leads to the determination of all the coefficients of the $\chi^{(2)}$ and $\chi^{(3)}$ tensors, the signs of which can be reversed thanks to the various possible scenarios of rotation of the crystal structure and knowing the links between the microscopic and macroscopic scales. Angular distributions are also presented and analyzed in terms of symmetry for the real and imaginary parts of the refractive index, as well as for phase-matching and quasi-phase-matching, categorized by optical class. This monograph makes use of group theory in the theoretical development, but it limits the use of higher mathematics to only what is necessary.

I wish that I had a book of this sort available to me when I was learning nonlinear optics.

Robert W. Boyd
Professor of Optics and Physics at the University of Rochester and Professor of Physics and Electrical and Computer Engineering at the University of Ottawa

Preface

Since ancient times, light has revealed the beauty of crystals, and these crystals have awakened mankind to the perception of symmetry. Light, crystal, symmetry—this triptych constitutes the thematic core of this book. It is dedicated to crystal optics, both in the linear and nonlinear regimes, focusing on coherent light, three-dimensional crystals, and symmetries of spatial or thermodynamic origin.

Our idea is to connect between mainly three "bibles" in physics: Born and Wolf for electromagnetism and linear optics [1980. Principle of Optics, Pergamon Press Ed., sixth edition, New York]; Boyd for nonlinear optics [2008. Nonlinear Optics, Academic Press, third edition], and Nye for symmetry [1985. Physical properties of crystals. Oxford University Press].

The gravity center of this book is on electric susceptibility, which lies at the heart of all optical properties through the electric constitutive relation and thus Maxwell's equations. We strive to cover areas complementary to those addressed in the three aforementioned books, with a particular emphasis on tensor representations and the angular distributions of several relevant optical parameters. These include the real and imaginary parts of the refractive index, phase-matching and quasi-phase-matching loci, and the effective coefficient, as well as their symmetry analysis. We provide full demonstrations, and many aspects discussed are based on our own research over these last decades, providing a deep level of completeness of the proposed description.

In addition to the three reference books mentioned earlier, our work complements that of other authors who have made major contributions to the field and have been influential to us: these include Bloembergen [1965. Nonlinear optics. New York: Benjamin], Butcher and Cotter [1990. The elements of nonlinear optics. Cambridge series in modern optics. Cambridge University Press], Yariv and Yeh [2002. Optical waves in crystals. New York: Wiley], Rottwitt and Tidemand-Lichtenberg [2015. Nonlinear Optics, CRC Press], Zyss [1994. Molecular nonlinear optics: materials, physics and devices. Quantum electronic principles and applications series. New York: Academic Press], Byer [1973. Treatise in quantum electronics, edited by H. Rabin & C. L. Tang. New York: Academic Press], Ebrahimzadeh and Dunn [2000. Optical parametric oscillators, in Handbook of optics, Vol. IV, pp. 2201–2272. New York: McGraw-Hill], Grynberg, Aspect and Fabre [2010. Introduction to Quantum Optics - From the Semi-classical Approach to Quantized Light, Cambridge University Press], and Checkhova and Banzer [2021. Polarization of Light in Classical, Quantum, and Nonlinear Optics, Series De Gruyter Textbook].

This book has been designed as a course for Masters and PhD students, but it is also intended for physicists who want to immerse themselves in the fundamentals of linear and nonlinear crystal optics, and for chemists who are interested in optics and want to identify the relevant parameters for designing or characterizing new crystals. We hope you enjoy reading it!

About the Authors

Benoît BOULANGER received his Ph.D. degree in material sciences from the University of Nancy in 1989. He was researcher at the National Center for Scientific Research (CNRS) from 1989 to 2000 at the University of Nancy, Stanford University, and the University of Bourgogne. In 2000, he joined the University of Grenoble-Alpes as professor. He does his research at Institut Néel CNRS where he is a group leader. He held a position of distinguished professor at Tianjin University of Technology from 2018 to 2021. His work is at the frontiers between nonlinear crystal optics, quantum optics, and material engineering with achievements in frequency conversion and generation of new quantum states of light. He has authored over 350 papers in international peer-reviewed journals and conference proceedings. He teaches laser physics, nonlinear optics, crystallography, and tensors, and he also regularly gives tutorials on crystal parametric nonlinear optics in international conferences and summer schools. He served as associate editor for *Optics Letters* from 2014 to 2020. He was general co-chair of the international conferences Non Linear Optics (Hawaii) in 2013 and Advanced Solid State Lasers (Boston) in 2018, and will be general co-chair of the French-Israel Symposium on Non Linear and Quantum Optics from 2026.

Simon JOLY is an associate professor at the University of Bordeaux in the IMS laboratory (Talence, France) since 2013 and is on delegation at the CEA of Pessac. He received his Ph.D. in physics from the University of Grenoble in 2009. From 2010 to 2013, he undertook three postdoctoral positions at the Institute of Molecular Science in Okazaki, Japan, at the CROMA laboratory in Chambéry, and then at CEA LETI in Grenoble. He has been involved in research activities in nonlinear optics, micro-lasers, terahertz spectroscopy, and infrared imaging. His current research focuses on two main areas: the development of photonic components for sensing applications and reliability concerns within the ONDES Research Group at the IMS Laboratory, and powerful and tunable sources for multispectral terahertz imaging at CEA Tech in Pessac.

Patricia SEGONDS received her Ph.D. in physics from the University of Grenoble in 1989. During her postdoctoral fellowship at Ohio State University, she was appointed associate professor at the University of Bordeaux. She conducted research on ultrafast all-optical switching in glasses at the Centre de Physique Moléculaire et d'Optique Hertzienne (CPMOH) until joining the University of Grenoble-Alpes in 1998. She is currently a professor at the NEEL Institute, where her research deals with optics and frequency conversion in nonlinear crystals and their characterization and implementation in optical parametric oscillators. She has authored over 230 papers in peer-reviewed journals and conference proceedings. She teaches practical courses in linear and nonlinear optics and has recorded tutorials in various areas of physics. Since 2013, she has served the European Optical Society (EOS), 2022–2024 as president, 2024–2026 as responsible for the JEOS-RP journal. She has organized several EOS and OPTICA conferences.

Yannick PETIT received his Ph.D. degree in physics of condensed matter and radiations from the University of Grenoble in 2007. He was a postdoctoral researcher in terawatt femtosecond laser applications at the University of Geneva, Switzerland, from 2007 to 2010. In 2010, he joined the University of Bordeaux as associate professor. He does his research as a physicist at the Institute of Chemistry and Condensed Matter of Bordeaux (ICMCB). His work deals with femtosecond laser material structuring and 3D-printing for the development of photonic functionalities and integrated photonic circuits. He has authored over 100 papers in refereed journals and conference proceedings. He teaches wave optics, nonlinear optics, and optics of solids at the University of Bordeaux.

1 Phenomenological Introduction

1.1 CLASSICAL DESCRIPTION OF THE COUPLING BETWEEN LIGHT AND MATTER

The classical modeling used here provides an initial introduction to linear and nonlinear optics while establishing a link with the medium in which the fields propagate.

1.1.1 The Involved Forces

The Lorentz model gives a good qualitative description of light-matter coupling. It assumes an electric field of light $\vec{E}(t)$ oscillating as a function of time at one or more circular frequencies and interacting with an ideal gas of electrons where the kinetic theory applies. Then the motion of each electron from its equilibrium position can be described classically (Feynman, Leighton, Sands 2005) [1]. To start, we consider here a one-dimensional (1D) interaction, which means that the electric field of light is oriented in a single direction, and that the electron of matter can move only along that direction, labelled the x-axis, as shown in Figure 1.1. Along this direction, all the vectors $\vec{U}(t)$ involved hereafter are defined regarding a unity vector $\vec{u}$, and then write:

$$\vec{U}(t) = U(t)\ \vec{u} \tag{1.1}$$

where $U(t)$ is a scalar quantity corresponding to the magnitude of vector $\vec{U}(t)$ projected along the x-axis at the time t.

Three forces apply to an electron under the electric field vector, $\vec{E}(t)$: the electric force, the restoring force, and the damping force, as shown in Figure 1.1. The electric force is always oriented in an opposite direction compared to the two others as explained hereafter.

1.1.1.1 The Electric Force

The electric force $\vec{F}_c$, also called the Coulomb force, involves the charge q of the electron and the electric field $\vec{E}(t)$ along the x-axis. It writes:

$$\vec{F}_c(t) = F_c(t)\ \vec{u} \tag{1.2}$$

with the magnitude given by:

$$F_c(t) = qE(t) \tag{1.3}$$

The unit of the magnitude $E(t)$ is the Volt per meter [V/m], and that of $F_c(t)$ is the Newton [N]. The charge q is a constant which unit is the Coulomb [C]. Its value

DOI: 10.1201/9780429091735-1

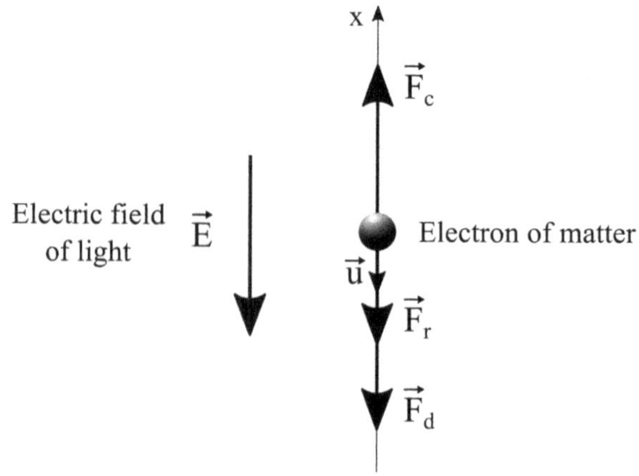

FIGURE 1.1 Forces involved in the Lorentz model when an electron interacts with an electric field $\vec{E}(t)$ oriented along the x-axis. $\vec{F}_c(t)$ is the Coulomb force, $\vec{F}_r(t)$ the restoring force, and $\vec{F}_d(t)$ the damping force. They are all parallel to the unit vector $\vec{u}$ along the motion direction of the electron.

is $q = -1.6 \times 10^{-19}$ C. Since $q < 0$, $\vec{F}_c(t)$ is oriented with the opposite direction of $\vec{E}(t)$ as shown in Figure 1.1.

The consequence of the electric force is that a microscopic electric dipolar moment $\vec{\mu}(t)$ is created at the time t along the x-axis, which writes:

$$\vec{\mu}(t) = \mu(t)\ \vec{u} \tag{1.4}$$

where

$$\mu(t) = q\ x(t) \tag{1.5}$$

The quantity $x(t)$ stands for the motion of the electron from its equilibrium position $x = 0$ at the time t.

By considering locally N independent electrons per volume unit moving along the x-axis, all the corresponding microscopic electric dipolar moments $\vec{\mu}(t)$ are identical. Then they can be simply added at the macroscopic scale, giving rise to the macroscopic induced polarization, that is expressed per volume unit as (Born, Wolf 1999) [2]:

$$\vec{P}(t) = N\ \vec{\mu}\ (t) = N\ q\ x(t)\ \vec{u} = P(t)\ \vec{u} \tag{1.6}$$

with

$$P(t) = N\ q\ x(t) \tag{1.7}$$

The goal of the present modeling is to be able to calculate the macroscopic induced polarization $\vec{P}(t)$, which thus requires to know the instantaneous displacement $x(t)$ of each electron. It depends not only on the Coulomb force but also on the restoring and damping forces, as will be shown hereafter.

1.1.1.2 The Restoring Force

In matter, due to the interactions between the electrons and the nuclei on the one hand, and between the electrons themselves on the other hand, these electrons are placed in a potential well, of energy V. It leads to a force that will oppose the Coulomb force, as shown in Figure 1.1. It is called the restoring force, $\vec{F}_r(t)$, expressed as:

$$\vec{F}_r(t) = F_r(t)\ \vec{u} \tag{1.8}$$

with

$$F_r(t) = \frac{\partial V}{\partial x} \tag{1.9}$$

At this level, it is interesting to consider two general situations according to the symmetry of the chemical bond in which the electrons are involved. For simplicity we consider here diatomic bonds in which the electrons can only move along the bond axis.

The first situation corresponds to the case of a homopolar bond, which means that the two atoms are the same, as depicted in Figure 1.2.

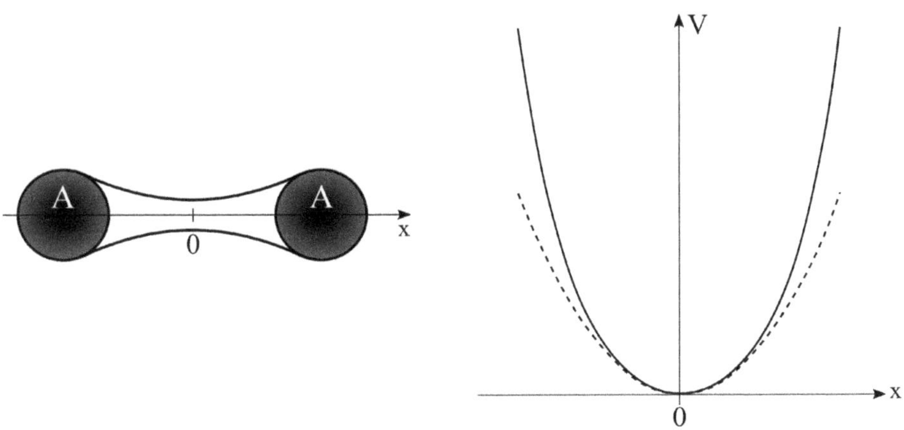

FIGURE 1.2 (Left) Homopolar bond linking two identical atoms A, the x-axis being the bond axis. (Right) Corresponding centrosymmetric potential energy V as a function of the electron motion x; the continuous line stands for its expansion as a Taylor power series with only even exponents of x; the dashed line is when exponent 2, namely the quadratic contribution with respect to position x, is the only one to be considered.

In that case, the potential energy V is centrosymmetric, so that it is a centrosymmetric function of the electron motion $x(t)$, *i.e.* $V(-x,t) = V(x,t)$ as shown in Figure 1.2. It means that the electrons "see" the same environment from both sides of their equilibrium position 0. A good description of V is then given by its expansion as a Taylor power series of $x(t)$, with only even exponents of $x(t)$ as following (Bloembergen 1996) [3]:

$$V(t) = \frac{1}{2} m\omega_0^2 x^2(t) + \frac{1}{4} mbx^4(t) + \frac{1}{6} mdx^6(t) + ... \tag{1.10}$$

The term $\frac{1}{2} m\omega_0^2 x^2(t)$ corresponds to the so-called "harmonic oscillator" depicted by the dashed line of Figure 1.2, the quantity ω_0 being the Eigen resonance angular frequency of the harmonic oscillator. The higher order terms $\frac{1}{4} mbx^4(t) + \frac{1}{6} mdx^6(t) + ...$ lead to a deviation from the harmonic oscillator; it is said to bring "anharmonicity". As a consequence, the full expression of Eq. (1.10) corresponds to the so-called "anharmonic oscillator".

Note that the magnitudes of ω_0, b, and d depend on matter. The amplitudes of b and d are all the greater as the electrons can be moved far away from their equilibrium position. The highest values are then obtained with electron gases in semiconductors or with aromatic cycles, or double and triple bonds in organic media. These materials are then qualified as having a "hyperpolarizability".

The restoring force being proportional to the first derivative of the potential energy, its magnitude is also a Taylor power series, but with only odd exponents of $x(t)$. Then according to Eq. (1.8) to Eq. (1.10), it comes:

$$\vec{F}_r(t) = F_r(t)\ \vec{u} = \left[m\omega_0^2 x(t) + mbx^3(t) + mdx^5(t) + ... \right] \vec{u} \tag{1.11}$$

The second situation corresponds to the case of two different atoms, A and B, forming an asymmetrical chemical bond, as shown in Figure 1.3.

The potential energy is thus a non-centrosymmetric function of the electron motion, *i.e.* $V(-x,t) \neq -V(x,t)$, as shown by the continuous line plotted in Figure 1.3. Then V can be expanded in a new Taylor power series of $x(t)$, where even and odd exponents have to be considered as following:

$$V(t) = \frac{1}{2} m\omega_0^2 x^2(t) + \frac{1}{3} max^3(t) + \frac{1}{4} mbx^4(t) + ... \tag{1.12}$$

As explained previously, the parameters a and b of the anharmonic term depend on the capability of the electrons to be delocalized.

According to Eq. (1.8), Eq. (1.9), and Eq. (1.12), it comes for the restoring force:

$$\vec{F}_r(t) = F_r(t)\ \vec{u} = \left[m\omega_0^2 x(t) + max^2(t) + mbx^3(t) + ... \right] \vec{u} \tag{1.13}$$

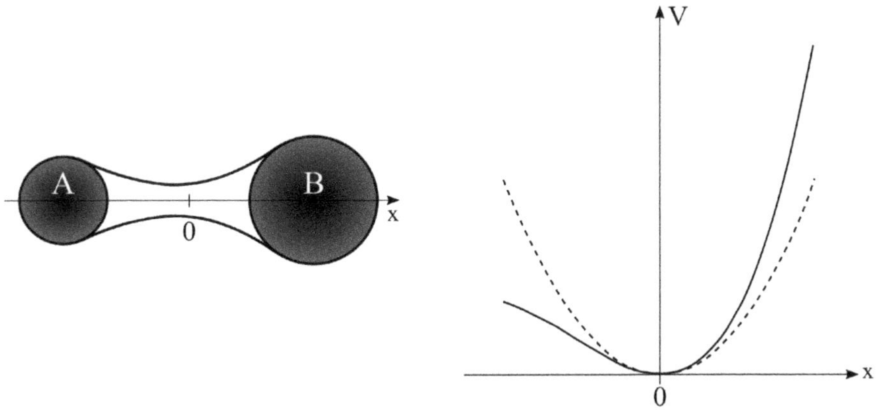

FIGURE 1.3 (Left) Heteropolar bond constituted by two distinct atoms A and B, the x-axis being the bond axis. (Right) Corresponding non-centrosymmetric potential energy V as a function of the electron motion x; the continuous line stands for its expansion as a Taylor power series with odd and even exponents of x; the dashed line is when the quadratic contribution is the only one to be considered.

Figure 1.2 and Figure 1.3 well show that the harmonic term $\frac{1}{2}m\omega_0^2x^2(t)$ well describes the potential energy around the close vicinity of the equilibrium position of the electron only. It will be seen later on that it corresponds to a situation in which the electric field of light is small compared with the electric field seen by the electron. For higher values of the electric field of light, the potential energy of the electron involves both the harmonic and anharmonic terms, leading to a potential distortion with respect to the harmonic case.

1.1.1.3 The Damping Force

The damping force is also opposed to the Coulomb force. It is due to any energy loss in matter like the friction of electrons. It writes:

$$\vec{F}_d(t) = F_d(t)\ \vec{u} \tag{1.14}$$

with:

$$F_d(t) = m\ \Gamma\ \mathrm{v}(t) \tag{1.15}$$

where $m \simeq 9\times10^{-31}kg$ is the electron mass, and Γ is the damping coefficient which unity is $[s^{-1}]$. The quantity $\mathrm{v}(t)$ is the magnitude of the electron velocity, in $[m.s^{-1}]$, expressed as:

$$\mathrm{v}(t) = \frac{dx(t)}{dt} \tag{1.16}$$

The quantity Γ depends on matter. It is assumed here that $\frac{1}{\Gamma}$ is negligible compared to the duration over which the electric field of light is applied. The unit of $F_d(t)$ is the Newton $[N]$.

1.1.2 Dynamics Motion of the Electrons

From the expressions of all the forces applied to the electron, it is possible to determine its motion using the law of Newtonian dynamics, *i.e.* (Feynman, Leighton, Sands 2005) [1]:

$$\vec{F}_c(t)+\vec{F}_r(t)+\vec{F}_d(t)=m\ \vec{\gamma}(t) \tag{1.17}$$

where m is electron mass and $\vec{\gamma}$, in $[m.s^{-2}]$, is the electron acceleration vector that writes:

$$\vec{\gamma}\ (t)=\gamma\ (t)\ \vec{u}=\frac{d\vec{v}(t)}{dt}=\frac{d^2x(t)}{dt^2}\vec{u} \tag{1.18}$$

The three forces that are involved, as well as the acceleration, have to be projected along the working axis, *i.e.* the x-axis. Then from Eq. (1.2), Eq. (1.3), Eq. (1.8), Eq. (1.14), Eq. (1.15), Eq. (1.16), and Eq. (1.18), Eq. (1.17) reduces to the following scalar expression:

$$qE(t)-m\Gamma\frac{dx(t)}{dt}-F_r\ (t)=m\frac{d^2x(t)}{dt^2} \tag{1.19}$$

According to Eq. (1.11) describing $F_r(t)$ in the case of a homopolar bond, Eq. (1.19) writes:

$$\frac{d^2x(t)}{dt^2}+\Gamma\frac{dx(t)}{dt}+\omega_0^2x(t)+bx^3(t)+dx^5(t)+...=\frac{q}{m}E(t) \tag{1.20}$$

And in the case of a heteropolar bond, Eq. (1.19) writes from Eq. (1.13):

$$\frac{d^2x(t)}{dt^2}+\Gamma\frac{dx(t)}{dt}+\omega_0^2x(t)+ax^2(t)+bx^3(t)+...=\frac{q}{m}E(t) \tag{1.21}$$

The integration of Eq. (1.20) and Eq. (1.21) gives access to $x(t)$ and then to the macroscopic induced polarization $P(t)$ according to Eq. (1.7). It is the aim of the two next sections. Two operating regimes are traditionally considered: the one governed only by the harmonic oscillator, *i.e.* ω_0^2x, and the one additionally requiring the anharmonic oscillator to be taken into account: $ax^2+bx^3+....$ These two regimes describe respectively the linear and nonlinear optical properties of matter. The nonlinear optical properties are called "quadratic" when the quadratic anharmonic term is considered, *i.e.* ax^2; they are called "cubic" when it is the cubic anharmonic term, *i.e.* bx^3.

The linear optical effects are predominant when the magnitude of the electric field of light is typically 10^{-5} times lower that the one seen by the electrons of matter, which is of about 3×10^8 V.cm^{-1}: that corresponds to light intensity typically below 0.25 W.cm^{-2}. The nonlinear optical effects can be observed above this intensity, knowing that the linear optical effects still exist in the nonlinear regime.

1.1.3 The Harmonic Oscillator and the Associated Linear Optical Effects

In this case, the potential energy is governed only by the harmonic term, the chemical bond being homopolar as well as heteropolar. Then Eq. (1.20) and Eq. (1.21) both reduce to:

$$\frac{d^2x(t)}{dt^2} + \Gamma\frac{dx(t)}{dt} + \omega_0^2 x(t) = \frac{q}{m}E(t) \tag{1.22}$$

Equation (1.22) can be solved exactly. An electric field $\vec{E}(t)$ oscillating at one circular frequency ω is considered here. The complex notation is used to express $E(t)$. Since this field is a real quantity, it comes:

$$E(t) = \frac{1}{2}\left[E(\omega)e^{i\omega t} + E(-\omega)e^{-i\omega t}\right] \tag{1.23}$$

where the amplitude $E(-\omega) = E^*(\omega)$ stands for the complex conjugate of $E(\omega)$ as reminded by the symbol *.

By inserting Eq. (1.23) into Eq. (1.22), it is easy to calculate the instantaneous motion of the electron, which is written $x^{(1)}(t)$ and expressed as:

$$x^{(1)}(t) = \frac{q}{2m}\left[\frac{E(\omega)}{\left[\omega_0^2 - \omega^2 + i\Gamma\omega\right]}e^{i\omega t} + \frac{E(-\omega)}{\left[\omega_0^2 - \omega^2 - i\Gamma\omega\right]}e^{-i\omega t}\right] \tag{1.24}$$

Equation (1.24) shows that under the linear regime, the electron swings back and forth permanently from its equilibrium position, and close to that position.

In the linear regime, the induced polarization is called the "linear induced polarization". It can be calculated by inserting Eq. (1.24) in Eq. (1.7), which gives:

$$P^{(1)}(t) = \frac{1}{2}\left[P^{(1)}(\omega)e^{i\omega t} + P^{(1)}(-\omega)e^{-i\omega t}\right] \tag{1.25}$$

with

$$P^{(1)}(\pm\omega) = \frac{Nq^2}{m}\frac{E(\pm\omega)}{\left[\omega_0^2 - \omega^2 \pm i\Gamma\omega\right]} \tag{1.26}$$

Equation (1.26) shows that the magnitude of the induced polarization $P^{(1)}$ is proportional to the optical electric field E, which depicts the linearity of the regime. Equation (1.25) and Eq. (1.26) clearly show that under linear optical effects, $P^{(1)}$ oscillates at the same circular frequency ω as that of the applied optical electric field. It is one important signature of the linear regime. The spectrum of $P^{(1)}$ is shown in Figure 1.4 in which the positive circular frequency is the only one which is considered.

The positive circular frequency will be the only one to be considered in order to lighten the expressions given hereafter, the negative ones being in the complex conjugate.

From Eq. (1.26) it comes:

$$P^{(1)}(\omega) = \varepsilon_0 \chi^{(1)}(\omega) E(\omega) \tag{1.27}$$

where ε_0 is the vacuum permittivity, and $\chi^{(1)}(\omega)$ the first-order electric susceptibility expressed as:

$$\chi^{(1)}(\omega) = \frac{Nq^2}{m\,\varepsilon_0\left[\omega_0^2 - \omega^2 + i\Gamma\omega\right]} \tag{1.28}$$

Note that $\chi^{(1)}(\omega)$ is dimensionless since $P^{(1)}(\omega)$ and $\varepsilon_0 E(\omega)$ have the same dimension.

The first-order electric susceptibility is obviously a complex number, and Eq. (1.28) also stands for $\chi^{(1)}(-\omega) = \chi^{(1)}(\omega)^*$ by exchanging ω for $-\omega$.

By multiplying both the upper part and the lower part of Eq. (1.28) by $\left[\omega_0^2 - \omega^2 - i\Gamma\omega\right]$, $\chi^{(1)}(\omega)$ writes:

$$\chi^{(1)}(\omega) = \frac{Nq^2}{m\,\varepsilon_0\left[\left(\omega_0^2 - \omega^2\right)^2 + \Gamma^2\omega^2\right]}\left[(\omega_0^2 - \omega^2) - i\Gamma\omega\right] \tag{1.29a}$$

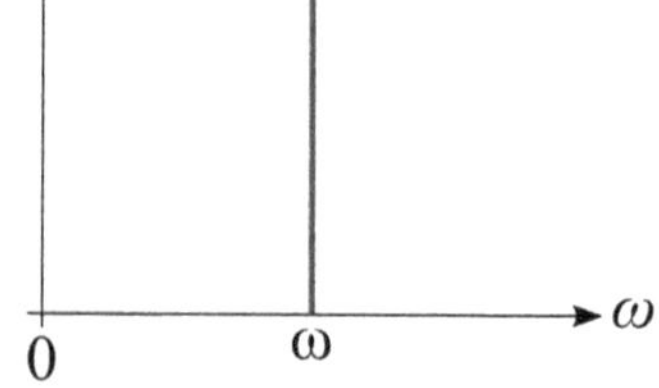

FIGURE 1.4 Spectrum of the induced polarization, $P = P^{(1)}$, when the applied optical electric field $E(t)$ is oscillating at the circular frequency ω.

Equation (1.29a) has the advantage of being able to separate the real and imaginary parts of the first-order electric susceptibility. The real part $\mathrm{Re}\{\chi^{(1)}(\omega)\}$ governs the propagation of light while the second one $\mathrm{Im}\{\chi^{(1)}(\omega)\}$ leads to absorption of light, which is called "linear absorption".

The imaginary term defined previously is due to the damping force defined in Part 1.1.1.3, which is the standard framework of the Lorentz model. But there may also be a driving force instead of a damping force in the model, which corresponds to the case of amplification of light. In that case, all the previous calculations have to be performed by replacing $-\Gamma\omega$ with $+\Gamma\omega$ in Eq. (1.14). At the end, it comes easily the following expression of the first-order electric susceptibility, where $\left[(\omega_0^2-\omega^2)-i\Gamma\omega\right]$ in Eq. (1.29a) is replaced by $\left[(\omega_0^2-\omega^2)+i\Gamma\omega\right]$, *i.e.*:

$$\chi^{(1)}(\omega)=\frac{Nq^2}{m\ \varepsilon_0\left[\left(\omega_0^2-\omega^2\right)^2+\Gamma^2\omega^2\right]}\left[(\omega_0^2-\omega^2)+i\Gamma\omega\right] \tag{1.29b}$$

The imaginary part of Eq. (1.29b) is relevant for describing the spontaneous and stimulated emissions that will be defined in Part 1.3.

As will be seen hereafter in Part 1.1.4 and Part 1.1.5, the modeling of the quadratic and cubic anharmonic oscillators implies also to take into account the harmonic oscillator contribution. The corresponding calculations will be performed in the following by using a damping force in the model, *i.e.* given by Eq. (1.14) and Eq. (1.15), which corresponds to usual situations.

1.1.4 The Quadratic Anharmonic Oscillator and the Associated Nonlinear Optical Effects

1.1.4.1 Single Excitation

The electric field of light considered here oscillates at one circular frequency ω, as in the previous section describing the linear regime. Then $\vec{E}(t)$ is described by Eq. (1.23).

As shown by Eq. (1.21), the quadratic anharmonic term of the restoring force only exists for heteropolar bonds. In that case, it is relevant to focus only on the quadratic nonlinear interactions; then Eq. (1.21) reduces to (Bloembergen 1996) [3]:

$$\frac{d^2x(t)}{dt^2}+\Gamma\frac{dx(t)}{dt}+\omega_0^2x(t)+ax^2(t)=\frac{q}{m}E(t) \tag{1.30}$$

No exact resolution of such a second-order differential equation is possible. But an approximate solution can be found using a perturbation theory in which the general solution is taken as (Boyd 2008) [4]:

$$x(t)=x^{(1)}(t)+x^{(2)}(t) \tag{1.31}$$

where $x^{(1)}(t)$ is the exact solution of the harmonic oscillator given by Eq. (1.24).

Then Eq. (1.30) can be split into the two following equations using Eq. (1.31):

$$\frac{d^2x^{(1)}(t)}{dt^2}+\Gamma\frac{dx^{(1)}(t)}{dt}+\omega_0^2x^{(1)}(t)=\frac{q}{m}E(t) \tag{1.32a}$$

$$\frac{d^2x^{(2)}(t)}{dt^2}+\Gamma\frac{dx^{(2)}(t)}{dt}+\omega_0^2x^{(2)}(t)=-a\left[x^{(1)}(t)\right]^2 \tag{1.32b}$$

When inserting the solution of Eq. (1.32a), *i.e.* Eq. (1.24), into Eq. (1.32b), it comes:

$$\frac{d^2x^{(2)}(t)}{dt^2}+\Gamma\frac{dx^{(2)}(t)}{dt}+\omega_0^2x^{(2)}(t)=-\frac{a}{4}\frac{q^2}{m^2}\left[\frac{E(\omega)}{\left[\omega_0^2-\omega^2+i\Gamma\omega\right]}e^{i\omega t}+C.C.\right]^2 \tag{1.33}$$

where C.C. stands for the complex conjugate. The solution of Eq. (1.33) is then:

$$x^{(2)}(t)=-\frac{a}{4}\frac{q^2}{m^2}\left[\begin{array}{l}\dfrac{E^2(\omega)e^{2i\omega t}}{\left[\omega_0^2-4\omega^2+i\Gamma 2\omega\right]\left[\omega_0^2-\omega^2+i\Gamma\omega\right]^2}\\ +\dfrac{E(\omega)E(-\omega)e^{i(\omega-\omega)t}}{\omega_0^2\left[\omega_0^2-\omega^2+i\Gamma\omega\right]\left[\omega_0^2-\omega^2-i\Gamma\omega\right]}\\ +C.C.\end{array}\right] \tag{1.34}$$

From the sum of Eq. (1.24) and Eq. (1.34), we get the approximate solution of $x(t)$ according to Eq. (1.31), which describes the quadratic anharmonic oscillation of the electron. The spectrum of $x(t)$ includes the circular frequency of the excitation light, *i.e.* ω, arising from $x^{(1)}(t)$. But two new ones appear, at 0 and 2ω, thanks to $x^{(2)}(t)$. It is a first signature of the nonlinear regime: the circular frequency at 0, which is a static electric field created in the matter and corresponds to an effect that is called "optical rectification"; the component at 2ω, which is an optical circular frequency, corresponds to "second-harmonic generation" (SHG).

The induced polarization can be then expressed by inserting Eq. (1.24) and Eq. (1.34) into Eq. (1.7), which gives:

$$P(t)=P^{(1)}(t)+P^{(2)}(t) \tag{1.35}$$

$P^{(1)}(t)$ is the induced linear polarization given by Eq. (1.25) and Eq. (1.26), and $P^{(2)}(t)$ is the induced quadratic polarization expressed as:

$$P^{(2)}(t)=\frac{1}{2}\varepsilon_0\left[\begin{array}{l}\chi^{(2)}(2\omega=\omega+\omega)E^2(\omega)e^{2i\omega t}\\ +\chi^{(2)}(0=\omega-\omega)E(\omega)E(-\omega)e^{i(\omega-\omega)t}\\ +C.C.\end{array}\right] \tag{1.36}$$

Equation (1.36) can also write:

$$P^{(2)}(t) = \frac{1}{2}\left[P^{(2)}(2\omega)e^{i2\omega t} + P^{(2)}(0) + C.C.\right] \tag{1.37}$$

$P^{(2)}(2\omega)$ and $P^{(2)}(0)$ are the two spectrum components of the induced quadratic polarization expressed respectively as:

$$P^{(2)}(2\omega) = \varepsilon_0 \chi^{(2)}(2\omega = \omega + \omega)E^2(\omega) \tag{1.38}$$

for the second-harmonic circular frequency, and

$$P^{(2)}(0) = \varepsilon_0 \chi^{(2)}(0 = \omega - \omega)E(\omega)E(-\omega) \tag{1.39}$$

for the static electric field component.

The corresponding spectrum of the induced polarization P is shown in Figure 1.5.

The quantities $\chi^{(2)}(2\omega = \omega + \omega)$ and $\chi^{(2)}(0)$ are the two components of the second-order electric susceptibility that write according to Eq. (1.34):

$$\chi^{(2)}(2\omega = \omega + \omega) = -\frac{aNq^3}{2m_e^2\varepsilon_0\left[\omega_0^2 - 2\omega + i\Gamma 2\omega\right]\left[\omega_0^2 - \omega + i\Gamma\omega\right]^2} \tag{1.40}$$

and

$$\chi^{(2)}(0 = \omega - \omega) = -\frac{aNq^3}{2m_e^2\varepsilon_0\omega_0^2\left[\omega_0^2 - \omega + i\Gamma\omega\right]\left[\omega_0^2 - \omega - i\Gamma\omega\right]} \tag{1.41}$$

Note that Eq. (1.40) and Eq. (1.41) are complex numbers, so that Eq. (1.40) also stands for $\chi^{(1)}(-2\omega = -\omega - \omega)$ after exchanging 2ω and ω for -2ω and $-\omega$, respectively. Similarly, Eq. (1.41) also stands for $\chi^{(1)}(0 = -\omega + \omega)$ after exchanging ω for $-\omega$.

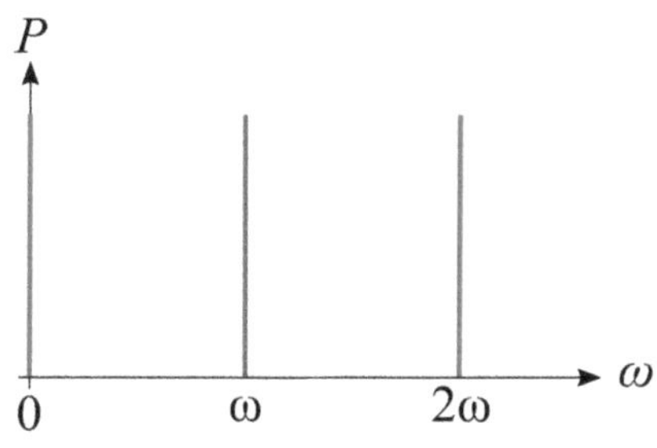

FIGURE 1.5 Spectrum of the induced polarization P when the excitation optical electric field is oscillating at the circular frequency ω. The blue peak corresponds to the contribution of $P^{(1)}$ while the red ones is for the contributions of $P^{(2)}$.

The dimension of $P^{(2)}(2\omega)$ and $P^{(2)}(0)$ being the same one as that of $\varepsilon_0 E(\omega)$, their unit is in [m/V]. Their magnitude is of about 10^{-12} m/V.

Equation (1.38) and Eq. (1.39) highlight that $P^{(2)}(2\omega)$ and $P^{(2)}(0)$ are both quadratic as a function of the magnitude of the excitation optical electric field since they are proportional to $E^2(\omega)$ and $E(\omega)E(-\omega)$, respectively.

1.1.4.2 Double Excitation

The excitation optical electric field $\vec{E}(t)$ that is considered now is oscillating at two circular frequencies ω_1 and ω_2, *i.e.*:

$$E(t) = \frac{1}{2}\left[E(\omega_1)e^{i\omega_1 t} + E(\omega_2)e^{i\omega_2 t} + C.C.\right] \tag{1.42}$$

The insertion of Eq. (1.42) into Eq. (1.32a) leads to the exact solution $x^{(1)}(t)$ describing the simultaneous harmonic motion of the electron at both the circular frequencies ω_1 and ω_2. It writes:

$$x^{(1)}(t) = \frac{1}{2}\frac{q}{m}\sum_{p=1}^{2}\left[\frac{E(\omega_p)}{\left[\omega_0^2 - \omega_p + i\Gamma\omega_p\right]}e^{i\omega_p t} + C.C.\right] \tag{1.43}$$

The term $x^{(2)}(t)$ can be found by inserting Eq. (1.43) into Eq. (1.32b). Then it comes:

$$x^{(2)}(t) = -\frac{aq^2}{4m^2}\left[\begin{array}{l} \dfrac{E^2(\omega_1)e^{i2\omega_1 t}}{D(2\omega_1)D^2(\omega_1)} + \dfrac{E(\omega_1)E(-\omega_1)e^{i(\omega_1-\omega_1)t}}{D(0)D(\omega_1)D(-\omega_1)} \\ + \dfrac{E^2(\omega_2)e^{i2\omega_2 t}}{D(2\omega_2)D^2(\omega_2)} + \dfrac{E(\omega_2)E(-\omega_2)e^{i(\omega_2-\omega_2)t}}{D(0)D(\omega_2)D(-\omega_2)} \\ + \dfrac{E(\omega_1)E(\omega_2)e^{i(\omega_1+\omega_2)t}}{D(\omega_1+\omega_2)D(\omega_1)D(\omega_2)} + \dfrac{E(-\omega_1)E(\omega_2)e^{i(\omega_2-\omega_1)t}}{D(\omega_2-\omega_1)D(-\omega_1)D(\omega_2)} \\ + C.C. \end{array}\right] \tag{1.44}$$

where

$$D(\pm u) = \omega_0^2 - u^2 \pm i\Gamma u \tag{1.45}$$

with $u = \omega_1, \omega_2, 2\omega_1, 2\omega_2, \omega_1 + \omega_2$ and $\omega_2 - \omega_1$, and *C.C.* standing for the sum of the complex conjugates of the different terms in brackets.

The spectrum of $x(t)$ is very rich since seven circular frequencies exist. We find the same ones than in the previous section, *i.e.* at ω_p, $2\omega_p$, and $\omega_p - \omega_p = 0$ with $(p = 1,2)$, corresponding to the linear optical effect, SHG and optical rectification, respectively, but two new circular frequencies appear, at $\omega_1 + \omega_2$ and $\omega_2 - \omega_1$: the two associated nonlinear optical phenomena are the "quadratic sum-frequency generation" (quadratic SFG) and the "quadratic difference-frequency generation" (quadratic DFG), respectively (Bloembergen 1996) [3]. Note that SHG and optical rectification are the degenerate cases in frequency of SFG and DFG, respectively.

By inserting Eq. (1.31), Eq. (1.43), Eq. (1.44), and Eq. (1.45) into Eq. (1.7) and Eq. (1.35), it comes:

$$P^{(1)}(t)=\frac{1}{2}\varepsilon_0\left[\chi^{(1)}(\omega_1)E(\omega_1)e^{i\omega_1 t}+\chi^{(1)}(\omega_2)E(\omega_2)e^{i\omega_2 t}+C.C.\right] \quad (1.46)$$

and

$$P^{(2)}(t)=\frac{1}{2}\varepsilon_0\left[\begin{array}{l}\chi^{(2)}(0=\omega_1-\omega_1)E(\omega_1)E(-\omega_1)e^{i(\omega_1-\omega_1)t}\\+\chi^{(2)}(0=\omega_2-\omega_2)E(\omega_2)E(-\omega_2)e^{i(\omega_2-\omega_2)t}\\+\chi^{(2)}(2\omega_1=\omega_1+\omega_1)E^2(\omega_1)e^{2i\omega_1 t}\\+\chi^{(2)}(2\omega_2=\omega_2+\omega_2)E^2(\omega_2)e^{2i\omega_2 t}\\+\chi^{(2)}(\omega_3=\omega_1+\omega_2)E(\omega_1)E(\omega_2)e^{i(\omega_1+\omega_2)t}\\+\chi^{(2)}(\omega'_3=\omega_2-\omega_1)E(-\omega_1)E(\omega_2)e^{i(\omega_2-\omega_1)t}\\+C.C.\end{array}\right] \quad (1.47)$$

Equation (1.47) can also be expressed as:

$$P^{(2)}(t)=\frac{1}{2}\left[\begin{array}{l}P^{(2)}(2\omega_1=\omega_1+\omega_1)e^{i2\omega_1 t}+P^{(2)}(2\omega_2=\omega_2+\omega_2)e^{i2\omega_2 t}\\+P^{(2)}(0=\omega_1-\omega_1)+P^{(2)}(0=\omega_2-\omega_2)\\+P^{(2)}(\omega_3=\omega_1+\omega_2)e^{i(\omega_3)t}+P^{(2)}(\omega'_3=\omega_2-\omega_1)e^{i(\omega'_3)t}\\+C.C.\end{array}\right] \quad (1.48)$$

The induced quadratic polarization at $2\omega_1$ and $2\omega_2$ are both given by Eq. (1.38), and the two quadratic polarization contributions to optical rectification at 0 are both given by Eq. (1.39). Additionally, the two other components are expressed as:

$$P^{(2)}(\omega_3)=\varepsilon_0\chi^{(2)}(\omega_3=\omega_1+\omega_2)E(\omega_1)E(\omega_2) \quad (1.49a)$$

and

$$P^{(2)}(\omega_3')=\varepsilon_0\chi^{(2)}(\omega_3'=\omega_2-\omega_1)E(\omega_2)E(-\omega_1) \quad (1.49b)$$

The electric susceptibilities $\chi^{(1)}(\omega_p)$ with $p=1,2$, $\chi^{(2)}(2\omega_p=\omega_p+\omega_p)$, and $\chi^{(2)}(0=\omega_p-\omega_p)$ are given by Eq. (1.29a), Eq. (1.29b), Eq. (1.40), and Eq. (1.41), respectively. For $\chi^{(2)}(\omega_3=\omega_1+\omega_2)$ and $\chi^{(2)}(\omega'_3=\omega_2-\omega_1)$, it comes:

$$\chi^{(2)}(\omega_3=\omega_2+\omega_1)=-\frac{aNq^3}{2m_e^2\varepsilon_0 D(\omega_2+\omega_1)D(\omega_2)D(\omega_1)} \quad (1.50a)$$

and

$$\chi^{(2)}(\omega_3' = \omega_2 - \omega_1) = -\frac{aNq^3}{2m_e^2\varepsilon_0 D(\omega_2 - \omega_1)D(\omega_2)D(-\omega_1)} \tag{1.50b}$$

where $D(u = \pm\omega_1, \omega_2, \omega_2 \pm \omega_1)$ are given by Eq. (1.45).

The corresponding spectrum of the induced polarization is shown in Figure 1.6.

It will be seen in Part 1.2 that the induced polarization leads to the radiation of an optical field at the same frequency. It is therefore entirely conceivable that this radiated field could be coupled with itself or with pre-existing frequencies. Thus, by this way numerous new circular frequencies could be potentially created from the initial spectrum of Figure 1.6 thanks to other optical quadratic effects. For example: $2\omega_1 + 2\omega_1 = 4\omega_1$, $2\omega_2 + 2\omega_2 = 4\omega_2$, $\omega_3 + \omega_1 = 2\omega_1 + \omega_2$, $\omega_3' - \omega_1 = \omega_2 - 2\omega_1$, and so on.

These kinds of recombination can also lead to the same circular frequencies than those initially created: $\omega_3' + \omega_1 = \omega_2$, $\omega_3 - \omega_1 = \omega_2$, or $\omega_3 - \omega_2 = \omega_1$ for example. The circular frequencies of the induced quadratic polarization corresponding to the two later cases that are DFG processes are then:

$$P^{(2)}(\omega_2) = \varepsilon_0 \chi^{(2)}(\omega_2 = \omega_3 - \omega_1)E(\omega_3)E(-\omega_1) \tag{1.51a}$$

and

$$P^{(2)}(\omega_1) = \varepsilon_0 \chi^{(2)}(\omega_1 = \omega_3 - \omega_2)E(\omega_3)E(-\omega_2) \tag{1.51b}$$

with

$$\chi^{(2)}(\omega_2 = \omega_3 - \omega_1) = -\frac{aNq^3}{2m_e^2\varepsilon_0 D(\omega_2)D(\omega_3)D(-\omega_1)} \tag{1.52a}$$

and

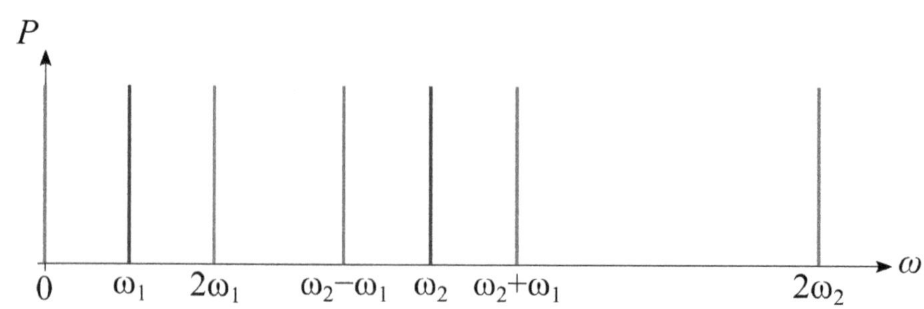

FIGURE 1.6 Spectrum of the induced polarization P when the excitation optical electric field is simultaneously oscillating at two circular frequencies ω_1 and ω_2. The peaks at ω_1 and ω_2 correspond to the contributions of $P^{(1)}$ while the other ones stand for the contributions of $P^{(2)}$.

$$\chi^{(2)}(\omega_1=\omega_3-\omega_2)=-\frac{aNq^3}{2m_e^2\varepsilon_0 D(\omega_1)D(\omega_3)D(-\omega_2)} \tag{1.52b}$$

where $D(u=\pm\omega_1,\pm\omega_2,\omega_3)$ is given by Eq. (1.45).

1.1.4.3 Relations between the First-Order and Second-Order Electric Susceptibilities

A comparison between the first-order electric susceptibility $\chi^{(1)}(\omega_p)$ from Eq. (1.28) with $p=1,2,3$, and the second-order electric susceptibility $\chi^{(2)}(\omega_2\pm\omega_1)$ from Eq. (1.50a) and Eq. (1.50b) allows us to establish that (Miller 1964) [5]:

$$\frac{\chi^{(2)}(\omega_2\pm\omega_1)}{\chi^{(1)}(\omega_1)\chi^{(1)}(\omega_2)\chi^{(1)}(\omega_2\pm\omega_1)}=\delta \tag{1.53}$$

with

$$\delta=-2a\frac{m_e\varepsilon_0^2}{N^2q^3} \tag{1.54}$$

The parameter δ is called the Miller's index. It is independent of circular frequency, which can be interesting in the framework of crystal engineering (Chemla, Zyss 1987, Zyss 1994) [6,7].

1.1.5 The Cubic Anharmonic Oscillator and the Associated Nonlinear Optical Effects

1.1.5.1 Single Excitation

The electric field of the incident light $\vec{E}(t)$ oscillates at one circular frequency ω, so that it is described by Eq. (1.23).

We are interested here in the nonlinear regime involving the cubic term bx^3 in the motion equation of the electrons. It can exist for both homopolar and heteropolar bonds, as shown by Eq. (1.20) and Eq. (1.21). Since the cubic nonlinear effects are targeted here, the starting equation is obtained by retaining only the term in bx^3 in the nonlinear part of the two later equations, which corresponds to (Bloembergen 1996) [3]:

$$\frac{d^2x(t)}{dt^2}+\Gamma\frac{dx(t)}{dt}+\omega_0^2x(t)+bx^3(t)=\frac{q}{m}E(t) \tag{1.54}$$

As in the case of quadratic effects, there is no exact resolution of such a differential equation, so that an approximate solution is considered, of form:

$$x(t)=x^{(1)}(t)+x^{(3)}(t) \tag{1.55}$$

where $x^{(1)}(t)$ is the exact solution of the harmonic oscillator given by Eq. (1.24).

Then Eq. (1.54) can be split into the two following equations:

$$\frac{d^2x^{(1)}(t)}{dt^2}+\Gamma\frac{dx^{(1)}(t)}{dt}+\omega_0^2x^{(1)}(t)=\frac{q}{m_e}E(t) \tag{1.56a}$$

$$\frac{d^2x^{(3)}(t)}{dt^2}+\Gamma\frac{dx^{(3)}(t)}{dt}+\omega_0^2x^{(3)}(t)=-b\left[x^{(1)}(t)\right]^3 \tag{1.56b}$$

When inserting Eq. (1.24) into Eq. (1.56b), it becomes:

$$x^{(3)}(t)=-\frac{b}{8}\frac{q^3}{m_e^3}\left[\frac{E^3(\omega)e^{3i\omega t}}{D(3\omega)D^3(\omega)}+3\frac{E^2(\omega)E(-\omega)e^{i\omega t}}{D^3(\omega)D(-\omega)}+C.C.\right] \tag{1.57}$$

where $D(u=\pm\omega,3\omega)$ are given by Eq. (1.45).

By inserting Eq. (1.24) and Eq. (1.57) into Eq. (1.55) and then into Eq. (1.7), the magnitude of the cubic induced polarization writes:

$$P(t)=P^{(1)}(t)+P^{(3)}(t) \tag{1.58}$$

where $P^{(1)}(t)$ is the induced linear polarization given by Eq. (1.25) and Eq. (1.26), and $P^{(3)}(t)$ the induced cubic polarization expressed as:

$$P^{(3)}(t)=\frac{1}{2}\varepsilon_0\left[\begin{array}{l}\chi^{(3)}(3\omega=\omega+\omega+\omega)E^3(\omega)e^{3i\omega t}\\+\chi^{(3)}(\omega=\omega+\omega-\omega)E^2(\omega)E(-\omega)e^{i\omega t}\\+C.C.\end{array}\right] \tag{1.59}$$

Equation (1.59) can also be written:

$$P^{(3)}(t)=\frac{1}{2}\left[P^{(3)}(3\omega)e^{i3\omega t}+P^{(3)}(\omega)e^{i\omega t}+C.C.\right] \tag{1.60}$$

with

$$P^{(3)}(3\omega)=\varepsilon_0\chi^{(3)}(3\omega=\omega+\omega+\omega)E^3(\omega) \tag{1.61}$$

where

$$\chi^{(3)}(3\omega=\omega+\omega+\omega)=-\frac{bNq^4}{4m_e^3\varepsilon_0D(3\omega)D^3(\omega)} \tag{1.62}$$

and

$$P^{(3)}(\omega)=\varepsilon_0\chi^{(3)}(\omega=\omega+\omega-\omega)E^2(\omega)E\ (-\omega) \tag{1.63}$$

where

$$\chi^{(3)}(\omega=\omega+\omega-\omega)=-\frac{3bNq^4}{4m_e^3\varepsilon_0 D^3(\omega)D(-\omega)} \quad (1.64)$$

The quantity $D(u=\pm\omega,3\omega)$ is given by Eq. (1.45).

Equation (1.61) and Eq. (1.62) describe "third-harmonic generation (THG)", while Eq. (1.63) and Eq. (1.64) correspond to the "optical Kerr effect".

The corresponding spectrum of the induced polarization is shown in Figure 1.7.

It is important to notice that the circular frequency ω has two origins: the linear optical effect on the one hand, and the optical Kerr effect $(\omega=\omega+\omega-\omega)$ on the other hand.

According to Eq. (1.62) and Eq. (1.64), the unit of the third-order electric susceptibility is in $[m^2/V^2]$. It has an order of magnitude of 10^{-21} m^2/V^2.

1.1.5.2 Triple Excitation

In that case the excitation optical electric field $E(t)$ writes:

$$E(t)=\frac{1}{2}\sum_{p=1,2,3}\left[E(\omega_p)e^{i\omega_p t}+C.C.\right] \quad (1.65)$$

The relation of order $\omega_1 \leq \omega_2 \leq \omega_3$ is taken as the convention in the following.

Using exactly the same approach as that used in Part 1.1.4.2 for the quadratic effects, it appears that the spectrum of the cubic polarization induced by three excitation spectral components is extremely rich. These components can be split into two categories.

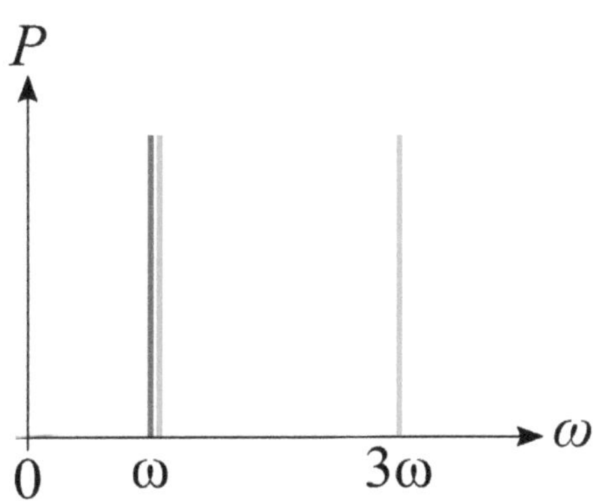

FIGURE 1.7 Spectrum of the induced polarization P when the excitation optical electric field is oscillating at the circular frequency ω. The blue peak corresponds to the contribution of $P^{(1)}$ while the green peaks stand for the two contributions of $P^{(3)}$.

The first category is of same kind as that of the quadratic nonlinear effects with up to 14 schemes and 11 new distinct spectral components; it can be described from THG, SFG, and DFG, *i.e.*:

- $3\omega_1 = \omega_1 + \omega_1 + \omega_1$, $3\omega_2 = \omega_2 + \omega_2 + \omega_2$, and $3\omega_3 = \omega_3 + \omega_3 + \omega_3$ from THG,
- $2\omega_1 + \omega_2 = \omega_1 + \omega_1 + \omega_2, 2\omega_1 + \omega_3 = \omega_1 + \omega_1 + \omega_3, 2\omega_2 + \omega_1 = \omega_2 + \omega_2 + \omega_1$, $2\omega_2 + \omega_3 = \omega_2 + \omega_2 + \omega_3$, $2\omega_3 + \omega_1 = \omega_3 + \omega_3 + \omega_1$, $2\omega_3 + \omega_2 = \omega_3 + \omega_3 + \omega_2$, and $\omega_4 = \omega_1 + \omega_2 + \omega_3$ from SFG,
- $\omega_3 - 2\omega_1 = \omega_3 - \omega_1 - \omega_1$, $\omega_3 - 2\omega_2 = \omega_3 - \omega_2 - \omega_2$, $\omega_2 - 2\omega_1 = \omega_2 - \omega_1 - \omega_1$, and $\omega_4 = \omega_3 - \omega_1 - \omega_2$ from DFG.

Note that THG is the full degenerate case of SFG.

The second category of spectral components is of a new type with up to 18 schemes and 9 new distinct spectral components; it proceeds by a kind of mixing between SFG and DFG, *i.e.*:

- $\omega_1 = \omega_1 + \omega_1 - \omega_1$, $\omega_2 = \omega_2 + \omega_2 - \omega_2$, $\omega_3 = \omega_3 + \omega_3 - \omega_3$, $\omega_1 = \omega_1 + \omega_2 - \omega_2$, $\omega_1 = \omega_1 + \omega_3 - \omega_3$, $\omega_2 = \omega_2 + \omega_1 - \omega_1$, $\omega_2 = \omega_2 + \omega_3 - \omega_3$, $\omega_3 = \omega_3 + \omega_1 - \omega_1$, $\omega_3 = \omega_3 + \omega_2 - \omega_2$,
- $2\omega_3 - \omega_1 = \omega_3 + \omega_3 - \omega_1$, $2\omega_3 - \omega_2 = \omega_3 + \omega_3 - \omega_2$, $2\omega_2 - \omega_1 = \omega_2 + \omega_2 - \omega_1$, $2\omega_1 - \omega_2 = \omega_1 + \omega_1 - \omega_2$, $2\omega_2 - \omega_3 = \omega_2 + \omega_2 - \omega_3$, $2\omega_1 - \omega_3 = \omega_1 + \omega_1 - \omega_3$,
- $\omega_4 = \omega_3 + \omega_2 - \omega_1$, $\omega'_4 = \omega_3 + \omega_1 - \omega_2$, $\omega''_4 = \omega_1 + \omega_2 - \omega_3$.

These components arise from cubic nonlinear effects called "Four-Wave Mixing (FWM)", the optical Kerr effect corresponding to the full degeneracy in energy, *i.e.* $\omega_p = \omega_p + \omega_p - \omega_p$ with $p = 1,2,3,4$.

As said for optical quadratic effects in the previous section, the primary components listed can be recombined, giving rise to new components. For example, the component at ω_4 coming from the SFG ($\omega_4 = \omega_1 + \omega_2 + \omega_3$) can be recombined with the components at ω_2 and ω_3 under DFG leading to a component at ω_1 from ($\omega_1 = \omega_4 - \omega_2 - \omega_3$).

At the end, the cubic nonlinear effects can be divided into two categories characterized by the two following general relations: $\omega_4 = \omega_1 + \omega_2 + \omega_3$ with all the possible permutations, and $\omega_4 = -\omega_1 + \omega_2 + \omega_3$ with all the possible permutations.

The components of the induced cubic polarization corresponding to $\omega_4 = \omega_1 + \omega_2 + \omega_3$ are:

$$
\begin{aligned}
P^{(3)}(\omega_4) &= \varepsilon_0 \chi^{(3)}(\omega_4 = \omega_1 + \omega_2 + \omega_3) E(\omega_1) E(\omega_2) E(\omega_3) \\
P^{(3)}(\omega_1) &= \varepsilon_0 \chi^{(2)}(\omega_1 = \omega_4 - \omega_3 - \omega_2) E(\omega_4) E(-\omega_3) E(-\omega_2) \\
P^{(3)}(\omega_2) &= \varepsilon_0 \chi^{(2)}(\omega_2 = \omega_4 - \omega_3 - \omega_1) E(\omega_4) E(-\omega_3) E(-\omega_1) \\
P^{(3)}(\omega_3) &= \varepsilon_0 \chi^{(2)}(\omega_3 = \omega_4 - \omega_2 - \omega_1) E(\omega_4) E(-\omega_2) E(-\omega_1)
\end{aligned}
\tag{1.66}
$$

where the third-order electric susceptibilities are:

$$
\begin{aligned}
\chi^{(3)}(\omega_4 = \omega_3 + \omega_2 + \omega_1) &= -\frac{aNq^4}{4m_e^3 \varepsilon_0 D(\omega_4) D(\omega_1) D(\omega_2) D(\omega_3)} \\
\chi^{(3)}(\omega_1 = \omega_4 - \omega_3 - \omega_2) &= -\frac{aNq^4}{4m_e^3 \varepsilon_0 D(\omega_1) D(\omega_4) D(-\omega_3) D(-\omega_2)} \\
\chi^{(3)}(\omega_2 = \omega_4 - \omega_3 - \omega_1) &= -\frac{aNq^4}{4m_e^3 \varepsilon_0 D(\omega_2) D(\omega_4) D(-\omega_3) D(-\omega_1)} \\
\chi^{(3)}(\omega_3 = \omega_4 - \omega_2 - \omega_1) &= -\frac{aNq^4}{4m_e^3 \varepsilon_0 D(\omega_3) D(\omega_4) D(-\omega_2) D(-\omega_1)}
\end{aligned}
\tag{1.67}
$$

$D(\pm u)$ being given by Eq. (1.45).

In the same way, the components of induced cubic polarization corresponding to $\omega_4 = -\omega_1 + \omega_2 + \omega_3$ are:

$$
\begin{aligned}
P^{(3)}(\omega_1) &= \varepsilon_0 \chi^{(3)}(\omega_1 = -\omega_4 + \omega_2 + \omega_3) E(-\omega_4) E(\omega_2) E(\omega_3) \\
P^{(3)}(\omega_2) &= \varepsilon_0 \chi^{(2)}(\omega_2 = \omega_4 + \omega_1 - \omega_3) E(\omega_4) E(\omega_1) E(-\omega_3) \\
P^{(3)}(\omega_3) &= \varepsilon_0 \chi^{(2)}(\omega_3 = \omega_4 + \omega_1 - \omega_2) E(\omega_4) E(\omega_1) E(-\omega_2) \\
P^{(3)}(\omega_4) &= \varepsilon_0 \chi^{(2)}(\omega_4 = -\omega_1 + \omega_2 + \omega_3) E(-\omega_1) E(\omega_2) E(\omega_3)
\end{aligned}
\tag{1.68}
$$

with

$$
\begin{aligned}
\chi^{(3)}(\omega_1 = -\omega_4 + \omega_2 + \omega_3) &= -\frac{aNq^4}{4m_e^3 \varepsilon_0 D(\omega_1) D(-\omega_4) D(\omega_2) D(\omega_3)} \\
\chi^{(3)}(\omega_2 = \omega_4 + \omega_1 - \omega_3) &= -\frac{aNq^4}{4m_e^3 \varepsilon_0 D(\omega_2) D(\omega_4) D(\omega_1) D(-\omega_3)} \\
\chi^{(3)}(\omega_3 = \omega_4 + \omega_1 - \omega_2) &= -\frac{aNq^4}{4m_e^3 \varepsilon_0 D(\omega_3) D(\omega_4) D(\omega_1) D(-\omega_2)} \\
\chi^{(3)}(\omega_4 = -\omega_1 + \omega_2 + \omega_3) &= -\frac{aNq^4}{4m_e^3 \varepsilon_0 D(\omega_4) D(-\omega_1) D(\omega_2) D(\omega_3)}
\end{aligned}
\tag{1.69}
$$

$D(\pm u)$ being given by Eq. (1.45).

Note that the name "four-wave-mixing (FWM)" that is classically used is not well suited because any cubic process, whether the circular frequency relation is written $\omega_4 = \omega_1 + \omega_2 + \omega_3$ or $\omega_4 = -\omega_1 + \omega_2 + \omega_3$, corresponds to an interaction between four waves.

1.1.5.3 Relations between the First-Order and Third-Order Electric Susceptibilities

A comparison between the first-order electric susceptibility $\chi^{(1)}(\omega_p)$ from Eq. (1.28) with $p = 1,2,3,4$, and the third-order electric susceptibility from Eq. (1.67) and Eq. (1.69) shows that:

$$\frac{\chi^{(3)}(\omega_4 = \mp\omega_1 + \omega_2 + \omega_3)}{\chi^{(1)}(\omega_1)\chi^{(1)}(\omega_2)\chi^{(1)}(\omega_3)\chi^{(1)}(\omega_4)} = \delta \tag{1.70}$$

with

$$\delta = \frac{bm_e\varepsilon_0^3}{4N^3q^4} \tag{1.71}$$

As in the case of quadratic nonlinear optics, this parameter is of prime importance in the framework of crystal engineering (Chemla, Zyss 1987, Zyss 1994) [6,7].

1.2 LIGHT PROPAGATION IN THE LINEAR AND NONLINEAR REGIMES

1.2.1 Algorithm of Light-Matter Interaction

The description given in the previous sections allows us to have the main lines of the phenomenology of excitation of matter by an optical electric field. Actually, the incident optical electric field gives birth to an induced polarization that contains linear and nonlinear characteristics, which induces a rich spectrum. This excitation step is followed by a second one in which the polarization radiates an optical electric field over the same spectrum. The radiated optical field then becomes a source of excitation for the volume of adjacent matter, and so on. This sequence, valid in both time and spectral domain, constitutes an algorithm whose interaction length is typically the wavelength of light. It is depicted in Figure 1.8.

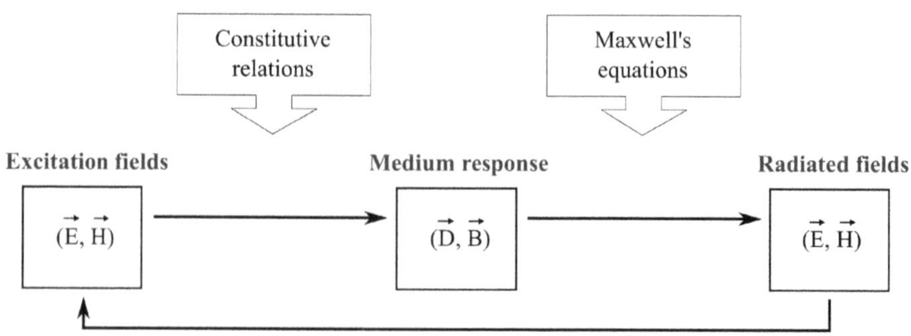

FIGURE 1.8 Algorithm describing the interaction between light and matter. $(\vec{E}, \vec{H})$ are respectively the electric field and magnetic field of the excitation or radiated light, while $(\vec{D}, \vec{B})$ are respectively the electric displacement and magnetic induction characterizing the excited matter.

The position vector had not been considered in the expressions of the electric field and induced polarization that have been given in the previous section because the description of the light-matter interaction was local. It is not the case in the present section where the target is now to model light propagation. Then a position vector $\vec{r}$ is added by now in all the considered quantities in the following. In particular, one has to consider the spatial phase term $e^{\pm i(\vec{k}.\vec{r})}$, where $\vec{k}$ is the wave vector, in any oscillating field.

1.2.2 Modeling in the Time Domain

1.2.2.1 Constitutive Equations

The constitutive equations deal with matter that is excited by an electro-magnetic wave. More precisely, they concern the coupling between the electric and magnetic properties of light and electrons. As a consequence, there are two constitutive relations: one dealing with the electric properties, the other one with the magnetic properties.

The applied optical electric field $\vec{E}(t,\vec{r})$ and the induced polarization of matter $\vec{P}(t,\vec{r})$ define the electric displacement vector $\vec{D}(t,\vec{r})$ from the following relation that is the "electric constitutive equation", $\vec{r}$ being the position vector (Bloembergen 1996) [3]:

$$\vec{D}(t,\vec{r}) = \varepsilon_0 \vec{E}(t,\vec{r}) + \vec{P}(t,\vec{r}) \tag{1.72}$$

with

$$\vec{P}(t,\vec{r}) = \vec{P}^{L}(t,\vec{r}) + \vec{P}^{NL}(t,\vec{r}) \tag{1.73}$$

where

$$\vec{P}^{L}(t,\vec{r}) = \vec{P}^{(1)}(t,\vec{r}) \tag{1.74}$$

and

$$\vec{P}^{NL}(t,\vec{r}) = \vec{P}^{(2)}(t,\vec{r}) + \vec{P}^{(3)}(t,\vec{r}) + \vec{P}^{(4)}(t,\vec{r}) + \dots. \tag{1.75}$$

for non-centrosymmetric media, where odd and even orders are allowed as seen in Part 1.1.4, and

$$\vec{P}^{NL}(t,\vec{r}) = \vec{P}^{(3)}(t,\vec{r}) + \vec{P}^{(5)}(t,\vec{r}) + \vec{P}^{(7)}(t,\vec{r})\dots \tag{1.76}$$

for centrosymmetric media, where only odd orders are allowed, as explained in Part 1.1.5.

The "magnetic constitutive equation" is formally analogous to the electrical constitutive relation given by Eq. (1.72) which writes (Born, Wolf 1999) [2]:

$$\vec{B}(t,\vec{r}) = \mu_0 \overrightarrow{H}(t,\vec{r}) + \mu_0 \overrightarrow{M}(t,\vec{r}) \tag{1.77}$$

μ_0 is the permeability of free space and $\vec{M}(t,\vec{r})$ the magnetization.

$\vec{M}(t,\vec{r})=0$ for non-magnetic media, which is the case in the present book. Then the magnetic constitutive relation reduces to:

$$\vec{B}(t,\vec{r})=\mu_0\vec{H}(t,\vec{r}) \tag{1.78}$$

1.2.2.2 Propagation Equation

Maxwell's equations describe the propagation of electro-magnetic waves in matter. In the time domain, and considering non-magnetic and non-conducting media, they are given by the following set of four equations (Rottwitt, Tidemand-Lichtenberg 2015) [8]:

$$div\left[\vec{D}(t,\vec{r})\right]=0 \tag{1.79}$$

$$\overrightarrow{rot}\left[\vec{E}(t,\vec{r})\right]=-\frac{\partial\vec{B}(t,\vec{r})}{\partial t} \tag{1.80}$$

$$div\left[\vec{B}(t,\vec{r})\right]=0 \tag{1.81}$$

$$\overrightarrow{rot}\left[\vec{H}(t,\vec{r})\right]=\frac{\partial\vec{D}(t,\vec{r})}{\partial t} \tag{1.82}$$

div and $\overrightarrow{rot}$ are spatial differential operators defined by:

$$div\left[\vec{A}\right]=\frac{\partial A_u}{\partial u}+\frac{\partial A_v}{\partial v}+\frac{\partial A_w}{\partial w} \tag{1.83}$$

and

$$\overrightarrow{rot}\left[\vec{A}\right]=\begin{pmatrix} rot_u\left[\vec{A}\right]=\dfrac{\partial A_w}{\partial v}-\dfrac{\partial A_v}{\partial w} \\ rot_v\left[\vec{A}\right]=\dfrac{\partial A_u}{\partial w}-\dfrac{\partial A_w}{\partial u} \\ rot_w\left[\vec{A}\right]=\dfrac{\partial A_v}{\partial u}-\dfrac{\partial A_u}{\partial v} \end{pmatrix} \tag{1.84}$$

where (A_u,A_v,A_w) are the Cartesian coordinates of vector $\vec{A}$ in an orthonormal frame (O,u,v,w).

It is important to notice that the fields $\vec{D}$ and $\vec{B}$ in Maxwell's equations correspond to the excited matter while $\vec{E}$ and $\vec{H}$ are the radiated fields of light, which corresponds exactly to the second step of the algorithm shown in Figure 1.8. Note also that the integration of these coupled equations can lead to the space and time evolutions of $\vec{E}$ and $\vec{H}$, which constitutes the propagation equation. An easy way to proceed is to combine Eq. (1.80) and Eq. (1.82), and to integrate the resulting equation. The combination can be done by applying the $\overrightarrow{rot}$ operator to Eq. (1.80), and by using Eq. (1.78) and Eq. (1.82), which gives:

$$\overrightarrow{rot}\left[\overrightarrow{rot}\left[\vec{E}(t,\vec{r})\right]\right]=-\mu_0\frac{\partial^2\vec{D}(t,\vec{r})}{\partial t^2} \tag{1.85}$$

After inserting Eq. (1.72) in Eq. (1.85), we get:

$$\overrightarrow{rot}\left[\overrightarrow{rot}\left[\vec{E}(t,\vec{r})\right]\right]=-\mu_0\varepsilon_0\frac{\partial^2\vec{E}(t,\vec{r})}{\partial t^2}-\mu_0\frac{\partial^2\vec{P}(t,\vec{r})}{\partial t^2} \tag{1.86}$$

Then by using $\frac{1}{c^2}=\mu_0\varepsilon_0$ where c is the magnitude of the velocity of light in vacuum, Eq. (1.86) becomes:

$$\overrightarrow{rot}\left[\overrightarrow{rot}\left[\vec{E}(t,\vec{r})\right]\right]+\frac{1}{c^2}\frac{\partial^2\vec{E}(t,\vec{r})}{\partial t^2}=-\mu_0\frac{\partial^2\vec{P}(t,\vec{r})}{\partial t^2} \tag{1.87}$$

Equation (1.87) is the conventional propagation equation of light in matter. In order to address the linear and nonlinear regimes, it is necessary to express the induced polarization by Eq. (1.73), which gives:

$$\overrightarrow{rot}\left[\overrightarrow{rot}\left[\vec{E}(t,\vec{r})\right]\right]+\frac{1}{c^2}\frac{\partial^2\vec{E}(t,\vec{r})}{\partial t^2}=-\mu_0\frac{\partial^2\vec{P}^L(t,\vec{r})}{\partial t^2}-\mu_0\frac{\partial^2\vec{P}^{NL}(t,\vec{r})}{\partial t^2} \tag{1.88}$$

Note that the $\overrightarrow{rot}$ operator can be also written as a function of the vector operator Nabla $\vec{\nabla}$. Then Eq. (1.88) is written:

$$\vec{\nabla}\times\vec{\nabla}\times\vec{E}(t,\vec{r})+\frac{1}{c^2}\frac{\partial^2\vec{E}(t,\vec{r})}{\partial t^2}=-\mu_0\frac{\partial^2\vec{P}^L(t,\vec{r})}{\partial t^2}-\mu_0\frac{\partial^2\vec{P}^{NL}(t,\vec{r})}{\partial t^2} \tag{1.89}$$

These two possible writings will be used in Chapter 5 to Chapter 8.

1.2.3 Modeling in the Frequency Domain

1.2.3.1 Fourier Transform

In nonlinear optics, it is especially interesting to consider the spectrum of the optical field that can be described by the superposition of monochromatic electromagnetic waves with a spectral distribution. Then all the fields relative to the interaction between light and matter, *i.e.* $\overrightarrow{A}(t,\vec{r})$ with $\overrightarrow{A}=(\overrightarrow{D},\overrightarrow{B},\overrightarrow{E},\overrightarrow{H})$, can be written:

$$\overrightarrow{A}(t,\vec{r})=\int_{-\infty}^{+\infty}\overrightarrow{A}(\omega,\vec{r})e^{-i\omega t}d\omega \tag{1.90}$$

where $\vec{A}(\omega,\vec{r})$ corresponds to any circular frequency ω of the field $\vec{A}(t,\vec{r})$. It is given by the Fourier transform of $\vec{A}(t,\vec{r})$, *i.e.*:

$$\vec{A}(\omega,\vec{r}) = \frac{1}{2\pi}\int_{-\infty}^{+\infty} \vec{A}(t,\vec{r})e^{i\omega t}\,dt \tag{1.91}$$

with

$$\vec{A}(-\omega,\vec{r}) = \vec{A}^*(\omega,\vec{r}) \tag{1.92}$$

1.2.3.2 Constitutive Equations

The frequency components of all the quantities depending on time and introduced in Part 1.2.2 can then be defined using Eq. (1.90), which gives for the induced polarization:

$$\vec{P}(\omega,\vec{r}) = \vec{P}^{L}(\omega,\vec{r}) + \vec{P}^{NL}(\omega,\vec{r}) \tag{1.93}$$

where

$$\vec{P}^{L}(\omega,\vec{r}) = \vec{P}^{(1)}(\omega,\vec{r}) \tag{1.94}$$

and

$$\vec{P}^{NL}(\omega,\vec{r}) = \vec{P}^{(2)}(\omega,\vec{r}) + \vec{P}^{(3)}(\omega,\vec{r}) + \vec{P}^{(4)}(\omega,\vec{r}) + \dots. \tag{1.95}$$

for non-centrosymmetric media, and

$$\vec{P}^{NL}(\omega,\vec{r}) = \vec{P}^{(3)}(\omega,\vec{r}) + \vec{P}^{(5)}(\omega,\vec{r}) + \vec{P}^{(7)}(\omega,\vec{r})\dots \tag{1.96}$$

for centrosymmetric media.

Furthermore, the two constitutive relations write:

$$\vec{D}(\omega,\vec{r}) = \varepsilon_0\vec{E}(\omega,\vec{r}) + \vec{P}(\omega,\vec{r}) \tag{1.97}$$

and

$$\vec{B}(\omega,\vec{r}) = \mu_0\vec{H}(\omega,\vec{r}) \tag{1.98}$$

1.2.3.3 Propagation Equation

The Fourier transform can be also applied to Eq. (1.80) until Eq. (1.83), which gives access to Maxwell's equations relative to each frequency component of the optical field. The Fourier transform does not affect the spatial operators, $div\left[\vec{A}(t,\vec{r})\right]$ and $\overrightarrow{rot}\left[\vec{A}(t,\vec{r})\right]$, but only the time differential for which it comes:

$$\frac{\partial\vec{A}(t,\vec{r})}{\partial t}=\int_{-\infty}^{+\infty}\vec{A}(\omega,\vec{r})\frac{\partial}{\partial t}[e^{-i\omega t}]d\omega=-i\omega\vec{A}(t,\vec{r}) \tag{1.99}$$

Then we get the following set of the four Maxwell's equations for each spectral component ω:

$$div\left[\vec{D}(\omega,\vec{r})\right]=0 \tag{1.100}$$

$$\overrightarrow{rot}\left[\vec{E}(\omega,\vec{r})\right]=i\omega\vec{B}(\omega,\vec{r}) \tag{1.101}$$

$$div\left[\vec{B}(\omega,\vec{r})\right]=0 \tag{1.102}$$

$$\overrightarrow{rot}\left[\vec{H}(\omega,\vec{r})\right]=-i\omega\vec{D}(\omega,\vec{r}) \tag{1.103}$$

Following the same way as in Part 1.2.2, it is easy to establish the propagation equation relative to each frequency component, *i.e.*:

$$\overrightarrow{rot}\left[\overrightarrow{rot}\left[\vec{E}(\omega,\vec{r})\right]\right]-\frac{\omega^2}{c^2}\vec{E}(\omega,\vec{r})=\omega^2\mu_0\vec{P}^L(\omega,\vec{r})+\omega^2\mu_0\vec{P}^{NL}(\omega,\vec{r}) \tag{1.104}$$

1.3 CLASSIFICATION OF LINEAR AND NONLINEAR OPTICAL EFFECTS USING A QUANTUM PICTURE

Here we use somewhat simplistic representations, but they allow us to approach the phenomena from a corpuscular point of view and to classify them.

1.3.1 Linear Optical Effects

Einstein had proposed a famous quantum picture for the description of linear absorption, spontaneous emission, and stimulated emission of light in matter, which are three linear optical effects mentioned in Part 1.1.3. This description is based on the interaction between photons and atoms, and has been at the heart of quantum and laser physics (Cohen Tannoudji 2004) [9]. Each photon carries a quantum of energy $\hbar\omega$, where $\hbar=\frac{h}{2\pi}$ is the reduced Planck constant. The atoms are characterized by two energy levels: the fundamental level E_f and the excited level $E_e>E_f$, separated by a bandgap $\Delta=E_f-E_e$. In the following, two kinds of situations will be considered: $\hbar\omega=\Delta$, which corresponds to what is called a "resonant process", and $\hbar\omega\neq\Delta$ corresponding to a "non-resonant process".

In all the following figures, the circles linked to horizontal arrows stand for propagating photons. The black horizontal lines denote the energy levels of each atom: the bottom one is the fundamental level, *i.e.* at E_f, and the top one is the excited one, *i.e.* at E_e. Vertical upward and downward arrows correspond to quanta of energy $\hbar\omega$ received or released by the atom. A black dot on the fundamental level means that the atom is in the fundamental state. It is in the excited state when the dot is on the excited level.

The first to be considered is an incident photon interacting with an atom initially in the fundamental state. If $\hbar\omega = \Delta$, the atom can reach the excited state thanks to absorption, as shown in Figure 1.9.

If now the atom is in its excited state, thanks to absorption or any other kind of excitation, it can spontaneously reach the fundamental state by emitting a photon carrying a quantum of energy that corresponds exactly to the gap, *i.e.* $\hbar\omega = \Delta$, as shown in Figure 1.10. This is the spontaneous emission of light. Distinct excited emitters in the considered medium behave independently and randomly such as incoherent emitters, so that there is no phase relationship between the distinct spontaneously emitted photons. Independent emitters thus behave as local sources, so that the resulting spontaneous emission is not directional.

The third situation imagined by Einstein is the stimulated emission. It is still a resonant elementary process corresponding to the interaction between an atom initially in the excited state. But in that case, the incident photon, with the quantum energy $\hbar\omega = \Delta$, can lead to the forced emission of an identical photon at $\hbar\omega$ by the atom reaching its fundamental state as shown in Figure 1.11. In this case, distinct emitters in the considered medium can behave in a synchronized way, such as coherent emitters, so that the distinct stimulated photons emitted from the same initial seed photon show the same phase. Emitters are no more independent emitters, so that the resulting stimulated emission is directional.

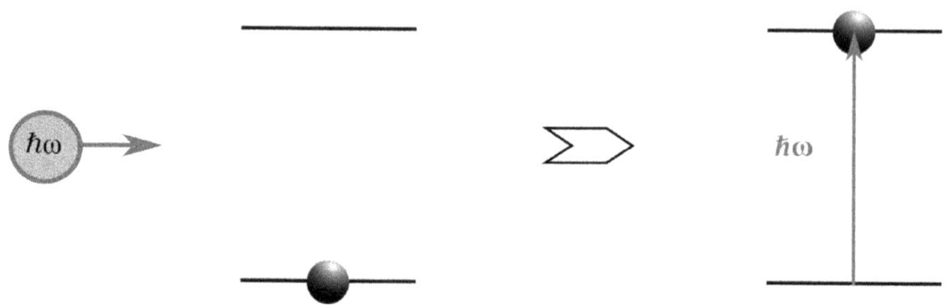

FIGURE 1.9 Linear absorption of light.

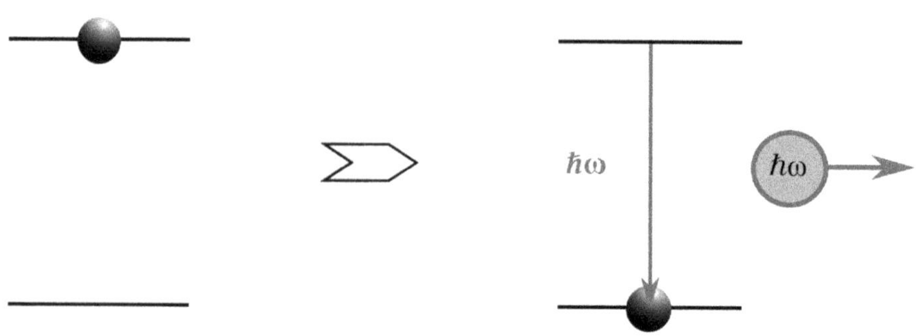

FIGURE 1.10 Spontaneous emission of light.

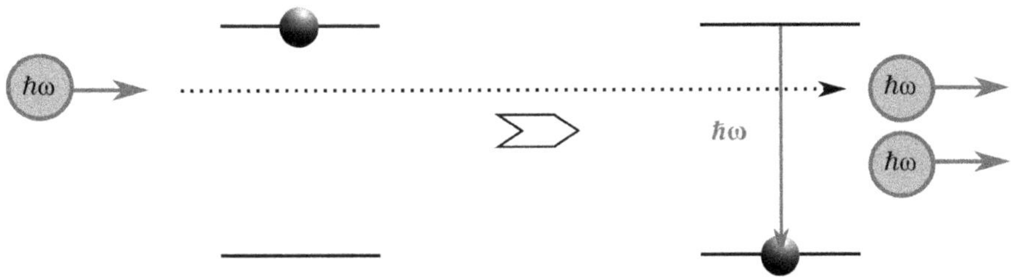

FIGURE 1.11 Stimulated emission of light.

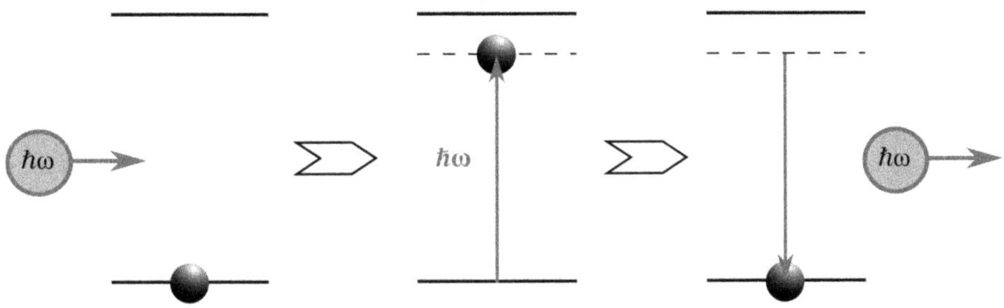

FIGURE 1.12 Linear propagation of light in a transparent medium.

Figure 1.11 well shows that two photons at $\hbar\omega$ are generated, which then corresponds to an amplification of the number of the incident photons, which is at the origin of the LASER effect (Cohen Tannoudji 2004) [9].

Linear absorption and stimulated emission are induced by an optical electric field, which is directly the framework of the Lorentz model described in Part 1.1. The case of spontaneous emission is more difficult to model since it is necessary to involve the quantum fluctuation of a vacuum as the excitation field (Cohen Tannoudji 2004) [9].

It is less common to see Einstein's representation used to describe the propagation of light in a transparent medium. That is what is being done hereafter. Thus, this situation corresponds to a non-resonant process: *i.e.* $\hbar\omega \neq \Delta$. If a photon at $\hbar\omega < \Delta$ is sent onto an atom in the fundamental state, as shown in Figure 1.12, it can transfer its energy to the atom which "sees" its own energy incremented by the quantum $\hbar\omega$. It is a virtual energy level for the atom, which is signified in Figure 1.12 by using a dashed line. The second step is the emission of light at $\hbar\omega$ correlatively to the return of the atom to its fundamental level. Then Figure 1.12 corresponds to the combination of Figure 1.9 and Figure 1.10 in which the real excitation level is replaced by a virtual one. Note that this quantum picture in two steps exactly corresponds to the classical picture at the macroscopic scale described by the algorithm presented in Figure 1.8.

In the case of a non-fluctuating medium, which is usually the case of a crystal, distinct oscillators are forced coherently, namely by an electric field at a given optical frequency and with a given phase distribution, so that the response of these oscillators leads to a coherent response that sustains light propagation and the associated directionality. But if a medium shows microscopic space-time random fluctuations, then forced oscillators partially behave as independent oscillators, so that the medium response also carries an incoherent response. In this case, the partially randomly phased relaxation from virtual states leads to a loss of phase relationship during the propagation light, resulting in the partial loss of light directionality. Such linear scattering process, also depicted by Figure 1.12, corresponds to the so-called Rayleigh scattering.

It is important to remind here that the three resonant elementary processes, *i.e.* the linear absorption as well as the spontaneous and stimulated emissions, are associated with the imaginary part of the first-order electric susceptibility, *i.e.* $\mathrm{Im}\left\{\chi^{(1)}\right\}$ where $\chi^{(1)}$ is given by Eq. (1.29a) for absorption, and Eq. (1.29b) for spontaneous and stimulated emissions. On the contrary, the propagation, which is a non-resonant process, is related to the real part of the first-order electric susceptibility, *i.e.* $\mathrm{Re}\left\{\chi^{(1)}\right\}$. Another major difference between these two kinds of situations concerns the dynamics: it is very quick—at the velocity of light—in the case of propagation, while it is much slower in the case of absorption and emission due to resonance with the energy levels of matter. These four linear optical interactions will be described in more detail in Chapter 5 and Chapter 6.

The goal of the two next sections is to describe quadratic and cubic nonlinear optical effects through spontaneous and stimulated elementary processes, as was done for the linear optical effects. The idea is to try to have a unified picture of all the optical interactions. This undertaking requires the use of the concepts of photon fusion and photon spontaneous or stimulated splitting that do not have their equivalent in the linear regime.

1.3.2 Quadratic Nonlinear Optical Effects

1.3.2.1 Fusion of Two Photons

In the following figures, the circles linked to horizontal arrows stand for propagating photons. Vertical upward and downward arrows correspond respectively to quanta of energies received and released by the atom.

Two incident photons at $\hbar\omega_1 < \Delta$ and $\hbar\omega_2 < \Delta$, with $(\hbar\omega_1 + \hbar\omega_2) < \Delta$, are first considered here. There is a probability that these two photons simultaneously transfer their energies to the atom initially in its fundamental state thanks to the excitation configuration shown in Figure 1.13. The two horizontal dashed lines stand for the two corresponding virtual levels.

The atom is thus brought at an energy $(\hbar\omega_1 + \hbar\omega_2)$, and it can then emit a photon with higher energy at:

$$\hbar\omega_3 = \hbar\omega_1 + \hbar\omega_2 \tag{1.105}$$

Equation (1.105) represents nothing more than the conservation of the energy of the photons that are involved.

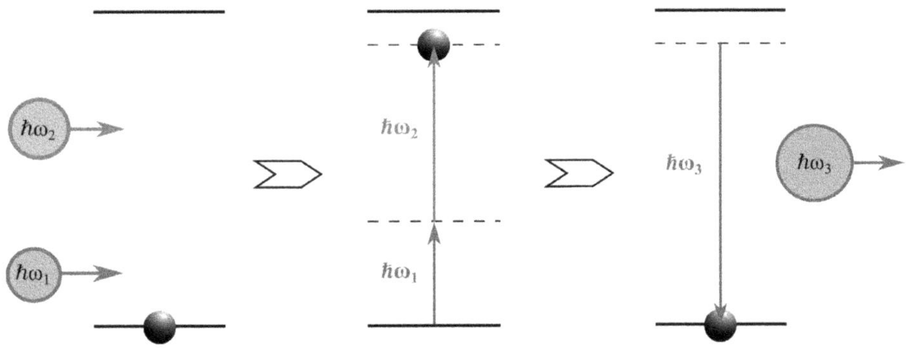

FIGURE 1.13 Fusion of two incident photons at $\hbar\omega_1$ and $\hbar\omega_2$ giving birth to a higher energy photon at $\hbar\omega_3 = \hbar\omega_1 + \hbar\omega_2$. It corresponds to quadratic SFG.

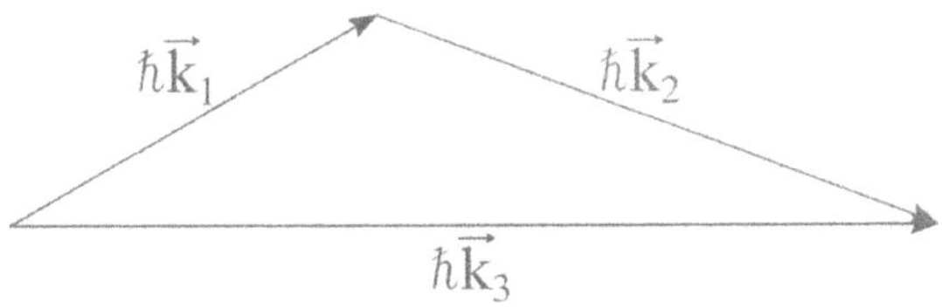

FIGURE 1.14 Momentum conservation corresponding to the fusion of two incident photons carrying the momenta $\hbar\vec{k_1}$ and $\hbar\vec{k_2}$, and giving birth to a higher energy photon such as $\hbar\vec{k_3} = \hbar\vec{k_1} + \hbar\vec{k_2}$.

The probability of occurrence of such an elementary process will be all the greater if the photon momentum is conserved, *i.e.*:

$$\hbar\vec{k_3} = \hbar\vec{k_1} + \hbar\vec{k_2} \tag{1.106}$$

where the $\vec{k_i}$, with i = 1, 2, 3, correspond to the wave vectors of the interacting photons. This condition has an equivalent in the wave approach, as will be seen and detailed in Chapter 7.

Figure 1.14 gives the representation of Eq. (1.106), the $\vec{k_i}$'s not necessarily being collinear.

This quantum description corresponds to the quadratic sum-frequency generation (SFG) previously introduced in Part 1.1.4.

Note that a fusion of two photons exhibiting the same quantum of energy $\hbar\omega$ will lead to the generation of a photon at $\hbar(2\omega)$, which is second-harmonic generation (SHG), as shown in Figure 1.15 (Franken 1961) [10].

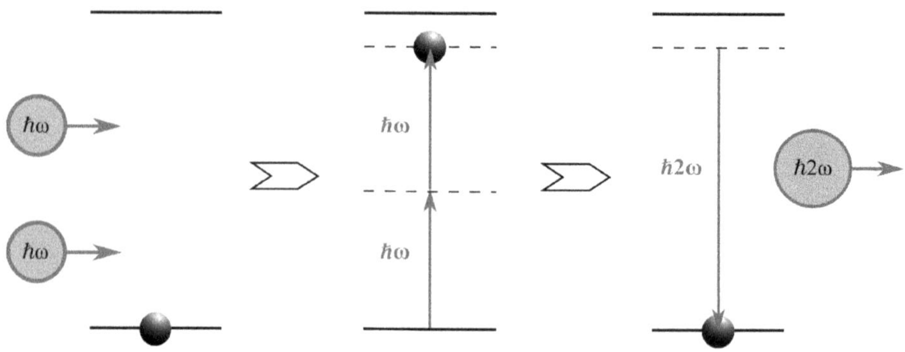

FIGURE 1.15 Fusion of two photons of same quantum of energy $\hbar\omega$ giving birth to a higher energy photon at $\hbar(2\omega) = \hbar\omega + \hbar\omega$. It corresponds to SHG.

This quantum picture of SHG shows that this process involves three entities, contrary to what the classical description performed in Part 1.1.4.1 might suggest, since a single incident excitation field was sufficient to lead to this phenomenon.

Note that SFG and SHG depend on the real part of the second-order electric susceptibility, *i.e.* $\mathrm{Re}\left\{\chi^{(2)}\right\}$, since they are non-resonant.

As for linear propagation in the case of a non-fluctuating medium, distinct oscillators are nonlinearly forced coherently so that the nonlinear response of these oscillators leads to a coherent SHG response that bears associated light directionality at the new radiated optical frequency. However, if a medium shows microscopic space-time random fluctuations, then nonlinearly forced oscillators partially behave as independent oscillators, so that the nonlinear medium response also carries an incoherent nonlinear response. In this case, the partially random-phase relaxation from virtual states leads to a loss of phase relationship during the light generation at double optical frequency, resulting in a partial loss of light directionality for this optical field. This nonlinear scattering process, also depicted by Figure 1.15, corresponds to the so-called Hyper-Rayleigh scattering, corresponding to the nonlinear counterpart of the Rayleigh scattering depicted in Figure 1.12 (Verbiest, Clays, Rodriguez 2009) [11].

1.3.2.2 Spontaneous and Stimulated Splitting of One Photon

The exact reverse of photon fusion is spontaneous photon splitting, which is shown in Figure 1.16. This process is called quadratic spontaneous parametric down-conversion (SPDC) or quadratic parametric fluorescence (PF) (Kulkarni 2022) [12].

Quadratic SPDC verifies the same photon energy and momentum conservations relations as SFG, *i.e.* Eq. (1.103) and Eq. (1.104). And as for photon fusion, the degenerate case in energy exists, which means that two photons at $\hbar\omega$ can be generated from the splitting of a single higher energy photon at $\hbar(2\omega)$. The two generated photons are the so-called "twin photons" often used in quantum optics (Aspect 1981) [13]. In the non-degenerate case, these photons are often called "signal" and "idler" photons, symbolized by "s" and "i", respectively. And usually, the signal photon is taken as the one with the higher energy, so that $(\omega_1, \omega_2) \equiv (\omega_i, \omega_s)$.

Note that the configuration depicted by Figure 1.16 corresponds to the propagation of light in the nonlinear quadratic regime, which is the "nonlinear equivalent" of the linear case described by Figure 1.12. It also corresponds to the algorithm of propagation shown in Figure 1.8.

The splitting of the incident photon at $\hbar\omega_3$ can be stimulated by a photon at $\hbar\omega_1$ or at $\hbar\omega_2$. A stimulation at $\hbar\omega_1$ is illustrated in Figure 1.17, which shows an amplification at $\hbar\omega_1$ as well as a generation at $\hbar\omega_2 = \hbar\omega_3 - \hbar\omega_1$.

Figure 1.17 shows that the splitting of the photon at $\hbar\omega_3$ stimulated by one photon at $\hbar\omega_1$ leads to an amplification at $\hbar\omega_1$, which is called "quadratic optical parametric amplification (OPA)", and to a new photon at $\hbar\omega_2 = \hbar\omega_3 - \hbar\omega_1$ corresponding to quadratic DFG (Byer 1968, Harris 1969) [14,15]. Then quadratic OPA and DFG are two dual optical nonlinear effects that correspond to the same elementary process.

As in the case of photon fusion, the spontaneous or stimulated photon splitting can be non-resonant, *i.e.* $\hbar\omega_3 < \Delta$ as described earlier, or resonant, *i.e.* $\hbar\omega_3 = \Delta$. The latter case is the nonlinear quadratic counterpart of the spontaneous or stimulated emission of the linear regime described in Part 1.3.1.

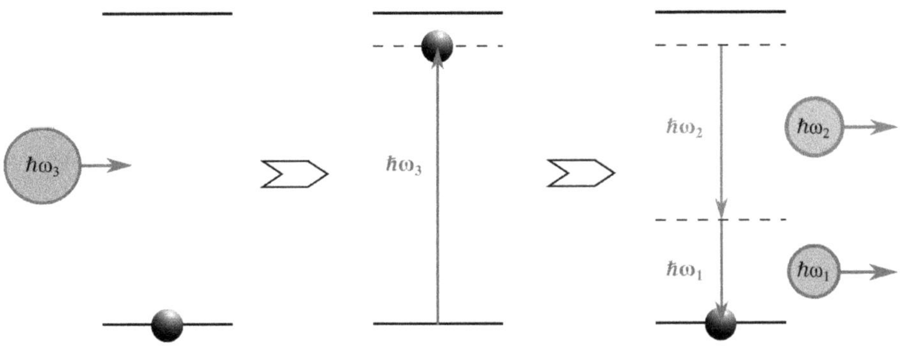

FIGURE 1.16 Spontaneous splitting of one photon at $\hbar\omega_3$ into two lower energy photons at $\hbar\omega_1$ and $\hbar\omega_2$. It corresponds to quadratic/second-order SPDC or PF.

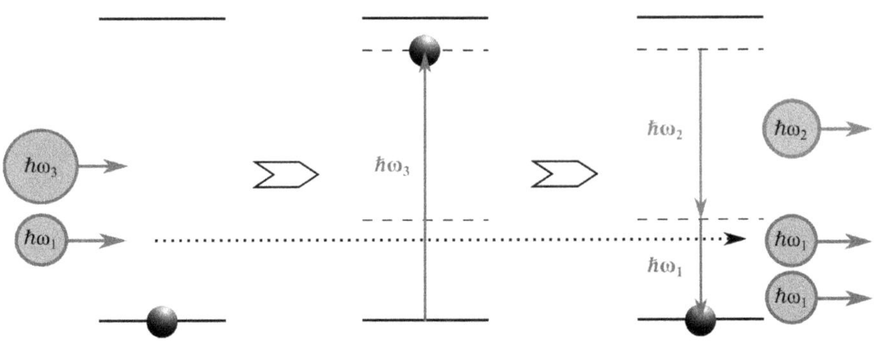

FIGURE 1.17 Splitting of one photon at $\hbar\omega_3$ stimulated by one photon at $\hbar\omega_1$; it leads to an OPA at $\hbar\omega_1$, and to a new photon at $\hbar\omega_2 = \hbar\omega_3 - \hbar\omega_1$ by quadratic DFG.

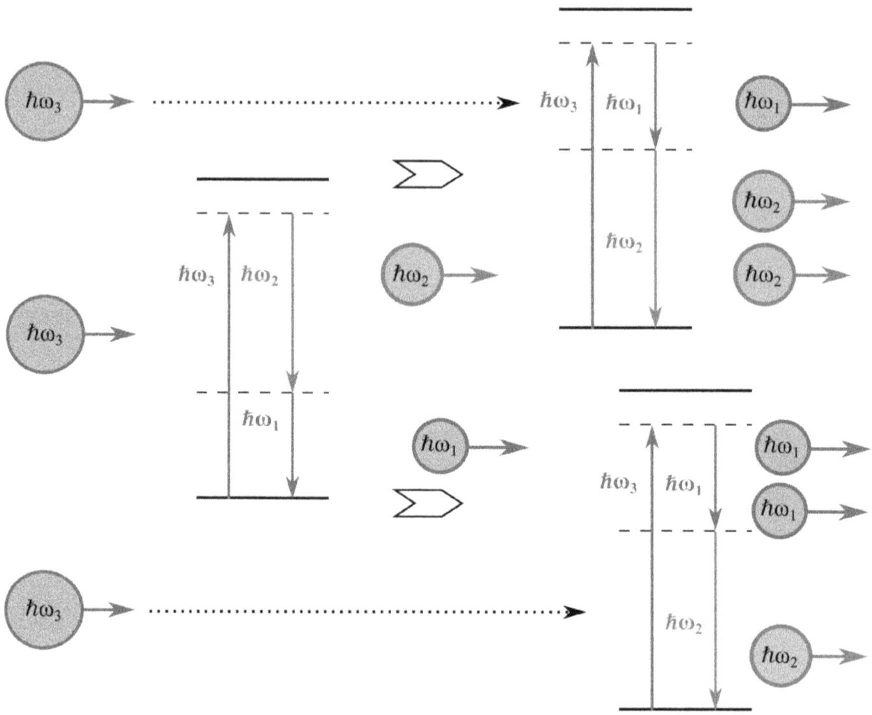

FIGURE 1.18 Cascading of spontaneous and stimulated photon splitting. It corresponds to quadratic OPG or OPO.

In realistic situations, the lighting of matter at $\hbar\omega_3$ leads in fact to a cascading of spontaneous and stimulated photon splitting, as shown in Figure 1.18. It is called "quadratic optical parametric generation (OPG)" for a single pass in matter, or "quadratic optical parametric oscillation (OPO)" when matter is inserted in an optical cavity in order to propagate light over several round trips, like the ones implemented for a LASER (Byer 1973, Ebrahimzadeh, Dunn 2000) [16,17].

The same kind of description can be applied to the cubic nonlinear effects, which is done hereafter.

1.3.3 Cubic Nonlinear Optical Effects

1.3.3.1 Fusion of Three Photons

The non-resonant fusion of three photons at $\hbar\omega_1$, $\hbar\omega_2$, and $\hbar\omega_3$, with $\hbar\omega_1 + \hbar\omega_2 + \hbar\omega_3 < \Delta$ is depicted in Figure 1.19 where we took $\omega_1 < \omega_2 < \omega_3 < \omega_4$. It is governed by the real part of the third-order electric susceptibility $\text{Re}\left\{\chi^{(3)}\right\}$, and it corresponds to the cubic SFG previously described in Part 1.1.5.2. It leads to the generation of a photon at $\hbar\omega_4$ with the corresponding photon energy conservation relation:

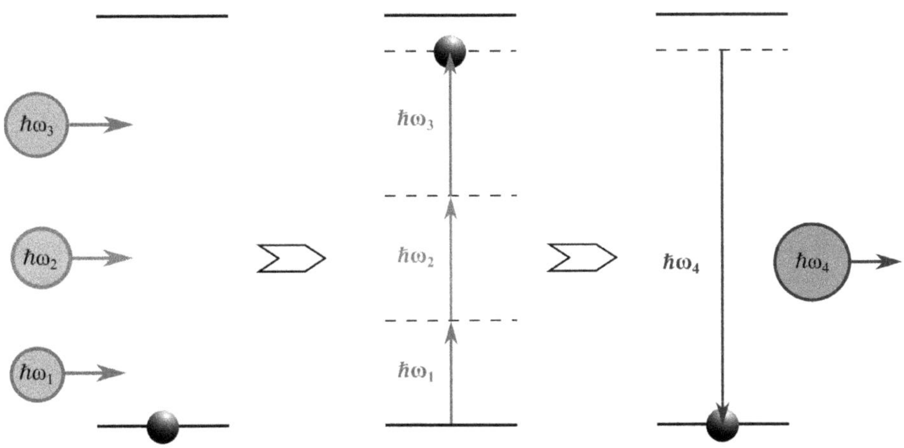

FIGURE 1.19 Fusion of three incident photons at $\hbar\omega_1$, $\hbar\omega_2$, and $\hbar\omega_3$ giving birth to a higher energy photon at $\hbar\omega_4 = \hbar\omega_1 + \hbar\omega_2 + \hbar\omega_3$. It corresponds to cubic SFG.

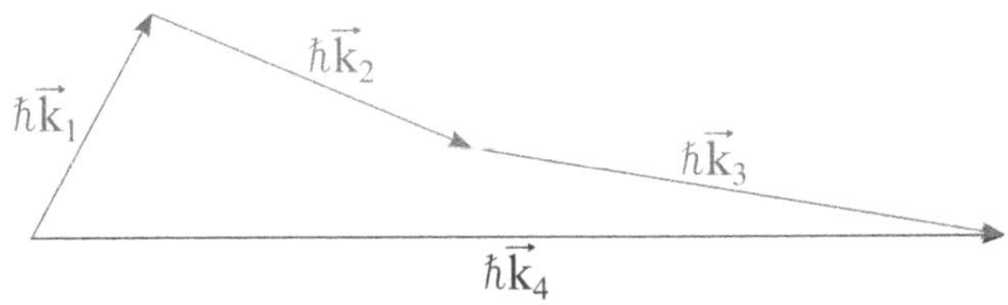

FIGURE 1.20 Momentum conservation corresponding to the fusion of two incident photons carrying the momenta $\hbar\vec{k_1}$, $\hbar\vec{k_2}$, and $\hbar\vec{k_3}$, and giving birth to a higher energy photon such as $\hbar\vec{k_4} = \hbar\vec{k_1} + \hbar\vec{k_2} + \hbar\vec{k_3}$.

$$\hbar\omega_4 = \hbar\omega_1 + \hbar\omega_2 + \hbar\omega_3 \tag{1.107}$$

The photon momentum conservation is then written:

$$\hbar\vec{k_4} = \hbar\vec{k_1} + \hbar\vec{k_2} + \hbar\vec{k_3} \tag{1.108}$$

It is depicted in Figure 1.20 in the general case in which the four wave vectors are not collinear.

The degenerate case in energy of the fusion of three photons corresponds to THG, as shown in Figure 1.21.

The three-photon fusion can be also resonant, *i.e.* $\hbar\omega_1 + \hbar\omega_2 + \hbar\omega_3 = \Delta$, so that it is linked to the imaginary part of the third-order electric susceptibility $\text{Im}\left\{\chi^{(3)}\right\}$.

1.3.3.2 Spontaneous and Stimulated Splitting of One Photon

Cubic SPDC, also called cubic parametric fluorescence, corresponds to the spontaneous splitting of one high energy photon at $\hbar\omega_4 < \Delta$ into three lower energy photons at $\hbar\omega_1$, $\hbar\omega_2$, and $\hbar\omega_3$. The corresponding photon energy and momentum conservations are given by Eq. (1.107) and Eq. (1.108) respectively, which is described in Figure 1.22.

The degenerate case of cubic SPDC leads to the generation of three photons at $\hbar\omega$ from a photon at $\hbar(3\omega)$. These photons are called "triple photons".

The photon splitting can be also stimulated, by one or two photons, as shown in Figure 1.23 and Figure 1.24, respectively (Bertrand 2025, Douady 2004) [18,19].

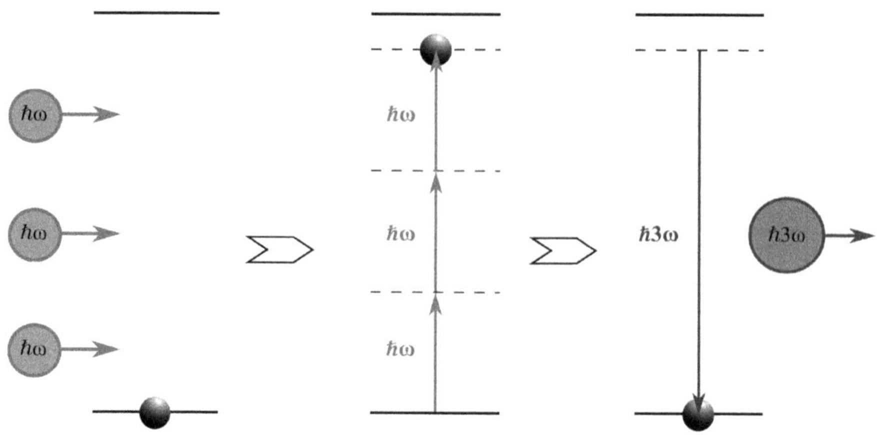

FIGURE 1.21 Fusion of three photons of same quantum of energy $\hbar\omega$ giving birth to a higher energy photon at $\hbar(3\omega) = \hbar\omega + \hbar\omega + \hbar\omega$. It corresponds to THG.

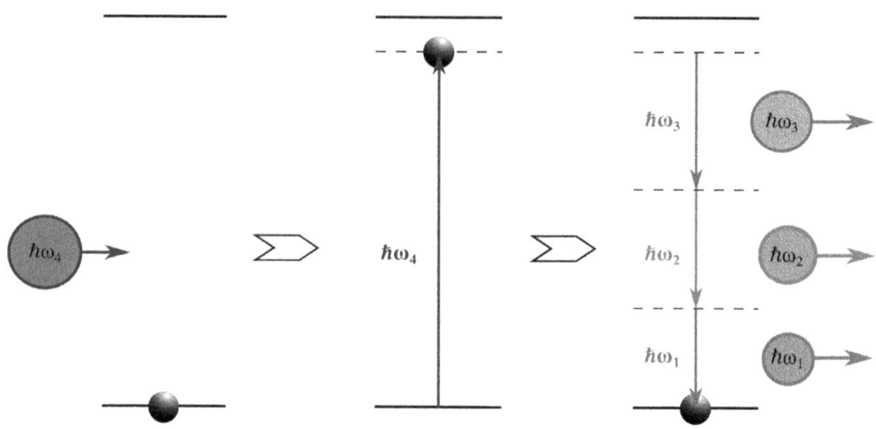

FIGURE 1.22 Spontaneous splitting of one photon at $\hbar\omega_4$ into three lower energy photons at $\hbar\omega_1$, $\hbar\omega_2$, and $\hbar\omega_3$. It corresponds to cubic SPDC or PF.

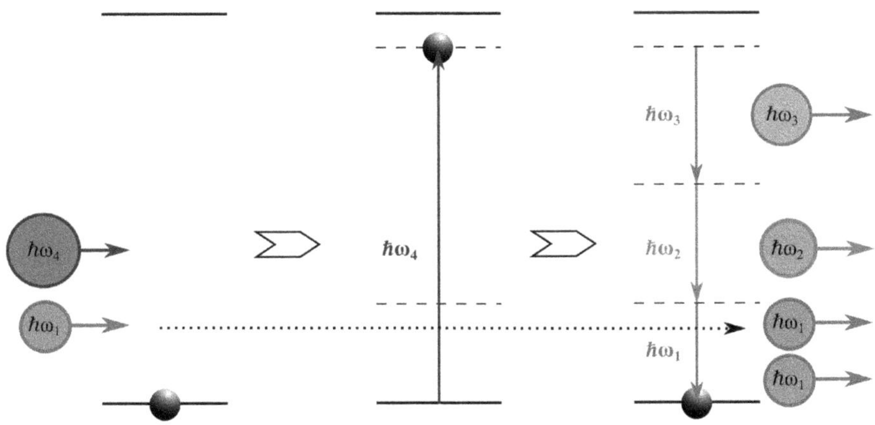

FIGURE 1.23 Splitting of one photon at $\hbar\omega_4$ stimulated by one photon at $\hbar\omega_1$ leading to an OPA at $\hbar\omega_1$, and to two new photons at $\hbar\omega_2 = \hbar\omega_4 - \hbar\omega_1 - \hbar\omega_3$ and $h\omega_3 = \hbar\omega_4 - \hbar\omega_1 - \hbar\omega_2$ by cubic DFG.

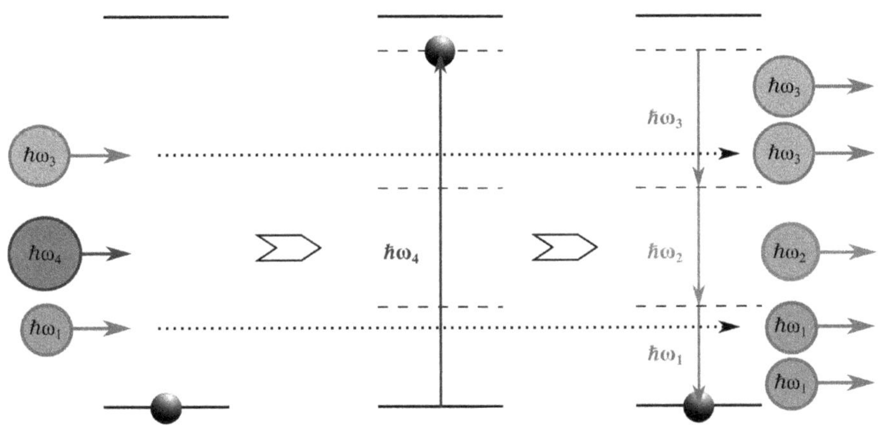

FIGURE 1.24 Splitting of one photon at $\hbar\omega_4$ simultaneously stimulated by one photon at $\hbar\omega_1$ and one photon at $\hbar\omega_3$ leading to an OPA at $\hbar\omega_1$ and $\hbar\omega_3$, and to a new photon at $\hbar\omega_2 = \hbar\omega_4 - \hbar\omega_1 - \hbar\omega_3$ by cubic DFG.

As in the case of the stimulated splitting of one photon into two photons, OPA and DFG are the two sides of the same coin. And there exists also the cascading of cubic SPDC and OPA, corresponding to cubic OPG or OPO.

1.3.3.3 Fusion-Scission of Two Photons Giving Birth to Two Photons

The previous cubic processes are of the same kinds as the quadratic ones, *i.e.* pure photon fusion or pure photon splitting. But as pointed out in the classical analysis performed in Part 1.1.5.2, there is a category of cubic interactions that have no equivalent in the quadratic ones, *i.e.* the so-called FWM. The corresponding photon energy conservation relation is written:

$$\hbar\omega_4 + \hbar\omega_1 = \hbar\omega_2 + \hbar\omega_3 \tag{1.109}$$

It was stated in Part 1.1.5.2 that these cubic interactions could be depicted by a combination of SFG and DFG. The quantum equivalent of such a description is a combination of a fusion of two photons followed by a splitting process giving birth to two new photons, as shown in Figure 1.25 in the non-resonant case, *i.e.* when $\hbar\omega_4 + \hbar\omega_1 = \hbar\omega_2 + \hbar\omega_3 < \Delta$.

The photon momentum conservation shown in Figure 1.26 is expressed as:

$$\hbar\vec{k_4} + \hbar\vec{k_1} = \hbar\vec{k_2} + \hbar\vec{k_3} \tag{1.110}$$

The same non-resonant interaction exists also when the four energies are equal, which is the fully degenerated optical Kerr effect depicted in Figure 1.27.

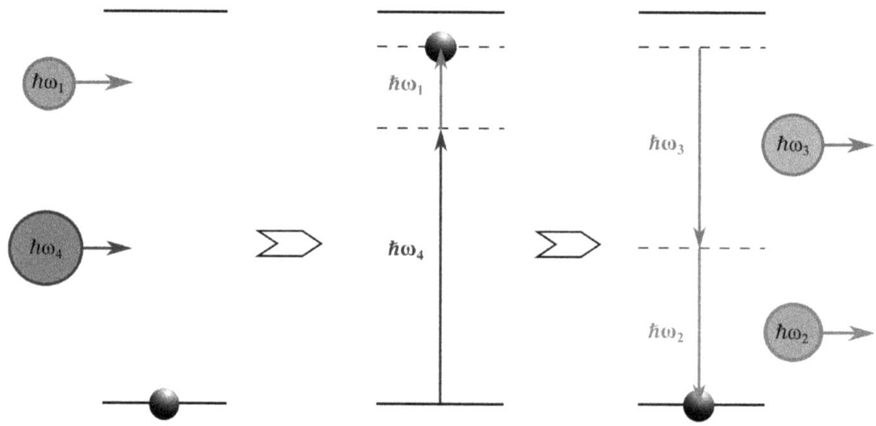

FIGURE 1.25 Fusion of two photons at $\hbar\omega_4$ and $\hbar\omega_1$ leading to the generation of two photons at $\hbar\omega_2$ and $\hbar\omega_3$ by FWM.

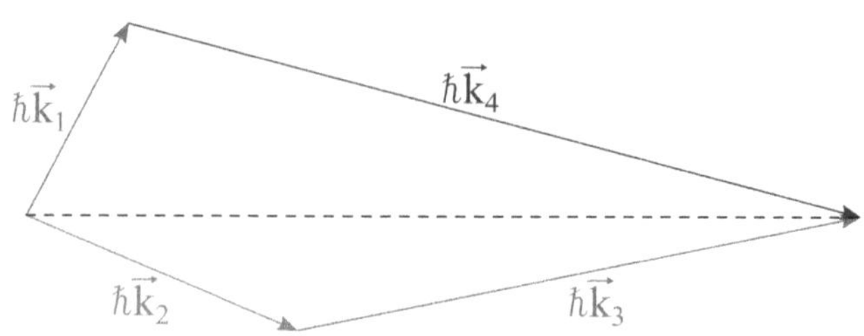

FIGURE 1.26 Momentum conservation corresponding to the fusion of two incident photons carrying the momenta $\hbar\vec{k_1}$ and $\hbar\vec{k_4}$, and giving birth to two higher energy photons carrying the momenta $\hbar\vec{k_2}$ and $\hbar\vec{k_3}$, such as $\hbar\vec{k_4} + \hbar\vec{k_1} = \hbar\vec{k_2} + \hbar\vec{k_3}$.

Figure 1.27 is the nonlinear equivalent of Figure 1.12 for the linear propagation. It then corresponds to the propagation of light in the nonlinear regime, and it is governed by $\mathrm{Re}\left\{\chi^{(3)}\right\}$.

The optical Kerr effect at resonance, *i.e.* when $\omega+\omega=\Delta$, can also exist, and it is governed by $\mathrm{Im}\left\{\chi^{(3)}\right\}$. This process is called "two-photon absorption (TPA)". The corresponding diagram is given in Figure 1.28. It can be directly compared with Figure 1.9 that corresponds to linear absorption.

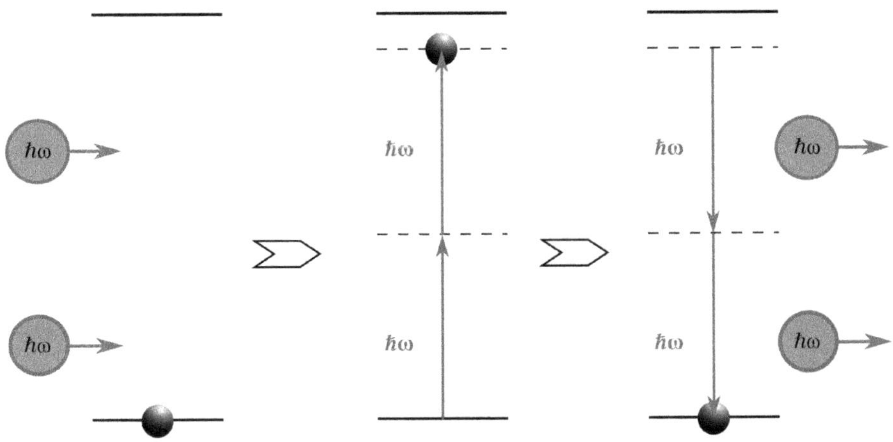

FIGURE 1.27 Fusion of two photons at $\hbar\omega$ leading to the generation of two photons at $\hbar\omega$ by the optical Kerr effect.

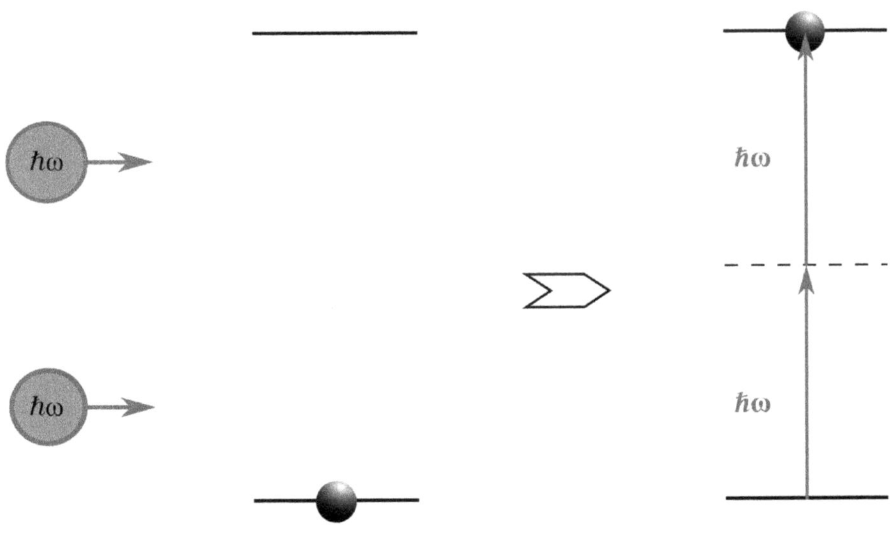

FIGURE 1.28 Absorption of two photons at $\hbar\omega$ by TPA.

Two other famous FWM processes are the Coherent Anti-Stokes Raman Scattering and the Coherent Stokes Raman Scattering described by Figure 1.29 and Figure 1.30, respectively.

Coherent Anti-Stokes Raman Scattering involves two incident photons at $\hbar\omega_P$, called the "pump", and one photon at $\hbar\omega_S$ that is the Stokes photon, as shown in Figure 1.29. The atom reaches a first virtual excited level thanks to the absorption at $\hbar\omega_P$, followed by a stimulated emission at $\hbar\omega_S$ allowing the atom to reach an intermediate excited state of matter from which it can reabsorb a photon at $\hbar\omega_P$. The atom is then in a new virtual excited state from which a spontaneous emission occurs at $\hbar\omega_{AS}$ which is the Anti-Stokes photon. Figure 1.29 also shows that there is an amplification at $\hbar\omega_S$.

In Coherent Stokes Raman Scattering, the incident photons are at $\hbar\omega_P$ and $\hbar\omega_{AS}$, leading to amplification at $\hbar\omega_{AS}$ and to the generation to a new photon at $\hbar\omega_S$ according to Figure 1.30.

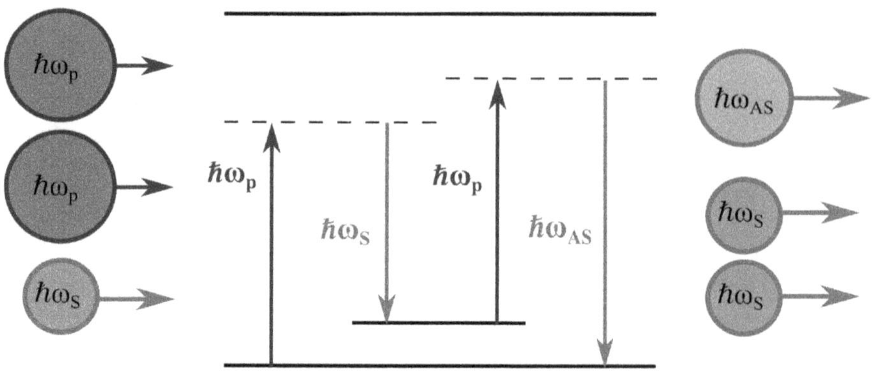

FIGURE 1.29 Coherent Anti-Stokes Raman Scattering.

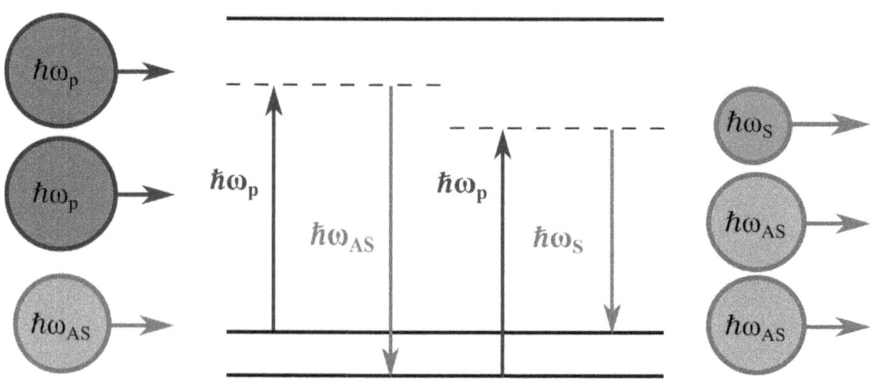

FIGURE 1.30 Coherent Stokes Raman Scattering.

It is interesting to notice that the Coherent Anti-Stokes Raman Scattering and the Coherent Stokes Raman Scattering are two FWM interactions ($\hbar\omega_4 + \hbar\omega_1 = \hbar\omega_2 + \hbar\omega_3$) that are partially degenerate in energy, *i.e.* $\omega_4 = \omega_p$, $\omega_1 = \omega_2 = \omega_S$, and $\omega_3 = \omega_{AS}$ for $\omega_4 = \omega_p$ Coherent Anti-Stokes Raman Scattering; and $\omega_4 = \omega_p$, $\omega_1 = \omega_2 = \omega_{AS}$, and $\omega_3 = \omega_S$ for Coherent Stokes Raman Scattering.

1.4 SUMMARY

We saw how important it is to combine the wave and quantum pictures in order to have a unified and complete description of the linear and nonlinear optical effects. Quadratic and cubic nonlinear optics then appeared as based on sum- and difference-frequency generations and their equivalents in terms of photon fusion and splitting respectively, the splitting being spontaneous or stimulated. Thus, nonlinear optics makes it possible to carry out optical frequency synthesis. The resulting light is often called "parametric light", and one speaks of "parametric nonlinear optics".

Contrary to the case of the LASER effect in which the involved elementary processes, *i.e.* linear absorption, spontaneous emission, and stimulated emission, are resonant, nonlinear optics offers the possibility to have frequency conversion out of resonance. It is a very good example showing that light can transfer energy to atoms even when the interaction is non-resonant. But in that case, the energy is not stored but instantaneously released under light. Whether the optical effects are linear or nonlinear, their efficiency depends on the imaginary part of the corresponding electrical susceptibility in the resonant case, or on its real part in the non-resonant case. On the other hand, the instantaneous nature of the non-resonant process means that there is a phase link between the excitation light and the emitted light. In particular, parametric light will retain the spatial and temporal coherence properties of laser light, the direction of the electric polarization of the excitation and emitted light playing a crucial role as will be seen in Chapter 4 to Chapter 8 (Chekhova, Banzer 2021) [20].

Finally, the strength of the Lorentz model using the harmonic and anharmonic regimes pictures is also to provide dispersion equations for the linear and nonlinear electric susceptibilities. It will be of prime importance in the mastering of the corresponding properties, particularly in the framework of the design of efficient linear and nonlinear crystals. This phenomenological model also brings to light the crucial role of symmetry in nonlinear optics, by establishing the impossibility to have any quadratic optical nonlinear effect in a centrosymmetric medium, while a non-centrosymmetric medium allows quadratic as well cubic optical nonlinear interactions. We will come back to this feature and other symmetry effects in Chapter 2 through Chapter 4 using the tensor formalism. Actually, it will be seen that the different orders of the electric susceptibility are described by tensors.

REFERENCES

[1] Feynman R. P., Leighton R. B., and Sands M. 2005. The Feynman Lectures on Physics, Addison, rev. and ext. ed., US: Wesley.

[2] Born M. and Wolf E. 1980. Principle of Optics, 6th ed., New York: Pergamon Press.

[3] Bloembergen N. 1996. Nonlinear Optics, 4th ed., New York and Amsterdam: W.A. Benjamin, Inc.
[4] Boyd R. W. 2008. Nonlinear Optics, 3rd ed., New York: Academic Press.
[5] Miller R. C. 1964. Optical second harmonic generation in piezoelectric crystals. Applied Physics Letters 5: 17–19.
[6] Chemla D. S. and Zyss J. 1987. Nonlinear Optical Properties of Organic Molecules and Crystals. Quantum Electronic Principles and Applications Series. New York: Academic Press.
[7] Zyss J. 1994. Editor. Molecular Nonlinear Optics: Materials, Physics and Devices. Quantum Electronic Principles and Applications Series. New York: Academic Press.
[8] Rottwitt K. and Tidemand-Lichtenberg P. 2015. Nonlinear Optics. CRC Press.
[9] Cohen Tannoudji C. 2004. Atoms in Electromagnetic Fields. World Scientific Series on Atomic, Molecular and Optical Physics, 2nd ed., World Scientific Publishing Co Pte Ltd.
[10] Franken P., Hill A. E., Peters C. W., and Weinreich G. 1961. Generation of optical harmonics. Physical Review Letters 7: 118.
[11] Verbiest T., Clays K., and Rodriguez V. 2009. Second-Order Nonlinear Optical Characterization Techniques. CRC Press.
[12] Kulkarni G., Rioux J., Braverman B., Chekhova M. V., and Boyd R. W. 2022. Classical model of spontaneous parametric down-conversion. Physical Review Research 4: 033098.
[13] Aspect A., Grangier P., and Roger R. 1981. Experimental test of realistic local theories via Bell's theorem. Physical Review Letters 47(7): 460–463.
[14] Byer R. L., Oshman M. K., Young J. F., and Harris S. E. 1968. Visible CW parametric oscillator. Applied Physics Letters 13: 109.
[15] Harris S. E. 1969. Tunable optical parametric oscillators. Proceedings IEEE 57(12): 2096–2113.
[16] Byer R. L. 1973. Treatise in Quantum Electronics, edited by H. Rabin & C. L. Tang. New York: Academic Press.
[17] Ebrahimzadeh M. and Dunn M. H. 2000. Optical parametric oscillators. In Handbook of Optics, vol. IV, pp. 2201–2272. New York: McGraw-Hill.
[18] Bertrand J., Boutou V., Felix C., Jegouso D., and Boulanger B. 2025. Experimental demonstration and modeling of near-infrared nonlinear third-order triple-photon generation stimulated over one mode. Applied Physics Letters Quantum. https://doi.org/10.1063/5.0254046.
[19] Douady J. and Boulanger B. 2004. Experimental demonstration of a pure third-order optical parametric downconversion process. Optics Letters 29(23): 2794–2796.
[20] Chekhova M. and Banzer P. 2021. Polarization of Light: In Classical, Quantum, and Nonlinear Optics. De Gruyter Textbook Series. Berlin, Boston: De Gruyter.

2 Mathematical Backgrounds in Orientation Symmetry and Tensors

2.1 ORIENTATION SYMMETRY

2.1.1 Symmetries of Finite Order

2.1.1.1 Symmetry Operations and Elements of Finite Order

The orientation symmetry of crystals is described on the basis of rotations around five n-fold axes, with $n = 1, 2, 3, 4, 6$, and the inversion written $\bar{1}$, and the four combinations of the inversion with the 2-, 3-, 4-, and 6-fold rotations (Fischer and Koch 2005) [1]. These combinations correspond to the so-called "roto-inversion operations" that are written: $\bar{2}$, $\bar{3}$, $\bar{4}$, and $\bar{6}$. That gives a total of ten symmetry operations describing the orientation symmetry of finite order. The numeral n is called the "order" of the symmetry operation or element.

A crystal having an n-fold axis is identically transformed to itself by a rotation of an angle of 360° / n around this axis. The effects of these symmetry operations are schematically shown in Figure 2.1 using the nomenclature of Hermann-Mauguin, which is also the one used in crystallography. Another one is the nomenclature of Schönflies commonly used in spectroscopy (Pechrasky and Zavalij 2009) [2]. These two nomenclatures hold for the symmetry operations as well as for the symmetry elements that generate these operations. Among the ten configurations depicted in Figure 2.1, it clearly appears that the 1-fold axis corresponds to identity, the $\bar{2}$-fold axis is equivalent to a mirror plane perpendicular to the $\bar{2}$-fold axis itself, and the $\bar{6}$-fold axis is equivalent to a 3-fold axis associated with a perpendicular mirror plane. In the following, the $\bar{2}$-fold axis will be often written m. It is also straightforward to notice from Figure 2.1 that among the five roto-inversion operations, only $\bar{1}$ and $\bar{3}$ exhibit an inversion center: it is obvious for $\bar{1}$, and it appears that $\bar{3}$ is equivalent to a 3-fold axis associated with an inversion center. Furthermore, Figure 2.1 allows us to know the "multiplicity" of each symmetry operation, defined by the number of objects generated by the operation, these objects being equivalent. We can notice that the multiplicity of an n-fold axis and a $\bar{n}$-fold axis is the same and equal to n, except for the case $\bar{3}$ in which 6 objects are generated.

It is not so difficult to have a three-dimensional (3D) picture of the effect of each of the ten elementary operations of orientation symmetry as shown in Figure 2.1. However, when several of these elementary operations are combined, as will be

DOI: 10.1201/9780429091735-2

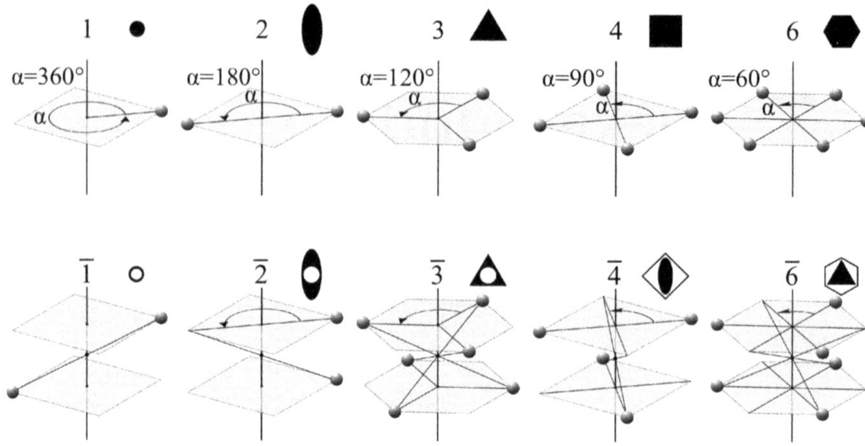

FIGURE 2.1 The ten symmetry operations describing orientation symmetry: 1, 2, 3, 4, and 6 stand for the rotation operations, and $\bar{1}$, $\bar{2}$, $\bar{3}$, $\bar{4}$, and $\bar{6}$ for the roto-inversion operations. The graphical black symbol at the top of each graph corresponds to the representation of each symmetry element. In each case, the grey spheres correspond to the symmetrically equivalent objects generated by the operation. α is the angle between two consecutive symmetrically equivalent objects.

done in Part 2.1.1.2 devoted to the thirty-two crystal classes, the 3D representation is very difficult or simply impossible. In this case, a 2D representation based on the "stereographic projection" is used (Pechrasky and Zavalij 2009) [2]. To achieve such a representation, an oriented sphere is considered, in which a north pole (N) and south pole (S) are defined as drawn in Figure 2.2. The plane perpendicular to the NS axis, and located at the middle of this axis, is called the "equatorial plane". Above and below this plane are the north and south hemispheres respectively. The stereographic projection of any point of the surface of the sphere is made on the equatorial plane according to the following rule depicted in Figure 2.2: the image "a" of any point "A" from the north hemisphere is given by the intersection between the line AS and the equatorial plane, the corresponding symbol being a cross "x"; and the image "b" of any point "B" from the south hemisphere is given by the intersection between the line BN and the equatorial plane, the corresponding symbol being a circle "o".

The example of the inversion operation $\bar{1}$ that transforms point A into point B or the reverse, with the inversion center being located at the center of the sphere, is depicted in Figure 2.3, showing the corresponding locations of points a and b. The center of the sphere is depicted in the stereographic projection by the graphical symbol ○ that represents the symmetry element of the inversion operation, as introduced in Figure 2.1.

Another example is given in Figure 2.4, corresponding to symmetry operation 3, the 3-fold rotation axis being taken as the NS axis. The center of the sphere is depicted in the stereographic projection by the graphical symbol ▲ that represents the symmetry element of the 3-fold rotation operation, as introduced in

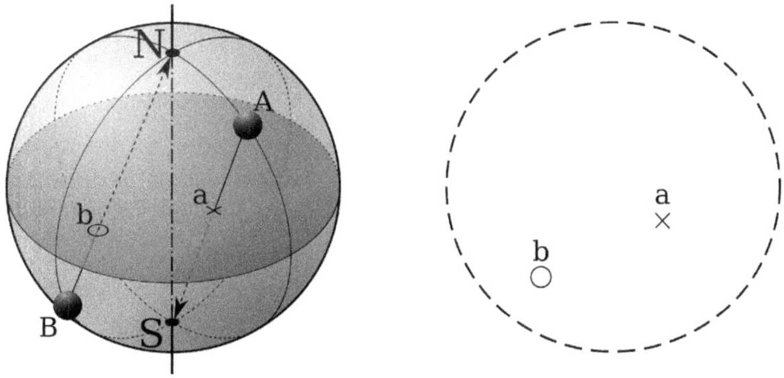

FIGURE 2.2 Principle of the stereographic projection: point "a" located in the equatorial plane, represented by the cross "x", is the projection of point A located at the surface of the sphere in the north hemisphere; point "b" located in the equatorial plane, represented by the circle "o", is the projection of point B located at the surface of the sphere in the south hemisphere.

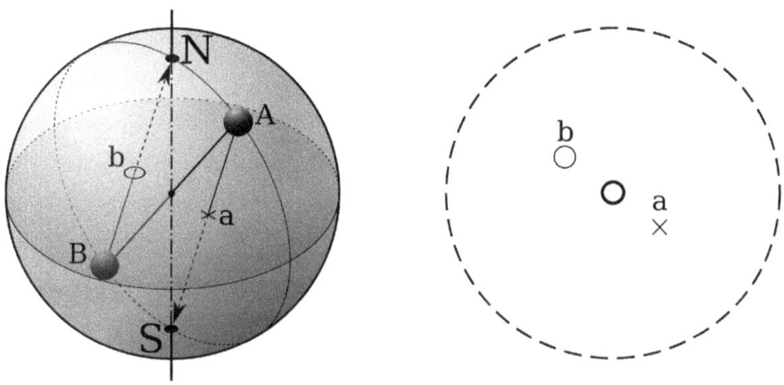

FIGURE 2.3 Example of the stereographic projection of the symmetry operation $\bar{1}$, the inversion center $\bar{1}$ being located at the center of the sphere.

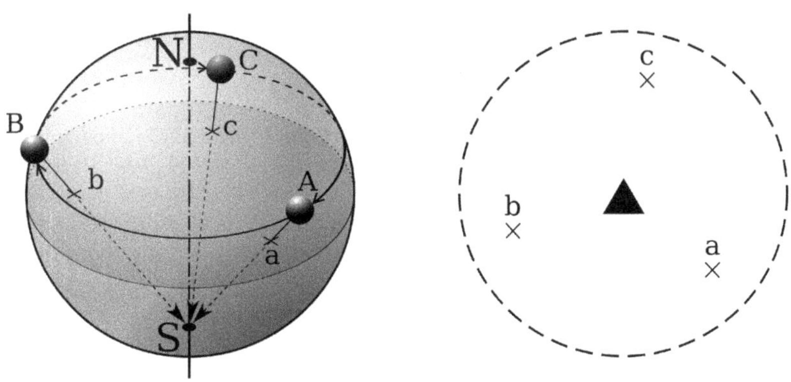

FIGURE 2.4 Example of the stereographic projection of the symmetry operation 3 in which the 3-fold axis is taken as the NS axis.

Figure 2.1. One can easily see that the 3-fold rotation around the NS axis that transforms A, B, and C one between each other also transforms a, b, and c one into the other ones, confirming the conservation of the angles by the stereographic projection.

It is also important to underline that there is a convention for the representation of the symmetry elements using the stereographic projection, as introduced with Figure 2.1 (Fischer and Koch 2005) [1], and illustrated by several examples shown in Figure 2.5.

The stereographic projections relative to the thirty-two crystal classes are given in Part 2.1.1.2.

To study the influence of these symmetry operations on the physical properties, which will be considered in Part 2.3, it is necessary to know the matrix representations of these operations. It can be easily done by considering the transformation

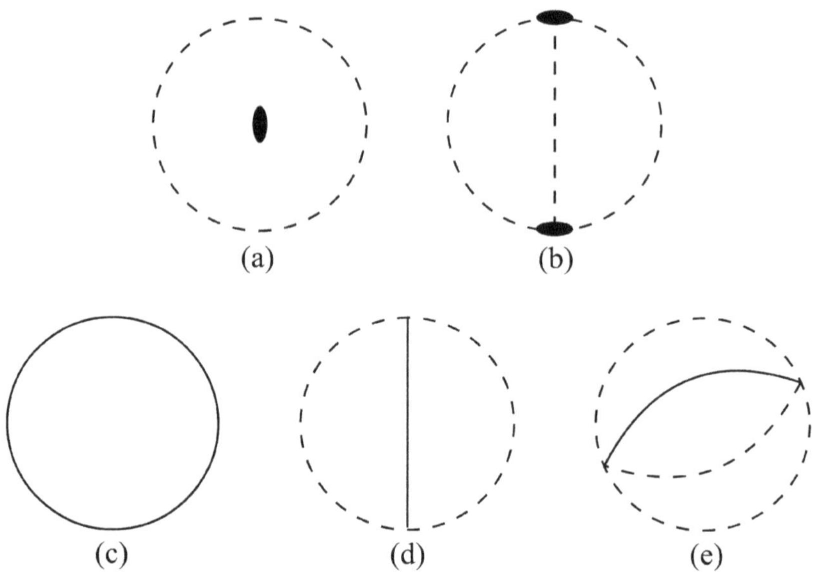

FIGURE 2.5 Representation of several symmetry elements using the stereographic projection: (a) 2-fold axis along the NS axis; (b) 2-fold axis in the equatorial plane; (c) mirror plane in the equatorial plane; (d) mirror plane perpendicular to the equatorial plane; (e) mirror plane making an arbitrary angle with the equatorial plane, the north part depicted by the continuous line and its south part by a dashed line. The equatorial plane is in a dashed line when it does not contain a mirror plane.

of an orthonormal frame, usually taken as a direct frame, by each of the symmetry operations that is considered.

Let $(\overrightarrow{OX_1},\overrightarrow{OX_2},\overrightarrow{OX_3})$ be the initial orthonormal frame, and $(\overrightarrow{OX_1'},\overrightarrow{OX_2'},\overrightarrow{OX_3'})$ the transformed frame by an n-fold rotation axis having an arbitrary orientation, as shown in Figure 2.6. Then the corresponding matrix transformation equation is given by (Nye 1985) [3]:

$$\begin{pmatrix} x_1' \\ x_2' \\ x_3' \end{pmatrix} = \begin{pmatrix} a_{11} & a_{12} & a_{13} \\ a_{21} & a_{22} & a_{23} \\ a_{31} & a_{32} & a_{33} \end{pmatrix} \begin{pmatrix} x_1 \\ x_2 \\ x_3 \end{pmatrix} \tag{2.1}$$

where (x_1', x_2', x_3') with $i=(1,2,3)$ are the Cartesian coordinates of a point in the transformed frame $(\overrightarrow{OX_1'}, \overrightarrow{OX_2'}, \overrightarrow{OX_3'})$, and corresponding to a point of coordinates (x_1,x_2,x_3) in the initial frame $(\overrightarrow{OX_1},\overrightarrow{OX_2},\overrightarrow{OX_3})$. The coefficients (a_{i1},a_{i2},a_{i3}) are the Cartesian coordinates of the axis $\overrightarrow{OX_i'}$ in $(\overrightarrow{OX_1},\overrightarrow{OX_2},\overrightarrow{OX_3})$.

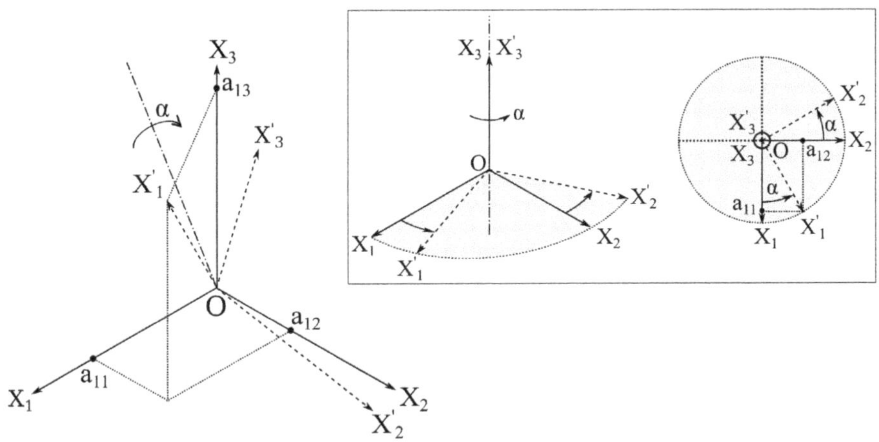

FIGURE 2.6 Transformation of the frame $(\overrightarrow{OX_1},\overrightarrow{OX_2},\overrightarrow{OX_3})$ into $(\overrightarrow{OX_1'}, \overrightarrow{OX_2'}, \overrightarrow{OX_3'})$ by a rotation of angle $\alpha = 360°/n$ around an n-fold axis depicted by the dotted-dashed axis. The coefficients (a_{11},a_{12},a_{13}) are the Cartesian coordinates of $\overrightarrow{OX_1'}$ in $(\overrightarrow{OX_1},\overrightarrow{OX_2},\overrightarrow{OX_3})$. The insert gives the example of rotation by an angle α around the $\overrightarrow{OX_3}$ axis.

From Eq. (2.1), the transformation matrix (a) is defined as:

$$(a)=\begin{pmatrix} a_{11} & a_{12} & a_{13} \\ a_{21} & a_{22} & a_{23} \\ a_{31} & a_{32} & a_{33} \end{pmatrix} \tag{2.2}$$

Equation (2.1) is developed as follows:

$$\begin{aligned} x'_1 &= a_{11}x_1 + a_{12}x_2 + a_{13}x_3 \\ x'_2 &= a_{21}x_1 + a_{22}x_2 + a_{23}x_3 \\ x'_3 &= a_{31}x_1 + a_{32}x_2 + a_{33}x_3 \end{aligned} \tag{2.3}$$

Equation (2.3) can be written in a more synthetic way:

$$x'_i = \sum_{j=1}^{3} a_{ij}x_j \tag{2.4}$$

where $(i, j) = (1,2,3)$.

Let's consider the insert of Figure 2.6 which gives a simple example of a rotation of angle $\alpha = 360° / n$ around an n-fold axis along $\overrightarrow{OX_3}$. The transformation matrix (a) directly rewrites as the following transformation matrix $(a)_n$:

$$(a)_n = \begin{pmatrix} \cos\alpha & \sin\alpha & 0 \\ -\sin\alpha & \cos\alpha & 0 \\ 0 & 0 & 1 \end{pmatrix} \tag{2.5}$$

Using Eq. (2.5), it is easy to establish the matrices associated with the 1-, 2-, 3-, 4-, and 6-fold rotation axis around $\overrightarrow{OX_3}$. In this case, α has the following values: 360° (n=1), 180° (n=2), 120° (n=3), 90° (n=4), and 60° (n=6) respectively.

From Eq. (2.1), it is also easy to establish the matrix $(a)_{\bar{1}}$ of the inversion $\bar{1}$:

$$(a)_{\bar{1}} = \begin{pmatrix} -1 & 0 & 0 \\ 0 & -1 & 0 \\ 0 & 0 & -1 \end{pmatrix} \tag{2.6}$$

The matrices of the different roto-inversions around the $\overrightarrow{OX_3}$ axis can be then calculated from the matrix product of Eq. (2.5) by Eq. (2.6) that commutes, *i.e.*:

$$(a)_{\bar{n}} = (a)_{\bar{1}} \bullet (a)_n = (a)_n \bullet (a)_{\bar{1}} \tag{2.7}$$

The calculation of the determinant $|a|$ of the rotation matrices (a) is an important issue for the following, as will be seen in Part 2.3 and Part 2.4. It is given by:

$$|a| = a_{11}(a_{22}a_{33} - a_{32}a_{23}) + a_{12}(a_{31}a_{23} - a_{21}a_{33}) + a_{13}(a_{21}a_{32} - a_{31}a_{22}) \quad (2.8)$$

According to Eq. (2.5) to Eq. (2.8), it appears that the matrix describing the rotation around a 1-, 2-, 3-, 4-, or 6-fold axis exhibits the same determinant equal to 1, whereas it is equal to –1 for the roto-inversions $\bar{1}$, $\bar{2}$, $\bar{3}$, $\bar{4}$, and $\bar{6}$. This feature will be used in Part 2.3 for the reduction of the tensors describing the physical properties of a crystal medium.

2.1.1.2 The Seven Crystal Systems and the Thirty-Two Crystal Classes

The ten orientation symmetry elements, *i.e.* $1,\bar{1},2,\bar{2},3,\bar{3},4,\bar{4},6$, and $\bar{6}$, can be combined according to specific geometrical rules in the framework of group theory only. There are thirty-two possible resulting combinations that define the 32 point groups of finite order, also called "point groups" or "orientation symmetry classes" or "crystal classes" (Hahn and Klapper 2005) [4]. All these names are equivalent.

These thirty-two crystal classes can be grouped into seven types of crystal lattices that are called "crystal systems". Each crystal system is characterized by a specific unit cell. The 3D unit cell is based on three vectors, written $\vec{a}$, $\vec{b}$, $\vec{c}$. They are called the "crystallographic axes" and constitute the "crystallographic frame" written $(\vec{a},\vec{b},\vec{c})$. These three vectors depict a direct trihedron, and the angles between the crystallographic axes are $\alpha = (\widehat{\vec{b},\vec{c}})$; $\beta = (\widehat{\vec{c},\vec{a}})$; $\gamma = (\widehat{\vec{a},\vec{b}})$, as shown in Figure 2.7.

The seven crystal systems are the following ones: triclinic, monoclinic, orthorhombic, trigonal, tetragonal, hexagonal and cubic. The corresponding unit cells are drawn in Figure 2.8.

The orientation of the crystallographic frame according to the orientation of symmetry elements, as well as the different conventions of nomenclature can be found in the International Tables for Crystallography (Fischer and Koch 2005) [1]. From a practical point of view, the crystallographic frame, including the orientation and the magnitude of its axes, are determined using X-rays diffraction (Pechrasky and Zavalij 2009) [2].

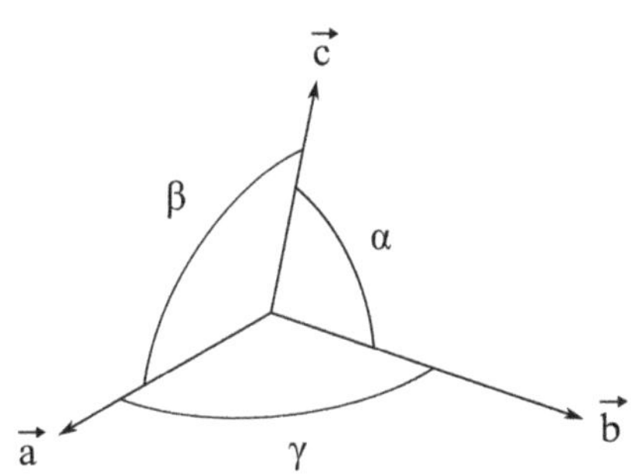

FIGURE 2.7 Description of the unit cell of a 3D crystal lattice. $(\vec{a},\vec{b},\vec{c})$ is the crystallographic frame and (α,β,γ) are the angles between the three crystallographic axes.

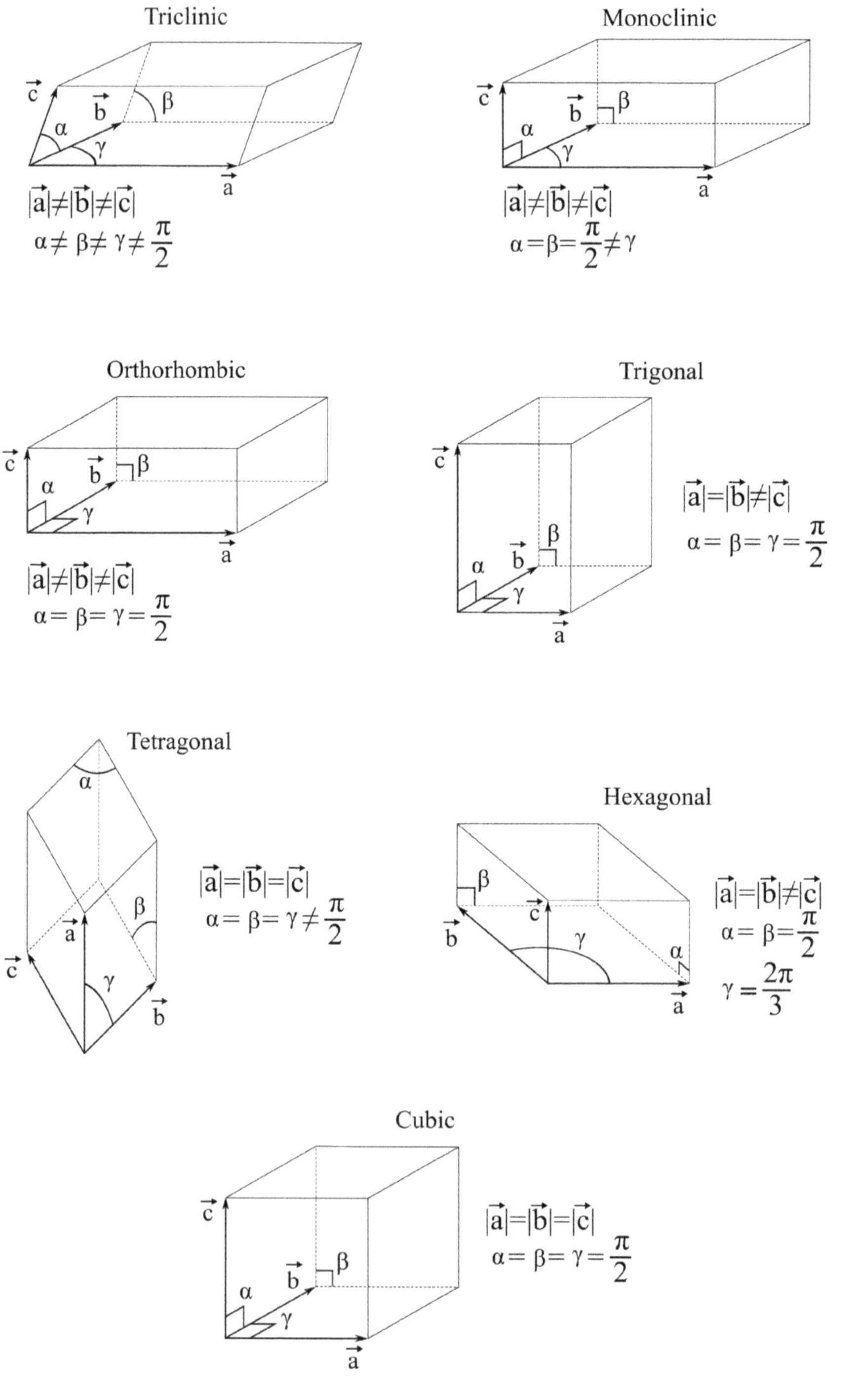

FIGURE 2.8 Unit cells of the seven crystal systems.

Using the standard international notation, the stereographic projections of the thirty-two crystal classes are given in Figure 2.9 to Figure 2.15. They are grouped into the seven crystal systems. The multiplicity M corresponds to the total number of symmetrically equivalent objects that can be generated by all the symmetry operations of the crystal class that can be considered.

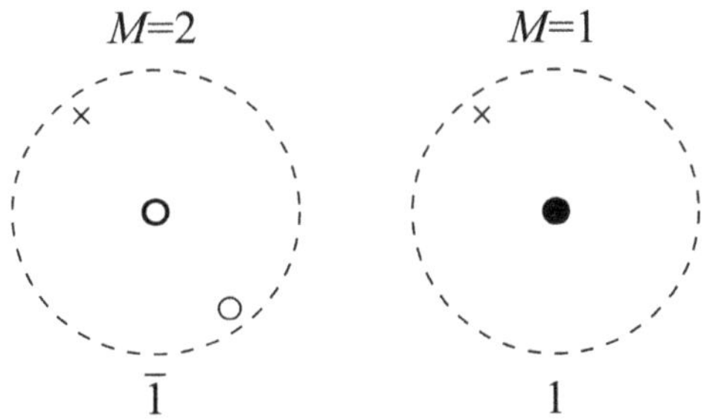

FIGURE 2.9 Stereographic projections of the two triclinic crystal classes 1 and $\bar{1}$, M corresponds to the multiplicity.

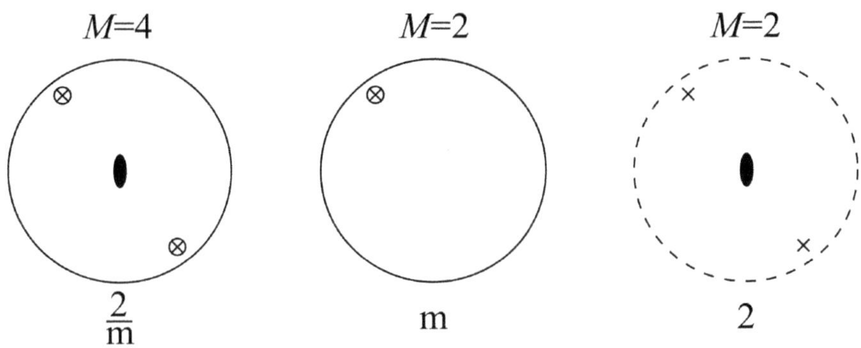

FIGURE 2.10 Stereographic projections of the three monoclinic crystal classes $\frac{2}{m}$, m and 2. M corresponds to the multiplicity.

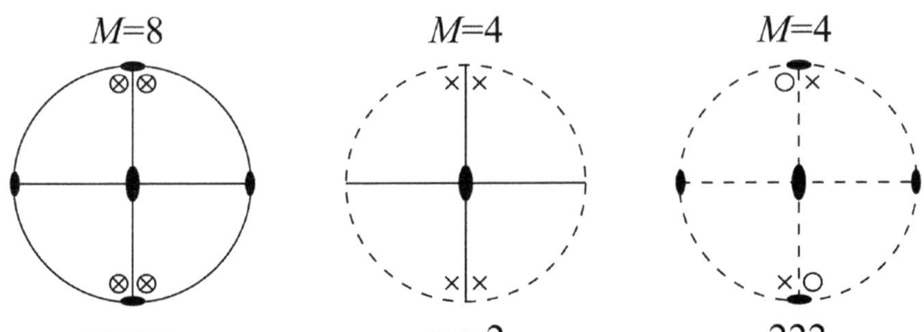

FIGURE 2.11 Stereographic projections of the three orthorhombic crystal classes *mmm*, *mm*2 and 222. M corresponds to the multiplicity.

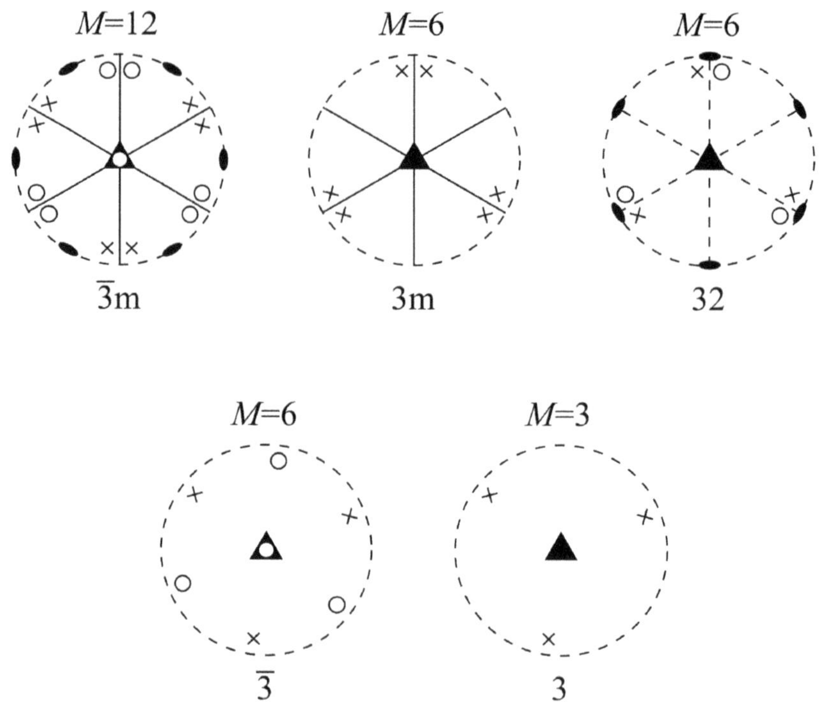

FIGURE 2.12 Stereographic projections of the five trigonal crystal classes $\bar{3}m$, $3m$, 32, $\bar{3}$ and 3. M corresponds to the multiplicity.

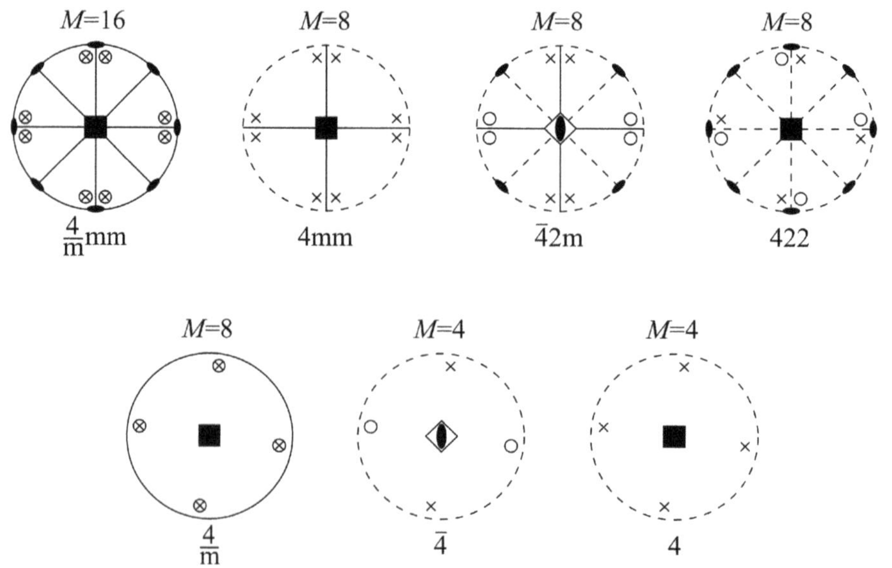

FIGURE 2.13 Stereographic projections of the seven tetragonal crystal classes $\frac{4}{m}mm$, $4mm$, $\bar{4}2m$, 422, $\frac{4}{m}$, $\bar{4}$ and 4. M corresponds to the multiplicity.

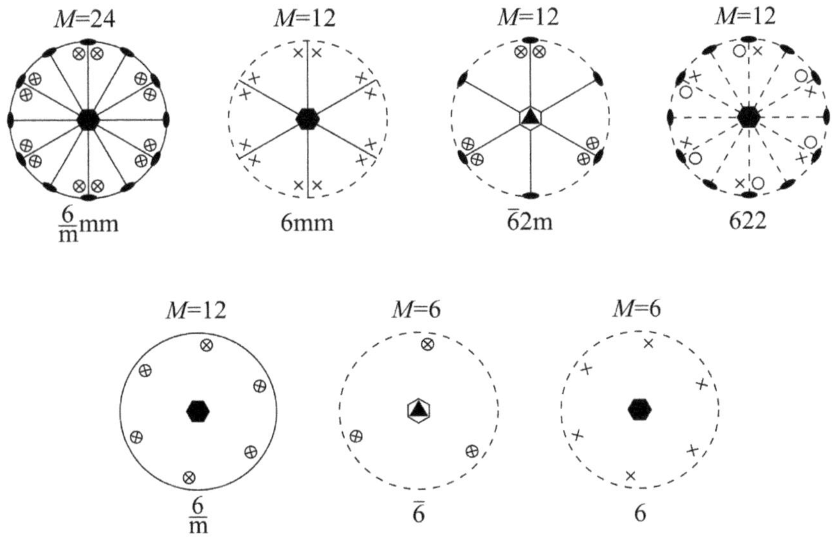

FIGURE 2.14 Stereographic projections of the seven hexagonal crystal classes $\frac{6}{m}mm$, $6mm$, $\bar{6}2m$, 622, $\frac{6}{m}$, $\bar{6}$ and 6. M corresponds to the multiplicity.

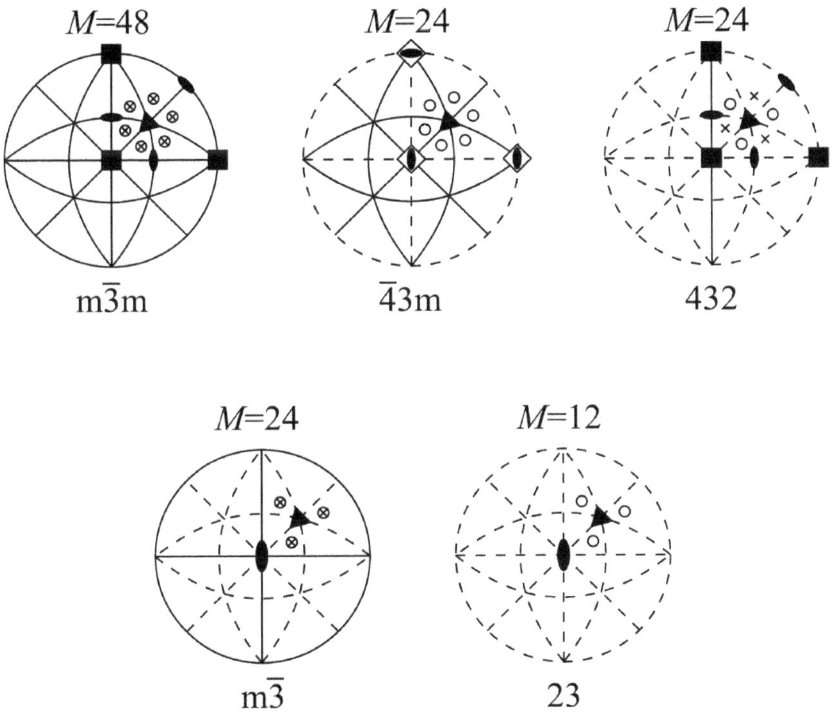

FIGURE 2.15 Stereographic projections of the five cubic crystal classes $m\bar{3}m$, $\bar{4}3m$, 432, $m\bar{3}$ and 23. M corresponds to the multiplicity. The symmetry elements and symmetrical objects are represented only in one fourth of the stereographic projection for more visibility.

The crystallographic frame $(\vec{a},\vec{b},\vec{c})$, that is defined according to the symmetry elements of the crystal (Pechrasky and Zavalij 2009) [2], is not so important in the framework of the tensor calculations carried out in crystal optics, or in crystal physics in general. Actually, the tensor representing a physical property has to be expressed in an orthonormal frame, called the "dielectric frame" or the "optical frame" when dealing with optics. The crucial point is then the orientation of this frame according to the orientation symmetry elements of the crystal, which will be properly defined in Chapter 3.

It is important to note that the crystal classes $\bar{1}$, $\frac{2}{m}$, mmm, $\bar{3}m$, $\frac{4}{m}mm$, $\frac{6}{m}mm$ and $m\bar{3}m$ have respectively the orientation symmetry of the triclinic, monoclinic, orthorhombic, trigonal, tetragonal, hexagonal and cubic unit cells depicted in Figure 2.8. These seven crystal classes are then holohedral crystal classes since they exhibit the same symmetry than that of the lattice. All the other crystal classes are then merohedral crystal classes, which means that they have only some symmetry elements of the lattice, not all. Note that for each crystal system, the holohedral crystal classes are the ones that necessarily show the largest values of multiplicity, while the merohedral classes show lower multiplicity values.

As shown in Figure 2.9 to Figure 2.15, the seven holohedral crystal classes have a center of inversion. These classes are called "centrosymmetric" crystal classes. And there are four merohedral centrosymmetric crystal classes that are also centrosymmetric, *i.e.* $\bar{3}$, $\frac{4}{m}$, $\frac{6}{m}$, and $m\bar{3}$. The twenty-one other crystal classes have no inversion center, so that they are non-centrosymmetric.

Finally, against intuition, Figure 2.15 shows that three cubic crystal classes are non-centrosymmetric: $\bar{4}3m$, 432 and 23.

It will be seen in Part 2.3 how the orientation symmetry of matter can act on the physical properties attached to the matter it-self.

2.1.2 Symmetries of Infinite Order

Orientation symmetry is not relevant for matter only. Indeed, it has also to be considered for the quantities involved in physical properties, as for example the electric and magnetic field of light propagating in matter. In these cases, it is often necessary to consider point groups having rotation axes of infinite order, which means that the object is equivalent to it-self by a rotation of any angle around this axis. The matrix associated with such a rotation around the $\overrightarrow{OX_3}$ axis is also given by Eq. (2.5), where α can take any value between 0° and 360°.

From group theory, it can be shown that there are seven such point groups, *i.e.* ∞m, $\frac{\infty}{m}m$, $\frac{\infty}{m}\frac{\infty}{m}$, ∞, $\infty 2$, $\frac{\infty}{m}$ and $\infty\infty$ in Schönflies nomenclature (De Graef and Mc Henry 2012) [5]. They are represented in Figure 2.16.

The three groups of the upper part of Figure 2.16 are represented by a static half-cone (∞m), a static full-cone that is equivalent to a static cylinder $\left(\frac{\infty}{m}m\right)$, and a static sphere $\left(\frac{\infty}{m}\frac{\infty}{m}\right)$. The half-cone corresponds to any vector, like the

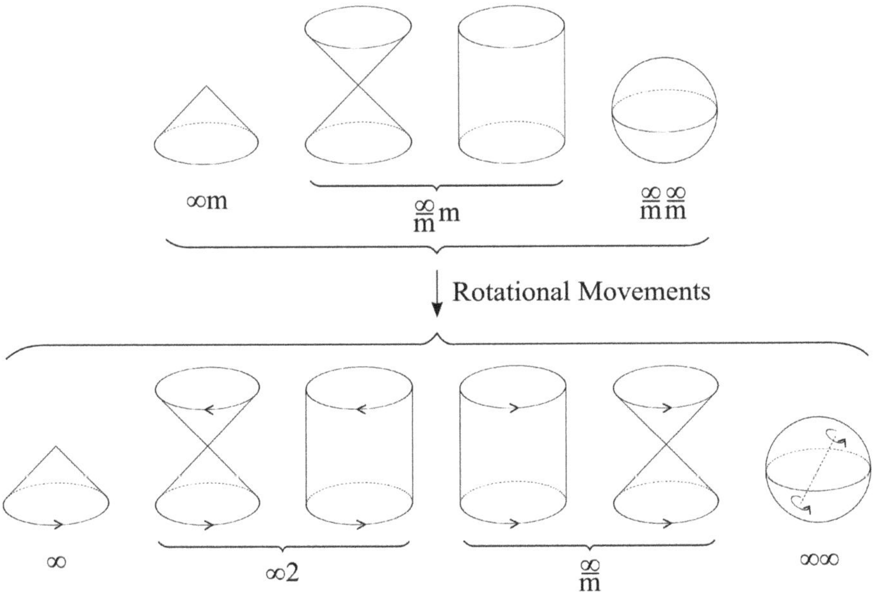

FIGURE 2.16 Representation of the seven point groups having a rotation axis of infinite order. The arrows correspond to rotational movements relatively to certain parts of the objects.

electric field or a gradient of potential for example: its symmetry elements are a rotation axis of infinite order and an infinite number of mirror planes containing the rotation axis. The full cone or cylinder can represent a uniaxial mechanical strain: there are the same symmetry elements than for the half-cone with in addition a mirror plane perpendicular to the rotation axis. The sphere corresponds to a scalar, like the temperature for example: it exhibits an infinite number of rotation axes of infinite order with an infinite number a mirror planes perpendicular to these rotation axes.

The lower part of Figure 2.16 concerns the point groups described by moving objects, the rotations of the objects, depicted by the arrows, having the effect of removing all or part of the mirror planes. The rotating half-cone (∞) has only the rotation axis of infinite order as symmetry element, like the combination of an electric field parallel to a magnetic field. When the two opposite faces of a full-cone or a full cylinder rotate in opposite directions (∞ 2), then the only remaining symmetry elements are the rotation axes: it is the case of the tensor of optical gyration. In the case where the two opposite faces turn in the same direction, the 2-fold axes disappear while a mirror plane normal to the rotation axis of infinite order appears: it corresponds to a "pseudo-vector", like the magnetic field. The last case described in Figure 2.16 is $(\infty\ \infty)$, corresponding to a sphere where each opposite surface element perpendicular to any diameter rotates in the same direction around the corresponding diameter: it would be the case of a sphere full of a liquid exhibiting optical activity.

2.2 TENSOR ALGEBRA

2.2.1 RANK AND MATRIX REPRESENTATION OF TENSORS

Any physical quantity must be described by a tensor in order to take into account the symmetry of matter and the dimensionality of the space N, $N = 0,1,2$ or 3. It is described by a matrix of Ω coefficients depending on N but also on the tensor rank R of the considered tensor, the latter being an intrinsic characteristic of the physical quantity that is considered. The number of coefficients Ω is given by:

$$\Omega = N^R \tag{2.9}$$

So in a 3D space, *i.e.* $N = 3$, a rank-1 tensor $(R = 1)$ has $\Omega = 3^1 = 3$ coefficients, a rank-2 tensor has $\Omega = 3^2 = 9$ coefficients, etc. Note that a rank-0 tensor $(R = 0)$ is always described by only one coefficient $(\Omega = 1)$ whatever the value of the dimensionality N, since $N^0 = 1$.

Some examples of tensor ranks and the corresponding properties are given hereafter: $R = 0$ for scalar quantities as temperature; $R = 1$ for vector quantities as an electric field; $R = 2,3$ and 4 for the first-, second-, and third-order electric susceptibilities respectively that have been defined in Chapter 1.

A tensor will be represented by a matrix $(T)_R$. Each coefficient of this matrix can be written $T_{ijkl...}$, where the indices $i, j, k, l, ...$ are Cartesian indices relative to an orthonormal frame in the considered space of dimension (N), the number of indices being exactly equal to the rank (R) of the tensor. These Cartesian indices can be numbers or letters. In the case of a 3D space for example, the numbers 1,2 and 3 are used when the orthonormal frame is written $(\overrightarrow{OX_1}, \overrightarrow{OX_2}, \overrightarrow{OX_3})$, but letters x, y and z are considered when the orthonormal frame writes $(\overrightarrow{Ox}, \overrightarrow{Oy}, \overrightarrow{Oz})$. Then the representative matrix $(T)_{R=1}$ of a rank-1 tensor can be written:

$$\vec{T} = (T)_{R=1} = \begin{pmatrix} T_1 \\ T_2 \\ T_3 \end{pmatrix} \text{ or } \vec{T} = (T)_{R=1} = \begin{pmatrix} T_x \\ T_y \\ T_z \end{pmatrix} \tag{2.10}$$

This (3×1) matrix, where 3 corresponds to the number of lines and 1 for the number of columns, describes a vector.

For a rank-2 tensor, the matrix $(T)_{R=2}$ is often written as a (3×3) matrix, *i.e.*:

$$(T)_{R=2} = \begin{pmatrix} T_{11} & T_{12} & T_{13} \\ T_{21} & T_{22} & T_{23} \\ T_{31} & T_{32} & T_{33} \end{pmatrix} \text{ or } (T)_{R=2} = \begin{pmatrix} T_{xx} & T_{xy} & T_{xz} \\ T_{yx} & T_{yy} & T_{yz} \\ T_{zx} & T_{zy} & T_{zz} \end{pmatrix} \tag{2.11}$$

The 3^2 coefficients T_{ij} are ordered in the matrix: the ordering sequence is 1,2 then 3 (or x, y then z) for the first index of each column from top to bottom and for the second index of each line from left to right. Such a (3×3) matrix will be largely considered in the following chapters since it describes the linear electric susceptibility $\chi^{(1)}$ that is at the heart of linear optics. Note that a (9×1) column matrix for $(T)_{R=2}$ can be also used as shown in Part 2.2.3. This will be useful for the calculation of the quadratic induced polarization.

Another widely used matrix in the following will be a (3×9) matrix. It represents the $3^3=27$ coefficients to describe a rank-3 tensor. It represents for example the second-order electric susceptibility $\chi^{(2)}$ that writes in the orthonormal frame $(\overrightarrow{Ox},\overrightarrow{Oy},\overrightarrow{Oz})$:

$$(T)_{R=3}=\begin{pmatrix} T_{xxx} & T_{xyy} & T_{xzz} & T_{xyz} & T_{xzy} & T_{xxz} & T_{xzx} & T_{xxy} & T_{xyx} \\ T_{yxx} & T_{yyy} & T_{yzz} & T_{yyz} & T_{yzy} & T_{yxz} & T_{yzx} & T_{yxy} & T_{yyx} \\ T_{zxx} & T_{zyy} & T_{zzz} & T_{zyz} & T_{zzy} & T_{zxz} & T_{zzx} & T_{zxy} & T_{zyx} \end{pmatrix} \tag{2.12}$$

Here also, it is important to carefully consider the ordering sequence of the Cartesian indices in the matrix for further tensor calculations.

Note that another matrix of 27 coefficients will be used in Part 2.2.3, Chapter 4 and Chapter 8: it will be a (1×27) matrix enabling the calculation of the cubic induced polarization.

The highest rank tensor considered in the present book is 4, which means $3^4=81$ coefficients. In the case of the third-order electric susceptibility, the tensor is described by the following (3×27) matrix composed of five blocks $[A]$, $[B]$, $[C]$, $[D]$ and $[E]$, *i.e.*:

$$(T)_{R=4}=([A][B][C][D][E]) \tag{2.13}$$

with

$$A=\begin{bmatrix} T_{xxxx} & T_{xyyy} & T_{xzzz} \\ T_{yxxx} & T_{yyyy} & T_{yzzz} \\ T_{zxxx} & T_{zyyy} & T_{zzzz} \end{bmatrix} \tag{2.14}$$

$$B=\begin{bmatrix} T_{xyzz} & T_{xzyz} & T_{xzzy} & T_{xyyz} & T_{xyzy} & T_{xzyy} \\ T_{yyzz} & T_{yzyz} & T_{yzzy} & T_{yyyz} & T_{yyzy} & T_{yzyy} \\ T_{zyzz} & T_{zzyz} & T_{zzzy} & T_{zyyz} & T_{zyzy} & T_{zzyy} \end{bmatrix} \tag{2.15}$$

$$C=\begin{bmatrix} T_{xxzz} & T_{xzxz} & T_{xzzx} & T_{xxxz} & T_{xxzx} & T_{xzxx} \\ T_{yxzz} & T_{yzxz} & T_{yzzx} & T_{yxxz} & T_{yxzx} & T_{yzxx} \\ T_{zxzz} & T_{zzxz} & T_{zzzx} & T_{zxxz} & T_{zxzx} & T_{zzxx} \end{bmatrix} \tag{2.16}$$

$$D=\begin{bmatrix} T_{xxyy} & T_{xyxy} & T_{xyyx} & T_{xxxy} & T_{xxyx} & T_{xyxx} \\ T_{yxyy} & T_{yyxy} & T_{yyyx} & T_{yxxy} & T_{yxyx} & T_{yyxx} \\ T_{zxyy} & T_{zyxy} & T_{zyyx} & T_{zxxy} & T_{zxyx} & T_{zyxx} \end{bmatrix} \tag{2.17}$$

$$E=\begin{bmatrix} T_{xxyz} & T_{xxzy} & T_{xyxz} & T_{xyzx} & T_{xzxy} & T_{xzyx} \\ T_{yxyz} & T_{yxzy} & T_{yyxz} & T_{yyzx} & T_{yzxy} & T_{yzyx} \\ T_{zxyz} & T_{zxzy} & T_{zyxz} & T_{zyzx} & T_{zzxy} & T_{zzyx} \end{bmatrix} \tag{2.18}$$

2.2.2 Polar and Axial Tensors

Two kind of tensors can describe physical properties: the polar tensors and the axial tensors also called "pseudo-tensors". They are defined according to their transformation by an orientation symmetry operation.

The coefficients of a rank-R polar tensor are transformed by a given symmetry operation as follows (Nye 1985) [3]:

$$T'_{ijkl...} = \sum_{mnop...} a_{im} a_{jn} a_{ko} a_{lp} \cdots T_{mnop...} \tag{2.19}$$

where $(i,j,k,l,m,n,o,p,...) = 1,2,3$ or x, y, z according to the chosen orthonormal frame.

$T'_{ijkl...}$ stands for any coefficient of the transformed tensor $(T')_R$, while $T_{mnop...}$ denotes any coefficient of the initial tensor $(T)_R$. The coefficients a_{xy} in Eq. (2.19) are the coefficients of the matrix (a) associated with the considered orientation symmetry operation, as defined by Eq. (2.2). Note that the rank R of the tensor is conserved through this transformation.

Quite similarly to Eq. (2.19), the definition of a rank-R axial tensor is (Nye 1985) [3]:

$$T'_{ijkl...} = |a| \sum_{mnop...} a_{im} a_{jn} a_{ko} a_{lp} \cdots T_{mnop...} \tag{2.20}$$

where $|a|$ is the determinant of the transformation matrix (a) that is given by Eq. (2.8).

Then according to Eq. (2.19) and Eq. (2.20), it appears that both a polar tensor and an axial tensor exhibit the same transformation by orientation symmetry associated with $|a|=1$. It is the case of the rotations 1, 2, 3, 4 and 6. On the contrary, when $|a|=-1$, which concerns the roto-inversions $\bar{1}$, $\bar{2}$, $\bar{3}$, $\bar{4}$ and $\bar{6}$, all the coefficients of an axial tensor change their sign compared to those of a polar tensor.

The example of the inversion $\bar{1}$ is particularly interesting since the calculations are easy to perform and this symmetry operation is strongly discriminant as it will be shown in Part 2.3.5. Indeed, according to Eq. (2.6) and Eq. (2.19), it is easy to show that any polar tensor of even rank is transformed identically to it-self by the inversion symmetry, *i.e.* $T'_{ij} = T_{ij}\ (R=2)$, $T'_{ijkl} = T_{ijkl}\ (R=4)$, etc. The situation is the opposite in the case of an odd rank polar tensor since it comes: $T'_i = -T_i\ (R=1)$, $T'_{ijk} = -T_{ijk}\ (R=3)$, etc. The case of an axial tensor transformed by the inversion $\bar{1}$ is exactly the reverse: $T'_{ij} = -T_{ij}\ (R=2)$, $T'_{ijkl} = -T_{ijkl}\ (R=4)$, and $T'_i = T_i\ (R=1)$, $T'_{ijk} = T_{ijk}\ (R=3)$ according to Eq. (2.6) and Eq. (2.20).

It is very useful for the following to notice that the transformation of a rank-R polar tensor is formally equivalent to the transformation of the products of R Cartesian coordinates. Then this transformation can be easily written from Eq. (2.4) (Nye 1985) [3]:

$$x'_i x'_j x'_k x'_l \ldots = \sum_{mnop\ldots} a_{im} a_{jn} a_{ko} a_{lp} \cdots x_m x_n x_o x_p \cdots \tag{2.21}$$

Equation (2.21) is identical to Eq. (2.19) if $x'_i x'_j x'_k x'_l \ldots$ is identified to $T'_{ijkl\ldots}$ on the one hand, and $x_m x_n x_o x_p \ldots$ to $T_{mnop\ldots}$ on the other hand.

The same considerations apply in the case of axial tensors if Eq. (2.21) is multiplied by the determinant of the symmetry operation that is considered, *i.e.* (Nye 1985) [3]:

$$x'_i x'_j x'_k x'_l \ldots = |a| \sum_{mnop\ldots} a_{im} a_{jn} a_{ko} a_{lp} \cdots x_m x_n x_o x_p \cdots \tag{2.22}$$

A direct application of this feature will be done in Part 2.3.5.

2.2.3 Contracted Product and Tensor Product

Two tensor operators will be used for the tensor calculations performed in this book: the contracted product and the tensor product.

2.2.3.1 Contracted Product

The contracted product involves two tensors with the same or different ranks.

The scalar product, which concerns two vectors, is the simplest example of such a contracted product. It is written:

$$\vec{U}.\vec{V} = \sum_i U_i V_i \tag{2.23}$$

Then the contracted product of two rank-1 tensors, *i.e.* two vectors, gives rise to a rank-0 tensor, *i.e.* a scalar, which writes:

$$(U)_{R=1} \cdot (V)_{R=1} = (W)_{R=1-1=0} \tag{2.24}$$

More generally, the contracted product of two tensors of any but equal rank gives rise to a scalar quantity. It writes:

$$(U)_{R=N} \cdot (V)_{R=N} = (W)_{R=N-N=0} \tag{2.25}$$

where N is an integer.

The cases where $N=3$ and $N=4$ will be considered in Chapter 7 and Chapter 8 for the calculation of the quadratic and cubic effective coefficients respectively, *i.e.* (Boulanger 1991, Boulanger 1993) [7,8]:

$$(U)_{R=3} \therefore (V)_{R=3} = (W)_{R=0} = \sum_{ijk} U_{ijk} V_{ijk} \tag{2.26}$$

and

$$(U)_{R=4} :: (V)_{R=4} = (W)_{R=0} = \sum_{ijkl} U_{ijkl} V_{ijkl} \tag{2.27}$$

The graphical symbols (.˙.) and (::) signify that the contracted products concern two rank-3 tensors and two rank-4 tensors respectively. Note that the simple dot (.) is sometimes used, as done in Eq. (2.25), in order to simplify the writing.

The contracted product of tensors of different ranks writes:

$$(U)_{R=M} \cdot (V)_{R=N \le M} = (W)_{R=M-N} \tag{2.28}$$

Where M and N are two integers, with the relation of order $M > N$. As previously, the graphical symbols (.) can also be used for the contracted product in general, but it can be useful to take a more explicit writing giving the rank of the tensor $(V)_{R=N \le M}$ located on the right side of the graphical symbol in Eq. (2.28). Then (:) or (.˙.) will be used to denote the fact that the right tensor is of rank-2 or -3 respectively.

Contracted products of tensors with $(M=2, N=1)$, $(M=3, N=2)$ and $(M=4, N=3)$ will concern the calculations of the linear induced polarization vector and the quadratic and cubic induced polarization vectors, respectively.

For $(M=2, N=1)$, it comes:

$$(U)_{R=2} \cdot (V)_{R=1} = (W)_{R=1} = \vec{W} \tag{2.29}$$

$$\sum_j U_{ij} V_j = W_i \tag{2.30}$$

Then it comes using Eqs. (2.10) and (2.11) in the frame $\overrightarrow{Ox}, \overrightarrow{Oy}, \overrightarrow{Oz}$:

$$\begin{pmatrix} U_{xx} & U_{xy} & U_{xz} \\ U_{yx} & U_{yy} & U_{yz} \\ U_{zx} & U_{zy} & U_{zz} \end{pmatrix} \cdot \begin{pmatrix} V_x \\ V_y \\ V_z \end{pmatrix} = \begin{pmatrix} W_x \\ W_y \\ W_z \end{pmatrix} \tag{2.31a}$$

And, according to Eq. (2.31a), the Cartesian coordinates of the resulting vector $\overrightarrow{W}$ also write:

$$W_i = U_{ix}V_x + U_{iy}V_y + U_{iz}V_z \tag{2.31b}$$

with $i = (x, y, z)$.

It is important to notice that the compatibility between the ordering sequence of the second Cartesian index of the coefficients in each line of tensor (U) and the ordering sequence of the coefficients of the column matrix (V) is necessary. Note also that Eq. (2.32) is formally identical to Eq. (2.1) corresponding to the change of frame where the transformation matrix (a) plays the role of $(W)_{R=2}$.

For $(M = 3, N = 2)$, the contracted product writes:

$$(U)_{R=3} : (V)_{R=2} = (W)_{R=1} = \overrightarrow{W} \tag{2.32}$$

$$\sum_{jk} U_{ijk}V_{jk} = W_i \tag{2.33}$$

According to Eq. (2.12), the matrix equation is:

$$\begin{pmatrix} U_{xxx} & U_{xyy} & U_{xzz} & U_{xyz} & U_{xzy} & U_{xxz} & U_{xzx} & U_{xxy} & U_{xyx} \\ U_{yxx} & U_{yyy} & U_{yzz} & U_{yyz} & U_{yzy} & U_{yxz} & U_{yzx} & U_{yxy} & U_{yyx} \\ U_{zxx} & U_{zyy} & U_{zzz} & U_{zyz} & U_{zzy} & U_{zxz} & U_{zzx} & U_{zxy} & U_{zyx} \end{pmatrix} : \begin{pmatrix} V_{xx} \\ V_{yy} \\ V_{zz} \\ V_{yz} \\ V_{zy} \\ V_{xz} \\ V_{zx} \\ V_{xy} \\ V_{yx} \end{pmatrix} = \begin{pmatrix} W_x \\ W_y \\ V_z \end{pmatrix} \tag{2.34}$$

Note that the second and third Cartesian indices of (U) correspond to the two Cartesian indices of (V), while the first Cartesian index corresponds to that of (W).

Then as previously, it gives for the three Cartesian coordinates of $\overrightarrow{W}$:

$$\begin{aligned} W_i = {} & U_{ixx}V_{xx} + U_{iyy}V_{yy} + U_{izz}V_{zz} + U_{iyz}V_{yz} + U_{izy}V_{zy} + U_{ixz}V_{xz} \\ & + U_{izx}V_{zx} + U_{ixy}V_{xy} + U_{ixy}V_{yx} \end{aligned} \tag{2.35}$$

with $i = (x, y, z)$.

For $(M = 4, N = 3)$, the contracted product is:

$$(U)_{R=4} \therefore (V)_{R=3} = (W)_{R=1} = \overrightarrow{W} \tag{2.36}$$

$$\sum_{jkl} U_{ijkl} V_{jkl} = W_i \tag{2.37}$$

Tensor (U) is represented by the (3 x 27) matrix given by Eq. (2.13) to Eq. (2.18), and (V) is described by a (1 x 27) column matrix. Then the Cartesian coordinates of the resulting vector $\overrightarrow{W}$ are given each by the sum of 27 terms, *i.e.*:

$$\begin{aligned} W_i = & U_{ixxx}V_{xxx} + U_{iyyy}V_{yyy} + U_{izzz}V_{zzz} \\ & + U_{iyzz}V_{yzz} + U_{izzy}V_{zzy} + U_{izyz}V_{zyz} + U_{iyyz}V_{yyz} + U_{izyy}V_{zyy} + U_{iyzy}V_{yzy} \\ & + U_{ixzz}V_{xzz} + U_{izzx}V_{zzx} + U_{izxz}V_{zxz} + U_{ixxz}V_{xxz} + U_{izxx}V_{zxx} + U_{ixzx}V_{xzx} \\ & + U_{ixyy}V_{xyy} + U_{iyyx}V_{yyx} + U_{iyxy}V_{yxy} + U_{ixxy}V_{xxy} + U_{iyxx}V_{yxx} + U_{ixyx}V_{xyx} \\ & + U_{ixyz}V_{xyz} + U_{ixzy}V_{xzy} + U_{iyxz}V_{yxz} + U_{iyzx}V_{yzx} + U_{izxy}V_{zxy} + U_{izyx}V_{zyx} \end{aligned} \tag{2.38}$$

with $i = (x, y, z)$.

Here also the three last Cartesian indices of (U) correspond to the three indices of (V), the first one being linked to (W).

From a general point of view, it is important to notice that the contracted product leads to a reduction of the number of tensor coefficients, *i.e.* 3^{M-N} compared with 3^M and $3^{N \le M}$, in a 3D space.

2.2.3.2 Tensor Product

Contrary to the contracted product, the tensor product increases of the resulting tensor rank and then of the associated number of tensor coefficients. In the framework of the present book tensor products between n vectors (with $n = 2, 3, 4$) will be considered leading to tensors of rank $R = 2, 3, 4$, or between a scalar and a vector.

In the case of two vectors $(n = 2)$, the tensor product writes:

$$\vec{U} \otimes \vec{V} = (U)_{R=1} \otimes (V)_{R=1} = (W)_{R=1+1=2} \tag{2.39}$$

The corresponding matrix equation is:

$$\begin{pmatrix} U_x \\ U_y \\ U_z \end{pmatrix} \otimes \begin{pmatrix} V_x \\ V_y \\ V_z \end{pmatrix} = \begin{pmatrix} W_{xx} = U_x V_x \\ W_{yy} = U_y V_y \\ W_{zz} = U_z V_z \\ W_{yz} = U_y V_z \\ W_{zy} = U_z V_y \\ W_{xz} = U_x V_z \\ W_{zx} = U_z V_x \\ W_{xy} = U_x V_y \\ W_{yx} = U_y V_x \end{pmatrix} \tag{2.40}$$

The $3^2 = 9$ coefficients correspond to the 3^2 possible combinations two by two between the coefficients of tensors $(U)_{R=1}$ and $(V)_{R=1}$. Note the same ordering of the Cartesian coordinates here for this tensor product as those also used for the vector (9×1) of Eq. (2.34).

In the same way, the tensor product of three vectors $(n = 3)$ and four vectors $(n = 4)$ are respectively given by:

$$\vec{U} \otimes \vec{V} \otimes \vec{W} = (U)_{R=1} \otimes (V)_{R=1} \otimes (W)_{R=1} = (X)_{R=1+1+1=3} \tag{2.41}$$

and

$$\vec{U} \otimes \vec{V} \otimes \vec{W} \otimes \vec{X} = (U)_{R=1} \otimes (V)_{R=1} \otimes (W)_{R=1} \otimes (X)_{R=1} = (Y)_{R=1+1+1+1=4} \tag{2.42}$$

The resulting tensors have respectively 3^3 and 3^4 coefficients corresponding to all the possible combinations between the coefficients of the initial tensors.

The product of a scalar by a vector is also a tensor product, *i.e.*:

$$U \otimes \vec{V} = (U)_{R=0} \otimes (V)_{R=1} = (W)_{R=0+1} \tag{2.43}$$

The corresponding matrix equation is:

$$U \otimes \begin{pmatrix} V_x \\ V_y \\ V_z \end{pmatrix} = \begin{pmatrix} W_x = U\ V_x \\ W_y = U\ V_y \\ W_z = U\ V_z \end{pmatrix} \tag{2.44}$$

Note that usually in that case, instead of writing $(\otimes)$ no graphical symbol is used since the tensor product $U \otimes \vec{V}$ is equivalent to the algebraic product $U\vec{V}$.

It is possible to learn more mathematics of tensor algebra and matrices by reading for example (Bourbaki 1998) [9].

2.2.4 Field Tensor and Matter Tensor

A physical equation involves both matter and field tensors that are defined below.

Any physical property is represented by a matter tensor $(U)_{R=M}$. The magnitude and relative signs of the coefficients of this tensor depends on the nature and arrangement of the atoms in the matter only. Then the symmetry of $(U)_{R=M}$ has a crucial importance at this level, as it will be seen below in Part 2.3.5. And it is important to note that the coefficients of the matter tensor do not depend on the direction of any property, as for example the direction of propagation of light or the direction of application of a mechanical strain.

The existence of the matter tensor $(U)_{R=M}$ is justified by the connection established between two field tensors only: that of the exciting field tensor $(V(\vec{r}))_{R=N}$, and that of the resulting field tensor $(W(\vec{r}))_{R=P}$, *i.e.* the matter response, these two field tensors depending on the direction $\vec{r}$. The exciting field can have any orientation and magnitude, but the orientation and magnitude of the resulting field depends on both the exciting field and the matter tensor according to the following contracted product:

$$\left(U\right)_{R=M} \cdot \left(V\left(\vec{r}\right)\right)_{R=N} = \left(W\left(\vec{r}\right)\right)_{R=P} \tag{2.45}$$

with $P = M - N$.

The left side of Eq. (2.45) can also be multiplied by one or several scalars. Below are listed some examples of tensor equations that will be used.

$$\left(S\right)_{R=0} \otimes \left(\left(U\right)_{R=2} \cdot \left(V\left(\vec{r}\right)\right)_{R=1}\right)_{R=1} = \left(W\left(\vec{r}\right)\right)_{R=1} \tag{2.46}$$

$$\left(S\right)_{R=0} \otimes \left(\left(U\right)_{R=3} : \left(\left(X\left(\vec{r}\right)\right)_{R=1} \otimes \left(Y\left(\vec{r}\right)\right)_{R=1}\right)_{R=2}\right)_{R=1} = \left(W\left(\vec{r}\right)\right)_{R=1} \tag{2.47}$$

$$\left(U\right)_{R=2} : \left(\left(X\left(\vec{r}\right)\right)_{R=1} \otimes \left(Y\left(\vec{r}\right)\right)_{R=1}\right)_{R=2} = \left(W\left(\vec{r}\right)\right)_{R=0} \tag{2.48}$$

$$\left(U\right)_{R=3} \therefore \left(\left(X\left(\vec{r}\right)\right)_{R=1} \otimes \left(Y\left(\vec{r}\right)\right)_{R=1} \otimes \left(Z\left(\vec{r}\right)\right)_{R=1}\right)_{R=3} = \left(W\left(\vec{r}\right)\right)_{R=0} \tag{2.49}$$

$$\left(S\right)_{R=0} \otimes \left(\left(U\right)_{R=4} \therefore \left(\left(X\left(\vec{r}\right)\right)_{R=1} \otimes \left(Y\left(\vec{r}\right)\right)_{R=1} \otimes \left(Z\left(\vec{r}\right)\right)_{R=1}\right)_{R=3}\right)_{R=1} = \left(W\left(\vec{r}\right)\right)_{R=1} \tag{2.50}$$

$$\left(U\right)_{R=4} :: \left(\left(X\left(\vec{r}\right)\right)_{R=1} \otimes \left(Y\left(\vec{r}\right)\right)_{R=1} \otimes \left(Z\left(\vec{r}\right)\right)_{R=1} \otimes \left(T\left(\vec{r}\right)\right)_{R=1}\right)_{R=4} = \left(W\left(\vec{r}\right)\right)_{R=0} \tag{2.51}$$

$$\left(U\right)_{R=3} \cdot \left(V\left(\vec{r}\right)\right)_{R=1} = \left(W\left(\vec{r}\right)\right)_{R=2} \tag{2.52}$$

$$\left(U\right)_{R=4} : \left(V\left(\vec{r}\right)\right)_{R=2} = \left(W\left(\vec{r}\right)\right)_{R=2} \tag{2.53}$$

Equation (2.46) corresponds to the first-order electric constitutive equations of electro-magnetism, where $\left(S\right)_{R=0}$ is the dielectric permittivity of vacuum, $\left(U\right)_{R=2}$ the linear electric susceptibility, $\left(V\left(\vec{r}\right)\right)_{R=1}$ the applied electric field of light, and $\left(W\left(\vec{r}\right)\right)_{R=1}$ the linear induced polarization.

Equation (2.47) corresponds to the second-order electric constitutive equations of electro-magnetism, where $\left(S\right)_{R=0}$ is the dielectric permittivity of vacuum, $\left(U\right)_{R=3}$ the quadratic electric susceptibility, $\left(X\left(\vec{r}\right)\right)_{R=1}$ and $\left(Y\left(\vec{r}\right)\right)_{R=1}$ the two applied electric fields of light, and $\left(W\left(\vec{r}\right)\right)_{R=1}$ the quadratic induced polarization.

In Eq. (2.48), $\left(U\right)_{R=2}$ corresponds to the linear electric susceptibility, $\left(X\left(\vec{r}\right)\right)_{R=1}$ is the applied electric fields of light, and $\left(Y\left(\vec{r}\right)\right)_{R=1}$ is the optical field that is radiated by the first order induced polarization. $\left(W\left(\vec{r}\right)\right)_{R=1}$ is the linear effective coefficient that governs the efficiency of the linear interactions.

In Eq. (2.49), $\left(U\right)_{R=3}$ corresponds to the quadratic electric susceptibility, $\left(X\left(\vec{r}\right)\right)_{R=1}$ and $\left(Y\left(\vec{r}\right)\right)_{R=1}$ to the two applied electric fields of light, and $\left(Z\left(\vec{r}\right)\right)_{R=1}$ to the optical field that is radiated by the quadratic induced polarization, and $\left(W\left(\vec{r}\right)\right)_{R=1}$ to the quadratic effective coefficient that governs the efficiency of the quadratic interactions.

Equation (2.50) corresponds to the third-order electric constitutive equations of electro-magnetism, where $\left(S\right)_{R=0}$ is the dielectric permittivity of vacuum, $\left(U\right)_{R=4}$ the cubic electric susceptibility, $\left(X\left(\vec{r}\right)\right)_{R=1}$, $\left(Y\left(\vec{r}\right)\right)_{R=1}$, $\left(Z\left(\vec{r}\right)\right)_{R=1}$ the three applied electric fields of light, and $\left(W\left(\vec{r}\right)\right)_{R=1}$ the cubic induced polarization.

In Eq. (2.51), $(U)_{R=4}$ corresponds to the cubic electric susceptibility, $\left(X(\vec{r})\right)_{R=1}$, $\left(Y(\vec{r})\right)_{R=1}$, $\left(Z(\vec{r})\right)_{R=1}$ to the three applied electric fields of light, $\left(T(\vec{r})\right)_{R=1}$ to the optical electric field that is radiated by the cubic induced polarization, and $\left(W(\vec{r})\right)_{R=0}$ corresponds to the cubic effective coefficient that governs the efficiency of the third-order nonlinear interactions.

Equation (2.52) corresponds to the Pockels effect, where $(U)_{R=3}$ corresponds to the linear electro-optical tensor, $\left(V(\vec{r})\right)_{R=1}$ to the static or quasi-static applied electric field, and $\left(W(\vec{r})\right)_{R=2}$ to the deformation of the index ellipsoid.

Finally, Eq. (2.53) corresponds to the photo-elastic effect, where $(U)_{R=4}$ corresponds to the photo-elastic tensor, $\left(V(\vec{r})\right)_{R=2}$ to the applied mechanical strain, and $\left(W(\vec{r})\right)_{R=2}$ to the deformation of the index ellipsoid.

From Eq. (2.46) to Eq. (2.53), the space variable $\vec{r}$ stands for the direction of propagation of the light or of the induced polarization, as well as for the direction of a mechanical strain.

2.3 REDUCTION OF TENSORS BY THE ORIENTATION SYMMETRY

2.3.1 THE NEUMANN PRINCIPLE

The Neumann principle (NP) establishes the conditions relative to the compatibility of symmetry of the medium and of the considered physical properties. More precisely, if matter remains invariant according to several orientation symmetry operations, then any physical property of matter has to be invariant according to the same orientation symmetry operations (Shuvalov 1981) [6].

As a consequence, the matter tensor $(T)_R$ corresponding to a given physical property has to be invariant according to all the orientation symmetry operations (a) of the point group of matter, which can be written according to Eqs. (2.19) and (2.20) (Nye 1985) [3]:

$$T'_{ijkl\ldots} = \sum_{mnop\ldots} a_{im} a_{jn} a_{ko} a_{lp} \cdots T_{mnop\ldots} \overset{NP}{=} T_{ijkl\ldots} \tag{2.54}$$

if the tensor is polar, and

$$T'_{ijkl\ldots} = |a| \sum_{mnop\ldots} a_{im} a_{jn} a_{ko} a_{lp} \cdots T_{mnop\ldots} \overset{NP}{=} T_{ijkl\ldots} \tag{2.55}$$

for an axial tensor.

In Eq. (2.54) and Eq. (2.55), the left equality stands for the considered transformation, and the right one, written $\overset{NP}{=}$, is imposed by the Neumann principle.

As mentioned earlier in Part 2.1.1, the greatest attention must be paid to the orientation of the considered orthonormal frame, in which the tensor is expressed, relatively to the orientation of the symmetry elements of matter. This important point will be explained in Chapter 3.

According to Eq. (2.54) and Eq. (2.55), the Neumann principle applied to tensors of ranks $R=1,2,3,4$ is written by the following set of equations:

$$T_i' = \sum_m a_{im} T_m \overset{NP}{=} T_i \tag{2.56}$$

$$T_{ij}' = \sum_{mn} a_{im} a_{jn} T_{mn} \overset{NP}{=} T_{ij} \tag{2.57}$$

$$T_{ijk}' = \sum_{mno} a_{im} a_{jn} a_{ko} T_{mno} \overset{NP}{=} T_{ijk} \tag{2.58}$$

$$T_{ijkl}' = \sum_{mnop} a_{im} a_{jn} a_{ko} a_{lp} T_{mnop} \overset{NP}{=} T_{ijkl} \tag{2.59}$$

for polar tensors, or

$$T_i' = |a| \sum_m a_{im} T_m \overset{NP}{=} T_i \tag{2.60}$$

$$T_{ij}' = |a| \sum_{mn} a_{im} a_{jn} T_{mn} \overset{NP}{=} T_{ij} \tag{2.61}$$

$$T_{ijk}' = |a| \sum_{mno} a_{im} a_{jn} a_{ko} T_{mno} \overset{NP}{=} T_{ijk} \tag{2.62}$$

$$T_{ijkl}' = |a| \sum_{mnop} a_{im} a_{jn} a_{ko} a_{lp} T_{mnop} \overset{NP}{=} T_{ijkl} \tag{2.63}$$

in the case of axial tensors.

The four following situations can occur, the tensors being polar or axial:

$$T_{ijkl...}' = T_{ijkl...} \overset{NP}{=} T_{ijkl...} \tag{2.64}$$

$$T_{ijkl...}' = -T_{ijkl...} \overset{NP}{=} T_{ijkl...} \tag{2.65}$$

$$T_{ijkl...}' = T_{uvwx...} \overset{NP}{=} T_{ijkl...} \tag{2.66}$$

$$T_{ijkl...}' = -T_{uvwx...} \overset{NP}{=} T_{ijkl...} \tag{2.67}$$

where i, j, k, l, u, v, x are Cartesian indices.

When Eq. (2.64) is satisfied, then the coefficient $T_{ijkl...}$ is not equal to zero. The only way to satisfy Eq. (2.65) is that $T_{ijkl...} = 0$. Furthermore, Eq. (2.66) and Eq. (2.67) provide relationships, as equality or opposite signs, between two different coefficients of the considered tensor $(T)_R$.

Equation (2.65) to Eq. (2.67) lead to a reduction of the number of independent coefficients. This feature is of prime importance from a practical point of view since it allows us to substantially lighten the tensor calculations, mostly when the tensor rank is high.

This work has to be done for each the symmetry operation (a) of matter and for the 3^R coefficients of the matter tensor $(T)_R$ of the considered physical property. It is thus a hard work: for example it gives up to $3^4 \times 23 = 1863$ operations in the case of a rank-4 tensor ($3^4 = 81$ coefficients) in the crystal class $m\bar{3}m$ where there are 23 orientation symmetry elements.

However, the calculations can be lightened by using the mathematical analogy with the Cartesian coordinates transformation described by Eq. (2.21) and Eq. (2.22). Then considering $(\overrightarrow{OX_1},\overrightarrow{OX_2},\overrightarrow{OX_3})$ as the orthonormal frame, the Neumann principle becomes:

$$x_i'x_j'x_k'x_l'\ldots = \sum_{mnop\ldots}^{NP} a_{im}a_{jn}a_{ko}a_{lp}\cdots x_m x_n x_o x_p \cdots = x_i x_j x_k x_l \cdots \tag{2.68}$$

or

$$x_i'x_j'x_k'x_l'\ldots = |a_{uv}| \sum_{mnop\ldots}^{NP} a_{im}a_{jn}a_{ko}a_{lp}\cdots x_m x_n x_o x_p \cdots = x_i x_j x_k x_l \cdots \tag{2.69}$$

for polar or axial tensors respectively.

2.3.2 Some Examples of Tensor Reduction

2.3.2.1 The Centrosymmetric Crystal Classes

The case of the eleven centrosymmetric crystal classes, *i.e.* having the inversion $\bar{1}$, as defined in Part 2.1.1.2, is of particular interest. Actually in these cases and according to Eq. (2.6), each Cartesian coordinate is transformed in its reverse, *i.e.*: $x_1' = -x_1$, $x_2' = -x_2$, $x_3' = -x_3$ when using the $(\overrightarrow{OX_1},\overrightarrow{OX_2},\overrightarrow{OX_3})$ frame, as shown in Figure 2.17.

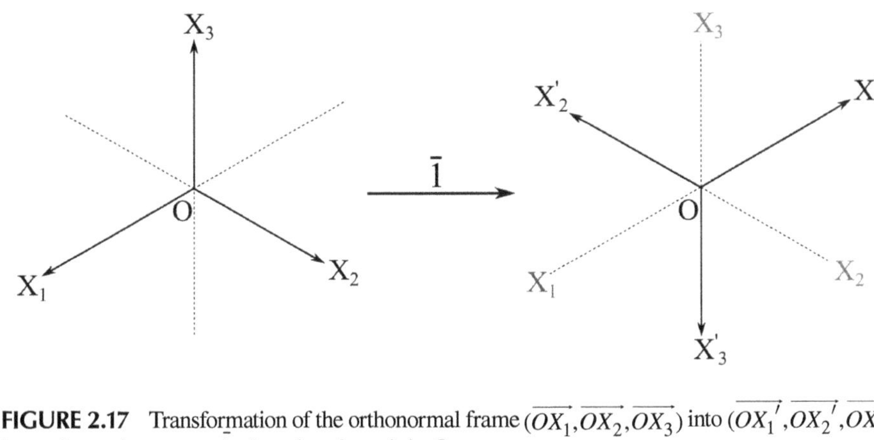

FIGURE 2.17 Transformation of the orthonormal frame $(\overrightarrow{OX_1},\overrightarrow{OX_2},\overrightarrow{OX_3})$ into $(\overrightarrow{OX_1'},\overrightarrow{OX_2'},\overrightarrow{OX_3'})$ by an inversion center $\bar{1}$ placed at the origin O.

Note that the transformed frame is no more direct.

Then in the case of a rank-3 polar tensor $(T)_{R=3}$, it comes according to Eq. (2.68):

$$x_i'x_j'x_k' = (-x_i)(-x_j)(-x_k) = -x_i x_j x_k \overset{NP}{=} x_i x_j x_k \tag{2.70}$$

It corresponds exactly to the situation described by Eq. (2.60): the only way to fulfill the right equality, imposed by the Neumann principle, is to have $x_i x_j x_k = 0$, which means that $T_{ijk} = 0$. Thus, in any centrosymmetric crystal class, all the 3^3 coefficients of this rank-3 polar tensor, like the quadratic electric susceptibility tensor for example, are nil. This result can be obviously generalized to any polar tensor of odd rank, since the product of an odd number of (-1) is necessary equal to -1. It is also very simple to establish that polar tensors of even rank are not concerned by this dramatic feature since the product of an even number of (-1) is equal to 1: it is the case for the cubic electric susceptibility tensor and the photo-elastic tensor, both having a rank $R = 4$.

In the same way, it is interesting to take the example of an axial rank-2 tensor $(T)_{R=2}$. Then according to Eq. (2.69), the Neumann principle gives:

$$x_i'x_j' = -(-x_i)(-x_j) = -x_i x_j \overset{NP}{=} x_i x_j \tag{2.71}$$

As previously, the consequence is $x_i x_j = 0$ leading to $T_{ij} = 0$. Thus, all the 3^2 coefficients of the axial tensor are nil, which is the case for example of the gyration tensor describing the optical activity effect. And here also, the result can be generalized to any other axial tensor of even rank, since $-(-1)^{R=2p} = -1$ with p any natural integer, while axial tensors of odd rank are not concerned since $-(-1)^{R=2p+1} = 1$.

2.3.2.2 The Holohedral Orthorhombic Crystal Class mmm

The stereographic projection of the *mmm* crystal class is given in Figure 2.11. The goal here is to reduce a rank-2 polar tensor $(T)_{R=2}$. The calculation will be carried out using $(\overrightarrow{Ox}, \overrightarrow{Oy}, \overrightarrow{Oz})$ as the orthonormal frame instead of $(\overrightarrow{OX_1}, \overrightarrow{OX_2}, \overrightarrow{OX_3})$, in order to well show that the denomination of the axes has no importance.

A starting and "natural" orientation for the orthonormal frame is with one principal axis in each mirror plane, as shown in Figure 2.18. The relevance of this choice will appear at the end of the calculation, given the required characteristics of this orientation according to the final shape of the reduced tensor that is desired in the framework of crystal optics, as it will be explained in Chapter 3.

The method consists in applying Eq. (2.63) for the 3^2 coefficients T_{ij} and the 3 mirror planes $m \perp \overrightarrow{Oz}$, $m \perp \overrightarrow{Ox}$, and $m \perp \overrightarrow{Oy}$. The transformation performed by $m \perp \overrightarrow{Oz}$ on $(\overrightarrow{Ox}, \overrightarrow{Oy}, \overrightarrow{Oz})$ is given in Figure 2.19.

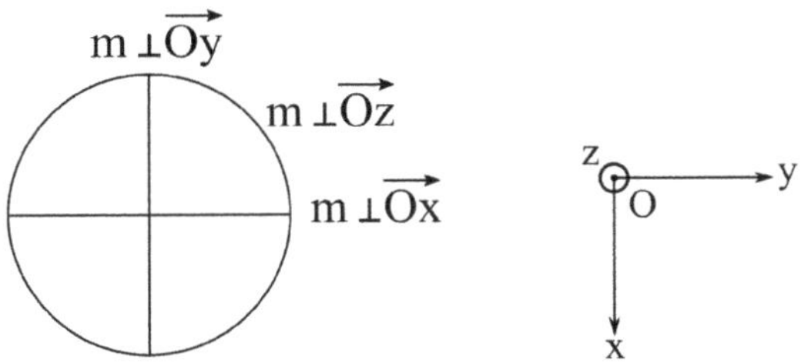

FIGURE 2.18 Orientation of the orthonormal frame $(\overrightarrow{Ox},\overrightarrow{Oy},\overrightarrow{Oz})$ according to the orientation symmetry elements in the *mmm* crystal class.

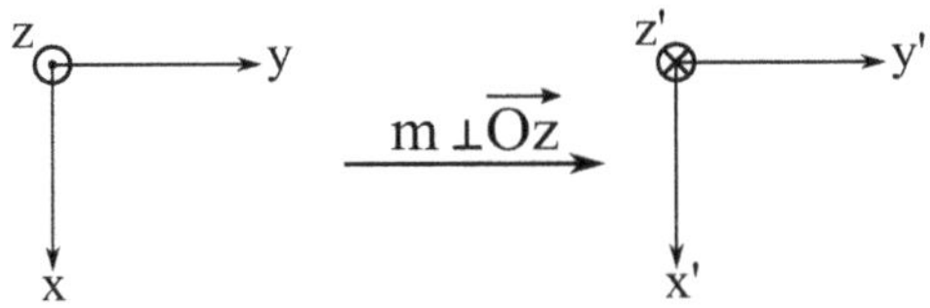

FIGURE 2.19 Transformation of $(\overrightarrow{Ox},\overrightarrow{Oy},\overrightarrow{Oz})$ into $(\overrightarrow{O'_x},\overrightarrow{O'_y},\overrightarrow{O'_z})$ by $m\perp\overrightarrow{Oz}$.

The effects of this mirror plane on the coefficients of the tensor $(T)_{R=2}$ are:

$$\begin{aligned}
x'x' &= xx \overset{NP}{=} xx \Rightarrow T_{xx} \neq 0\\
x'y' &= xy \overset{NP}{=} xy \Rightarrow T_{xy} \neq 0\\
x'z' &= x(-z) \overset{NP}{=} xz \Rightarrow T_{xz} = 0\\
y'x' &= yx \overset{NP}{=} yx \Rightarrow T_{yx} \neq 0\\
y'y' &= yy \overset{NP}{=} yy \Rightarrow T_{yy} \neq 0\\
y'z' &= y(-z) \overset{NP}{=} yz \Rightarrow T_{yz} = 0\\
z'x' &= (-z)x \overset{NP}{=} zx \Rightarrow T_{zx} = 0\\
z'y' &= (-z)y \overset{NP}{=} zy \Rightarrow T_{zy} = 0\\
z'z' &= (-z)(-z) \overset{NP}{=} zz \Rightarrow T_{zz} \neq 0
\end{aligned} \tag{2.72}$$

Then four coefficients become nil.

Thus, it will be sufficient to consider the five other coefficients for the next step, *i.e.* for example the effect of $m\perp\overrightarrow{Ox}$. The corresponding transformation on $(\overrightarrow{Ox},\overrightarrow{Oy},\overrightarrow{Oz})$ is depicted in Figure 2.20.

FIGURE 2.20 Transformation of $(\overrightarrow{Ox},\overrightarrow{Oy},\overrightarrow{Oz})$ into $(\overrightarrow{O'_x},\overrightarrow{O'_y},\overrightarrow{O'_z})$ by $m \perp \overrightarrow{Ox}$.

The effect on the of the remaining T_{ij} coefficients by $m \perp \overrightarrow{Ox}$ is:

$$\begin{aligned}
&x'x' = (-x)(-x) \overset{NP}{=} xx \Rightarrow T_{xx} \neq 0\\
&x'y' = (-x)y \overset{NP}{=} xy \Rightarrow T_{xy} = 0\\
&y'x' = y(-x) \overset{NP}{=} yx \Rightarrow T_{yx} = 0\\
&y'y' = yy \overset{NP}{=} yy \Rightarrow T_{yy} \neq 0\\
&z'z' = zz \overset{NP}{=} zz \Rightarrow T_{zz} \neq 0
\end{aligned} \tag{2.73}$$

There are three new cancellations, so that the action of the last mirror plane $m \perp \overrightarrow{Oy}$ has to be applied to the three remaining coefficients. Figure 2.21 shows the corresponding transformation on $(\overrightarrow{Ox},\overrightarrow{Oy},\overrightarrow{Oz})$:

The effect on the three last coefficients by $m \perp \overrightarrow{Oy}$ is:

$$\begin{aligned}
&x'x' = xx \overset{NP}{=} xx \Rightarrow T_{xx} \neq 0\\
&y'y' = (-y)(-y) \overset{NP}{=} yy \Rightarrow T_{yy} \neq 0\\
&z'z' = zz \overset{NP}{=} zz \Rightarrow T_{zz} \neq 0
\end{aligned} \tag{2.74}$$

This last step does not bring new cancellations. Then the corresponding matrix is diagonal in the case of the *mmm* crystal class, *i.e.*:

$$(T)_{R=2} = \begin{pmatrix} T_{xx} & 0 & 0 \\ 0 & T_{yy} & 0 \\ 0 & 0 & T_{zz} \end{pmatrix} \tag{2.75}$$

Note that the crystal class *mmm* is centrosymmetric, the inversion center $\bar{1}$ being a consequence of the combination of the three mirrors. Then Eq. (2.75) is compatible with the result obtained in Part 2.3.2.1 concerning the fact that a polar tensor with an odd rank can have non-zero coefficients in a centrosymmetric crystal class.

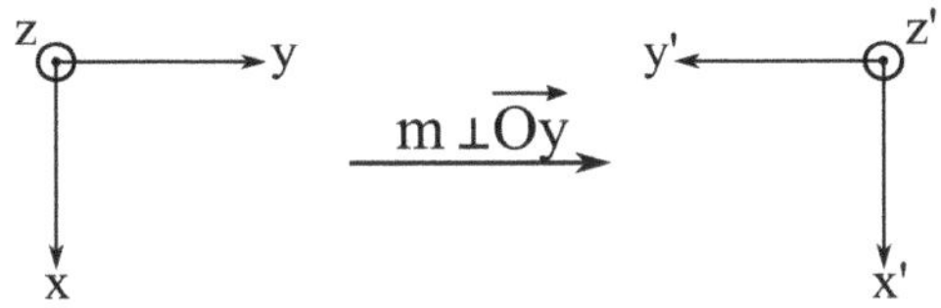

FIGURE 2.21 Transformation of $(\overrightarrow{Ox},\overrightarrow{Oy},\overrightarrow{Oz})$ into $(\overrightarrow{O'_x},\overrightarrow{O'_y},\overrightarrow{O'_z})$ by $m \perp \overrightarrow{Oy}$.

The last comment deals with the choice made for the orientation of $(\overrightarrow{Ox},\overrightarrow{Oy},\overrightarrow{Oz})$ in the example considered above: in crystal optics this orientation will be considered as the relevant one since the tensor of the first-order electric susceptibility (a rank-2 polar tensor) is diagonal, which obviously leads to a simplification of any calculation involving this physical property. The choice of the frame was easy in the case of the crystal class *mmm* . It will appear in Chapter 3 that it will not be always the case for other crystal classes.

2.4 SUMMARY

The media that are considered here are three-dimensional crystals: they are defined by a three-dimensional periodic arrangement of atoms, molecules or ions forming a pattern called the "unit cell". It is repeated in the space thanks to geometrical translations defining the "crystal lattice". The symmetry of the unit cell is not necessary the same than that of the crystal lattice, which defines two kinds of crystals: "holohedral crystals" where the two symmetries are the same, and "merohedral crystals" where they are different. Both symmetries are described by "symmetry operations of finite order" that are performed by "symmetry elements of finite order". At the macroscopic scale, the symmetry of any physical property of a crystal, *i.e.* the spatial symmetry of its angular distribution for example, depends on the "orientation symmetry" of this crystal, also called the crystallographic "point group: it takes into account the symmetry operations that are common to the unit cell and to the lattice. In the framework of group theory, it can be shown that there exist only thirty-two possible combinations of symmetry elements of finite order, defining the thirty-two crystallographic "point groups of finite order" also called the "crystal classes of orientation symmetry". If the orientation symmetry of a physical property of a crystal depends on the point group of this crystal, it is not necessary the case for the symmetry group of this property. In particular, it can exhibit "symmetry operations of infinite order", so that it can belong to one of the seven possible "point groups of infinite order". The physical properties are described by two kinds of tensors: the "polar tensors" and the "axial tensors". By mastering some elements of tensor algebra, and by using the Neumann principle, it becomes possible to take into account the orientation symmetry of matter at the level of each physical property exhibited by this matter. It is of prime importance for going deeply in the understanding of crystal optics. The stereographic projection and the transformation matrix are particularly useful in this context because they allow us to apply the Neumann principle in order to reduce the number of independent coefficients of the matter tensor representing the considered physical property.

A crucial point is the choice of the orientation of the orthonormal frame according to the orientation symmetry elements, this frame being the basis where is expressed any tensor of the crystal. All these aspects will be extensively used in Chapter 3. Tensor algebra is the mathematical heart of this book. The contracted product and the tensor product are the two relevant tools for mastering the calculation of the different orders of the induced polarization of matter performed in Chapter 4, as well as the determination of the angular distribution of any optical property, which constitutes the aim of Chapter 5 to Chapter 8.

REFERENCES

[1] Fischer W. and Koch E. 2005. Symmetry Operations, Part 11: International Tables for Crystallography, Vol. A: Space-Group Symmetry, edited by T. Hahn. Dordrecht, Netherlands: International Union of Crystallography, Kluwer Academic Publisher.

[2] Pechrasky V. and Zavalij P. 2009. Fundamental of Powder Diffraction and Structural Characterization of Materials. Heidelberg: Springer Verlag.

[3] Nye J. F. 1985. Physical Properties of Crystals. Oxford: Oxford University Press.

[4] Hahn Th. and Klapper H. 2005. Point Groups and Crystal Classes, Part 10: International Tables for Crystallography, Vol. A: Space-Group Symmetry, edited by T. Hahn. Dordrecht, Netherlands: International Union of Crystallography, Kluwer Academic Publisher.

[5] De Graef M. and Mc Henry M. E. 2012. Structure of Materials: An Introduction to Crystallography, Diffraction and Symmetry. Cambridge: Cambridge University Press.

[6] Shuvalov L. A. 1981. Modern Crystallography IV—Physical Properties of Crystals. Springer Series in Solid-State Sciences, No. 37. Heidelberg: Springer Verlag.

[7] Boulanger B. and Marnier G. 1991. Field factor calculation for the study of the relationships between all the 3-wave non-linear optical interactions in uniaxial and biaxial crystals. Journal of Physics: Condensed Matter 3: 8327–8350.

[8] Boulanger B., Fève J. P., and Marnier G. 1993. Field factor formalism for the study of the tensorial symmetry of the 4-wave non-linear optical parametric interactions in uniaxial and biaxial crystal classes. Physical Review E 48(6): 4730–4751.

[9] Bourbaki N. 1998. Algebra I: Chapters 1–3. Springer Science & Business Media.

3 Symmetries of the Electric Susceptibility

3.1 ELECTRIC SUSCEPTIBILITY TENSORS AND INDUCED POLARIZATIONS

The matrices that describe the linear, quadratic, and cubic electric susceptibility tensors correspond respectively to Eq. (2.11), Eq. (2.12), and Eq. (2.13) of Chapter 2, with $(T)_{R=2} = \chi^{(1)}$, $(T)_{R=3} = \chi^{(2)}$, and $(T)_{R=4} = \chi^{(3)}$. They are given hereafter in Figure 3.1, Figure 3.2, and Figure 3.3, respectively, using the symbol • for any coefficient of the matrices: the corresponding Cartesian indices x, y, or z can be read in the left-side vertical and top-side horizontal indications. For example, the coefficient of the second line and third column in Figure 3.1 corresponds to $\chi^{(1)}_{yz}$, that of the third line and sixth column in Figure 3.2 is $\chi^{(2)}_{zxz}$, and that of the first line and twenty-fourth column in Figure 3.3 corresponds to $\chi^{(3)}_{xyxz}$. In this chapter, the Cartesian indices are relative to the orthonormal frame $(\overrightarrow{Ox}, \overrightarrow{Oy}, \overrightarrow{Oz})$ defined in Chapter 2. Its orientation according to that of the orientation symmetry elements of the crystal will be first taken as an arbitrarily orientation in Part 3.1. But there exists a specific orientation of $(\overrightarrow{Ox}, \overrightarrow{Oy}, \overrightarrow{Oz})$ for which the $\chi^{(1)}$ tensor is diagonal, as will be shown in Part 3.2. Such a frame is called the "dielectric frame" or the "optical frame", and it is the one conventionally used in crystal optics.

In practical cases, the matrices described are simplified. Actually, a reduction of the number of independent coefficients of the first-, second-, and third-order electric susceptibility tensors can occur thanks to thermodynamics and spatial symmetry considerations of matter. The aim of this chapter is then to show how no absorption, no frequency dispersion, and equality between circular frequencies lead to what is called "intrinsic symmetries", and how the orientation symmetry governs the "extrinsic symmetry" by using Neumann's principle described in Part 3.3. The corresponding matrices will be given for the thirty-two crystal classes.

The linear, quadratic, and cubic induced polarizations vectors, $\overrightarrow{P^{(1)}}$, $\overrightarrow{P^{(2)}}$, and $\overrightarrow{P^{(3)}}$ respectively, are also involved in this chapter. It is then necessary at this level to write the general tensor equations that were introduced in Chapter 1 in a scalar form. And as explained in Chapter 1, $\chi^{(1)}$, $\chi^{(2)}$, and $\chi^{(3)}$ are complex quantities in the general case.

The position vector $\vec{r}$ is omitted in the following expressions of the electric field and induced polarizations since there is no consideration of propagation in the present chapter. Only circular frequency (ω) or time (t) will be considered.

The tensor equation of $\overrightarrow{P^{(1)}}$ has the form given by Eq. (2.29). Then from the scalar Eq. (1.27) it comes:

$$\overrightarrow{P^{(1)}}(\omega) = \varepsilon_0[\chi^{(1)}(\omega).\vec{E}(\omega)] \tag{3.1}$$

DOI: 10.1201/9780429091735-3

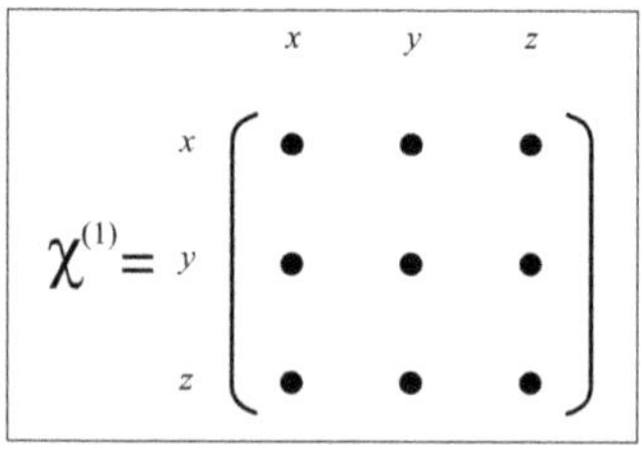

FIGURE 3.1 General form of the matrix describing the linear electric susceptibility tensor $\chi^{(1)}$ that is of rank $R = 2$.

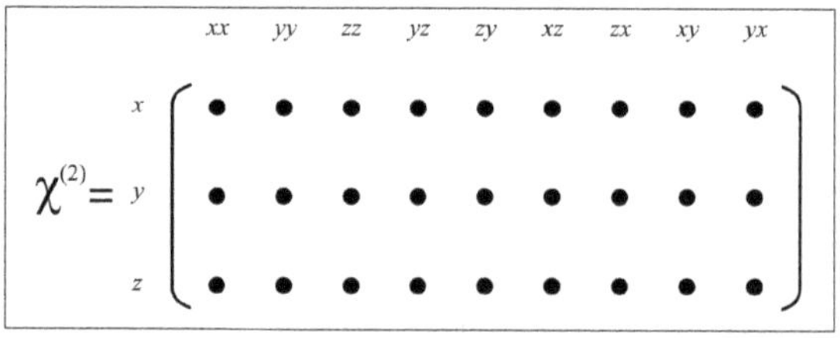

FIGURE 3.2 General form of the matrix describing the quadratic electric susceptibility tensor $\chi^{(2)}$ that is of rank $R = 3$.

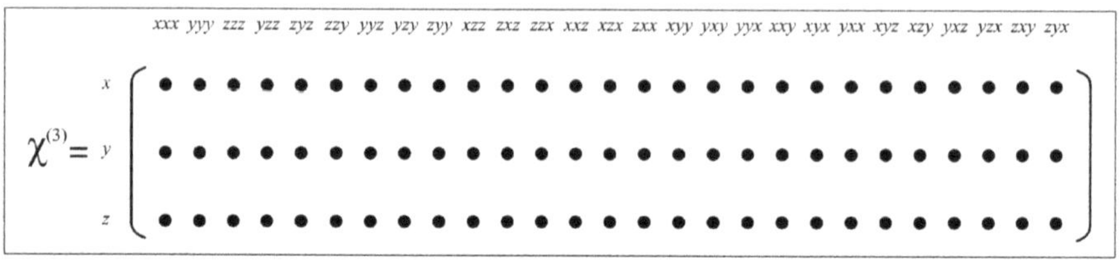

FIGURE 3.3 General form of the matrix describing the cubic electric susceptibility tensor $\chi^{(3)}$ that is of rank $R = 4$.

The tensor equation of $\overrightarrow{P^{(2)}}$ has the form given by Eq. (2.32). Then from the scalar Eq. (1.49a), Eq. (1.51a), and Eq. (1.51b) it comes for the three Fourier components at $(\omega_1, \omega_2, \omega_3)$ that are coupled with $\chi^{(2)}$:

$$\begin{aligned}
\overrightarrow{P^{(2)}}(\omega_1) &= \varepsilon_0[\chi^{(2)}(\omega_1 = \omega_3 - \omega_2) : \vec{E}(\omega_3) \otimes \vec{E}^*(\omega_2)] \\
\overrightarrow{P^{(2)}}(\omega_2) &= \varepsilon_0[\chi^{(2)}(\omega_2 = \omega_3 - \omega_1) : \vec{E}(\omega_3) \otimes \vec{E}^*(\omega_1)] \\
\overrightarrow{P^{(2)}}(\omega_3) &= \varepsilon_0[\chi^{(2)}(\omega_3 = \omega_1 + \omega_2) : \vec{E}(\omega_1) \otimes \vec{E}(\omega_2)]
\end{aligned} \tag{3.2}$$

where $\vec{E}^*(\omega_i) = \vec{E}(-\omega_i)$.

For $\overrightarrow{P^{(3)}}$, the tensor equation has the form given by Eq. (2.36). As explained in Chapter 1, two situations have to be considered corresponding to the two possible energy conservation relations: $\omega_4 = \omega_1 + \omega_2 + \omega_3$ and $\omega_4 + \omega_1 = \omega_2 + \omega_3$. Then from the scalar Eq. (1.66) and Eq. (1.68) respectively, it comes for the coupled Fourier components at $(\omega_1, \omega_2, \omega_3, \omega_4)$ that are coupled with $\chi^{(3)}$:

In the case $\omega_4 = \omega_1 + \omega_2 + \omega_3$

$$
\begin{aligned}
\overrightarrow{P^{(3)}}(\omega_1) &= \varepsilon_0[\chi^{(3)}(\omega_1 = \omega_4 - \omega_2 - \omega_3) \therefore \overrightarrow{E}(\omega_4) \otimes \overrightarrow{E}^*(\omega_2) \otimes \overrightarrow{E}^*(\omega_3)] \\
\overrightarrow{P^{(3)}}(\omega_2) &= \varepsilon_0[\chi^{(3)}(\omega_2 = \omega_4 - \omega_1 - \omega_3) \therefore \overrightarrow{E}(\omega_4) \otimes \overrightarrow{E}^*(\omega_1) \otimes \overrightarrow{E}^*(\omega_3)] \\
\overrightarrow{P^{(3)}}(\omega_3) &= \varepsilon_0[\chi^{(3)}(\omega_3 = \omega_4 - \omega_1 - \omega_2) \therefore \overrightarrow{E}(\omega_4) \otimes \overrightarrow{E}^*(\omega_1) \otimes \overrightarrow{E}^*(\omega_2)] \\
\overrightarrow{P^{(3)}}(\omega_4) &= \varepsilon_0[\chi^{(2)}(\omega_4 = \omega_1 + \omega_2 + \omega_3) \therefore \overrightarrow{E}(\omega_1) \otimes \overrightarrow{E}(\omega_2) \otimes \overrightarrow{E}(\omega_3)]
\end{aligned}
\tag{3.3}
$$

In the case $\omega_4 + \omega_1 = \omega_2 + \omega_3$

$$
\begin{aligned}
\overrightarrow{P^{(3)}}(\omega_1) &= \varepsilon_0[\chi^{(3)}(\omega_1 = -\omega_4 + \omega_2 + \omega_3) \therefore \overrightarrow{E}^*(\omega_4) \otimes \overrightarrow{E}(\omega_2) \otimes \overrightarrow{E}(\omega_3)] \\
\overrightarrow{P^{(3)}}(\omega_2) &= \varepsilon_0[\chi^{(3)}(\omega_2 = \omega_4 + \omega_1 - \omega_3) \therefore \overrightarrow{E}(\omega_4) \otimes \overrightarrow{E}(\omega_1) \otimes \overrightarrow{E}^*(\omega_3)] \\
\overrightarrow{P^{(3)}}(\omega_3) &= \varepsilon_0[\chi^{(3)}(\omega_3 = \omega_4 + \omega_1 - \omega_2) \therefore \overrightarrow{E}(\omega_4) \otimes \overrightarrow{E}(\omega_1) \otimes \overrightarrow{E}^*(\omega_2)] \\
\overrightarrow{P^{(3)}}(\omega_4) &= \varepsilon_0[\chi^{(3)}(\omega_4 = -\omega_1 + \omega_2 + \omega_3) \therefore \overrightarrow{E}^*(\omega_1) \otimes \overrightarrow{E}(\omega_2) \otimes \overrightarrow{E}(\omega_3)]
\end{aligned}
\tag{3.4}
$$

3.2 INTRINSIC SYMMETRIES OF THE ELECTRIC SUSCEPTIBILITY TENSORS

3.2.1 Symmetry in the Approximation of a Lossless Medium

3.2.1.1 First-Order Electric Susceptibility

A lossless medium is considered, that is to say without any absorption. In such a matter, when there is no entropy variation, and in the case of an infinitesimal reversible transformation of matter under the light interaction, the total differential of the instantaneous thermodynamic potential function $\phi(t)$ can be written in the time domain (Born and Wolf 1999) [1]:

$$
\left\langle d\phi^{(1)}(t) \right\rangle_t \tag{3.5}
$$

where $\vec{E}(t)$ is the electric field that is radiated by the induced polarization $\vec{P}(t)$. For the purpose of the following demonstrations, we will consider monochromatic waves, at the circular frequency ω. And since $\vec{E}(t)$ and $\vec{P}(t)$ are real quantities, they are simply:

$$\vec{P}(t) = \frac{1}{2}\left[\vec{P}(\omega)e^{-i\omega t} + \overrightarrow{P^*}(\omega)e^{i\omega t}\right] \quad (3.6)$$

and

$$\vec{E}(t) = \frac{1}{2}\left[\vec{E}(\omega)e^{-i\omega t} + \overrightarrow{E^*}(\omega)e^{i\omega t}\right] \quad (3.7)$$

where $\vec{E}(\omega)$ is the complex electric field that is radiated by the complex induced polarization $\vec{P}(\omega)$.

According to Eq. (3.6) and Eq. (3.7), the time averaged Eq. (3.5) depends only on non-oscillating contributions, which gives:

$$\left\langle d\phi(t)\right\rangle_t = -\frac{1}{4}\left[\vec{P}(\omega).d\overrightarrow{E^*}(\omega) + \overrightarrow{P^*}(\omega).d\vec{E}(\omega)\right] \quad (3.8)$$

In the linear regime, $\vec{P}(\omega)$ that is the first-order induced polarization, *i.e.* $\overrightarrow{P^{(1)}}(\omega)$, as defined in Chapter 1. Then according to Eq. (3.8), the corresponding time averaged thermodynamic potential, written $\left\langle d\phi^{(1)}(t)\right\rangle_t$, is given by:

$$\left\langle d\phi^{(1)}(t)\right\rangle_t = -\frac{1}{4}\left[\overrightarrow{P^{(1)}}(\omega).d\overrightarrow{E^*}(\omega) + \overrightarrow{P^{(1)*}}(\omega).d\vec{E}(\omega)\right] \quad (3.9)$$

According to Eq. (3.1), Eq. (2.30), and Eq. (2.43), Eq. (3.9) becomes:

$$\begin{aligned}\left\langle d\phi^{(1)}(t)\right\rangle_t &= -\frac{1}{4}\varepsilon_0\sum_i\left[\sum_j \chi_{ij}^{(1)}(\omega)E_j(\omega)dE_i^*(\omega) + \sum_j \chi_{ij}^{(1)*}(\omega)E_j^*(\omega)dE_i(\omega)\right]\\ &= -\frac{1}{4}\varepsilon_0\sum_{ij}\chi_{ij}^{(1)}(\omega)E_j(\omega)dE_i^*(\omega) + \chi_{ij}^{(1)*}(\omega)E_j^*(\omega)dE_i(\omega)\end{aligned} \quad (3.10)$$

where i and j stand for x, y, or z.

In the following, the quantity $\left\langle d\phi^{(1)}(t)\right\rangle_t$ will be written $d\phi$ in order to lighten the writing.

Then from Eq. (3.10) it comes:

$$\frac{\partial\phi^{(1)}}{\partial E_i^*(\omega)} = -\frac{1}{4}\varepsilon_0\sum_i \chi_{ij}^{(1)}(\omega)E_j(\omega) \quad (3.11)$$

And then:

$$\frac{\partial}{\partial E_j(\omega)}\left(\frac{\partial\phi^{(1)}}{\partial E_i^*(\omega)}\right) = \frac{\partial^2\phi^{(1)}}{\partial E_j(\omega)\partial E_i^*(\omega)} = -\frac{1}{4}\varepsilon_0\chi_{ij}^{(1)}(\omega) \quad (3.12)$$

The coefficient $\chi_{ji}^*(\omega)$ can be obtained in the same way, by permuting at first indices i and j in Eq. (3.10), *i.e.*:

$$d\phi^{(1)} = -\frac{1}{4}\varepsilon_0 \sum_{ji} \chi_{ji}^{(1)}(\omega)E_i(\omega)dE_j^*(\omega) + \chi_{ji}^{(1)*}(\omega)E_i^*(\omega)dE_j(\omega) \tag{3.13}$$

Then:

$$\frac{\partial \phi^{(1)}}{\partial E_j(\omega)} = -\frac{1}{4}\varepsilon_0 \sum_i \chi_{ji}^{(1)*}(\omega)E_i^*(\omega) \tag{3.14}$$

And finally:

$$\frac{\partial}{\partial E^*{}_i(\omega)}\left(\frac{\partial \phi^{(1)}}{\partial E_j(\omega)}\right) = \frac{\partial^2 \phi^{(1)}}{\partial E^*{}_i(\omega)\partial E_j(\omega)} = -\frac{1}{4}\varepsilon_0 \chi_{ji}^{(1)*}(\omega) \tag{3.15}$$

Since the mixed derivatives of the same function do not depend on the order of derivation, it comes:

$$\frac{\partial^2 \phi^{(1)}}{\partial E_j(\omega)\partial E_i^*(\omega)} = \frac{\partial^2 \phi^{(1)}}{\partial E_i^*(\omega)\partial E_j(\omega)} \tag{3.16}$$

And thus:

$$-\frac{1}{4}\varepsilon_0 \chi_{ij}^{(1)}(\omega) = -\frac{1}{4}\varepsilon_0 \chi_{ji}^{(1)*}(\omega) \tag{3.17}$$

Which is equivalent to:

$$\chi_{ij}^{(1)}(\omega) = \chi_{ji}^{(1)*}(\omega) \tag{3.18}$$

It means that in a lossless medium, $\chi^{(1)}$ is a Hermitian tensor.

As a consequence, since in a lossless medium $\chi^{(1)}$ is a pure real quantity as explained in Chapter 1. Then $\chi^{(1)} = \chi^{(1)*}$, so that Eq. (3.18) leads to:

$$\chi_{ij}^{(1)}(\omega) = \chi_{ji}^{(1)}(\omega) \tag{3.19}$$

As a consequence, $\chi^{(1)}$ is a symmetric tensor, as shown in Figure 3.4, in which the "bridge" between two dots means the equality between the two corresponding coefficients.

Figure 3.4 shows a $\chi^{(1)}$ tensor with 6 independent coefficients among 9 when the medium is transparent at the considered circular frequency. This approximation will be considered all throughout this chapter.

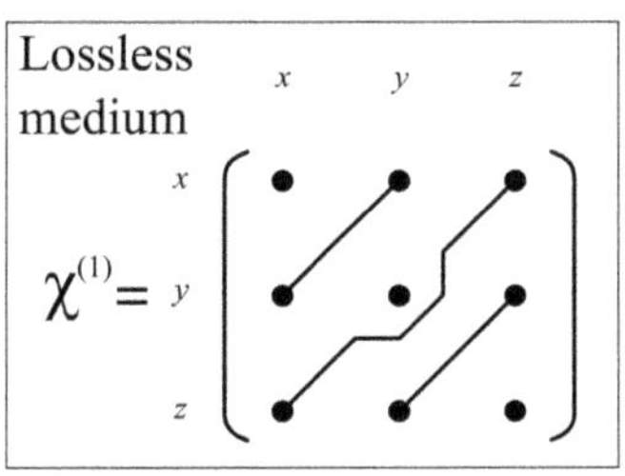

FIGURE 3.4 Matrix describing the symmetric first-order electric susceptibility tensor of a lossless medium.

3.2.1.2 Second-Order Electric Susceptibility—ABDP Symmetry

In the case of the second-order nonlinear regime, the induced polarization is $\overrightarrow{P^{(2)}}$. Then according to Eq. (3.8), the corresponding time averaged thermodynamic potential $\left\langle d\phi^{(2)}(t)\right\rangle_t$ is expressed as (Bloembergen 1996) [2]:

$$\left\langle d\phi^{(2)}(t)\right\rangle_t = -\frac{1}{4}\left[\overrightarrow{P^{(2)}}(\omega).d\overrightarrow{E^*}(\omega)+\overrightarrow{P^{(2)*}}(\omega).d\vec{E}(\omega)\right] \tag{3.20}$$

$\left\langle d\phi^{(2)}(t)\right\rangle_t$ will be written $d\phi^{(2)}$ in the following.

The second-order nonlinear regime involves three waves at three different circular frequencies $(\omega_1,\omega_2,\omega_3)$ linked by the energy conservation relation $\omega_3=\omega_1+\omega_2$, which is equivalent to $\omega_1=\omega_3-\omega_2$ and $\omega_2=\omega_3-\omega_1$. Therefore, the thermodynamic potential function is the same whatever the circular frequency considered, and writes:

$$\begin{aligned} d\phi^{(2)} &= -\frac{1}{4}\left[\overrightarrow{P^{(2)}}(\omega_3).d\overrightarrow{E^*}(\omega_3)+\overrightarrow{P^{(2)*}}(\omega_3).d\vec{E}(\omega_3)\right] \\ &= -\frac{1}{4}\left[\overrightarrow{P^{(2)}}(\omega_1).d\overrightarrow{E^*}(\omega_1)+\overrightarrow{P^{(2)*}}(\omega_1).d\vec{E}(\omega_1)\right] \\ &= -\frac{1}{4}\left[\overrightarrow{P^{(2)}}(\omega_2).d\overrightarrow{E^*}(\omega_2)+\overrightarrow{P^{(2)*}}(\omega_2).d\vec{E}(\omega_2)\right] \end{aligned} \tag{3.21}$$

For such calculations that aim to take into account the anisotropy of the medium, we took care of considering different summation indices in order to distinguish the different circular frequencies, *i.e.*, *i* for ω_3, *j* for ω_1, and *k* for ω_2, knowing that each index *i*, *j*, and *k* can stand for *x*, *y*, or *z*. Obviously we could write the three equations with the same index, but thanks to their differentiation we can obtain interesting results, as shown hereafter.

It is also interesting to develop each of the three parts of Eq. (3.21), *i.e.* at ω_3, ω_1, and ω_2.

In the case of ω_3, from Eq. (3.2), Eq. (2.33), and Eq. (2.43), it comes:

$$d\phi^{(2)} = -\frac{1}{4}\varepsilon_0\left[\begin{array}{l}\sum\limits_{ijk}\chi_{ijk}{}^{(2)}(\omega_3=\omega_1+\omega_2)E_j(\omega_1)E_k(\omega_2)dE_i^*(\omega_3) \\ +\chi_{ijk}{}^{(2)*}(\omega_3=\omega_1+\omega_2)E_j^*(\omega_1)E_k^*(\omega_2)dE_i(\omega_3)\end{array}\right] \tag{3.22}$$

And so:

$$\frac{\partial\phi^{(2)}}{\partial E_i^*(\omega_3)} = -\frac{1}{4}\varepsilon_0\sum_{jk}\chi_{ijk}{}^{(2)}(\omega_3=\omega_1+\omega_2)E_j(\omega_1)E_k(\omega_2) \tag{3.23}$$

$$\frac{\partial}{\partial E_j(\omega_1)}\left(\frac{\partial\phi^{(2)}}{\partial E_i^*(\omega_3)}\right) = -\frac{1}{4}\varepsilon_0\sum_{k}\chi_{ijk}{}^{(2)}(\omega_3=\omega_1+\omega_2)E_k(\omega_2) \tag{3.24}$$

$$\frac{\partial}{\partial E_k(\omega_2)}\left[\frac{\partial}{\partial E_j(\omega_1)}\left(\frac{\partial \phi^{(2)}}{\partial E_i^*(\omega_3)}\right)\right] = -\frac{1}{4}\varepsilon_0 \chi_{ijk}{}^{(2)}(\omega_3 = \omega_1 + \omega_2) \tag{3.25}$$

Which can be written:

$$\frac{\partial^3 \phi^{(2)}}{\partial E_k(\omega_2)\partial E_j(\omega_1)\partial E_i^*(\omega_3)} = -\frac{1}{4}\varepsilon_0 \chi_{ijk}{}^{(2)}(\omega_3 = \omega_1 + \omega_2) \tag{3.26}$$

The equivalent of Eq. (3.22) at ω_1 is:

$$\begin{aligned} d\phi^{(2)} &= -\frac{1}{4}\varepsilon_0 \sum\nolimits_j [\sum\nolimits_{ik} \chi_{jik}{}^{(2)}(\omega_1 = \omega_3 - \omega_2) E_i(\omega) E_k^*(\omega_2) dE_j^*(\omega_1) \\ &+ \sum\nolimits_{ik} \chi_{jik}{}^{(2)*}(\omega_1 = \omega_3 - \omega_2) E_i^*(\omega_3) E_k(\omega_2) dE_j(\omega_1)] \end{aligned} \tag{3.27}$$

By doing as previously in order to get $\chi_{jik}{}^{(2)*}(\omega_1)$, it comes:

$$\frac{\partial^3 \phi^{(2)}}{\partial E_i^*(\omega_3)\partial E_k(\omega_2)\partial E_j(\omega_1)} = -\frac{1}{4}\varepsilon_0 \chi_{jik}{}^{(2)*}(\omega_1 = \omega_3 - \omega_2) \tag{3.28}$$

And finally, the same work performed at ω_2 leads to the following equations:

$$d\phi^{(2)} = -\frac{1}{4}\varepsilon_0 \left[\begin{aligned} &\sum_{kij} \chi_{kij}{}^{(2)}(\omega_2 = \omega_3 - \omega_1) E_i(\omega) E_j^*(\omega_1) dE_k^*(\omega_2) \\ &+ \chi_{kij}{}^{(2)*}(\omega_2 = \omega_3 - \omega_1) E_i^*(\omega_3) E_j(\omega_1) dE_k(\omega_2) \end{aligned}\right] \tag{3.29}$$

and

$$\frac{\partial^3 \phi^{(2)}}{\partial E_i^*(\omega_3)\partial E_j(\omega_1)\partial E_k(\omega_2)} = -\frac{1}{4}\varepsilon_0 \chi_{kij}{}^{(2)*}(\omega_2 = \omega_3 - \omega_1) \tag{3.30}$$

The next step deals with the crossing between Eq. (3.26), Eq. (3.28), and Eq. (3.30). The left hands of the three equalities are equal thanks to the commutativity of the order of derivation, *i.e.*:

$$\begin{aligned} \frac{\partial^3 \phi^{(2)}}{\partial E_k(\omega_2)\partial E_j(\omega_1)\partial E_i^*(\omega_3)} &= \frac{\partial^3 \phi^{(2)}}{\partial E_i^*(\omega_3)\partial E_k(\omega_2)\partial E_j(\omega_1)} \\ &= \frac{\partial^3 \phi^{(2)}}{\partial E_i^*(\omega_3)\partial E_j(\omega_1)\partial E_k(\omega_2)} \end{aligned} \tag{3.31}$$

which implies:

$$\chi_{ijk}{}^{(2)}(\omega_3 = \omega_1 + \omega_2) = \chi_{jik}{}^{(2)*}(\omega_1 = \omega_3 - \omega_2) = \chi_{kij}{}^{(2)*}(\omega_2 = \omega_3 - \omega_1) \tag{3.32}$$

Because $\chi^{(2)}$ is a real quantity in lossless matter, *i.e.* $\chi^{(2)} = \chi^{(2)*}$, then it comes:

$$\chi_{ijk}^{(2)}(\omega_3 = \omega_1 + \omega_2) = \chi_{jik}^{(2)}(\omega_1 = \omega_3 - \omega_2) = \chi_{kij}^{(2)}(\omega_2 = \omega_3 - \omega_1) \quad (3.33)$$

These equalities are verified under what is called the Armstrong-Bloembergen-Ducuing-Pershan (ABDP) symmetry (Armstrong 1962) [3]. These relations provide equalities between different coefficients of $\chi^{(2)}$ at different circular frequencies, which will be very useful for the establishment of the propagation equation in the nonlinear regime as shown in Chapter 7. But there is no reduction of the number of the independent coefficients of the tensor at a given circular frequency.

3.2.1.3 Third-Order Electric Susceptibility—ABDP Symmetry

The previous approach can be applied for the third-order interactions, knowing that two situations have to be considered according to the two possible energy conservation relations: $\omega_4 = \omega_1 + \omega_2 + \omega_3$ and $\omega_4 + \omega_1 = \omega_2 + \omega_3$.

The relation $\omega_4 = \omega_1 + \omega_2 + \omega_3$ being equivalent to $\omega_1 = \omega_4 - \omega_2 - \omega_3$, $\omega_2 = \omega_4 - \omega_1 - \omega_3$, and $\omega_3 = \omega_4 - \omega_1 - \omega_2$, the time averaged thermodynamic potential function can be written with respect to any of the four circular frequencies, *i.e.*:

$$\begin{aligned} d\phi^{(3)} &= -\frac{1}{4}\left[\overrightarrow{P^{(3)}}(\omega_4).d\overrightarrow{E^*}(\omega_4) + \overrightarrow{P^{(3)*}}(\omega_4).d\vec{E}(\omega_4)\right] \\ &= -\frac{1}{4}\left[\overrightarrow{P^{(3)}}(\omega_1).d\overrightarrow{E^*}(\omega_1) + \overrightarrow{P^{(3)*}}(\omega_1).d\vec{E}(\omega_1)\right] \\ &= -\frac{1}{4}\left[\overrightarrow{P^{(3)}}(\omega_2).d\overrightarrow{E^*}(\omega_2) + \overrightarrow{P^{(3)*}}(\omega_2).d\vec{E}(\omega_2)\right] \\ &= -\frac{1}{4}\left[\overrightarrow{P^{(3)}}(\omega_3).d\overrightarrow{E^*}(\omega_3) + \overrightarrow{P^{(3)*}}(\omega_3).d\vec{E}(\omega_3)\right] \end{aligned} \quad (3.34)$$

Using Eq. (3.3), Eq. (2.38), and Eq. (2.43), and by taking care of considering different summation indices in order to distinguish the different circular frequencies, *i.e.*, i for ω_4, j for ω_1, k for ω_2, and l for ω_3, the following equalities appear:

$$\begin{aligned} &\chi_{ijkl}^{(3)}(\omega_4 = \omega_1 + \omega_2 + \omega_3) = \chi_{jikl}^{(3)*}(\omega_1 = \omega_4 - \omega_2 - \omega_3) \\ &= \chi_{kijl}^{(3)*}(\omega_2 = \omega_4 - \omega_1 - \omega_3) = \chi_{lijk}^{(3)*}(\omega_3 = \omega_4 - \omega_1 - \omega_2) \end{aligned} \quad (3.35)$$

And because $\chi^{(3)}$ is a real quantity in a lossless medium, *i.e.* $\chi^{(3)} = \chi^{(3)*}$, then Eq. (3.35) leads to:

$$\begin{aligned} &\chi_{ijkl}^{(3)}(\omega_4 = \omega_1 + \omega_2 + \omega_3) = \chi_{jikl}^{(3)}(\omega_1 = \omega_4 - \omega_2 - \omega_3) \\ &= \chi_{kijl}^{(3)}(\omega_2 = \omega_4 - \omega_1 - \omega_3) = \chi_{lijk}^{(3)}(\omega_3 = \omega_4 - \omega_1 - \omega_2) \end{aligned} \quad (3.36)$$

The same approach is now applied to the second case of the third-order processes. The relation $\omega_4 = -\omega_1 + \omega_2 + \omega_3$ being equivalent to $\omega_1 = -\omega_4 + \omega_2 + \omega_3$, $\omega_2 = \omega_4 + \omega_1 - \omega_3$, and $\omega_3 = \omega_4 + \omega_1 - \omega_2$, then it comes from Eq. (3.4), Eq. (2.38), and Eq. (2.43):

$$\begin{aligned} &\chi_{ijkl}^{(3)}(\omega_4 = -\omega_1 + \omega_2 + \omega_3) = \chi_{jikl}^{(3)}(\omega_1 = -\omega_4 + \omega_2 + \omega_3) \\ &= \chi_{kijl}^{(3)}(\omega_2 = \omega_4 + \omega_1 - \omega_3) = \chi_{lijk}^{(3)}(\omega_3 = \omega_4 + \omega_1 - \omega_2) \end{aligned} \tag{3.37}$$

As in the case of $\chi^{(2)}$, these ABDP relations do not affect the symmetry of $\chi^{(3)}$ at a given circular frequency, which means no reduction of the number of independent tensor coefficients, but they provide equalities that will be useful for the establishment of the propagation equation in the nonlinear regime, as shown in Chapter 8.

3.2.2 Symmetry in the Approximation of a Lossless and Non-Dispersive Medium

3.2.2.1 Second-Order Electric Susceptibility—Kleinman Symmetry

If the dispersion can be neglected over the spectral range containing the three circular frequencies that are involved, which is generally the case in the middle of the transparency window of any medium, then it comes $\chi^{(2)}(\omega_3) = \chi^{(2)}(\omega_1) = \chi^{(2)}(\omega_2)$, so that the ABDP relations given by Eq. (3.33) lead directly to the three following equalities that are valid at ω_3, ω_1, and ω_2.

$$\chi_{ijk}^{(2)} = \chi_{jik}^{(2)} = \chi_{kij}^{(2)} \tag{3.38}$$

More relations can be found by taking into account the non-dispersion of $\chi^{(2)}$ directly at the level of Eq. (3.31), which gives:

$$\frac{\partial^3 \phi^{(2)}}{\partial E_k \partial E_j \partial E_i^*} = \frac{\partial^3 \phi^{(2)}}{\partial E_i^* \partial E_k \partial E_j} = \frac{\partial^3 \phi^{(2)}}{\partial E_i^* \partial E_j \partial E_k} \tag{3.39}$$

Then according to Eq. (3.26), Eq. (3.28), or Eq. (3.30), it comes that $\chi^{(2)}$ is invariant by any permutation of the Cartesian indices independently on the frequency, *i.e.*:

$$\chi_{ijk}^{(2)} = \chi_{ikj}^{(2)} = \chi_{jik}^{(2)} = \chi_{jki}^{(2)} = \chi_{kij}^{(2)} = \chi_{kji}^{(2)} \tag{3.40}$$

As a result, the $\chi^{(2)}$ tensor is fully symmetric, which is called the Kleinman symmetry (Kleinman 1962) [4]. The corresponding (3×9) matrix is given hereafter, the "bridge" between two dots meaning the equality between the two corresponding coefficients. Figure 3.5 shows that the $\chi^{(2)}$ tensor has ten independent coefficients among twenty-seven under Kleinman symmetry.

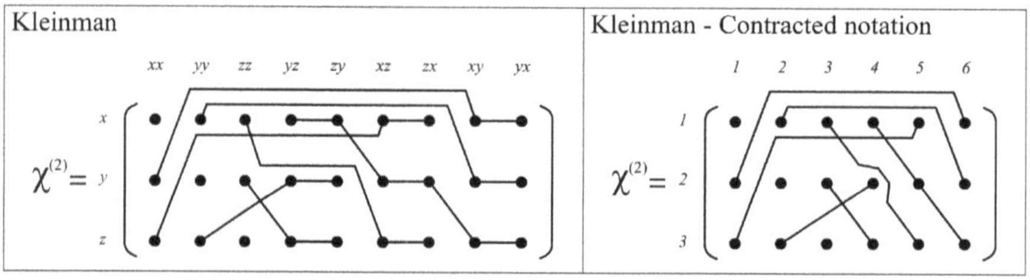

FIGURE 3.5 Symmetric second-order electric susceptibility tensor in the case of a lossless and non-dispersive medium using the general notation (left) and the contracted notation (right).

Because the symmetry of the $\chi^{(2)}$ tensor is full, a simplified notation can be used, in which the two last indices are contracted in a single index. In this case, any $\chi_{ijk}{}^{(2)}$ coefficient is written $\chi_{\mu m}{}^{(2)}$ with the following convention:

$$\begin{aligned}
&i \rightarrow \mu: \quad x \rightarrow 1 \quad y \rightarrow 2 \quad z \rightarrow 3 \\
&jk \rightarrow m: xx \rightarrow 1 \quad yy \rightarrow 2 \quad zz \rightarrow 3 \\
&yz = zy \rightarrow 4 \quad xz = zx \rightarrow 5 \quad xy = yx \rightarrow 6
\end{aligned} \tag{3.41}$$

Using the contracted notation, the corresponding (3x6) matrix exhibits ten independent coefficients as shown in Figure 3.5.

Note that Eq. (3.41) is the equivalent of Eq. (3.16). Actually, in the first-order case, the dispersion is obviously not relevant since a single circular frequency is involved. In the two cases, *i.e.* the first- and second-order cases, the consequence is a fully symmetrical tensor.

3.2.2.2 Third-Order Electric Susceptibility—Kleinman Symmetry

For the same reasons, if the medium is non-dispersive over the four circular frequencies $\omega_1, \omega_2, \omega_3$, and ω_4, then $\chi^{(3)}$ is symmetric in all indices at any of these circular frequencies, which leads to the matrices shown in Figure 3.6.

A contracted notation can also be used in that case. A $\chi_{ijkl}{}^{(3)}$ coefficient is written $\chi_{\mu m}{}^{(3)}$ with the following convention:

$$\begin{aligned}
&i \rightarrow \mu: \quad x \rightarrow 1 \quad y \rightarrow 2 \quad z \rightarrow 3 \\
&jkl \rightarrow m: xxx \rightarrow 1 \quad yyy \rightarrow 2 \quad zzz \rightarrow 3 \\
&yzz = zzy = zyz \rightarrow 4 \quad yyz = zyy = yzy \rightarrow 5 \\
&xzz = zzx = zxz \rightarrow 6 \quad xxz = zxx = xzx \rightarrow 7 \\
&xyy = yyx = yxy \rightarrow 8 \quad xxy = yxx = xyx \rightarrow 9 \\
&xyz = xzy = yxz = yzx = zxy = zyx \rightarrow 0
\end{aligned} \tag{3.42}$$

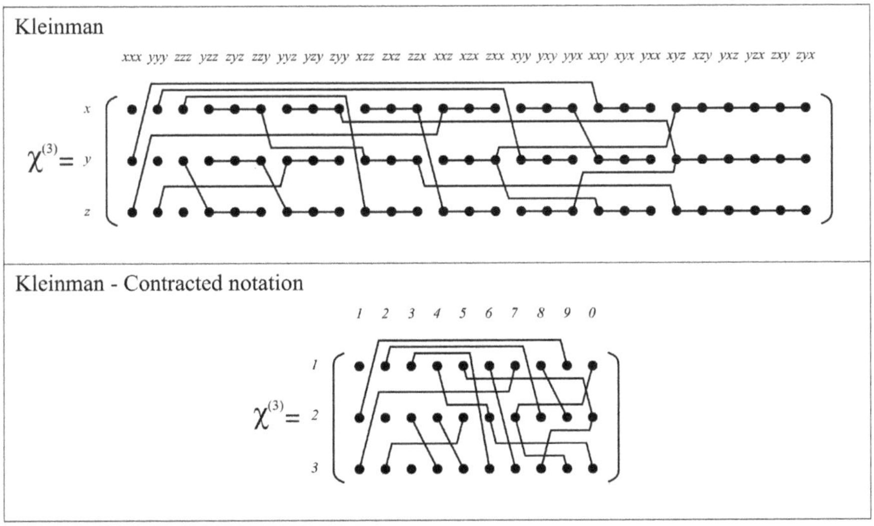

FIGURE 3.6 Symmetric third-order electric susceptibility tensor in the case of a lossless and non-dispersive medium using the general notation (top) and the contracted notation (bottom).

The (3×27) matrix becomes a (3×10) matrix when using the contracted notation given in Figure 3.6.

Due to the equalities between coefficients, $\chi^{(3)}$ exhibits thirteen independent coefficients, instead of eighty-one in the general case. It is a huge reduction. Note also that the matrices of Figure 3.6 are valid for the two possible energy conservation relations, *i.e.* $\omega_4 = \omega_1 + \omega_2 + \omega_3$ and $\omega_4 = \omega_1 + \omega_2 - \omega_3$.

3.2.3 Symmetry in the Case of a Lossless Medium and Equalities between Frequencies

3.2.3.1 Second-Order Electric Susceptibility with Equality between Frequencies

In that case the energy conservation is written $2\omega = \omega + \omega$, which is equivalent to $\omega = 2\omega - \omega$ where ω stands for $\omega_1 = \omega_2$ and 2ω for ω_3. Then the ABDP equalities given in Eq. 3.33 become:

$$\chi_{ijk}^{(2)}(2\omega = \omega + \omega) = \chi_{jik}^{(2)}(\omega = 2\omega - \omega) = \chi_{kij}^{(2)}(\omega = 2\omega - \omega) \qquad (3.43)$$

The second equality allows us to directly establish that $\chi^{(2)}(\omega = 2\omega - \omega)$ is symmetrical according to the first and third Cartesian indices that are the indices corresponding to the equal circular frequencies, *i.e.* $\omega_1 = \omega_2 = \omega$. The corresponding matrix, exhibiting eighteen independent coefficients, is shown in Figure 3.7.

Note that these equalities are part of Kleinman symmetry shown in Figure 3.5, which is not surprising since the dispersion is not relevant for the Cartesian indices relative to equal circular frequencies.

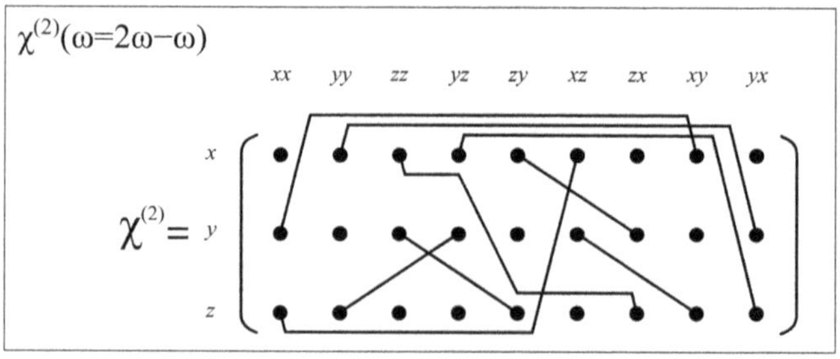

FIGURE 3.7 Second-order electric susceptibility tensor at $\omega = 2\omega - \omega$.

The symmetry of $\chi^{(2)}(2\omega = \omega + \omega)$, that governs second-harmonic generation (SHG), cannot be established from Eq. (3.43), but from Eq. (3.26) with $\omega_1 = \omega_2 (= \omega)$ and so $\omega_3 = 2\omega$, which gives:

$$\frac{\partial^3 \phi^{(2)}}{\partial E_k(\omega)\partial E_j(\omega)\partial E_i^*(2\omega)} = -\frac{1}{4}\varepsilon_0 \chi_{ijk}{}^{(2)}(2\omega = \omega + \omega) \tag{3.44}$$

The same kind of relation can be established for $\chi_{ikj}{}^{(2)}(2\omega = \omega + \omega)$, *i.e.*:

$$\frac{\partial^3 \phi^{(2)}}{\partial E_j(\omega)\partial E_k(\omega)\partial E_i^*(2\omega)} = -\frac{1}{4}\varepsilon_0 \chi_{ikj}{}^{(2)}(2\omega = \omega + \omega) \tag{3.45}$$

The commutativity of the order of derivation leads to the following equalities:

$$\frac{\partial^3 \phi^{(2)}}{\partial E_j(\omega)\partial E_k(\omega)\partial E_i^*(2\omega)} = \frac{\partial^3 \phi^{(2)}}{\partial E_k(\omega)\partial E_j(\omega)\partial E_i^*(2\omega)} \tag{3.46}$$

Then it can be deduced that:

$$\chi_{ijk}{}^{(2)}(2\omega = \omega + \omega) = \chi_{ikj}{}^{(2)}(2\omega = \omega + \omega) \tag{3.47}$$

Consequently $\chi^{(2)}(\omega = 2\omega - \omega)$ is symmetrical according to the indices corresponding to the two equal circular frequencies, *i.e.* the first and third ones, which leads to the matrix as shown in Figure 3.8.

Note that the contracted notation (3.41) can be used in that case.

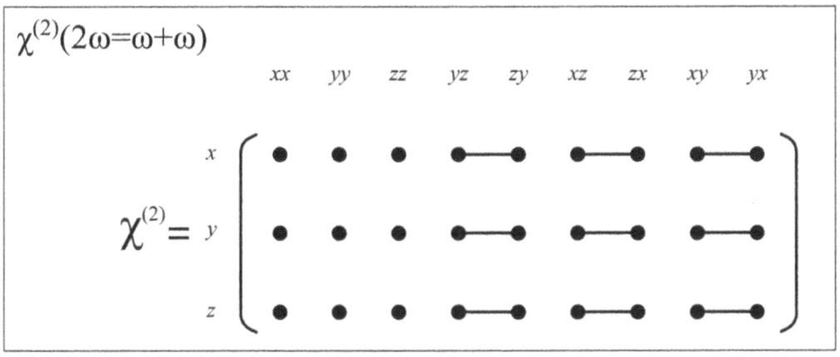

FIGURE 3.8 Second-order electric susceptibility tensor at $2\omega = \omega + \omega$.

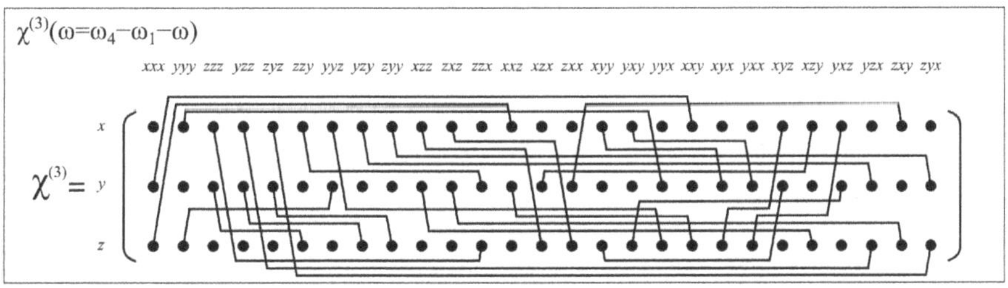

FIGURE 3.9 Third-order electric susceptibility tensor at $\omega = \omega_4 - \omega_1 - \omega$.

3.2.3.2 Third-Order Electric Susceptibility with Equality between Frequencies

It can also be established that $\chi^{(3)}$ is symmetrical in the indices corresponding to equal circular frequencies. Then for the two possible energy conservation relations, $\omega_4 = \omega_1 + \omega_2 + \omega_3$ or $\omega_4 + \omega_1 = \omega_2 + \omega_3$, the equalities can concern two, three, or four circular frequencies, these symmetries being included in those of Kleinman shown in Figure 3.6.

Next are given all the possible symmetry relations and the corresponding matrices. The (3×27) matrices of Figure 3.9 until Figure 3.12 exhibit fifty-four independent coefficients over the eighty-one and correspond to equalities between two circular frequencies ($\omega_2 = \omega_3$). Figure 3.13 and Figure 3.14 correspond to equalities between three frequencies ($\omega_1 = \omega_2 = \omega_3$), which lead to thirty-three independent coefficients. When the four circular frequencies are equal ($\omega_1 = \omega_2 = \omega_3 = \omega_4$) the $\chi^{(3)}$ tensor is exactly the same as that of Figure 3.6 having thirteen independent coefficients and corresponding to the symmetry of Kleinman.

1) $$\chi_{kijl}^{(3)}(\omega = \omega_4 - \omega_1 - \omega) = \chi_{lijk}^{(3)}(\omega = \omega_4 - \omega_1 - \omega) \tag{3.48}$$

2) $$\chi_{ijkl}^{(3)}(\omega_4 = \omega_1 + \omega + \omega) = \chi_{ijlk}^{(3)}(\omega_4 = \omega_1 + \omega + \omega) \tag{3.49}$$

$$\chi_{jikl}^{(3)}(\omega_1 = \omega_4 - \omega - \omega) = \chi_{jilk}^{(3)}(\omega_1 = \omega_4 - \omega - \omega) \tag{3.50}$$

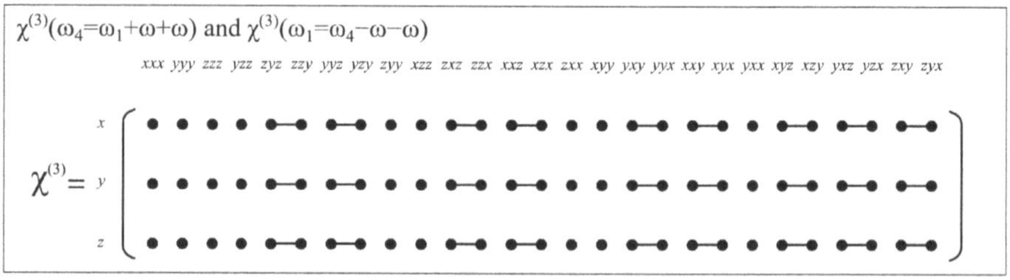

FIGURE 3.10 Third-order electric susceptibility tensor at $\omega_4 = \omega_1 + \omega + \omega$ and $\omega_1 = \omega_4 - \omega - \omega$.

$$3)\quad {\chi_{jikl}}^{(3)}(\omega = \omega_4 + \omega - \omega_3) = {\chi_{kijl}}^{(3)}(\omega = \omega_4 + \omega - \omega_3) \tag{3.51}$$

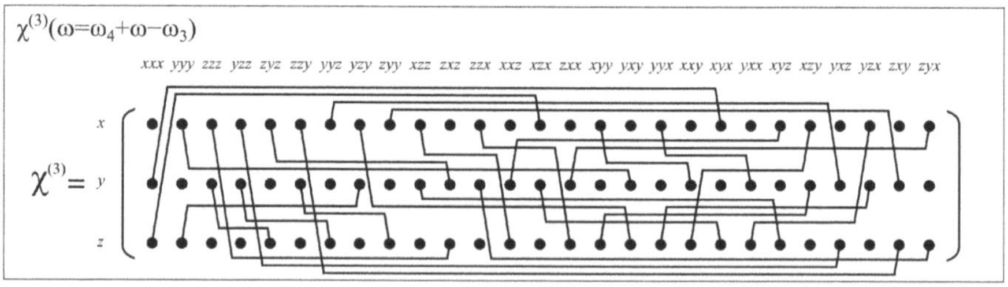

FIGURE 3.11 Third-order electric susceptibility tensor at $\omega = \omega_4 + \omega - \omega_3$.

$$4)\quad {\chi_{ijkl}}^{(3)}(\omega_4 = -\omega + \omega + \omega_3) = {\chi_{ikjl}}^{(3)}(\omega_4 = -\omega + \omega + \omega_3) \tag{3.52}$$

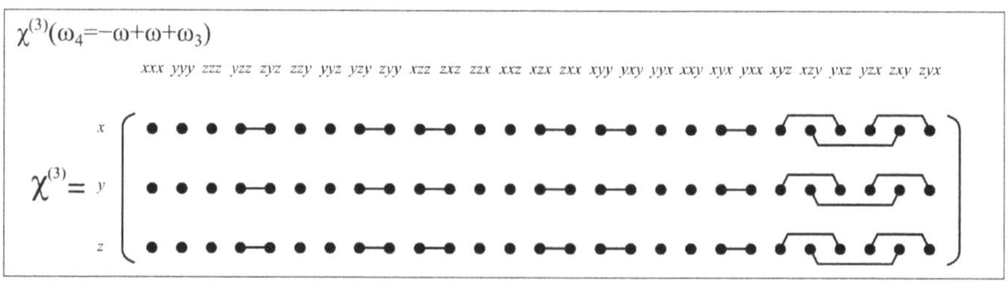

FIGURE 3.12 Third-order electric susceptibility tensor at $\omega_4 = -\omega + \omega + \omega_3$.

$$5)\quad {\chi_{kijl}}^{(3)}(\omega = 3\omega - \omega - \omega) = {\chi_{lijk}}^{(3)}(\omega = 3\omega - \omega - \omega) \tag{3.53}$$

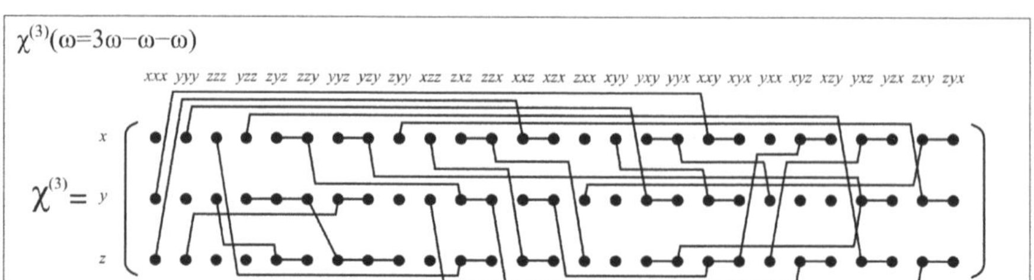

FIGURE 3.13 Third-order electric susceptibility tensor at $\omega = 3\omega - \omega - \omega$.

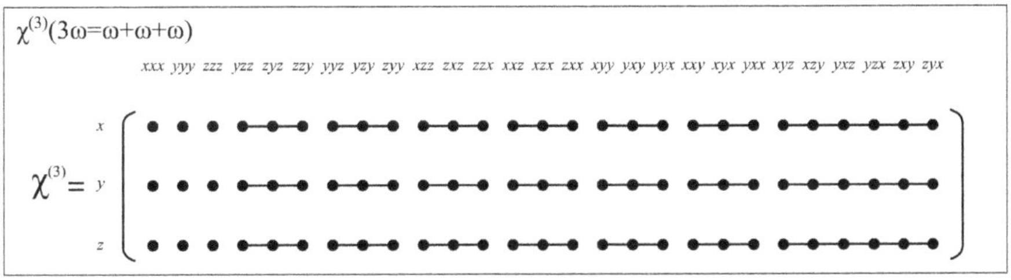

FIGURE 3.14 Third-order electric susceptibility tensor at $3\omega = \omega + \omega + \omega$.

6) $$\chi_{ijkl}^{(3)}(3\omega = \omega+\omega+\omega) = \chi_{ijlk}^{(3)}(3\omega = \omega+\omega+\omega)$$
$$= \chi_{ikjl}^{(3)}(3\omega = \omega+\omega+\omega) = \chi_{iklj}^{(3)}(3\omega = \omega+\omega+\omega) \quad (3.54)$$
$$= \chi_{iljk}^{(3)}(3\omega = \omega+\omega+\omega) = \chi_{ilkj}^{(3)}(3\omega = \omega+\omega+\omega)$$

Equation (3.54) corresponds to the case of third-harmonic generation (THG).
The contraction notation given by Eq. (3.42) can be used in that case.

7) $$\chi_{ijkl}^{(3)}(\omega = \omega+\omega-\omega) = \chi_{jikl}^{(3)}(\omega = \omega-\omega+\omega) \quad (3.55)$$
$$= \chi_{kijl}^{(3)}(\omega = \omega-\omega+\omega) = \chi_{ljki}^{(3)}(\omega = \omega+\omega-\omega)$$

According to Eq. (3.55) that corresponds to the Kerr effect, the tensor is fully symmetrical so that the corresponding matrix is that with Kleinman symmetry given in Figure 3.6.

3.3 EXTRINSIC SYMMETRIES—SPATIAL SYMMETRIES

3.3.1 First-Order Electric Susceptibility Tensors

3.3.1.1 Definition of the Dielectric Frame

All the calculations and matrices of Part 3.1 have been carried out in a $(\overrightarrow{Ox}, \overrightarrow{Oy}, \overrightarrow{Oz})$ orthonormal frame with an arbitrary orientation according to that of the orientation symmetry elements of the crystal. The corresponding matrix is given in Figure 3.4 in the case of lossless media. Since $\chi^{(1)}$ is a symmetrical tensor, there exists a specific orientation of $(\overrightarrow{Ox}, \overrightarrow{Oy}, \overrightarrow{Oz})$ for which it is represented by a diagonal matrix. This frame is called the "dielectric frame", also named the "optical frame", and it is the one that is used in crystal optics because it leads to the simplest form of the electric susceptibility tensor. It is true for $\chi^{(1)}$ obviously, but the form is also the simplest one, but not diagonal, for $\chi^{(2)}$ or $\chi^{(3)}$ as will be shown hereafter.

The orientation of the dielectric frame is not always linked to the orientation symmetry elements according to the crystal class that is considered.

As detailed in Part 2.3, Neumann's principle allows us to take into account the effect of the orientation symmetry on the matrix describing a given physical property. In the case of $\chi^{(1)}$, following the example given in Part 2.3.2.2, the application of Neumann's principle shows that the thirty-two crystal classes defined in Part 2.1.1.2 are arranged in three different diagonal matrix forms that defines the three optical classes, *i.e.*: biaxial, uniaxial, and isotropic. The corresponding matrices and the associated orientations of the dielectric frame are given in the following for the different crystal classes, from the lowest to the highest degrees of symmetry.

3.3.1.2 Biaxial Optical Class

The crystal classes of the triclinic, monoclinic, and orthorhombic crystal systems belong to the biaxial optical class.

When $(\overrightarrow{Ox},\overrightarrow{Oy},\overrightarrow{Oz})$ has an arbitrary orientation, then the matrix representing $\chi^{(1)}$ has six independent coefficients, as shown in Figure 3.15, the same ones as those in the general case shown in Figure 3.4. When $(\overrightarrow{Ox},\overrightarrow{Oy},\overrightarrow{Oz})$ is in coincidence with the dielectric frame, the $\chi^{(1)}$ matrix becomes diagonal, with three independent coefficients, $\chi^{(1)}_{xx} \neq \chi^{(1)}_{yy} \neq \chi^{(1)}_{zz}$, as shown in Figure 3.15.

The specificities of orientation of the dielectric frame is detailed in the following for the different crystal classes belonging to the biaxial optical class.

In the case of the triclinic classes, *i.e.* 1 and $\bar{1}$ with the stereographic projections given in Figure 2.9, there is absolutely no link between the crystallographic and dielectric frames. Consequently, it is a more complicated situation to find the orientation of the dielectric axes. Note also that the orientation of the dielectric frame is subject to change depending on any dispersive parameter of $\chi^{(1)}$, like for example the circular frequency, the direction of propagation of the light, or the temperature. Furthermore, the dielectric frame can never be in coincidence with the crystallographic frame since the first one is orthogonal contrary to the second one where $\alpha \neq \beta \neq \gamma \neq 90°$ as shown in Figure 2.8.

The stereographic projections of the three monoclinic crystal classes, *i.e.* $\frac{2}{m}$, m and 2, are shown in Figure 2.10. There exists a situation of partial reduction of the number of independent coefficients of the $\chi^{(1)}$ matrix when one axis, conventionally $\overrightarrow{Oy}$, is oriented along the 2-fold axis of classes 2 and $\frac{2}{m}$, and perpendicular

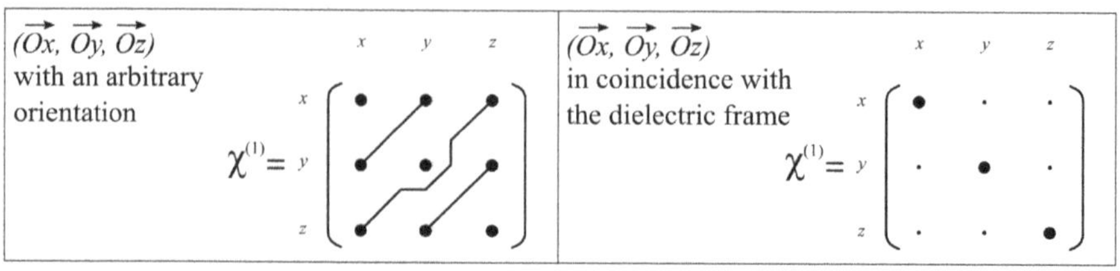

FIGURE 3.15 Matrices representing the first-order electric susceptibility tensor $\chi^{(1)}$ for the biaxial optical class and two different orientations of the orthonormal frame $(\overrightarrow{Ox},\overrightarrow{Oy},\overrightarrow{Oz})$: arbitrary orientation (left), and in the dielectric frame (right).

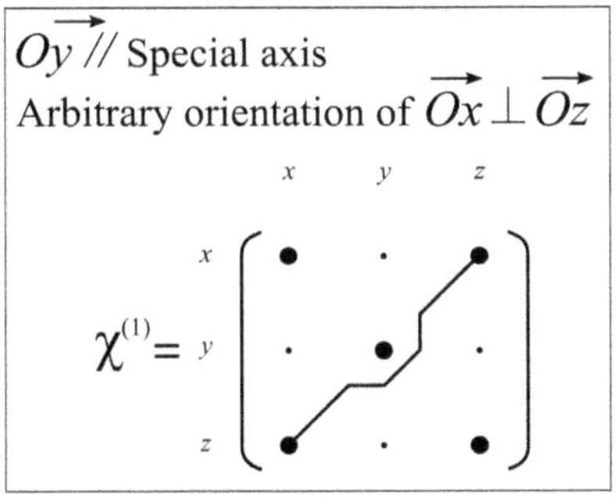

FIGURE 3.16 Matrix representing the first-order electric susceptibility tensor $\chi^{(1)}$ for monoclinic crystal classes when the $\overrightarrow{Oy}$ axis of the dielectric frame is taken along the special axis.

to the mirror in the case of the m class in which the two other axes $\overrightarrow{Ox}$ and $\overrightarrow{Oz}$ have arbitrary directions. This crystallographic axis is called the "special axis". The convention is to take the $\vec{b}$ axis as the special axis, the two other crystallographic $\vec{a}$- and $\vec{c}$-axes being in the plane perpendicular to $\vec{b}$ as shown in Figure 2.8. In that case, the matrix is that of Figure 3.16 in which the four independent coefficients are: $\chi_{xx}^{(1)} \neq \chi_{yy}^{(1)} \neq \chi_{zz}^{(1)} \neq \chi_{xz}^{(1)} = \chi_{zx}^{(1)}$.

Note that some authors do not put $\overrightarrow{Oy}$ along the special axis. Then the non-zero coefficients of the corresponding $\chi^{(1)}$ tensor write $\chi_{xx}^{(1)} \neq \chi_{yy}^{(1)} \neq \chi_{zz}^{(1)} \neq \chi_{yz}^{(1)} = \chi_{zy}^{(1)}$ when $\overrightarrow{Ox}$ is along the special axis, and $\chi_{xx}^{(1)} \neq \chi_{yy}^{(1)} \neq \chi_{zz}^{(1)} \neq \chi_{xy}^{(1)} = \chi_{yx}^{(1)}$ when it is $\overrightarrow{Oz}$. It is important and efficient to notice that these relations can be directly deduced from those established in the case in which $\overrightarrow{Oy}$ is along the special axis by replacing the Cartesian indices (x, y, z) by (z, x, y) for $\overrightarrow{Ox}$ along the special axis, or by (y, z, x) when it is $\overrightarrow{Oz}$. These two permutations correspond exactly to the change of orientation of the dielectric frame, *i.e.* $(\overrightarrow{Ox}, \overrightarrow{Oy}, \overrightarrow{Oz})$ becoming $(\overrightarrow{Oz}, \overrightarrow{Ox}, \overrightarrow{Oy})$ or $(\overrightarrow{Oy}, \overrightarrow{Oz}, \overrightarrow{Ox})$, respectively.

As mentioned earlier, the situation of full reduction is reached for an exact coincidence between $(\overrightarrow{Ox}, \overrightarrow{Oy}, \overrightarrow{Oz})$ and the dielectric frame, $\overrightarrow{Oy}$ being usually taken along the crystallographic axis $\vec{b}$ that is the special axis. Note that there is *a priori* no connection between $\overrightarrow{Ox}$ and $\overrightarrow{Oz}$ and the crystallographic axes $\vec{a}$ or $\vec{c}$, especially since the orientation of $\overrightarrow{Ox}$ and $\overrightarrow{Oz}$ can vary as a function of any dispersive parameter of $\chi^{(1)}$. And as for the triclinic classes, a full coincidence of the dielectric and crystallographic frames is never possible due to the existence of a non-right angle between two of the three crystallographic axes.

The orthorhombic crystal classes, *mmm*, *mm*2, and 222, are the biaxial crystals with the highest degree of symmetry, the corresponding stereographic projections being given in Figure 2.11. The three dielectric axes are in full coincidence with the symmetry elements: perpendicular to the three mirrors in the case of *mmm*, perpendicular to the two mirrors and along the 2-fold axis (it is conventionally $\overrightarrow{Oz}$), for *mm*2, and along the three 2-fold axes in the case of *mmm*. It means that $(\overrightarrow{Ox},\overrightarrow{Oy},\overrightarrow{Oz})$ is in strict coincidence with the crystallographic frame, so that X-ray diffraction is sufficient for determining the orientation of the dielectric frame.

3.3.1.3 Uniaxial Optical Class

This optical class brings the classes of the trigonal, tetragonal, and hexagonal crystal systems together. Neumann's principle does not lead to a reduction of the number of independent coefficients when $(\overrightarrow{Ox},\overrightarrow{Oy},\overrightarrow{Oz})$ has an arbitrary orientation. As a consequence, the $\chi^{(1)}$ matrix is that of the general case reminded in Figure 3.17. The dielectric and crystallographic frames are totally linked, and that even if they are not in strict coincidence: it is the case for the trigonal and hexagonal classes for which the crystallographic frame is not orthogonal. Due to the full link between the two frames, the dielectric frame cannot rotate as a function of dispersive parameters of $\chi^{(1)}$. The rotation axis of highest degree of symmetry is always one of the three dielectric axes: the convention is to take $\overrightarrow{Oz}$. Then Neumann's principle leads to a diagonal matrix with two independent coefficients $\chi_{xx}^{(1)} = \chi_{yy}^{(1)} \neq \chi_{zz}^{(1)}$, as shown in Figure 3.17.

The specificities of orientation of the dielectric frame are detailed here for the different uniaxial crystal classes.

1) $\overrightarrow{Oz}$ is along the 3-fold axis of the trigonal crystal classes: $\bar{3}m$, $3m$, 32, $\bar{3}$, and 3, the corresponding stereographic projections being shown in Figure 2.12. $\overrightarrow{Ox}$ is conventionally taken along the crystallographic axis $\vec{a}$ that is somewhere in the equatorial plane for classes 3 and $\bar{3}$; it is one of the 2-fold axes for classes 32 and $\bar{3}m$; and it is at the middle of two mirrors in the case of $3m$.
2) $\overrightarrow{Oz}$ is along the 4-fold axis for the tetragonal crystal classes: $\frac{4}{m}mm$, $4mm$, $\bar{4}2m$, 422, $\frac{4}{m}$, $\bar{4}$, and 4, the corresponding stereographic projections being

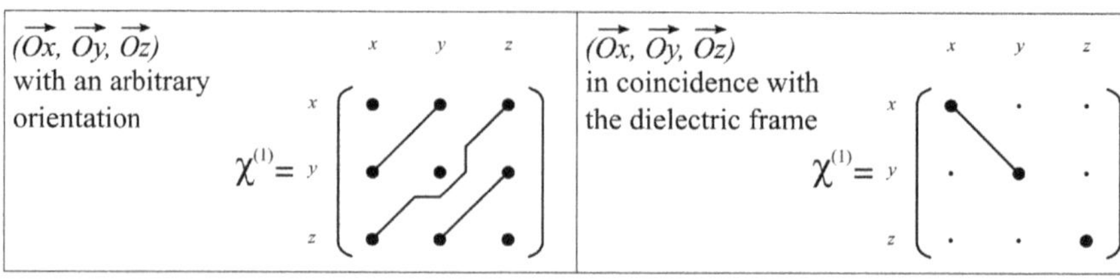

FIGURE 3.17 The forms of the matrix representing the first-order electric susceptibility tensor $\chi^{(1)}$ for the uniaxial optical class and two different orientations of the orthonormal frame $(\overrightarrow{Ox},\overrightarrow{Oy},\overrightarrow{Oz})$: arbitrary orientation (left), and in the dielectric frame (right).

shown in Figure 2.13. $\overrightarrow{Ox}$ is conventionally taken along the crystallographic axis $\vec{a}$, and $\overrightarrow{Oy}$ along $\vec{b}$: $\vec{a}$ and $\vec{b}$ are somewhere in the equatorial plane for classes 4, $\bar{4}$, and $\frac{4}{m}$, and along two perpendicular 2-fold axes for classes 422, $4mm$, $\bar{4}2m$, and $\frac{4}{m}mm$.

3) $\overrightarrow{Oz}$ is along the 6-fold axis of the hexagonal crystal classes: $\frac{6}{m}mm$, $6mm$, $\bar{6}2m$, 622, $\frac{6}{m}$, $\bar{6}$, and 6, the corresponding stereographic projections being shown in Figure 2.14. $\overrightarrow{Ox}$ is conventionally taken along the crystallographic axis $\vec{a}$: somewhere in the equatorial plane in the case of classes 6, $\bar{6}$, and $\frac{6}{m}$, along one of the 2-fold axes for classes 622, $\bar{6}2m$, and $\frac{6}{m}mm$, in one of the mirrors for class $6mm$, and at the middle of two mirrors in the case of $\bar{6}2m$.

3.3.1.4 Isotropic Optical Class

The cubic crystal classes, *i.e.* classes $m\bar{3}m$, $\bar{4}3m$, 432, $m\bar{3}$, and 23, belong to the isotropic optical class. The corresponding stereographic projections are given in Figure 2.15. As for the biaxial and uniaxial optical classes, there is no reduction of the number of independent coefficients of the $\chi^{(1)}$ matrix when the frame $(\overrightarrow{Ox},\overrightarrow{Oy},\overrightarrow{Oz})$ has an arbitrary orientation, as shown in Figure 3.18 (left). But when $(\overrightarrow{Ox},\overrightarrow{Oy},\overrightarrow{Oz})$ is in coincidence with the dielectric frame, the reduction is maximal, with only one independent coefficient, *i.e.* $\chi_{xx}^{(1)} = \chi_{yy}^{(1)} = \chi_{zz}^{(1)}$, giving the diagonal matrix of Figure 3.18.

This case is the simplest one since the dielectric and crystallographic frames exactly coincide. Then $(\overrightarrow{Ox},\overrightarrow{Oy},\overrightarrow{Oz})$ corresponds to $(\vec{a},\vec{b},\vec{c})$ that is along the three 2-fold axes in the cases of the 23 and $m\bar{3}$ classes, along the three 4-fold axes for $\bar{4}3m$, and along the three 4-fold axes in the cases of 432 and $m\bar{3}m$.

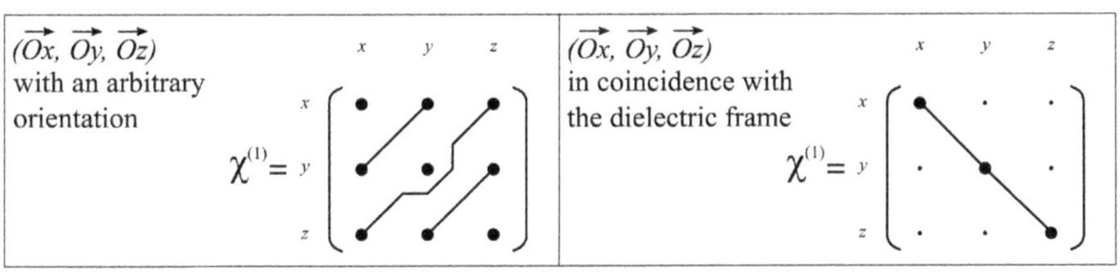

FIGURE 3.18 The forms of the matrix representing the first-order electric susceptibility tensor $\chi^{(1)}$ for the isotropic optical class and two different orientations of the orthonormal frame $(\overrightarrow{Ox},\overrightarrow{Oy},\overrightarrow{Oz})$: arbitrary orientation (left), and in the dielectric frame (right).

3.3.2 Second-Order Electric Susceptibility Tensors

The way to apply Neumann's principle to $\chi^{(2)}$ is detailed in Part 2.3.2.1 and Part 2.3.2.2, and the effect of Kleinman's symmetry is given in Part 3.2.2.1. All the calculations are performed in the orthonormal frame $(\overrightarrow{Ox},\overrightarrow{Oy},\overrightarrow{Oz})$ in exact coincidence with the dielectric frame, as defined in Part 3.3.2.

The application of Neumann's principle to $\chi^{(2)}$ leads to the same kinds of symmetry as for $\chi^{(1)}$, *i.e.* by annihilation of coefficient and by equality between coefficients, but with one more relation, *i.e.* $\chi^{(2)}_{ijk} = -\chi^{(2)}_{lmn}$. The matrix shown in Figure 3.19 summarizes these different relations and their symbols.

As demonstrated in Part 2.3.2.1, the 3^3 coefficients of $\chi^{(2)}$ are all equal to zero for the eleven centrosymmetric crystal classes: $\bar{1}$, $\frac{2}{m}$, mmm, $\bar{3}m$, $\bar{3}$, $\frac{4}{m}mm$, $\frac{4}{m}$, $\frac{6}{m}mm$, $\frac{6}{m}$, $m\bar{3}m$, and $m\bar{3}$. The (3×9) matrices representing $\chi^{(2)}$ for the twenty-one acentric crystal classes are given in the following from Figure 3.20 to Figure 3.39. Two matrices *per* crystal class are shown: that resulting from the application of Neumann's principle only, and that combining Neumann's principle and Kleinmann's symmetry.

Several singularities appear in the matrices by simply observing Figure 3.20 to Figure 3.35. Firstly, different crystal classes can exhibit the same matrix. There are eighteen forms of matrix after applying Neumann's principle, the three following couples of classes exhibiting each the same form: 422 (Figure 3.30) and 622 (Figure 3.35); 4 (Figure 3.32) and 6 (Figure 3.37); 23 and $\bar{4}3m$ (Figure 3.39). It also occurs under Kleinman's symmetry for classes 222 (Figure 3.24), $\bar{4}2m$

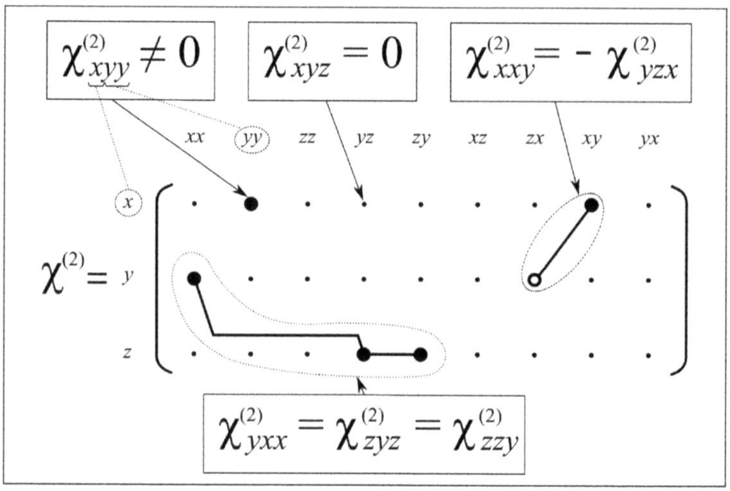

FIGURE 3.19 Scheme of matrix of the second-order electric susceptibility indicating the different relationships between coefficients and their symbols.

(Figure 3.39). $\bar{4}3m$ (Figure 3.39), and 23 (Figure 3.39) on the one hand, and $4mm$ (Figure 3.28), 4 (Figure 3.32), $6mm$ (Figure 3.33), and 6 (Figure 3.37) on the other hand, which gives fifteen matrix forms.

Secondly, Kleinman's symmetry does not bring more symmetry than Neumann's principle alone in the cases of classes $\bar{6}2m$ (Figure 3.34), $\bar{6}$ (Figure 3.36), 23 (Figure 3.39), and $\bar{4}3m$ (Figure 3.39).

Thirdly, Kleinman's symmetry leads to the annihilation of all the coefficients for classes 422 (Figure 3.30), 622 (Figure 3.35), and 432 (Figure 3.38). Knowing that the symmetry due to equalities between circular frequencies is necessarily included in that of Kleinman, it is interesting to consider the effect of circular frequencies equalities alone concerning these three classes: it appears that the equality between the two circular frequencies corresponding to the two last Cartesian indices, leading to $\chi_{ijk}^{(2)}(2\omega = \omega + \omega) = \chi_{ikj}^{(2)}(2\omega = \omega + \omega)$ governing SHG, does not change the form of the matrix for classes 422 and 622, contrary to the case of class 432. It means that for class 432 the magnitudes of the $\chi^{(2)}$ coefficients will be strictly equal to zero in the degenerate case, *i.e.* $\omega_1 = \omega_2 = \omega$, and weak in the non-degenerate case, *i.e.* $\omega_1 \neq \omega_2$, when Kleinman's symmetry applies. They will be weak and not equal to zero in this latter case because Kleinman's symmetry lies on approximations (no absorption, no dispersion) so that the equalities between coefficients are not strict. For classes 422 and 622, the magnitudes will be weak for the degenerate case as well as for the non-degenerate one if Kleinman's applies.

3.3.2.1 Triclinic Crystal Classes

Figure 3.20 shows the matrix obtained from Neumann's principle only, and the matrix coming from Neumann's principle and Kleinman's symmetry in the case of the acentric triclinic crystal class.

3.3.2.2 Monoclinic Crystal Classes

The form of $\chi^{(2)}$ when $\overrightarrow{Ox}$ or $\overrightarrow{Oz}$ is taken along the special axis can be deduced from that established when it is $\overrightarrow{Oy}$ (Figure 3.21 and Figure 3.22) by simply replacing the Cartesian indices (x, y, z) with (z, x, y) for $\overrightarrow{Ox}$ along the special axis, or with (y, z, x) when $\overrightarrow{Oz}$ is along the special axis.

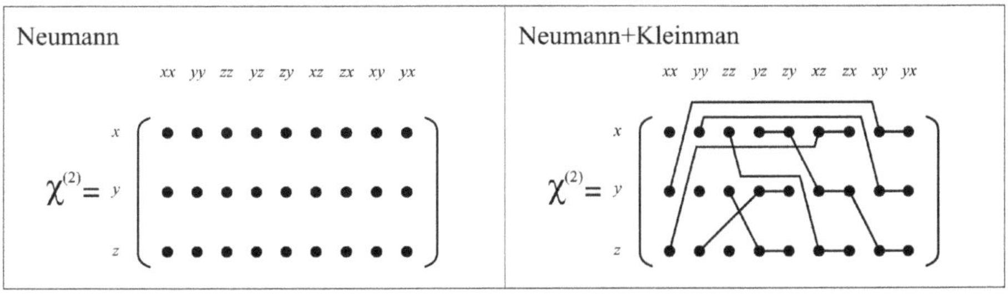

FIGURE 3.20 Matrices corresponding to the second-order electric susceptibility tensor $\chi^{(2)}$ for crystal class 1.

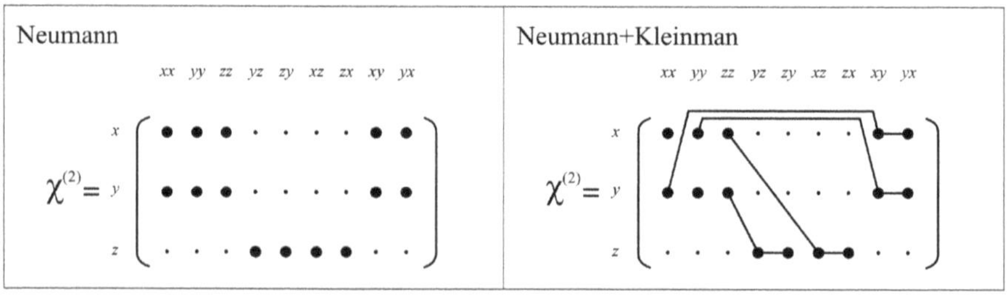

FIGURE 3.21 Matrices corresponding to the second-order electric susceptibility tensor $\chi^{(2)}$ for crystal class *m* with axis $\overrightarrow{Oy}$ perpendicular to mirror *m*.

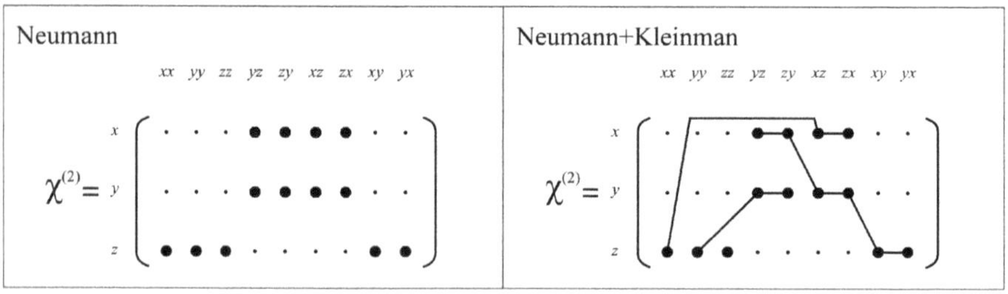

FIGURE 3.22 Matrices corresponding to the second-order electric susceptibility tensor $\chi^{(2)}$ for crystal class 2 with axis $\overrightarrow{Oy}$ along the 2-fold axis.

3.3.2.3 Orthorhombic Crystal Classes

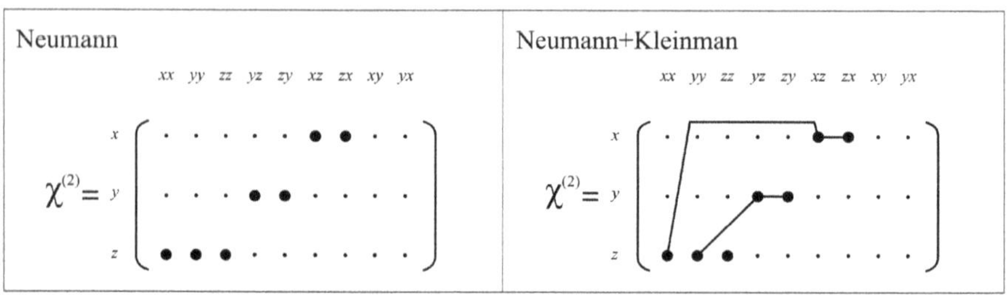

FIGURE 3.23 Matrices corresponding to the second-order electric susceptibility tensor $\chi^{(2)}$ for crystal class *mm*2.

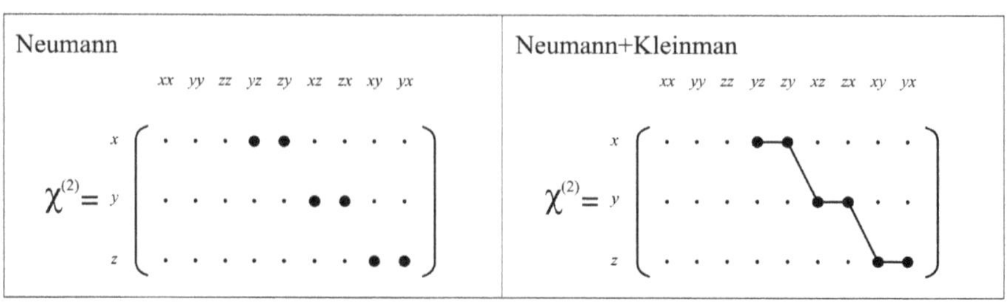

FIGURE 3.24 Matrices corresponding to the second-order electric susceptibility tensor $\chi^{(2)}$ for crystal class 222.

3.3.2.4 Trigonal Crystal Classes

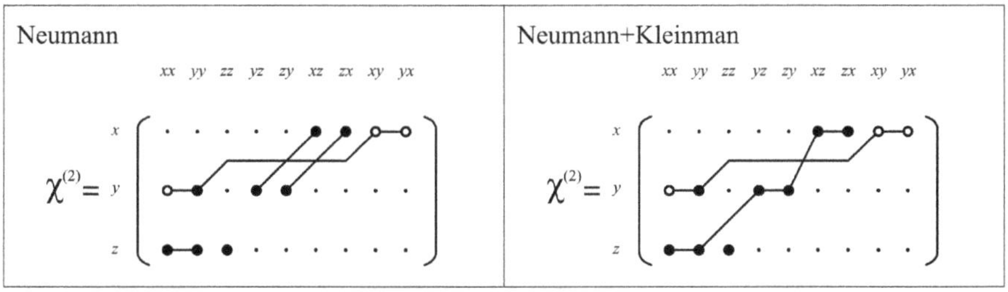

FIGURE 3.25 Matrices corresponding to the second-order electric susceptibility tensor $\chi^{(2)}$ for crystal class $3m$.

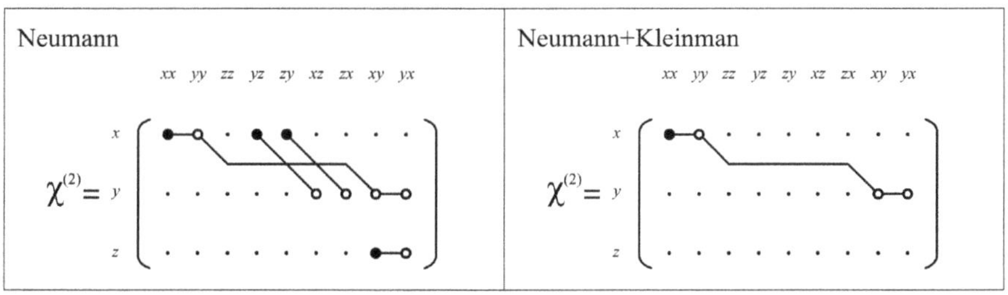

FIGURE 3.26 Matrices corresponding to the second-order electric susceptibility tensor $\chi^{(2)}$ for crystal class 32.

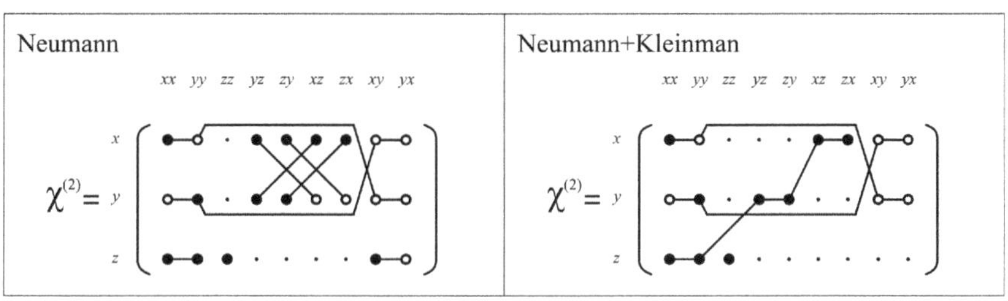

FIGURE 3.27 Matrices corresponding to the second-order electric susceptibility tensor $\chi^{(2)}$ for crystal class 3.

3.3.2.5 Tetragonal Crystal Classes

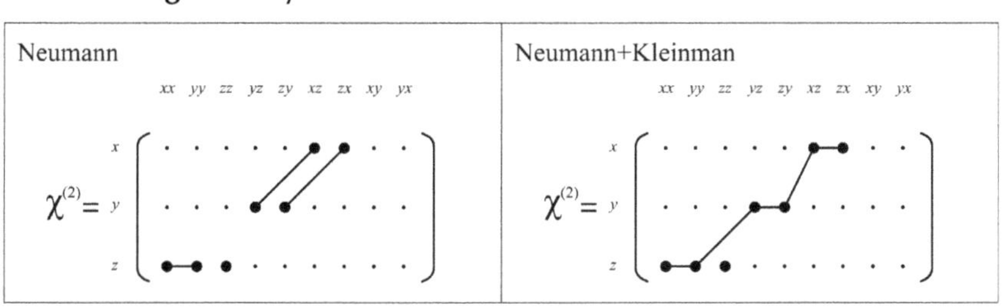

FIGURE 3.28 Matrices corresponding to the second-order electric susceptibility tensor $\chi^{(2)}$ for crystal class $4mm$.

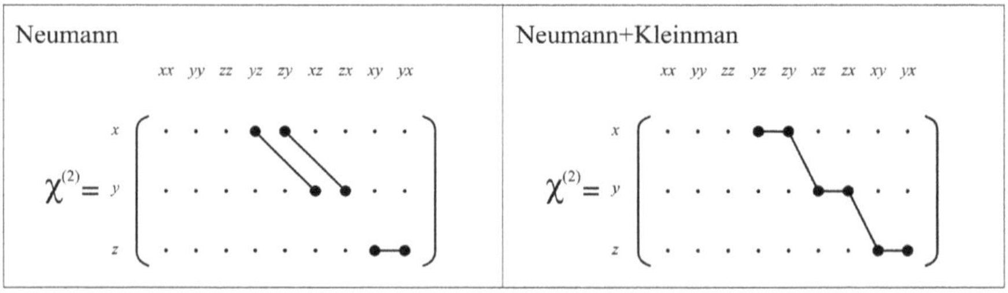

FIGURE 3.29 Matrices corresponding to the second-order electric susceptibility tensor $\chi^{(2)}$ for crystal class $\bar{4}2m$.

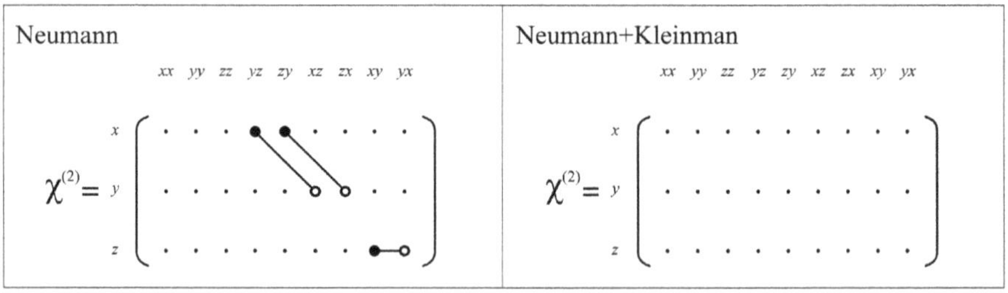

FIGURE 3.30 Matrices corresponding to the second-order electric susceptibility tensor $\chi^{(2)}$ for crystal class 422.

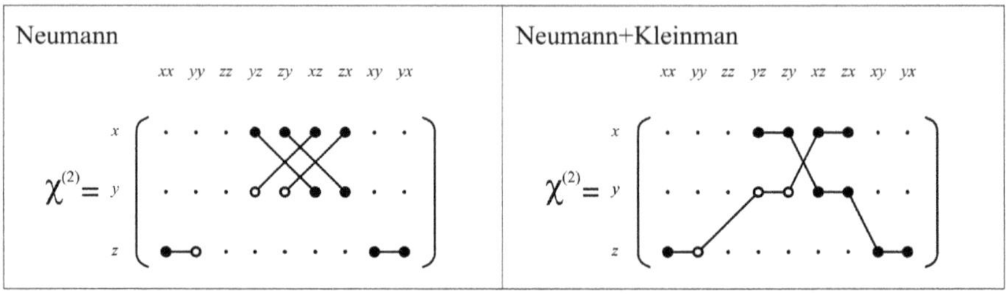

FIGURE 3.31 Matrices corresponding to the second-order electric susceptibility tensor $\chi^{(2)}$ for crystal class $\bar{4}$.

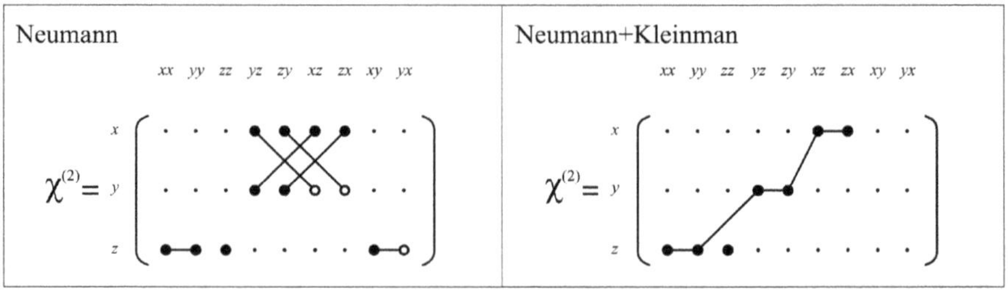

FIGURE 3.32 Matrices corresponding to the second-order electric susceptibility tensor $\chi^{(2)}$ for crystal class 4.

3.3.2.6 Hexagonal Crystal Classes

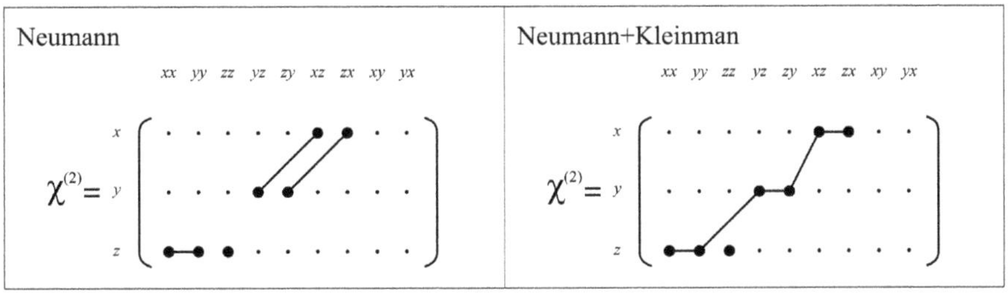

FIGURE 3.33 Matrices corresponding to the second-order electric susceptibility tensor $\chi^{(2)}$ for crystal class 6*mm*.

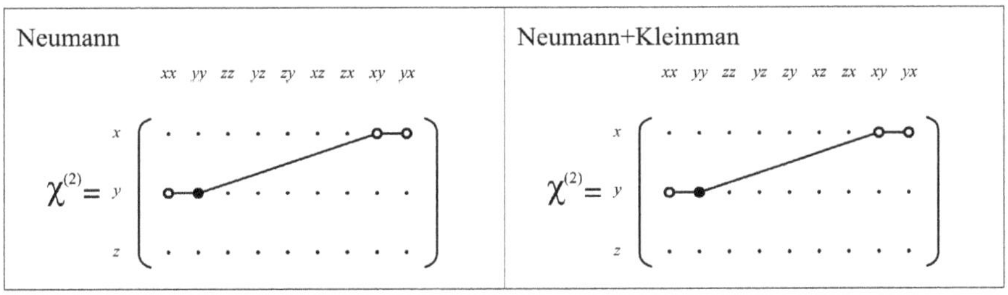

FIGURE 3.34 Matrices corresponding to the second-order electric susceptibility tensor $\chi^{(2)}$ for crystal class $\bar{6}2m$.

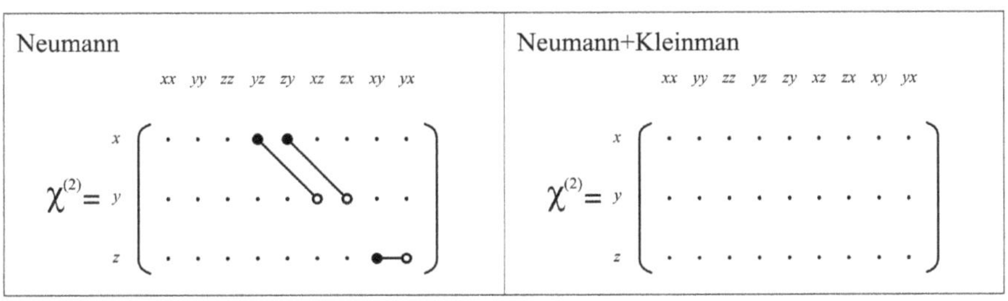

FIGURE 3.35 Matrices corresponding to the second-order electric susceptibility tensor $\chi^{(2)}$ for crystal class 622.

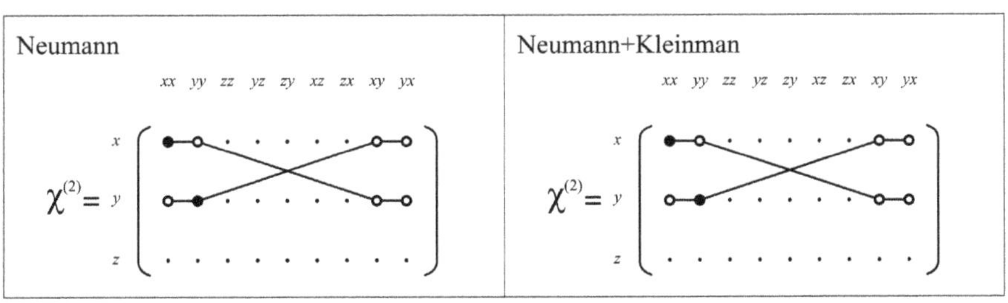

FIGURE 3.36 Matrices corresponding to the second-order electric susceptibility tensor $\chi^{(2)}$ for crystal class $\bar{6}$.

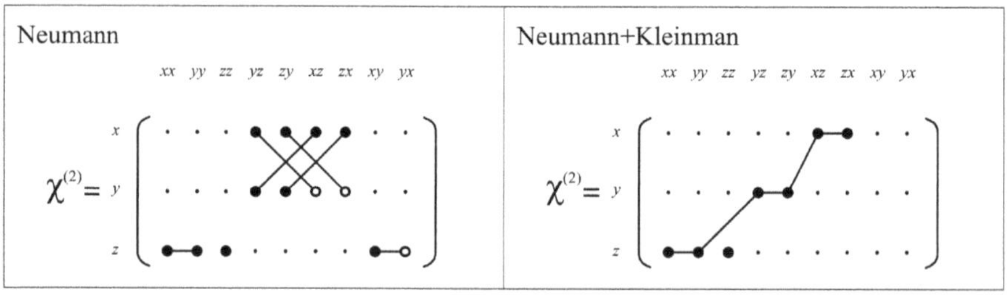

FIGURE 3.37 Matrices corresponding to the second-order electric susceptibility tensor $\chi^{(2)}$ for crystal class 6.

3.3.2.7 Cubic Crystal Classes

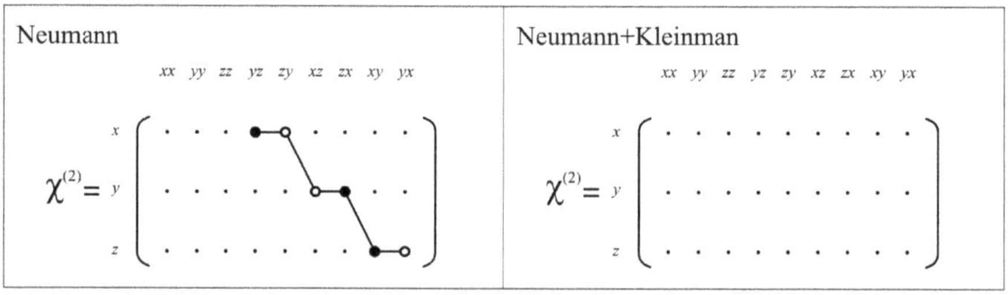

FIGURE 3.38 Matrices corresponding to the second-order electric susceptibility tensor $\chi^{(2)}$ for crystal class 432.

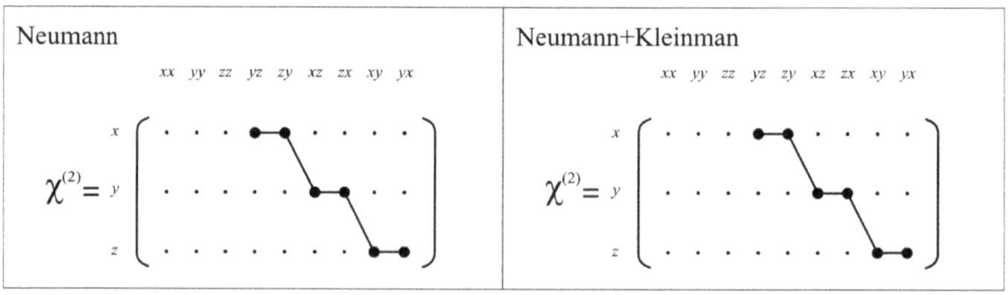

FIGURE 3.39 Matrices corresponding to the second-order electric susceptibility tensor $\chi^{(2)}$ for crystal classes 23 and $\bar{4}3m$.

3.3.3 Third-Order Electric Susceptibility Tensors

The matrices representing the $\chi^{(3)}$ tensors expressed in the dielectric frame are obtained after applying Neumann's principle alone on the one hand, and by combining Neumann's principle and Kleinmann's symmetry.

There is one more relation between coefficients than in the case of $\chi^{(2)}$, as for example: $\chi_{yzzz}^{(3)} = \chi_{zyzy}^{(3)} + \chi_{zzyy}^{(3)} + \chi_{zxzz}^{(3)}$. All the kinds of relations are shown in Figure 3.40.

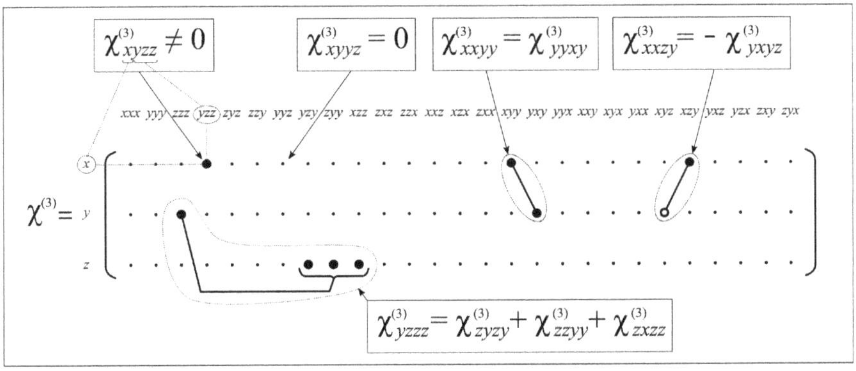

FIGURE 3.40 Scheme of matrix of the third-order electric susceptibility $\chi^{(3)}$ indicating the different relationships between coefficients and their symbols.

The twenty-one acentric crystal classes as well as the eleven centrosymmetric ones exhibit a $\chi^{(3)}$ tensor with non-zero coefficients as demonstrated in Part 2.3.2.1 on the basis of Neumann's principle, which means that all the crystal classes are "active" for $\chi^{(3)}$. The corresponding matrices are depicted from Figure 3.41 to Figure 3.51.

But despite a higher number of "active" crystal classes than for $\chi^{(2)}$, *i.e.* thirty-two instead of twenty-one, there are fewer matrix forms for $\chi^{(3)}$: eleven after having applied Neumann's principle alone, and nine when it is combined with Kleinmann's symmetry.

However, there is no complete cancellation of the coefficients due to Kleinman's symmetry or equalities between circular frequencies in $\chi^{(3)}$ contrary to $\chi^{(2)}$.

3.3.3.1 Triclinic Crystal Classes

The matrices corresponding to the two triclinic crystal classes are given in Figure 3.41.

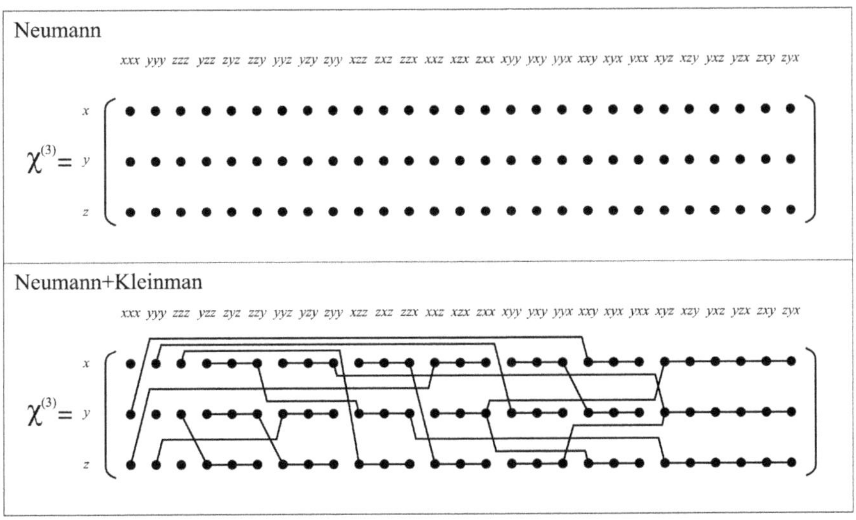

FIGURE 3.41 Matrices corresponding to the third-order electric susceptibility tensor $\chi^{(3)}$ for crystal classes 1 and $\bar{1}$.

3.3.3.2 Monoclinic Crystal Classes

When $\overrightarrow{Ox}$ or $\overrightarrow{Oz}$ are taken along the special axis, the form of tensor $\chi^{(3)}$ can be deduced from that established when $\overrightarrow{Oy}$ is along the special axis (Figure 3.42) by simply replacing the Cartesian indices (x,y,z) with (z,x,y) for $\overrightarrow{Ox}$ along the special axis, or with (y,z,x) when it is $\overrightarrow{Oz}$ along the special axis.

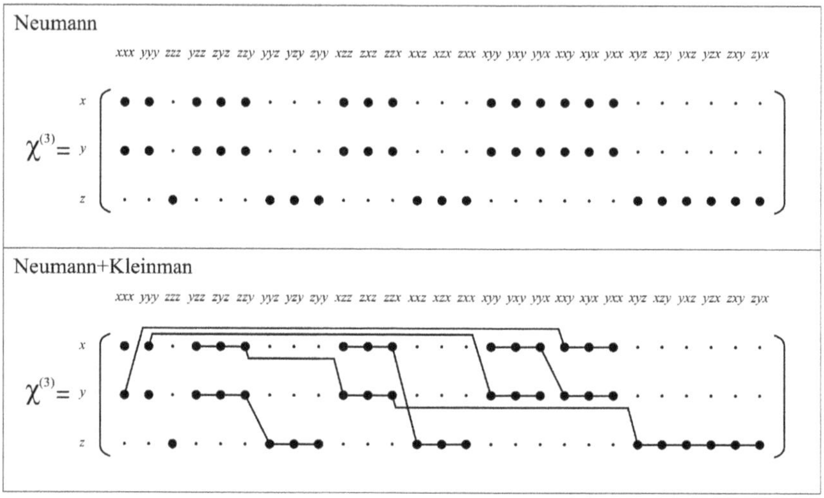

FIGURE 3.42 Matrices corresponding to the third-order electric susceptibility tensor $\chi^{(3)}$ for crystal classes m, 2, and $\frac{2}{m}$.

3.3.3.3 Orthorhombic Crystal Classes

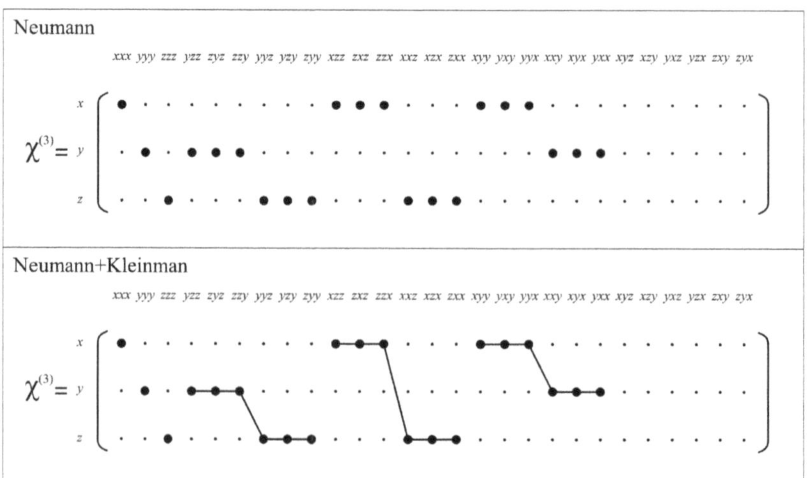

FIGURE 3.43 Matrices corresponding to the third-order electric susceptibility tensor $\chi^{(3)}$ for crystal classes $mm2$, 222, and mmm.

3.3.3.4 Trigonal Crystal Classes

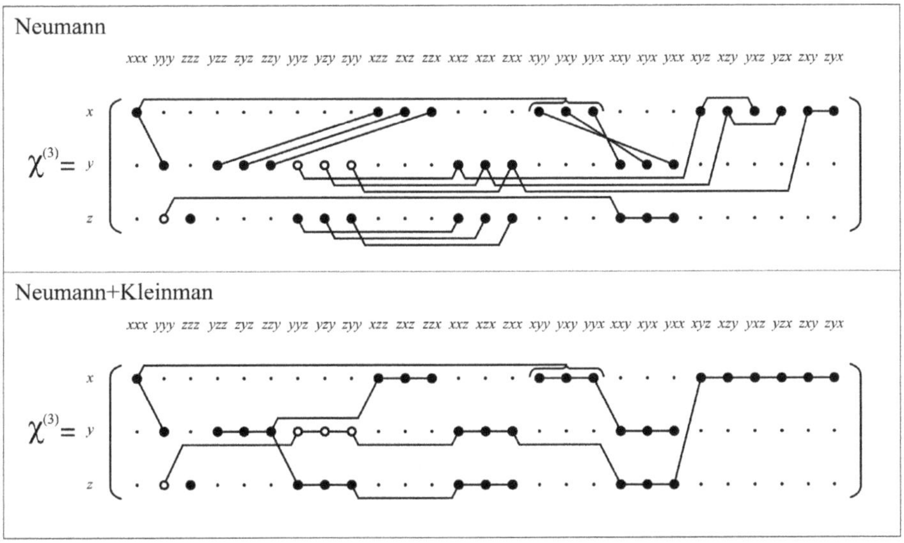

FIGURE 3.44 Matrices corresponding to the third-order electric susceptibility tensor $\chi^{(3)}$ for crystal classes $3m$, 32, and $\bar{3}m$.

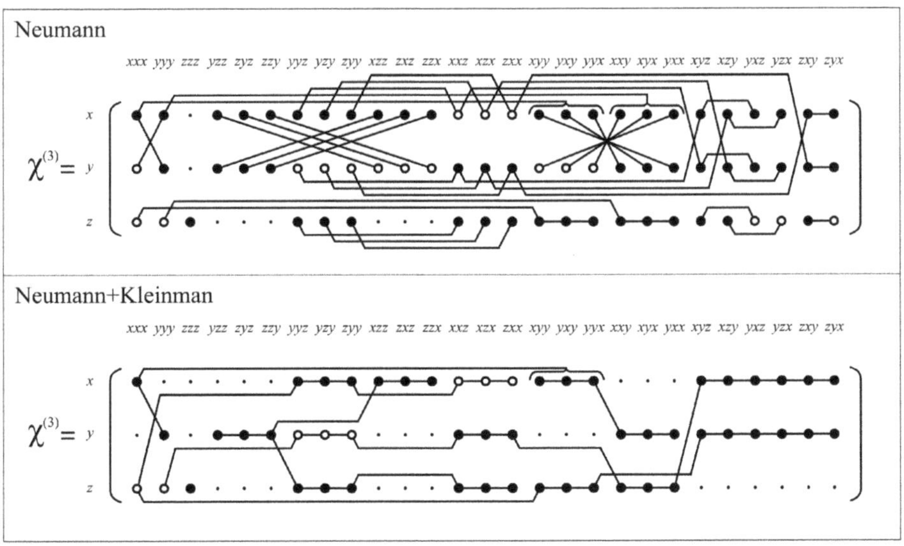

FIGURE 3.45 Matrices corresponding to the third-order electric susceptibility tensor $\chi^{(3)}$ for crystal classes 3 and $\bar{3}$.

3.3.3.5 Tetragonal Crystal Classes

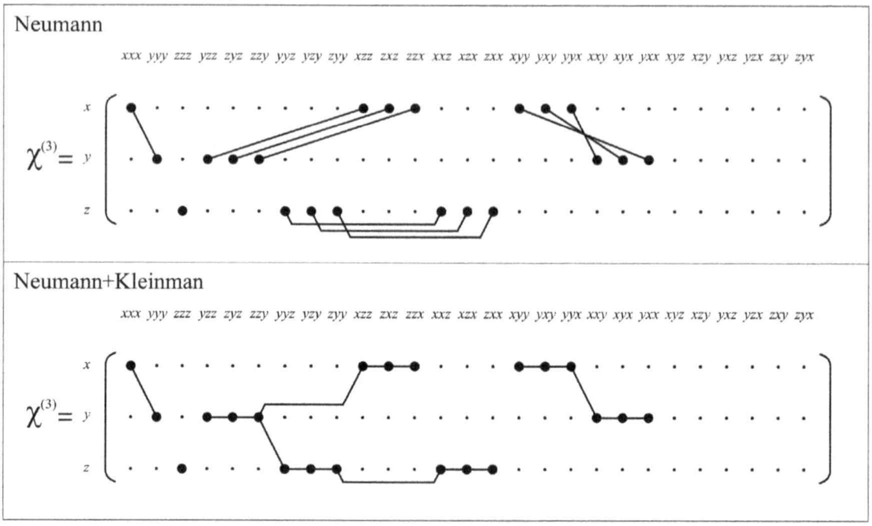

FIGURE 3.46 Matrices corresponding to the third-order electric susceptibility tensor $\chi^{(3)}$ for crystal classes $4mm$, $\bar{4}2m$, 422, and $\frac{4}{m}mm$.

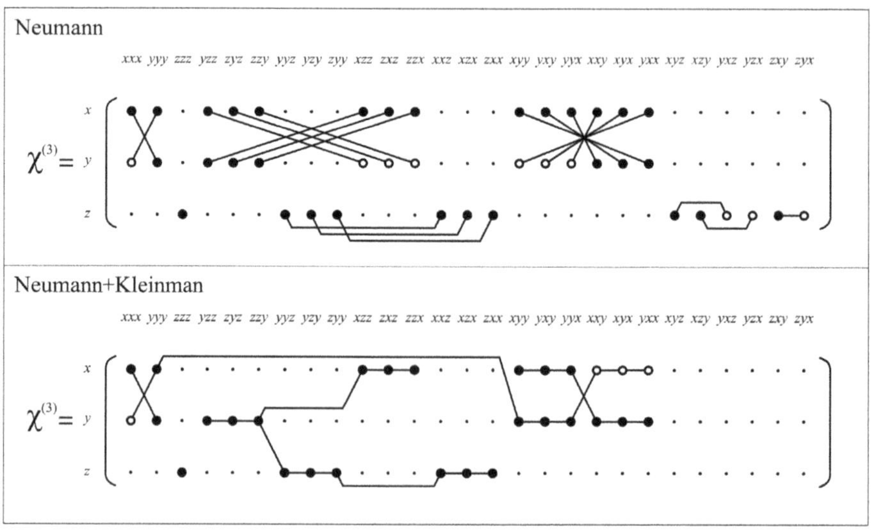

FIGURE 3.47 Matrices corresponding to the third-order electric susceptibility tensor $\chi^{(3)}$ for crystal classes $\bar{4}$, 4, and $\frac{4}{m}$.

3.3.3.6 Hexagonal Crystal Classes

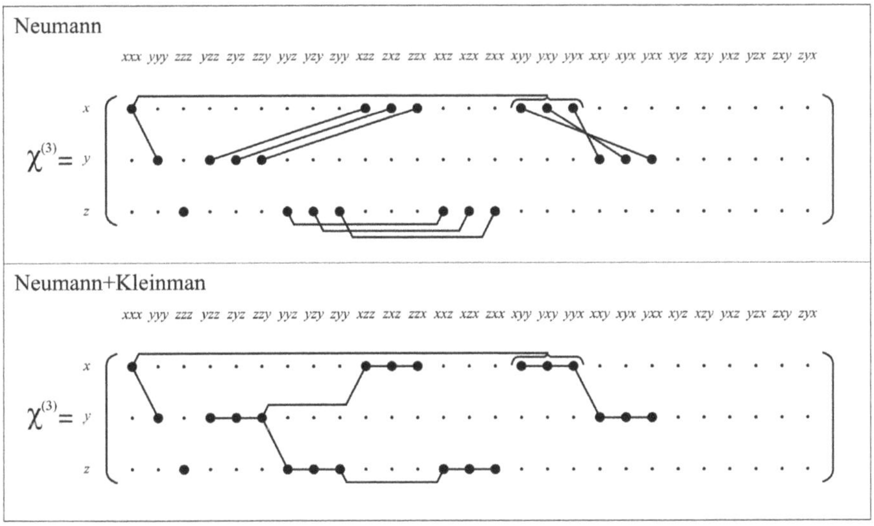

FIGURE 3.48 Matrices corresponding to the third-order electric susceptibility tensor $\chi^{(3)}$ for crystal classes $6mm$, $\bar{6}2m$, 622, and $\frac{6}{m}mm$.

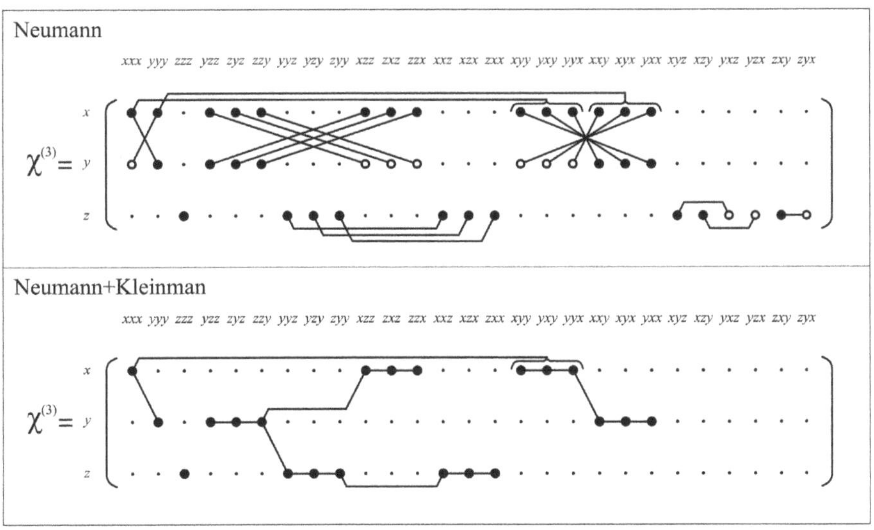

FIGURE 3.49 Matrices corresponding to the third-order electric susceptibility tensor $\chi^{(3)}$ for crystal classes $\bar{6}$, 6, and $\frac{6}{m}$.

3.3.3.7 Cubic Crystal Classes

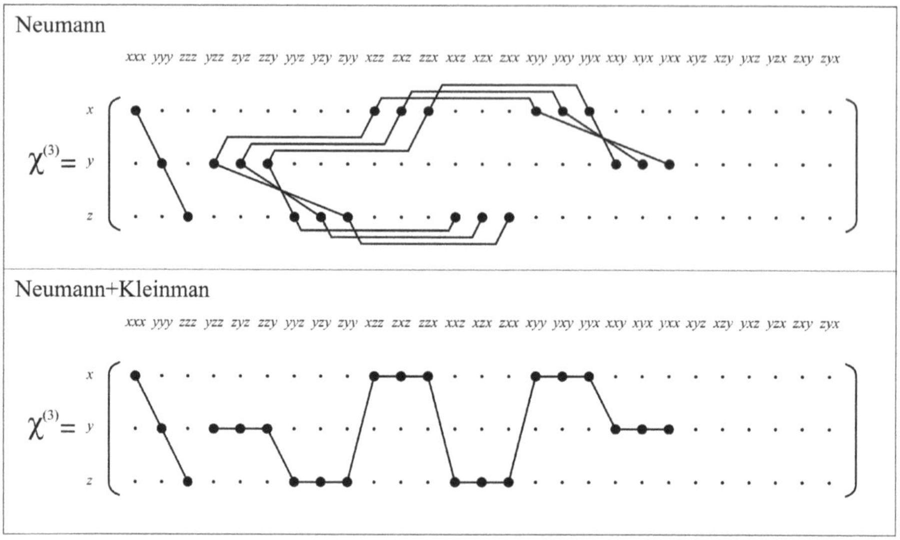

FIGURE 3.50 Matrices corresponding to the third-order electric susceptibility tensor $\chi^{(3)}$ for crystal classes 432, $\bar{4}3m$, and $m\bar{3}m$.

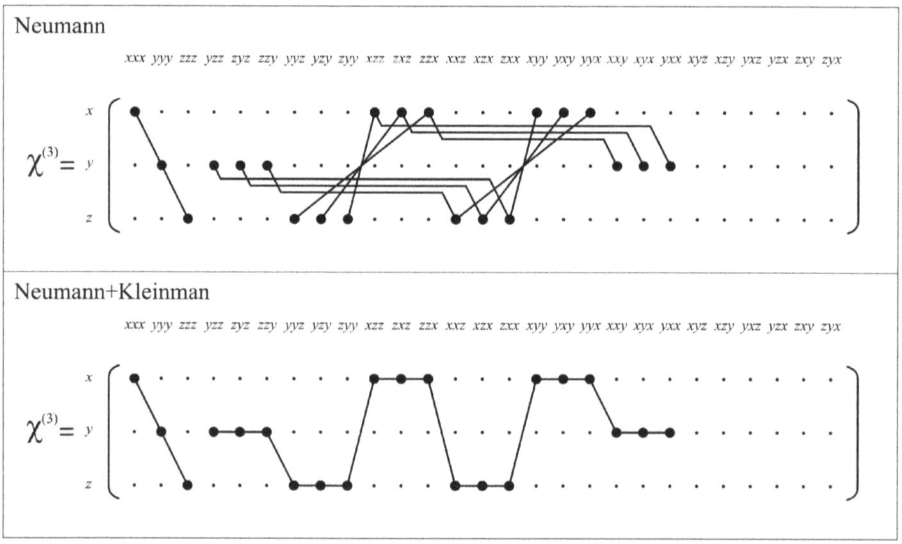

FIGURE 3.51 Matrices corresponding to the third-order electric susceptibility tensor $\chi^{(3)}$ for crystal classes 23 and $m\bar{3}$.

3.4 IMPACT OF THE RELATIVE ORIENTATION OF THE MOLECULES ON THE ELECTRIC SUSCEPTIBILITY USING THE BOND CHARGE MODEL

In Part 2.1, we considered N electrons that can move only along one axis, which allowed us to model in a simplifying way the magnitude of the different orders of the electric susceptibility at the microscopic scale. In real cases, electrons are involved in chemical bonds oriented and distributed in the three directions of space.

The bond charge model is very useful for a 3D modeling of the relative magnitudes of the coefficients of each order of the macroscopic electric susceptibility tensor (Levine 1974) [5]. This model is based on a decomposition of the macroscopic electric susceptibility into microscopic contributions which are the bond polarizabilities.

The calculation of the first-, second-, and third-order electric susceptibilities is performed hereafter at the level of the unit cell of the considered crystal, each chemical bond orientation being expressed in the dielectric frame, as shown in Figure 3.52.

According to Figure 3.52 and to the fact that the macroscopic first-order electric susceptibility is a rank-2 tensor, each $\chi_{ij}^{(1)}$ coefficients can be expressed as the oriented sum of the microscopic first-order dipolar moments, *i.e.* the linear microscopic polarizabilities, of all the chemical bonds contained in each unit cell, which gives (Levine 1974) [5]:

$$\chi_{ij}^{(1)} = \frac{1}{V_{uc}} \sum_{b=1}^{N_b} G_{ij}^{b} \mu_b^{(1)} \tag{3.56}$$

with

$$G_{ij}^{b} = \cos\theta_i^b \cos\theta_j^b \tag{3.57}$$

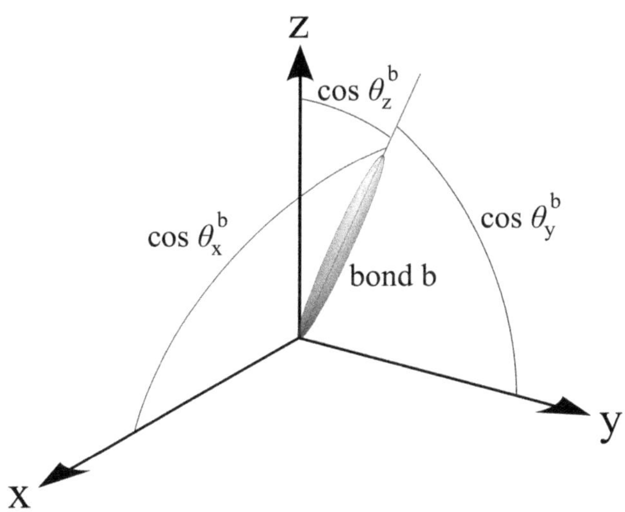

FIGURE 3.52 Orientation of a chemical bond (labeled *b*) in the dielectric frame (*x*, *y*, *z*) of the crystal.

N_b corresponds to the number of chemical bonds of type b, V_{uc} is the unit cell volume of the crystal, and $\theta_{i,j}^b$ is the angle between the axis of the b t[h] bond and the dielectric frame axis i (or j) where i and j stands for x, y, and z. The quantity $\mu_b^{(1)}$ is the first-order electric dipolar moment of the b t[h] bond that can be calculated from the plasma frequency and the energy band gap (Van Vechten 1969) [6].

Note that G_{ij}^b is one of the 3^2 coefficients of a rank-2 polar symmetrical tensor built from the cosine directors of the bond axis. Then it is a diagonal tensor in the dielectric frame, the three non-zero coefficients being G_{xx}^b, G_{yy}^b, and G_{zz}^b.

Similarly, each $\chi_{ijk}^{(2)}$ coefficient of the macroscopic second-order electric susceptibility tensor writes:

$$\chi_{ijk}^{(2)} = \frac{1}{V_{uc}} \sum_{b=1}^{N_b} G_{ijk}^b \mu_b^{(2)} \tag{3.58}$$

with

$$G_{ijk}^b = \cos\theta_i^b \cos\theta_j^b \cos\theta_k^b \tag{3.59}$$

and

$$\mu_b^{(2)} = \frac{1}{4} \frac{\partial \alpha_b^{(1)}}{\partial r} \frac{\alpha_b^{(1)}}{q_b} \tag{3.60}$$

where ∂r is the displacement of the bond charge q_b.

The microscopic quadratic polarizability $\mu_b^{(2)}$ of Eq. (3.60) is calculated by considering the modification of the microscopic linear polarizability $\mu_b^{(1)}$ due to the displacement of the bond charge under an applied electric field (Hansen 1988) [7]. And as explained in Chapter 1, the microscopic quadratic polarizability is non-zero if the corresponding electronic potential is non-centrosymmetric.

Each $\chi_{ijkl}^{(3)}$ coefficient of the macroscopic third-order electric susceptibility tensor can be written in the same way, the microscopic cubic polarizability $\mu_b^{(3)}$ being calculated from the microscopic linear and quadratic polarizability, $\mu_b^{(1)}$ and $\mu_b^{(2)}$ respectively, which gives (Boulanger 1999) [8]:

$$\chi_{ijkl}^{(3)} = \frac{1}{V_{uc}} \sum_{b=1}^{N_b} G_{ijkl}^b \mu_b^{(3)} \tag{3.61}$$

with

$$G_{ijkl}^b = \cos\theta_i^b \cos\theta_j^b \cos\theta_k^b \cos\theta_l^b \tag{3.62}$$

and

$$\mu_b^{(3)} = \frac{1}{4} \frac{\partial \alpha_b^{(2)}}{\partial r} \frac{\alpha_b^{(1)}}{q_b} \tag{3.63}$$

This simple geometrical approach has led to a satisfying agreement between calculations and measurements of the coefficients of the first-, second-, and third-order electric susceptibility tensors, especially for mineral crystals (Boulanger 1999) [8]. The case of organic crystals requires a more sophisticated model since the molecules are more complex, which implies to take into account the tensor nature of the microscopic polarizability (Zyss 1993) [9].

This approach is of huge interest for guiding the strategy in the framework of materials engineering. For example, Eq. (3.59) and Eq. (3.60) show that chemical bonds exhibiting a high microscopic quadratic polarizability can lead to a negligible second-order electric susceptibility at the macroscopic scale if these bonds constitute a centrosymmetric group. An illustration of this feature can be brought by considering a schematic situation of a crystal made of two kinds of chemical bonds, *i.e.* $b1$ and $b2$, exhibiting non-zero microscopic quadratic polarizability, *i.e.* $\mu_{b1}^{(2)}$ and $\mu_{b2}^{(2)}$ respectively. Then the corresponding macroscopic coefficients of the second-order electric susceptibility tensor can be expressed as:

$$\chi_{ijk}^{(2)} = \frac{1}{V_{uc}}\left[\sum_{b1=1}^{N_{b1}} G_{ijk}^{b1}\mu_{b1}^{(2)} + \sum_{b2=1}^{N_{b2}} G_{ijk}^{b2}\mu_{b2}^{(2)}\right] \tag{3.64}$$

with

$$G_{ijk}^{b1} = \cos\theta_i^{b1}\cos\theta_j^{b1}\cos\theta_k^{b1} \tag{3.65}$$

and

$$G_{ijk}^{b2} = \cos\theta_i^{b2}\cos\theta_j^{b2}\cos\theta_k^{b2} \tag{3.66}$$

We consider now the case in which the group of bonds $b1$ is centrosymmetric, with for example four bonds, *i.e.* $N_{b1} = 4$, with the axes lying along the diagonal of a square, as depicted in Figure 3.53. The group of bonds $b2$ is considered to be non-centrosymmetric and is not drawn here.

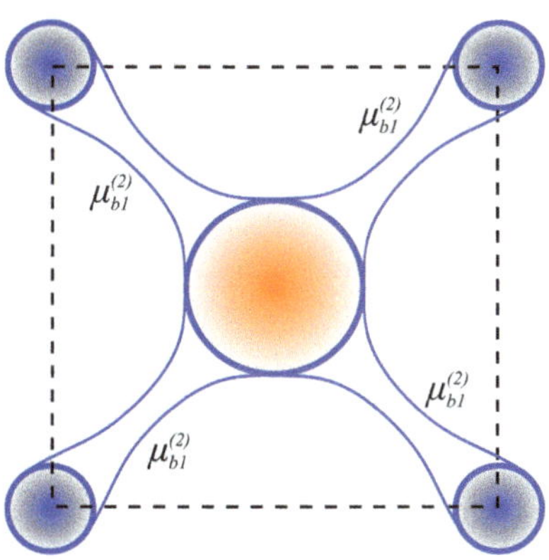

FIGURE 3.53 Non-centrosymmetric group of bonds $b1$, between each blue atom and the yellow one, exhibiting a microscopic quadratic polarizability $\mu_{b1}^{(2)}$.

According to Neumann's principle, the summation over the $b1$ bonds vanishes to zero, *i.e.*:

$$\sum_{b1=1}^{N_{b1}} G_{ijk}^{b1} \mu_{b1}^{(2)} = 0 \tag{3.67}$$

Then the macroscopic second-order electric susceptibility is only due to bonds $b2$, which gives:

$$\chi_{ijk}^{(2)} = \frac{1}{V_{uc}} \sum_{b2=1}^{N_{b2}} G_{ijk}^{b2} \mu_{b2}^{(2)} \tag{3.68}$$

From a more general point of view, it is easy to understand that a group of chemical bonds having a weak geometrical factor contributes weakly to the macroscopic electric susceptibility despite a high microscopic polarizability, which applies for any order of the electric susceptibility.

3.5 SUMMARY

The intrinsic symmetries concern lossless media. This approximation is usually valid even with an absorption coefficient of a few tens of percent. In that case $\chi^{(1)}$ can be considered as a real quantity and as a symmetrical tensor. Furthermore, ABDP and Kleinmann's symmetries can apply to $\chi^{(2)}$ and $\chi^{(3)}$ as described in detail in this chapter.

The electric susceptibility is the original optical property since it governs the link between the electric field of light and the polarization induced in matter. When expressed in the dielectric frame, each coefficient of the electric susceptibility tensor that is considered links given Cartesian components of the electric field of light with a given Cartesian component of the induced polarization, knowing that each Cartesian coefficient has to match a given circular frequency of light. For example, $\chi_{ijkl}^{(3)}(\omega_4 = \omega_1 + \omega_2 + \omega_3)$ couples $P_i^{(3)}(\omega_4)$ with the product $E_j(\omega_1)E_k(\omega_2)E_l(\omega_3)$.

Symmetry, whether intrinsic or extrinsic, plays a crucial role in the different orders of the electric susceptibility since it allows the corresponding tensors to be reduced in terms of the number of independent coefficients. Therefore, symmetry leads to significant simplifications for calculations. But it must be understood that the "shape" of the tensors is inseparable from the nonlinear optical process that is chosen to measure the different tensor coefficients. This means that the tensor at a given circular frequency is different depending on whether it is measured in a non-degenerate or degenerate situation regarding the circular frequencies: for example in a crystal belonging to $mm2$ and for which Kleinman's symmetry would not apply, it comes: $\chi_{ijk}^{(2)}(\omega_3 = \omega_a + \omega_b) \neq \chi_{ikj}^{(2)}(\omega_3 = \omega_a + \omega_b)$ while $\chi_{ijk}^{(2)}(\omega_3 = \omega_a + \omega_a) = \chi_{ikj}^{(2)}(\omega_3 = \omega_a + \omega_a)$. It is therefore necessary to specify the nonlinear process that was considered for the measurement when communicating the values of the tensor coefficients.

Both linear and nonlinear optical interactions will be described in the so-called dielectric frame, which is the frame in which the real part of the first-order electric susceptibility tensor is diagonal. It has been seen in this chapter that the dielectric frame is linked with the crystallographic frame for all the orientation symmetry classes except for the monoclinic and triclinic ones. For these later classes, it is important to notice that the orientations of the dielectric axes that are not linked with crystallographic axes necessarily depends on the light frequency since that results from the diagonalization of a non-diagonal tensor in an arbitrary reference frame whose coefficients depend on the frequency, which is a strong level of difficulty in the implementation of the optical properties.

REFERENCES

[1] Born M. and Wolf E. 1980. Principle of Optics, 6th ed., New York: Pergamon Press.

[2] Bloembergen N. 1996. Nonlinear Optics, 4th ed., New York and Amsterdam: W.A. Benjamin, Inc.

[3] Armstrong J. A., Bloembergen N., Ducuing J., and Pershan P. S. 1962. Interactions between light waves in a nonlinear dielectric. The Physical Review 127: 1918–1939.

[4] Kleinman D. A. 1962. Nonlinear dielectric polarization in optical media. Physical Review 126: 1977–1979.

[5] Levine B. F. 1974. Origin of the unusual dependence of the nonlinear optical susceptibility on bond length for ionic ferroelectrics. Physical Review B 10: 1655–1664.

[6] Van Vechten J. A. 1969. Quantum dielectric theory of electronegativity in covalent systems. I: Electronic dielectric constant. Physical Review 182: 891–905.

[7] Hansen N. K., Protas J. P., and Marnier G. 1988. The electron-density distribution in $KTiOPO_4$. Comptes rendus de l'Académie des Sciences, Paris 307: 475–478.

[8] Boulanger B., Fève J. P., Delarue P., Rousseau I., and Marnier G. 1999. Cubic optical nonlinearity of $KTiOPO_4$. Journal of Physics B: Atomic, Molecular & Optical Physics 32: 475–488.

[9] Zyss J. 1993. Molecular engineering implications of rotational invariance in quadratic nonlinear optics: From dipolar to octupolar molecules and materials. The Journal of Chemical Physics 98(9): 6583–6599.

4 Calculation of the Induced Polarization

4.1 FIRST-ORDER INDUCED POLARIZATION

The matrix equation corresponding to the tensor Eq. (3.1) giving the first-order polarization $\overrightarrow{P^{(1)}}$ induced by the electric field $\vec{E}$ and expressed in the dielectric frame is (Nye 1985) [1]:

$$\begin{pmatrix} P_x^{(1)} \\ P_y^{(1)} \\ P_z^{(1)} \end{pmatrix} = \varepsilon_o \begin{pmatrix} \chi_{xx}^{(1)} & 0 & 0 \\ 0 & \chi_{yy}^{(1)} & 0 \\ 0 & 0 & \chi_{zz}^{(1)} \end{pmatrix} \begin{pmatrix} E_x \\ E_y \\ E_z \end{pmatrix} \tag{4.1}$$

The Cartesian coordinates of $\overrightarrow{P^{(1)}}$ can be easily calculated according to Eq. (3.31) and Eq. (3.32), which gives:

$$\begin{cases} P_x^{(1)} = \varepsilon_o \chi_{xx}^{(1)} E_x \\ P_y^{(1)} = \varepsilon_o \chi_{yy}^{(1)} E_y \\ P_z^{(1)} = \varepsilon_o \chi_{zz}^{(1)} E_z \end{cases} \tag{4.2}$$

Since $(\chi_{xx}^{(1)} \neq \chi_{yy}^{(1)} \neq \chi_{zz}^{(1)})$, $(\chi_{xx}^{(1)} = \chi_{yy}^{(1)} \neq \chi_{zz}^{(1)})$, and $(\chi_{xx}^{(1)} = \chi_{yy}^{(1)} = \chi_{zz}^{(1)})$ according to the optical class that is considered, *i.e.* biaxial, uniaxial, and isotropic respectively, it appears from Eq. (4.2) that $\overrightarrow{P^{(1)}}$ and $\vec{E}$ are collinear in the isotropic case, contrary to the uniaxial and biaxial cases. Thus it appears that the dipolar moment that has been created in uniaxial and biaxial crystals is not collinear with the exciting electric field. This is here the first manifestation of the anisotropy of the optical properties, which was not taken into account in Chapter 2. This feature is not really intuitive and is directly connected with the orientation symmetry of matter.

4.2 SECOND-ORDER INDUCED POLARIZATION

The second-order induced polarization $\overrightarrow{P^{(2)}}$ is expressed in the dielectric frame. The matrix equation corresponding to the tensor Eq. (3.2) is (Boulanger & Zyss 2013) [2]:

DOI: 10.1201/9780429091735-4

$$\begin{pmatrix} P_x^{(1)} \\ P_y^{(1)} \\ P_z^{(1)} \end{pmatrix} = \varepsilon_o \begin{pmatrix} \chi_{xxx}^{(2)} & \chi_{xyy}^{(2)} & \chi_{xzz}^{(2)} & \chi_{xyz}^{(2)} & \chi_{xzy}^{(2)} & \chi_{xxz}^{(2)} & \chi_{xzx}^{(2)} & \chi_{xxy}^{(2)} & \chi_{xyx}^{(2)} \\ \chi_{yxx}^{(2)} & \chi_{yyy}^{(2)} & \chi_{yzz}^{(2)} & \chi_{yyz}^{(2)} & \chi_{yzy}^{(2)} & \chi_{yxz}^{(2)} & \chi_{yzx}^{(2)} & \chi_{yxy}^{(2)} & \chi_{yyx}^{(2)} \\ \chi_{zxx}^{(2)} & \chi_{zyy}^{(2)} & \chi_{zzz}^{(2)} & \chi_{zyz}^{(2)} & \chi_{zzy}^{(2)} & \chi_{zxz}^{(2)} & \chi_{zzx}^{(2)} & \chi_{zxy}^{(2)} & \chi_{zyx}^{(2)} \end{pmatrix} \begin{pmatrix} E_{x1}E_{x2} \\ E_{y1}E_{y2} \\ E_{z1}E_{z2} \\ E_{y1}E_{z2} \\ E_{z1}E_{y2} \\ E_{x1}E_{z2} \\ E_{z1}E_{x2} \\ E_{x1}E_{y2} \\ E_{y1}E_{x2} \end{pmatrix} \quad (4.3)$$

Equation (4.3) is totally general and can be used for the three circular frequencies ω_1, ω_2, and ω_3 that are involved, *i.e.*:

$$\left\{\overrightarrow{P^{(2)}}(\omega_3), \overrightarrow{E_1}(\omega_1), \overrightarrow{E_2}(\omega_2)\right\}, \left\{\overrightarrow{P^{(2)}}(\omega_1), \overrightarrow{E_1}(\omega_3), \overrightarrow{E_2}(-\omega_2)\right\}, \left\{\overrightarrow{P^{(2)}}(\omega_1), \overrightarrow{E_1}(\omega_3), \overrightarrow{E_2}(-\omega_1)\right\},$$

with $\omega_3 = \omega_1 + \omega_2$, $\overrightarrow{E_2}(-\omega_{1,2}) = \overrightarrow{E_2^*}(\omega_{1,2})$, and $\omega_1 = \omega_2$ or $\omega_1 \neq \omega_2$. Note that the indices 1 and 2 at the level of the products of electric fields are here in order to well distinguish the two fields, which will be of particular interest when $\omega_1 = \omega_2$ as will be seen later on.

The calculation of $\overrightarrow{P^{(2)}}$ from Eq. (4.3) has to be performed following Eq. (3.34) and Eq. (3.35) and taking the $\chi^{(2)}$ matrices of Figure 3.20 through Figure 3.39 according to the crystal class that is considered. The example of the crystal class $mm2$ is detailed hereafter for $P^{(2)}(\omega_3)$ using the matrices of Figure 3.23.

In the general case, when $\omega_1 \neq \omega_2$ and Kleinmann's symmetry does not apply, it comes:

$$\begin{aligned} P_x^{(2)}(\omega_3) &= \varepsilon_o[\chi_{xxz}^{(2)}(\omega_3)E_{x1}(\omega_1)E_{z2}(\omega_2) + \chi_{xzx}^{(2)}(\omega_3)E_{z1}(\omega_1)E_{x2}(\omega_2)] \\ P_y^{(2)}(\omega_3) &= \varepsilon_o[\chi_{yyz}^{(2)}(\omega_3)E_{y1}(\omega_1)E_{z2}(\omega_2) + \chi_{yzy}^{(2)}(\omega_3)E_{z1}(\omega_1)E_{y2}(\omega_2)] \\ P_z^{(2)}(\omega_3) &= \varepsilon_o[\chi_{zxx}^{(2)}(\omega_3)E_{x1}(\omega_1)E_{x2}(\omega_2) + \chi_{zyy}^{(2)}(\omega_3)E_{y1}(\omega_1)E_{y2}(\omega_2) \\ &\quad + \chi_{zzz}^{(2)}(\omega_3)E_{z1}(\omega_1)E_{z2}(\omega_2)] \end{aligned} \quad (4.4)$$

In the case when $\omega_1 \neq \omega_2$ and Kleinmann's symmetry applies, *i.e.* $\chi_{xxz}^{(2)}(\omega_3) = \chi_{xzx}^{(2)}(\omega_3) = \chi_{zzx}^{(2)}(\omega_3) \neq \chi_{yyz}^{(2)}(\omega_3) = \chi_{yzy}^{(2)}(\omega_3) = \chi_{zzy}^{(2)}(\omega_3) \neq \chi_{zzz}^{(2)}(\omega_3)$, then Eq. (4.4) reduces to:

$$\begin{aligned} P_x^{(2)}(\omega_3) &= \varepsilon_o\chi_{xxz}^{(2)}(\omega_3)[E_{x1}(\omega_1)E_{z2}(\omega_2) + E_{z1}(\omega_1)E_{x2}(\omega_2)] \\ P_y^{(2)}(\omega_3) &= \varepsilon_o\chi_{yyz}^{(2)}(\omega_3)[E_{y1}(\omega_1)E_{z2}(\omega_2) + E_{z1}(\omega_1)E_{y2}(\omega_2)] \\ P_z^{(2)}(\omega_3) &= \varepsilon_o[\chi_{zxx}^{(2)}(\omega_3)E_{x1}(\omega_1)E_{x2}(\omega_2) + \chi_{zyy}^{(2)}(\omega_3)E_{y1}(\omega_1)E_{y2}(\omega_2) \\ &\quad + \chi_{zzz}^{(2)}(\omega_3)E_{z1}(\omega_1)E_{z2}(\omega_2)] \end{aligned} \quad (4.5)$$

In the case of SHG, *i.e.* $\omega_1 = \omega_2 \equiv \omega$ and so $\omega_3 = 2\omega$, there is often a confusion in a lot of articles and reference books between the symmetry of $\chi^{(2)}$ and that at the level of the fields at ω. It is indeed important to notice that the degeneracy at the level of the circular frequencies, *i.e.* $\omega_1 = \omega_2 \equiv \omega$, does not lead to a degeneracy of the electric fields at ω. Actually, two cases can exist: $\overrightarrow{E_1}(\omega) = \overrightarrow{E_2}(\omega)$, which corresponds to the degeneracy at the level of the fields called type I SHG as will be seen in Chapter 7, and $\overrightarrow{E_1}(\omega) \neq \overrightarrow{E_2}(\omega)$ that is the non-degenerate case corresponding to type II SHG.

- In the case of type II SHG ($\overrightarrow{E_1}(\omega) \neq \overrightarrow{E_2}(\omega)$) with or without Kleinman's symmetry, Eq. (4.4) is written as Eq. (4.5), *i.e.*:

$$
\begin{aligned}
P_x^{(2)}(2\omega) &= \varepsilon_o \chi_{xxz}^{(2)}(2\omega)[E_{x1}(\omega)E_{z2}(\omega) + E_{z1}(\omega)E_{x2}(\omega)] \\
P_y^{(2)}(2\omega) &= \varepsilon_o \chi_{yyz}^{(2)}(2\omega)[E_{y1}(\omega)E_{z2}(\omega) + E_{z1}(\omega)E_{y2}(\omega)] \\
P_z^{(2)}(\omega_3) &= \varepsilon_o [\chi_{zxx}^{(2)}(2\omega)E_{x1}(\omega)E_{x2}(\omega) + \chi_{zyy}^{(2)}(2\omega)E_{y1}(\omega)E_{y2}(\omega) \\
&\qquad + \chi_{zzz}^{(2)}(2\omega)E_{z1}(\omega)E_{z2}(\omega)]
\end{aligned}
\tag{4.6}
$$

- In the case of type I SHG ($\overrightarrow{E_1}(\omega) = \overrightarrow{E_2}(\omega) \equiv \overrightarrow{E}(\omega)$) with or without Kleinman's symmetry, Eq. (4.6) reduces to:

$$
\begin{aligned}
P_x^{(2)}(2\omega) &= 2\varepsilon_o \chi_{xxz}^{(2)}(2\omega)E_x(\omega)E_z(\omega) \\
P_y^{(2)}(2\omega) &= 2\varepsilon_o \chi_{yyz}^{(2)}(2\omega)E_y(\omega)E_z(\omega) \\
P_z^{(2)}(\omega_3) &= \varepsilon_o [\chi_{zxx}^{(2)}(2\omega)E_x(\omega)^2 + \chi_{zyy}^{(2)}(2\omega)E_y(\omega)^2 + \chi_{zzz}^{(2)}(2\omega)E_z(\omega)^2]
\end{aligned}
\tag{4.7}
$$

Kleinman's symmetry only brings the two equalities $\chi_{xxz}^{(2)}(2\omega) = \chi_{zxx}^{(2)}(2\omega)$ and $\chi_{yyz}^{(2)}(2\omega) = \chi_{zyy}^{(2)}(2\omega)$.

4.3 THIRD-ORDER INDUCED POLARIZATION

The third-order induced polarization $\overrightarrow{P^{(3)}}$ is expressed in the dielectric frame. The matrix equations corresponding to the tensor Eq. (3.3) and Eq. (3.4) are not written here due to the numerous coefficients. The generic tensor equations are Eq. (2.36) and Eq. (2.37). Then according to Eq. (2.38), it comes for each $P_i^{(3)}$ Cartesian coordinates with $i = x, y, z$:

$$
\begin{aligned}
P_i^{(3)} = \varepsilon_0 [&\chi_{ixxx}^{(3)} V_{xxx} + \chi_{iyyy}^{(3)} V_{yyy} + \chi_{izzz}^{(3)} V_{zzz} \\
&+ \chi_{iyzz}^{(3)} V_{yzz} + \chi_{izzy}^{(3)} V_{zzy} + \chi_{izyz}^{(3)} V_{zyz} + \chi_{iyyz}^{(3)} V_{yyz} + \chi_{izyy}^{(3)} V_{zyy} + \chi_{iyzy}^{(3)} V_{yzy} \\
&+ \chi_{ixzz}^{(3)} V_{xzz} + \chi_{izzx}^{(3)} V_{zzx} + \chi_{izxz}^{(3)} V_{zxz} + \chi_{ixxz}^{(3)} V_{xxz} + \chi_{izxx}^{(3)} V_{zxx} + \chi_{ixzx}^{(3)} V_{xzx} \\
&+ \chi_{ixyy}^{(3)} V_{xyy} + \chi_{iyyx}^{(3)} V_{yyx} + \chi_{iyxy}^{(3)} V_{yxy} + \chi_{ixxy}^{(3)} V_{xxy} + \chi_{iyxx}^{(3)} V_{yxx} + \chi_{ixyx}^{(3)} V_{xyx} \\
&+ \chi_{ixyz}^{(3)} V_{xyz} + \chi_{ixzy}^{(3)} V_{xzy} + \chi_{iyxz}^{(3)} V_{yxz} + \chi_{iyzx}^{(3)} V_{yzx} + \chi_{izxy}^{(3)} V_{zxy} + \chi_{izyx}^{(3)} V_{zyx}]
\end{aligned}
\tag{4.8}
$$

where $V_{abc} = E_{a1}E_{b2}E_{c3}$ with $(a,b,c) = x, y,$ or z.

Equation (4.8) is valid for the two types of energy conservation and all the possible corresponding circular frequency components of $\overrightarrow{P^{(3)}}$, *i.e.*:

- for $\omega_4 = \omega_1 + \omega_2 + \omega_3$

$$\left\{\overrightarrow{P^{(3)}}(\omega_4), \overrightarrow{E_1}(\omega_1), \overrightarrow{E_2}(\omega_2), \overrightarrow{E_3}(\omega_3)\right\}, \left\{\overrightarrow{P^{(3)}}(\omega_1), \overrightarrow{E_1}(\omega_4), \overrightarrow{E_2}(-\omega_2), \overrightarrow{E_3}(-\omega_3)\right\},$$
$$\left\{\overrightarrow{P^{(3)}}(\omega_2), \overrightarrow{E_1}(\omega_4), \overrightarrow{E_2}(-\omega_1), \overrightarrow{E_3}(-\omega_3)\right\}, \left\{\overrightarrow{P^{(3)}}(\omega_3), \overrightarrow{E_1}(\omega_4), \overrightarrow{E_1}(-\omega_1), \overrightarrow{E_2}(-\omega_2)\right\}$$

- for $\omega_4 + \omega_1 = \omega_2 + \omega_3$

$$\left\{\overrightarrow{P^{(3)}}(\omega_4), \overrightarrow{E_1}(-\omega_1), \overrightarrow{E_2}(\omega_2), \overrightarrow{E_3}(\omega_3)\right\}, \left\{\overrightarrow{P^{(3)}}(\omega_1), \overrightarrow{E_1}(-\omega_4), \overrightarrow{E_2}(\omega_2), \overrightarrow{E_3}(\omega_3)\right\},$$
$$\left\{\overrightarrow{P^{(3)}}(\omega_2), \overrightarrow{E_1}(\omega_4), \overrightarrow{E_2}(\omega_1), \overrightarrow{E_3}(-\omega_3)\right\}, \left\{\overrightarrow{P^{(3)}}(\omega_3), \overrightarrow{E_1}(\omega_4), \overrightarrow{E_1}(\omega_1), \overrightarrow{E_2}(-\omega_2)\right\}$$

with $\omega_1 \neq \omega_2 \neq \omega_3$, $\omega_1 = \omega_2 \neq \omega_3$, $\omega_1 = \omega_2 = \omega_3$, and $\overrightarrow{E_a}(-\omega_b) = \overrightarrow{E_a}^*(\omega_b)$.

The calculation specific to each class can be done using the matrices of Figure 3.41 to Figure 3.51.

The example of crystals belonging to classes 23 and $m\bar{3}$ (Figure 3.51) is detailed here for $\overrightarrow{P^{(3)}}(\omega_4 = \omega_1 + \omega_2 + \omega_3)$ as well as for $\overrightarrow{P^{(3)}}(\omega_4 = -\omega_1 + \omega_2 + \omega_3)$.

In the general case, when $\omega_1 \neq \omega_2 \neq \omega_3$ and Kleinmann's symmetry does not apply, it comes:

$$\begin{aligned}
P_x^{(3)}(\omega_4) = \varepsilon_0[&\chi_{xxxx}^{(3)}(\omega_4)E_{x1}(\pm\omega_1)E_{x2}(\omega_2)E_{x3}(\omega_3) + \chi_{xxzz}^{(3)}(\omega_4)E_{x1}(\pm\omega_1)E_{z2}(\omega_2)E_{z3}(\omega_3) \\
&+ \chi_{xzzx}^{(3)}(\omega_4)E_{z1}(\pm\omega_1)E_{z2}(\omega_2)E_{x3}(\omega_3) + \chi_{xzxz}^{(3)}(\omega_4)E_{z1}(\pm\omega_1)E_{x2}(\omega_2)E_{z3}(\omega_3) \\
&+ \chi_{xxyy}^{(3)}(\omega_4)E_{x1}(\pm\omega_1)E_{y2}(\omega_2)E_{y3}(\omega_3) + \chi_{xyyx}^{(3)}(\omega_4)E_{y1}(\pm\omega_1)E_{y2}(\omega_2)E_{x3}(\omega_3) \\
&+ \chi_{xyxy}^{(3)}(\omega_4)E_{y1}(\pm\omega_1)E_{x2}(\omega_2)E_{y3}(\omega_3)] \\
P_y^{(3)}(\omega_4) = \varepsilon_0[&\chi_{yyyy}^{(3)}(\omega_4)E_{y1}(\pm\omega_1)E_{y2}(\omega_2)E_{y3}(\omega_3) + \chi_{yyzz}^{(3)}(\omega_4)E_{y1}(\pm\omega_1)E_{z2}(\omega_2)E_{z3}(\omega_3) \\
&+ \chi_{yzzy}^{(3)}(\omega_4)E_{z1}(\pm\omega_1)E_{z2}(\omega_2)E_{y3}(\omega_3) + \chi_{yzyz}^{(3)}(\omega_4)E_{z1}(\pm\omega_1)E_{y2}(\omega_2)E_{z3}(\omega_3) \\
&+ \chi_{yxxy}^{(3)}(\omega_4)E_{x1}(\pm\omega_1)E_{x2}(\omega_2)E_{y3}(\omega_3) + \chi_{yyxx}^{(3)}(\omega_4)E_{y1}(\pm\omega_1)E_{x2}(\omega_2)E_{x3}(\omega_3) \\
&+ \chi_{yxyx}^{(3)}(\omega_4)E_{x1}(\pm\omega_1)E_{y2}(\omega_2)E_{x3}(\omega_3)] \\
P_z^{(3)}(\omega_4) = \varepsilon_0[&\chi_{zzzz}^{(3)}(\omega_4)E_{z1}(\pm\omega_1)E_{z2}(\omega_2)E_{z3}(\omega_3) + \chi_{zyyz}^{(3)}(\omega_4)E_{y1}(\pm\omega_1)E_{y2}(\omega_2)E_{z3}(\omega_3) \\
&+ \chi_{zzyy}^{(3)}(\omega_4)E_{z1}(\pm\omega_1)E_{y2}(\omega_2)E_{y3}(\omega_3) + \chi_{zyzy}^{(3)}(\omega_4)E_{y1}(\pm\omega_1)E_{z2}(\omega_2)E_{y3}(\omega_3) \\
&+ \chi_{zxxz}^{(3)}(\omega_4)E_{x1}(\pm\omega_1)E_{x2}(\omega_2)E_{z3}(\omega_3) + \chi_{zzxx}^{(3)}(\omega_4)E_{z1}(\pm\omega_1)E_{x2}(\omega_2)E_{x3}(\omega_3) \\
&+ \chi_{zxzx}^{(3)}(\omega_4)E_{x1}(\pm\omega_1)E_{z2}(\omega_2)E_{x3}(\omega_3)]
\end{aligned} \tag{4.9}$$

In the case in which $\omega_1 \neq \omega_2 \neq \omega_3$ and Kleinmann's symmetry applies, then Eq. (4.9) reduces to:

$$\begin{aligned}
P_x^{(3)}(\omega_4) &= \varepsilon_0 [\chi_{xxxx}^{(3)}(\omega_4) E_{x1}(\pm\omega_1) E_{x2}(\omega_2) E_{x3}(\omega_3) \\
&+ \chi_{xxzz}^{(3)}(\omega_4) \begin{Bmatrix} E_{x1}(\pm\omega_1) E_{z2}(\omega_2) E_{z3}(\omega_3) \\ + E_{z1}(\pm\omega_1) E_{z2}(\omega_2) E_{x3}(\omega_3) + E_{z1}(\pm\omega_1) E_{x2}(\omega_2) E_{z3}(\omega_3) \end{Bmatrix} \\
&+ \chi_{xxyy}^{(3)}(\omega_4) \begin{Bmatrix} E_{x1}(\pm\omega_1) E_{y2}(\omega_2) E_{y3}(\omega_3) \\ + E_{y1}(\pm\omega_1) E_{y2}(\omega_2) E_{x3}(\omega_3) + E_{y1}(\pm\omega_1) E_{x2}(\omega_2) E_{y3}(\omega_3) \end{Bmatrix}] \\
P_y^{(3)}(\omega_4) &= \varepsilon_0 [\chi_{yyyy}^{(3)}(\omega_4) E_{y1}(\pm\omega_1) E_{y2}(\omega_2) E_{y3}(\omega_3) \\
&+ \chi_{yyzz}^{(3)}(\omega_4) \begin{Bmatrix} E_{y1}(\pm\omega_1) E_{z2}(\omega_2) E_{z3}(\omega_3) \\ + E_{z1}(\pm\omega_1) E_{z2}(\omega_2) E_{y3}(\omega_3) + \chi E_{z1}(\pm\omega_1) E_{y2}(\omega_2) E_{z3}(\omega_3) \end{Bmatrix} \\
&+ \chi_{yxxy}^{(3)}(\omega_4) \begin{Bmatrix} E_{x1}(\pm\omega_1) E_{x2}(\omega_2) E_{y3}(\omega_3) \\ + E_{y1}(\pm\omega_1) E_{x2}(\omega_2) E_{x3}(\omega_3) + E_{x1}(\pm\omega_1) E_{y2}(\omega_2) E_{x3}(\omega_3) \end{Bmatrix}] \\
P_z^{(3)}(\omega_4) &= \varepsilon_0 [\chi_{zzzz}^{(3)}(\omega_4) E_{z1}(\pm\omega_1) E_{z2}(\omega_2) E_{z3}(\omega_3) \\
&+ \chi_{zyyz}^{(3)}(\omega_4) \begin{Bmatrix} E_{y1}(\pm\omega_1) E_{y2}(\omega_2) E_{z3}(\omega_3) \\ + E_{z1}(\pm\omega_1) E_{y2}(\omega_2) E_{y3}(\omega_3) + E_{y1}(\pm\omega_1) E_{z2}(\omega_2) E_{y3}(\omega_3) \end{Bmatrix} \\
&+ \chi_{zxxz}^{(3)}(\omega_4) \begin{Bmatrix} E_{x1}(\pm\omega_1) E_{x2}(\omega_2) E_{z3}(\omega_3) \\ + E_{z1}(\pm\omega_1) E_{x2}(\omega_2) E_{x3}(\omega_3) + E_{z1}(\pm\omega_1) E_{x2}(\omega_2) E_{z3}(\omega_3) \end{Bmatrix}]
\end{aligned} \tag{4.10}$$

As in the case of $\chi^{(2)}$, it is necessary to be careful with the degeneracy regarding the circular frequency, *i.e.* partial $(\omega_1 \neq \omega_2 = \omega_3)$ and full $(\omega_1 = \omega_2 = \omega_3)$. Note that the full degeneracy corresponds to THG $(\omega_4 = 3\omega = \omega + \omega + \omega)$ or to the Kerr effect $(\omega_4 = \omega = -\omega + \omega + \omega)$. The comments made in Part 4.2 can be done here, but the number of situations regarding the electric fields is higher. Actually in the case $(\omega_1 \neq \omega_2 = \omega_3 \equiv \omega)$, there are the possibilities of non-degeneracy $(\overrightarrow{E_2}(\omega) \neq \overrightarrow{E_3}(\omega))$ or degeneracy $(\overrightarrow{E_2}(\omega) = \overrightarrow{E_3}(\omega))$; in the case $(\omega_1 = \omega_2 = \omega_3 \equiv \omega)$, there are the following possibilities: $\overrightarrow{E_1}(\omega) = \overrightarrow{E_2}(\omega) \equiv \vec{E}(\omega) \neq \overrightarrow{E_3}(\omega)$, $\overrightarrow{E_1}(\omega) \neq \overrightarrow{E_2}(\omega) = \overrightarrow{E_3}(\omega) \equiv \vec{E}(\omega)$, $\overrightarrow{E_1}(\omega) = \overrightarrow{E_3}(\omega) \equiv \vec{E}(\omega) \neq \overrightarrow{E_2}(\omega)$, $\overrightarrow{E_1}(\omega) = \overrightarrow{E_2}(\omega) = \overrightarrow{E_3}(\omega) \equiv \vec{E}(\omega)$.

Here are only detailed the examples of THG and the Kerr effect.

- In the cases $(\overrightarrow{E_1}(\omega) \neq \overrightarrow{E_2}(\omega) \neq \overrightarrow{E_3}(\omega))$, Kleinmann's symmetry applying or not, Eq. (4.9) has the same form as Eq. (4.10), *i.e.*:

$$\begin{aligned}
P_x^{(3)}(\omega_4) &= \varepsilon_0 [\chi_{xxxx}^{(3)}(\omega_4) E_{x1}(\pm\omega) E_{x2}(\omega) E_{x3}(\omega) \\
&+ \chi_{xxzz}^{(3)}(\omega_4) \begin{Bmatrix} E_{x1}(\pm\omega) E_{z2}(\omega) E_{z3}(\omega) \\ + E_{z1}(\pm\omega) E_{z2}(\omega) E_{x3}(\omega) + E_{z1}(\pm\omega) E_{x2}(\omega) E_{z3}(\omega) \end{Bmatrix}
\end{aligned}$$

$$
\begin{aligned}
&+\chi^{(3)}_{xxyy}(\omega_4)\begin{Bmatrix} E_{x1}(\pm\omega)E_{y2}(\omega)E_{y3}(\omega) \\ +E_{y1}(\pm\omega)E_{y2}(\omega)E_{x3}(\omega)+E_{y1}(\pm\omega)E_{x2}(\omega)E_{y3}(\omega)\end{Bmatrix}] \\
P^{(3)}_y(\omega_4) &= \varepsilon_0[\chi^{(3)}_{yyyy}(\omega_4)E_{y1}(\pm\omega)E_{y2}(\omega)E_{y3}(\omega) \\
&+\chi^{(3)}_{yyzz}(\omega_4)\begin{Bmatrix} E_{y1}(\pm\omega)E_{z2}(\omega)E_{z3}(\omega) \\ +E_{z1}(\pm\omega)E_{z2}(\omega)E_{y3}(\omega)+E_{z1}(\pm\omega)E_{y2}(\omega)E_{z3}(\omega)\end{Bmatrix} \\
&+\chi^{(3)}_{yxxy}(\omega_4)\begin{Bmatrix} E_{x1}(\pm\omega)E_{x2}(\omega)E_{y3}(\omega) \\ +E_{y1}(\pm\omega)E_{x2}(\omega)E_{x3}(\omega)+E_{x1}(\pm\omega)E_{y2}(\omega)E_{x3}(\omega)\end{Bmatrix}] \\
P^{(3)}_z(\omega_4) &= \varepsilon_0[\chi^{(3)}_{zzzz}(\omega_4)E_{z1}(\pm\omega)E_{z2}(\omega)E_{z3}(\omega) \\
&+\chi^{(3)}_{zyyz}(\omega_4)\begin{Bmatrix} E_{y1}(\pm\omega)E_{y2}(\omega)E_{z3}(\omega) \\ +E_{z1}(\pm\omega)E_{y2}(\omega)E_{y3}(\omega)+E_{y1}(\pm\omega)E_{z2}(\omega)E_{y3}(\omega)\end{Bmatrix} \\
&+\chi^{(3)}_{zxxz}(\omega_4)\begin{Bmatrix} E_{x1}(\pm\omega)E_{x2}(\omega)E_{z3}(\omega) \\ +E_{z1}(\pm\omega)E_{x2}(\omega)E_{x3}(\omega)+E_{x1}(\pm\omega)E_{z2}(\omega)E_{x3}(\omega)\end{Bmatrix}]
\end{aligned}
\tag{4.11}
$$

with $\omega_4 = 3\omega$ when it is $\vec{E_3}(\omega)$ (THG), and $\omega_4 = \omega$ when it is $\vec{E_3}(-\omega)$ (Kerr effect).

- In the cases of THG with $(\vec{E_1}(\omega) \neq \vec{E_2}(\omega) = \vec{E_3}(\omega) \equiv \vec{E}(\omega))$, Kleinmann's symmetry applying or not, Eq. (4.11) reduces to:

$$
\begin{aligned}
P^{(3)}_x(\omega_4) &= \varepsilon_0[\chi^{(3)}_{xxxx}(\omega_4)E_{x1}(\pm\omega)E_x(\omega)^2 \\
&+\chi^{(3)}_{xxzz}(\omega_4)\left\{E_{x1}(\pm\omega)E_z(\omega)^2+2E_{z1}(\pm\omega)E_z(\omega)E_x(\omega)\right\} \\
&+\chi^{(3)}_{xxyy}(\omega_4)\left\{E_{x1}(\pm\omega)E_y(\omega)^2+2E_{y1}(\pm\omega)E_y(\omega)E_x(\omega)\right\}] \\
P^{(3)}_y(\omega_4) &= \varepsilon_0[\chi^{(3)}_{yyyy}(\omega_4)E_{y1}(\pm\omega)E_y(\omega)^2 \\
&+\chi^{(3)}_{yyzz}(\omega_4)\left\{E_{y1}(\pm\omega)E_z(\omega)^2+2E_{z1}(\pm\omega)E_z(\omega)E_y(\omega)\right\} \\
&+\chi^{(3)}_{yxxy}(\omega_4)\left\{E_{y1}(\pm\omega)E_x(\omega)^2+2E_{x1}(\pm\omega)E_y(\omega)E_x(\omega)\right\}] \\
P^{(3)}_z(\omega_4) &= \varepsilon_0[\chi^{(3)}_{zzzz}(\omega_4)E_{z1}(\pm\omega)E_z(\omega)^2 \\
&+\chi^{(3)}_{zyyz}(\omega_4)\left\{E_{z1}(\pm\omega)E_y(\omega)^2+2E_{y1}(\pm\omega)E_z(\omega)E_y(\omega)\right\} \\
&+\chi^{(3)}_{zxxz}(\omega_4)\left\{E_{z1}(\pm\omega)E_x{}^2(\omega)+2E_{x1}(\pm\omega)E_z(\omega)E_x(\omega)\right\}]
\end{aligned}
\tag{4.12}
$$

with $\omega_4 = 3\omega$ when it is $\vec{E_3}(\omega)$ (THG), and $\omega_4 = \omega$ when it is $\vec{E_3}(-\omega)$ (Kerr effect).

- In the cases of THG, with $(\overrightarrow{E_1}(\omega) = \overrightarrow{E_2}(\omega) = \overrightarrow{E_3}(\omega) \equiv \vec{E}(\omega))$, Kleinmann's symmetry applying or not, Eq. (4.9) reduces to its simplest form, *i.e.*:

$$\begin{cases} P_x^{(3)}(3\omega) = \varepsilon_0[\chi_{xxxx}^{(3)}(3\omega)E_x(\omega)^3 + 3\chi_{xxzz}^{(3)}(3\omega)E_x(\omega)E_z(\omega)^2 \\ \qquad\qquad + 3\chi_{xxyy}^{(3)}(3\omega)E_x(\omega)E_y(\omega)^2] \\ P_y^{(3)}(3\omega) = \varepsilon_0[\chi_{yyyy}^{(3)}(3\omega)E_y(\omega)^3 + 3\chi_{yyzz}^{(3)}(3\omega)E_y(\omega)E_z(\omega)^2 \\ \qquad\qquad + 3\chi_{yxxy}^{(3)}(3\omega)E_x(\omega)^2 E_y(\omega)] \\ P_z^{(3)}(3\omega) = \varepsilon_0[\chi_{zzzz}^{(3)}(3\omega)E_z(\omega)^3 + 3\chi_{zyyz}^{(3)}(3\omega)E_y(\omega)^2 E_z(\omega) \\ \qquad\qquad + 3\chi_{zxxz}^{(3)}(3\omega)E_x(\omega)^2 E_z(\omega)] \end{cases} \tag{4.13}$$

Kleinman's symmetry does not change the form of Eq. (4.9) to Eq. (4.13), but it leads to the three following equalities between nonlinear coefficients: $\chi_{xxzz}^{(3)} = \chi_{zxxz}^{(3)}, \chi_{xxyy}^{(3)} = \chi_{yxxy}^{(3)}, \chi_{yyzz}^{(3)} = \chi_{zyyz}^{(3)}$.

4.4 SUMMARY

The calculations performed in this chapter well show the influence of orientation symmetry of matter on the different orders of the induced polarization. In particular, it was established that the linear induced polarization is not necessarily collinear with the exciting electric field in the case of an anisotropic medium.

As seen in Chapter 1, the induced polarization plays a key role in Maxwell equations. It is then of prime importance for the establishment of the propagation equation, which is the aim of Chapter 5 wherein both lossless and absorbing media will be considered.

REFERENCES

[1] Nye J. F. 1985. Physical Properties of Crystals. Oxford: Oxford University Press.

[2] Boulanger B. and Zyss J. 2013. Non-linear optical properties—chapter 1.7. In International Tables for Crystallography, Vol. D: Physical Properties of Crystals, edited by A. Authier, pp. 181–222. Dordrecht, Netherlands: International Union of Crystallography, Kluwer Academic Publisher.

5 Linear Optical Effects Associated with the Real Part of the First-Order Electric Susceptibility

5.1 LINEAR PROPAGATION EQUATION IN THE DIELECTRIC FRAME

When exciting matter by an optical electric field $\vec{E}(\omega,\vec{r})$ at the circular frequency ω in the linear regime, the induced nonlinear polarization is zero by definition, *i.e.* $\vec{P}^{NL}(\omega,\vec{r}) = 0$, so that Eq. (1.104) is only governed by the linear induced polarization $\vec{P}^{L}(\omega,\vec{r}) = \vec{P}^{(1)}(\omega,\vec{r})$, which gives:

$$\overrightarrow{rot}\left[\overrightarrow{rot}\left[\vec{E}(\omega,\vec{r})\right]\right] - \frac{\omega^2}{c^2}\vec{E}(\omega,\vec{r}) = \omega^2\mu_0\vec{P}^{(1)}(\omega,\vec{r}) \tag{5.1}$$

where μ_0 is the vacuum permittivity, and $c = \dfrac{1}{\sqrt{\mu_0\varepsilon_0}}$ the velocity of light in a vacuum.

By combining Eq. (3.1) and Eq. (5.1), it comes:

$$\overrightarrow{rot}\left[\overrightarrow{rot}\left[\vec{E}(\omega,\vec{r})\right]\right] - \frac{\omega^2}{c^2}\vec{E}(\omega,\vec{r}) = \omega^2\mu_0\varepsilon_0\chi^{(1)}(\omega)\cdot\vec{E}(\omega,\vec{r}) \tag{5.2}$$

The dot stands for the contracted product, ε_0 is the vacuum permittivity, and $\chi^{(1)}(\omega)$ is the dimensionless first-order electric susceptibility defined in Chapter 1.

As described in Chapter 3, $\chi^{(1)}(\omega)$ is a symmetric second-rank polar tensor in a lossless medium, so that any of its coefficients are real. It is then described by a (3 × 3) matrix with 6 independent coefficients among 9, as shown in Figure 3.4. When considering the orthonormal frame coinciding with the dielectric frame and labeled $(\overrightarrow{Ox},\overrightarrow{Oy},\overrightarrow{Oz})$, the simplest form for $\chi^{(1)}(\omega)$ is accessible since it is described by a diagonal matrix depicted in Figure 3.15 and in Eq. (4.1), *i.e.*:

$$\chi^{(1)}(\omega) = \begin{pmatrix} \chi^{(1)}_{xx}(\omega) & 0 & 0 \\ 0 & \chi^{(1)}_{yy}(\omega) & 0 \\ 0 & 0 & \chi^{(1)}_{zz}(\omega) \end{pmatrix} \tag{5.3}$$

DOI: 10.1201/9780429091735-5

By inserting Eq. (5.3) in Eq. (5.2), it comes:

$$\overrightarrow{rot}\left[\overrightarrow{rot}\left[\vec{E}(\omega,\vec{r})\right]\right]-\frac{\omega^2}{c^2}\left[Id+\chi^{(1)}(\omega)\right]\cdot\vec{E}(\omega,\vec{r})=\vec{0} \tag{5.4}$$

where Id stands for the (3 × 3) identity matrix:

$$Id=\begin{pmatrix}1&0&0\\0&1&0\\0&0&1\end{pmatrix} \tag{5.5}$$

As shown in Chapter 4, the thirty-two crystal classes are spread out in the three so-called optical classes: isotropic, uniaxial, and biaxial.

- For the isotropic optical class, $\chi_{xx}^{(1)}(\omega)=\chi_{yy}^{(1)}(\omega)=\chi_{zz}^{(1)}(\omega)$, as shown in Figure 3.18. The three coefficients will be written $\chi(\omega)$ in the following.
- For the uniaxial optical class, $\chi_{xx}^{(1)}(\omega)=\chi_{yy}^{(1)}(\omega)\neq\chi_{zz}^{(1)}(\omega)$, as shown in Figure 3.17. The following writing convention is used: $\chi_{xx}^{(1)}(\omega)=\chi_{yy}^{(1)}(\omega)\equiv\chi_{o}^{(1)}(\omega)$ and $\chi_{zz}^{(1)}(\omega)\equiv\chi_{e}^{(1)}(\omega)$, where the subscripts o and e stand for "ordinary" and "extraordinary", respectively. There exists two kinds of uniaxial crystals in nature according to the two possible relations of orders: $\chi_{o}^{(1)}(\omega)<\chi_{e}^{(1)}(\omega)$, which defines a positive uniaxial crystal, and $\chi_{e}^{(1)}(\omega)<\chi_{o}^{(1)}(\omega)$ for a negative uniaxial crystal.
- For the biaxial optical class, the three coefficients are completely independent, *i.e.*: $\chi_{xx}^{(1)}(\omega)\neq\chi_{yy}^{(1)}(\omega)\neq\chi_{zz}^{(1)}(\omega)$, as shown in Figure 3.15. As in the case of uniaxial crystals, there are two categories of biaxial crystals that are characterized by the two following relations defining the optical sign of a biaxial crystal: $\chi_{xx}^{(1)}(\omega)<\chi_{yy}^{(1)}(\omega)<\chi_{zz}^{(1)}(\omega)$ and $\chi_{zz}^{(1)}(\omega)<\chi_{yy}^{(1)}(\omega)<\chi_{xx}^{(1)}(\omega)$. By continuity of the uniaxial class, the biaxial class resulting from a lifting of the degeneracy of the uniaxial class, *i.e.* $\chi_{xx}^{(1)}=\chi_{yy}^{(1)}\rightarrow\chi_{xx}^{(1)}\neq\chi_{yy}^{(1)}$, crystals described by the first relation of order are named positive biaxial crystals while they are negative biaxial crystals when considering the second relation.

In Eq. (5.4) appears the relative dielectric permittivity:

$$\varepsilon_r(\omega)=Id+\chi^{(1)}(\omega) \tag{5.6}$$

Then Eq. (5.4) can be rewritten as:

$$\overrightarrow{rot}\left[\overrightarrow{rot}\left[\vec{E}(\omega,\vec{r})\right]\right]-\frac{\omega^2}{c^2}\varepsilon_r(\omega)\cdot\vec{E}(\omega,\vec{r})=\vec{0} \tag{5.7}$$

According to Eq. (5.3) and Eq. (5.5), $\varepsilon_r(\omega)$ is also a polar rank-2 tensor depicted by a (3 × 3) matrix that is diagonal in the dielectric frame:

$$\varepsilon_r(\omega)=\begin{pmatrix}\varepsilon_{r_{xx}}(\omega) & 0 & 0\\ 0 & \varepsilon_{r_{yy}}(\omega) & 0\\ 0 & 0 & \varepsilon_{r_{zz}}(\omega)\end{pmatrix} \tag{5.8}$$

The coefficients $\varepsilon_{r_{ii}}(\omega)$, with $i=x,y,z$, are the three principal values of the real part of the relative dielectric permittivity. Regarding the optical class, it comes:

- for the isotropic optical class, $\varepsilon_{r_{xx}}(\omega)=\varepsilon_{r_{yy}}(\omega)=\varepsilon_{r_{zz}}(\omega)\equiv\varepsilon_r(\omega)$;
- for the uniaxial optical class, $\varepsilon_{r_{xx}}(\omega)=\varepsilon_{r_{yy}}(\omega)\equiv\varepsilon_{r_o}(\omega)<\varepsilon_{r_{zz}}(\omega)\equiv\varepsilon_{r_e}(\omega)$ or $\varepsilon_{r_{xx}}(\omega)=\varepsilon_{r_{yy}}(\omega)\equiv\varepsilon_{r_o}(\omega)>\varepsilon_{r_{zz}}(\omega)\equiv\varepsilon_{r_e}(\omega)$;
- for the biaxial optical class, $\varepsilon_{r_{xx}}(\omega)<\varepsilon_{r_{yy}}(\omega)<\varepsilon_{r_{zz}}(\omega)$ or $\varepsilon_{r_{xx}}(\omega)>\varepsilon_{r_{yy}}(\omega)>\varepsilon_{r_{zz}}(\omega)$.

The plane wave is a possible solution of Eq. (5.7) and it is largely used since it enables a very good description of light propagation. In that case, the electric field $\vec{E}(\omega,\vec{r})$ at the position $\vec{r}$ is given by:

$$\vec{E}(\omega,\vec{r})=\vec{e}(\omega)E_0(\omega)e^{i\vec{k}(\omega)\cdot\vec{r}} \tag{5.9}$$

The unit electric field vector $\vec{e}(\omega)$, the wave vector $\vec{k}(\omega)$, and the position $\vec{r}$ have the following Cartesian coordinates in the dielectric frame $(\overrightarrow{Ox},\overrightarrow{Oy},\overrightarrow{Oz})$:

$$\vec{e}(\omega)=\begin{pmatrix}e_x\\ e_y\\ e_z\end{pmatrix}\ \vec{k}(\omega)=\begin{pmatrix}k_x\\ k_y\\ k_z\end{pmatrix}\ \vec{r}=\begin{pmatrix}x\\ y\\ z\end{pmatrix} \tag{5.10}$$

Using Eq. (5.9), the light propagation equation Eq. (5.7) writes:

$$\overrightarrow{rot}\left[\overrightarrow{rot}\left[E_0(\omega)e^{i\vec{k}(\omega)\cdot\vec{r}}\vec{e}(\omega)\right]\right]-\frac{\omega^2}{c^2}E_0(\omega)e^{i\vec{k}(\omega)\cdot\vec{r}}\varepsilon_r(\omega)\cdot\vec{e}(\omega)=\vec{0} \tag{5.11}$$

The dot in Eq. (5.11) stands for the contracted product.

By dividing both previous terms of Eq. (5.11) by the electric field amplitude $E_0(\omega)$, it comes:

$$\overrightarrow{rot}\left[\overrightarrow{rot}\left[e^{i\vec{k}(\omega)\cdot\vec{r}}\vec{e}(\omega)\right]\right]-\frac{\omega^2}{c^2}e^{i\vec{k}(\omega)\cdot\vec{r}}\varepsilon_r(\omega)\cdot\vec{e}(\omega)=\vec{0} \tag{5.12}$$

According to Eq. (1.84), $\overrightarrow{rot}\left[e^{i\vec{k}(\omega)\cdot\vec{r}}\vec{e}(\omega,\vec{u})\right]$ develops as:

$$\overrightarrow{rot}\left[e^{i\vec{k}(\omega,\vec{u})\cdot\vec{r}}\vec{e}(\omega)\right]=\vec{\nabla}\wedge\left[e^{i\vec{k}(\omega,\vec{u})\cdot\vec{r}}\vec{e}(\omega)\right] \tag{5.13}$$

where $\wedge$ stands for the vector product and $\vec{\nabla}$ for the vector operator Nabla. Then, from Eq. (5.10) and by using the Cartesian coordinates, Eq. (5.13) writes:

$$\overrightarrow{rot}\left[e^{i\vec{k}(\omega)\cdot\vec{r}}\vec{e}(\omega)\right]=\begin{pmatrix}\dfrac{\partial}{\partial x}\\ \dfrac{\partial}{\partial y}\\ \dfrac{\partial}{\partial z}\end{pmatrix}\wedge\begin{pmatrix}e^{i(k_x x+k_y y+k_z z)}e_x\\ e^{i(k_x x+k_y y+k_z z)}e_y\\ e^{i(k_x x+k_y y+k_z z)}e_z\end{pmatrix} \tag{5.14}$$

where $\dfrac{\partial}{\partial m}$, with $m = x, y, z$, are partial derivatives.

Equation (5.14) develops as:

$$\overrightarrow{rot}\left[e^{i\vec{k}(\omega)\cdot\vec{r}}\vec{e}(\omega)\right]=\begin{pmatrix}\dfrac{\partial e^{i(k_x x+k_y y+k_z z)}}{\partial y}e_z-\dfrac{\partial e^{i(k_x x+k_y y+k_z z)}}{\partial z}e_y\\ \dfrac{\partial e^{i(k_x x+k_y y+k_z z)}}{\partial z}e_x-\dfrac{\partial e^{i(k_x x+k_y y+k_z z)}}{\partial x}e_z\\ \dfrac{\partial e^{i(k_x x+k_y y+k_z z)}}{\partial x}e_y-\dfrac{\partial e^{i(k_x x+k_y y+k_z z)}}{\partial y}e_x\end{pmatrix}$$
$$=ie^{i(k_x x+k_y y+k_z z)}\begin{pmatrix}k_y e_z-k_z e_y\\ k_z e_x-k_x e_z\\ k_x e_y-k_y e_x\end{pmatrix} \tag{5.15}$$

Equation (5.15) is thus equivalent to:

$$\overrightarrow{rot}\left[e^{i\vec{k}(\omega)\cdot\vec{r}}\vec{e}(\omega)\right]=ie^{i\vec{k}(\omega)\cdot\vec{r}}\left[\vec{k}(\omega)\wedge\vec{e}(\omega)\right] \tag{5.16}$$

The application of $\overrightarrow{rot}$ on Eq. (5.16) gives:

$$\overrightarrow{rot}\left[\overrightarrow{rot}\left[e^{i\vec{k}(\omega)\cdot\vec{r}}\vec{e}(\omega)\right]\right]=i\vec{\nabla}\wedge\left(e^{i\vec{k}(\omega)\cdot\vec{r}}\left[\vec{k}(\omega)\wedge\vec{e}(\omega)\right]\right) \tag{5.17}$$

By developing Eq. (5.17), it comes:

$$\overrightarrow{rot}\left[\overrightarrow{rot}\left[e^{i\vec{k}(\omega)\cdot\vec{r}}\vec{e}(\omega)\right]\right]=i\begin{pmatrix}\dfrac{\partial}{\partial x}\\ \dfrac{\partial}{\partial y}\\ \dfrac{\partial}{\partial z}\end{pmatrix}\wedge\begin{pmatrix}e^{i(k_x x+k_y y+k_z z)}(k_y e_z-k_z e_y)\\ e^{i(k_x x+k_y y+k_z z)}(k_z e_x-k_x e_z)\\ e^{i(k_x x+k_y y+k_z z)}(k_x e_y-k_y e_x)\end{pmatrix}$$

$$=$$

$$i\begin{pmatrix}\dfrac{\partial e^{i(k_x x+k_y y+k_z z)}}{\partial y}(k_x e_y-k_y e_x)-\dfrac{\partial e^{i(k_x x+k_y y+k_z z)}}{\partial z}(k_z e_x-k_x e_z)\\ \dfrac{\partial e^{i(k_x x+k_y y+k_z z)}}{\partial z}(k_y e_z-k_z e_y)-\dfrac{\partial e^{i(k_x x+k_y y+k_z z)}}{\partial x}(k_x e_y-k_y e_x)\\ \dfrac{\partial e^{i(k_x x+k_y y+k_z z)}}{\partial x}(k_z e_x-k_x e_z)-\dfrac{\partial e^{i(k_x x+k_y y+k_z z)}}{\partial y}(k_y e_z-k_z e_y)\end{pmatrix} \tag{5.18}$$

$$=i^2 e^{i(k_x x+k_y y+k_z z)}\begin{pmatrix}k_y(k_x e_y-k_y e_x)-k_z(k_z e_x-k_x e_z)\\ k_z(k_y e_z-k_z e_y)-k_x(k_x e_y-k_y e_x)\\ k_x(k_z e_x-k_x e_z)-k_y(k_y e_z-k_z e_y)\end{pmatrix}$$

According to Eq. (5.16), and knowing that $i^2=-1$, then Eq. (5.18) writes:

$$\begin{aligned}\overrightarrow{rot}\left[\overrightarrow{rot}\left[e^{i\vec{k}(\omega)\cdot\vec{r}}\vec{e}(\omega)\right]\right]&=\vec{\nabla}\wedge\left[\vec{\nabla}\wedge\left[e^{i\vec{k}(\omega)\cdot\vec{r}}\vec{e}(\omega)\right]\right]\\ &=-e^{i\vec{k}(\omega)\cdot\vec{r}}\left\{\vec{k}(\omega)\wedge\left[\vec{k}(\omega)\wedge\vec{e}(\omega)\right]\right\}\end{aligned} \tag{5.19}$$

Using Eq. (5.19), Eq. (5.12) becomes:

$$-e^{i\vec{k}(\omega)\cdot\vec{r}}\left\{\vec{k}(\omega)\wedge\left[\vec{k}(\omega)\wedge\vec{e}(\omega)\right]\right\}-\frac{\omega^2}{c^2}e^{i\vec{k}(\omega)\cdot\vec{r}}\varepsilon_r(\omega)\cdot\vec{e}(\omega)=\vec{0} \tag{5.20}$$

Finally, after dividing both terms of Eq. (5.20) by $e^{i\vec{k}(\omega)\cdot\vec{r}}$, it comes:

$$-\left\{\vec{k}(\omega)\wedge\left[\vec{k}(\omega)\wedge\vec{e}(\omega)\right]\right\}-\frac{\omega^2}{c^2}\varepsilon_r(\omega)\cdot\vec{e}(\omega)=\vec{0} \tag{5.21}$$

It is interesting now to introduce in the equations the fact that the modulus of the wave vector, *i.e.* $\left\|\vec{k}(\omega)\right\|$, may vary as a function of the direction of propagation. If $\vec{u}$ is the unit vector of the direction of propagation, then the wave vector can be written as:

$$\vec{k}(\omega,\vec{u})=\left\|\vec{k}(\omega,\vec{u})\right\|\vec{u} \tag{5.22}$$

The Cartesian coordinates of $\vec{u}$ are also expressed in the dielectric frame, *i.e.*:

$$\vec{u} = \begin{pmatrix} u_x \\ u_y \\ u_z \end{pmatrix} \tag{5.23}$$

with

$$\left\|\vec{u}\right\| = \sqrt{u_x^2 + u_y^2 + u_z^2} = 1 \tag{5.24}$$

The modulus of the wave vector can also be expressed as a function of the phase velocity of light $\mathrm{v}(\omega,\vec{u})$ by the following relation:

$$\left\|\vec{k}(\omega,\vec{u})\right\| = \frac{\omega}{\mathrm{v}(\omega,\vec{u})} \tag{5.25}$$

The phase velocity is governed by the refractive index $n(\omega,\vec{u})$ that is the prominent physical quantity of linear optics. It is defined as:

$$\mathrm{v}(\omega,\vec{u}) = \frac{c}{n(\omega,\vec{u})} \tag{5.26}$$

Note that $n(\omega,\vec{u})$ is a dimensionless scalar quantity, namely a real quantity in the present case.

When inserting Eq. (5.22), Eq. (5.23), Eq. (5.25), and Eq. (5.26) in Eq. (5.21), and after simplification by $\frac{\omega^2}{c^2}$, it comes:

$$-n^2(\omega,\vec{u})\left\{\vec{u} \wedge \left[\vec{u} \wedge \vec{e}(\omega,\vec{u})\right]\right\} - \varepsilon_r^{(1)}(\omega)\cdot\vec{e}(\omega,\vec{u}) = \vec{0} \tag{5.27}$$

Using the vector relation $\vec{A} \wedge \left[\vec{B} \wedge \vec{C}\right] = \left(\vec{A}\cdot\vec{C}\right)\vec{B} - \left(\vec{A}\cdot\vec{B}\right)\vec{C}$, Eq. (5.27) becomes:

$$-n^2(\omega,\vec{u})\left\{\left(\vec{u}\cdot\vec{e}(\omega,\vec{u})\right)\vec{u} - (\vec{u}\cdot\vec{u})\vec{e}(\omega,\vec{u})\right\} - \varepsilon_r^{(1)}(\omega)\cdot\vec{e}(\omega,\vec{u}) = \vec{0} \tag{5.28}$$

Since $\vec{u}.\vec{u} = 1$ according to Eq. (5.22), then Eq. (5.28) reduces to:

$$n^2(\omega,\vec{u})\left[\vec{e}(\omega,\vec{u}) - \left(\vec{u}\cdot\vec{e}(\omega,\vec{u})\right)\vec{u}\right] - \varepsilon_r^{(1)}(\omega)\cdot\vec{e}(\omega,\vec{u}) = \vec{0} \tag{5.29}$$

When expressing $\varepsilon_r^{(1)}(\omega)$, $\vec{e}(\omega,\vec{u})$, and $\vec{u}$ by their Cartesian coordinates using Eq. (5.8), Eq. (5.10), and Eq. (5.23), Eq. (5.29) is thus equivalent to the following system of three scalar equations:

$$\begin{cases} \left[n^2(\omega,\vec{u})(1-u_x^2) - \varepsilon_{r_{xx}}(\omega)\right]e_x - n^2(\omega,\vec{u})u_xu_ye_y - n^2(\omega,\vec{u})u_xu_ze_z = 0 \\ -n^2(\omega,\vec{u})u_yu_xe_x + \left[n^2(\omega,\vec{u})(1-u_y^2) - \varepsilon_{r_{yy}}(\omega)\right]e_y - n^2(\omega,\vec{u})u_yu_ze_z = 0 \\ -n^2(\omega,\vec{u})u_zu_xe_x - n^2(\omega,\vec{u})u_zu_ye_y + \left[n^2(\omega,\vec{u})(1-u_z^2) - \varepsilon_{r_{zz}}(\omega)\right]e_z = 0 \end{cases} \tag{5.30}$$

Equation (5.30) is the linear propagation equation expressed as a function of the refractive index and projected onto the three principal axes of the dielectric frame. This equation is of prime importance because it will lead to the distributions of the refractive index, of the phase velocity, and of the unit electric field vector as a function of the direction of propagation, which will be done hereafter in Part 5.2 and Part 5.3.

5.2 INDEX SURFACE AND PHASE VELOCITY SURFACE

5.2.1 Index Surface

5.2.1.1 General Equation

Equation (5.30) is relative to a given direction of propagation $\vec{u}$. It corresponds to a system of three equations with three unknowns (e_x, e_y, e_z) that are the three Cartesian coordinates of the unit electric field vector. This system can be written in the form of the following matrix equation:

$$M\begin{pmatrix} e_x \\ e_y \\ e_z \end{pmatrix} = 0 \tag{5.31}$$

where M is a (3×3) matrix expressed as:

$$M = \begin{pmatrix} n^2(\omega,\vec{u})(1-u_x^2)-\varepsilon_{r_{xx}}(\omega) & -n^2(\omega,\vec{u})u_xu_y & -n^2(\omega,\vec{u})u_xu_z \\ -n^2(\omega,\vec{u})u_xu_y & n^2(\omega,\vec{u})(1-u_y^2)-\varepsilon_{r_{yy}}(\omega) & -n^2(\omega,\vec{u})u_yu_z \\ -n^2(\omega,\vec{u})u_xu_z & -n^2(\omega,\vec{u})u_yu_z & n^2(\omega,\vec{u})(1-u_z^2)-\varepsilon_{r_{zz}}(\omega) \end{pmatrix} \tag{5.32}$$

Equation (5.31) will have non-zero solutions if the determinant of M is equal to zero, *i.e.*:

$$\begin{vmatrix} n^2(\omega,\vec{u})(1-u_x^2)-\varepsilon_{r_{xx}}(\omega) & -n^2(\omega,\vec{u})u_xu_y & -n^2(\omega,\vec{u})u_xu_z \\ -n^2(\omega,\vec{u})u_yu_x & n^2(\omega,\vec{u})(1-u_y^2)-\varepsilon_{r_{yy}}(\omega) & -n^2(\omega,\vec{u})u_yu_z \\ -n^2(\omega,\vec{u})u_zu_x & -n^2(\omega,\vec{u})u_zu_y & n^2(\omega,\vec{u})(1-u_z^2)-\varepsilon_{r_{zz}}(\omega) \end{vmatrix} = 0 \tag{5.33}$$

Equation (5.33) is made easier by writing it as a combination of three (2×2) determinants as follows:

$$\left[n^2(\omega,\vec{u})(1-u_x^2)-\varepsilon_{r_{xx}}(\omega)\right]\times$$

$$\begin{vmatrix} n^2(\omega,\vec{u})(1-u_y^2)-\varepsilon_{r_{yy}}(\omega) & -n^2(\omega,\vec{u})u_yu_z \\ -n^2(\omega,\vec{u})u_zu_y & n^2(\omega,\vec{u})(1-u_z^2)-\varepsilon_{r_{zz}}(\omega) \end{vmatrix}$$

$$
\begin{aligned}
&+n^2(\omega,\vec{u})u_x u_y \times \begin{vmatrix} -n^2(\omega,\vec{u})u_y u_x & -n^2(\omega,\vec{u})u_y u_z \\ -n^2(\omega,\vec{u})u_z u_x & n^2(\omega,\vec{u})(1-u_z^2)-\varepsilon_{r_{zz}}(\omega) \end{vmatrix} \\
&-n^2(\omega,\vec{u})u_x u_z \times \begin{vmatrix} -n^2(\omega,\vec{u})u_y u_x & n^2(\omega,\vec{u})(1-u_y^2)-\varepsilon_{r_{yy}}(\omega) \\ -n^2(\omega,\vec{u})u_z u_x & -n^2(\omega,\vec{u})u_z u_y \end{vmatrix} \\
&=0
\end{aligned}
\tag{5.34}
$$

The calculation of Eq. (5.34) directly leads to:

$$
\begin{aligned}
&\left[n^2(\omega,\vec{u})(1-u_x^2)-\varepsilon_{r_{xx}}(\omega)\right] \\
&\left\{\left[n^2(\omega,\vec{u})(1-u_y^2)-\varepsilon_{r_{yy}}(\omega)\right]\left[n^2(\omega,\vec{u})(1-u_z^2)-\varepsilon_{r_{zz}}(\omega)\right]-n^4(\omega,\vec{u})u_y^2u_z^2\right\} \\
&+n^2(\omega,\vec{u})u_x u_y\left\{-n^2(\omega,\vec{u})u_x u_y\left[n^2(\omega,\vec{u})(1-u_z^2)-\varepsilon_{r_{zz}}(\omega)\right]-n^4(\omega,\vec{u})u_x u_y u_z^2\right\} \\
&-n^2(\omega,\vec{u})u_x u_z\left\{n^4(\omega,\vec{u})u_x u_y^2 u_z+n^2(\omega,\vec{u})u_x u_z\left[n^2(\omega,\vec{u})(1-u_y^2)-\varepsilon_{r_{yy}}(\omega)\right]\right\} \\
&=0
\end{aligned}
\tag{5.35}
$$

Finally, after developing Eq. (5.35), it comes:

$$
\begin{aligned}
&\left[n^2(\omega,\vec{u})-\varepsilon_{r_{xx}}(\omega)\right]\left[n^2(\omega,\vec{u})-\varepsilon_{r_{yy}}(\omega)\right]\left[n^2(\omega,\vec{u})-\varepsilon_{r_{zz}}(\omega)\right] \\
&-n^2(\omega,\vec{u})u_x^2\left[n^2(\omega,\vec{u})-\varepsilon_{r_{yy}}(\omega)\right]\left[n^2(\omega,\vec{u})-\varepsilon_{r_{zz}}(\omega)\right] \\
&-n^2(\omega,\vec{u})u_y^2\left[n^2(\omega,\vec{u})-\varepsilon_{r_{xx}}(\omega)\right]\left[n^2(\omega,\vec{u})-\varepsilon_{r_{zz}}(\omega)\right] \\
&-n^2(\omega,\vec{u})u_z^2\left[n^2(\omega,\vec{u})-\varepsilon_{r_{xx}}(\omega)\right]\left[n^2(\omega,\vec{u})-\varepsilon_{r_{yy}}(\omega)\right]=0
\end{aligned}
\tag{5.36}
$$

It is possible at this step to easily solve Eq. (5.36) for a propagation in a principal axis of the dielectric frame, *i.e.* $\overrightarrow{Ox}$, $\overrightarrow{Oy}$, or $\overrightarrow{Oz}$, which leads to two possible values in each case.

- Along $\overrightarrow{Ox}$ the coordinates of the unit wave vector $\vec{u}$ are $(1,0,0)$ so that Eq. (5.36) reduces to:

$$
-\varepsilon_{r_{xx}}(\omega)\left[n^2(\omega,\overrightarrow{Ox})-\varepsilon_{r_{yy}}(\omega)\right]\left[n^2(\omega,\overrightarrow{Ox})-\varepsilon_{r_{zz}}(\omega)\right]=0
\tag{5.37}
$$

Since $\varepsilon_{r_{xx}}(\omega)\neq 0$, the two solutions of Eq. (5.37) are $n(\omega,\overrightarrow{Ox})=\sqrt{\varepsilon_{r_{yy}}(\omega)}$ and $n(\omega,\overrightarrow{Ox})=\sqrt{\varepsilon_{r_{zz}}(\omega)}$.

- Along $\overrightarrow{Oy}$ the coordinates of the unit wave vector $\vec{u}$ are $(0,1,0)$ so that Eq. (5.36) reduces to:

$$-\varepsilon_{r_{yy}}(\omega)\left[n^2(\omega,\overrightarrow{Oy})-\varepsilon_{r_{xx}}(\omega)\right]\left[n^2(\omega,\overrightarrow{Oy})-\varepsilon_{r_{zz}}(\omega)\right]=0 \tag{5.38}$$

Since $\varepsilon^{(1)}_{r_{yy}}(\omega)\neq 0$, the two solutions of Eq. (5.38) are $n(\omega,\overrightarrow{Oy})=\sqrt{\varepsilon_{r_{xx}}(\omega)}$ and $n(\omega,\overrightarrow{Oy})=\sqrt{\varepsilon_{r_{zz}}(\omega)}$.

- Along $\overrightarrow{Oz}$ the coordinates of the unit wave vector $\vec{u}$ are $(0,0,1)$ so that Eq. (5.36) reduces to:

$$-\varepsilon_{r_{zz}}(\omega)\left[n^2(\omega,\overrightarrow{Oz})-\varepsilon_{r_{xx}}(\omega)\right]\left[n^2(\omega,\overrightarrow{Oz})-\varepsilon_{r_{yy}}(\omega)\right]=0 \tag{5.39}$$

Since $\varepsilon^{(1)}_{r_{zz}}(\omega)\neq 0$, the two solutions of Eq. (5.39) are $n(\omega,\overrightarrow{Oz})=\sqrt{\varepsilon_{r_{xx}}(\omega)}$ and $n(\omega,\overrightarrow{Oz})=\sqrt{\varepsilon_{r_{yy}}(\omega)}$.

The three quantities $\sqrt{\varepsilon_{r_{xx}}(\omega)}$, $\sqrt{\varepsilon_{r_{yy}}(\omega)}$, and $\sqrt{\varepsilon_{r_{zz}}(\omega)}$ are called the principal refractive indices, and are written with the following convention:

$$n_x(\omega)=\sqrt{\varepsilon_{r_{xx}}(\omega)} \qquad n_y(\omega)=\sqrt{\varepsilon_{r_{yy}}(\omega)} \qquad n_z(\omega)=\sqrt{\varepsilon_{r_{zz}}(\omega)} \tag{5.40}$$

Then it comes regarding the optical class:

- $n_x(\omega)=n_y(\omega)=n_z(\omega)\equiv n(\omega)$ for the isotropic optical class;
- $n_x(\omega)=n_y(\omega)\equiv n_o(\omega)<n_z(\omega)\equiv n_e(\omega)$ or $n_x(\omega)=n_y(\omega)\equiv n_o(\omega)>n_z(\omega)\equiv n_e(\omega)$ for the uniaxial optical class where $n_o(\omega)$ and $n_e(\omega)$ are the ordinary and extraordinary principal refractive indices, respectively;
- $n_x(\omega)<n_y(\omega)<n_z(\omega)$ or $n_x(\omega)>n_y(\omega)>n_z(\omega)$ for the biaxial optical class.

Equation (5.36) also allows us to calculate the refractive index n as a function of any other direction of propagation $\vec{u}$. The goal now is to put this equation into a form that can be directly exploited to find easily the solutions n. That can be done as follows: firstly, multiply the three-term product appearing in the first line of Eq. (5.36) by $(u_x^2+u_x^2+u_x^2)$, which is equal to 1 since $\vec{u}$ is a unit vector; secondly factorize the whole; and thirdly use the writing convention defined by Eq. (5.40). Finally, it comes:

$$\begin{aligned}&u_x^2n_x^2(\omega)\left[n^2(\omega,\vec{u})-n_y^2(\omega)\right]\left[n^2(\omega,\vec{u})-n_z^2(\omega)\right]\\&+u_y^2n_y^2(\omega)\left[n^2(\omega,\vec{u})-n_x^2(\omega)\right]\left[n^2(\omega,\vec{u})-n_z^2(\omega)\right]\\&+u_z^2n_z^2(\omega)\left[n^2(\omega,\vec{u})-n_x^2(\omega)\right]\left[n^2(\omega,\vec{u})-n_y^2(\omega)\right]=0\end{aligned} \tag{5.41}$$

By performing the products, dividing by $n_x^2(\omega)n_y^2(\omega)n_z^2(\omega)$, and ordering, Eq. (5.41) reduces to the following second-degree equation:

$$AX^2 + BX + C = 0 \tag{5.42}$$

with

$$\begin{cases} X = n^2(\omega,\vec{u}) \\ A = n_y^{-2}(\omega)n_z^{-2}(\omega)u_x^2 + n_x^{-2}(\omega)n_z^{-2}(\omega)u_y^2 + n_x^{-2}(\omega)n_y^{-2}(\omega)u_z^2 \\ B = -u_x^2\left[n_y^{-2}(\omega)+n_z^{-2}(\omega)\right] - u_y^2\left[n_x^{-2}(\omega)+n_z^{-2}(\omega)\right] - u_z^2\left[n_x^{-2}(\omega)+n_y^{-2}(\omega)\right] \\ C = 1 \end{cases} \tag{5.43}$$

The unknown quantity X being a real quantity since a lossless medium is considered, then Eq. (5.42) has solutions if its discriminant Δ is positive or nil, *i.e.*:

$$\Delta = B^2 - 4AC \geq 0 \tag{5.44}$$

Thus the two general solutions are:

$$X^{\pm} = \frac{-B \pm \sqrt{\Delta}}{2A} \tag{5.45}$$

Then it appears here that in any direction of propagation $\vec{u}$, there are two possible values for the refractive index as it had been previously established in the case of propagation along the principal axes of the dielectric frame. These two general solutions are written $n^-(\omega,\vec{u})$ and $n^+(\omega,\vec{u})$ and are expressed as:

$$n^{\pm}(\omega,\vec{u}) = \sqrt{\frac{-B \pm \sqrt{B^2 - 4AC}}{2A}} \tag{5.46}$$

The parameters A , B, and C are those defined by Eq. (5.43).

Note that $n^+(\omega,\vec{u}) > n^-(\omega,\vec{u})$ since $B < 0$.

The geometric representation of $n^{\pm}(\omega,\vec{u})$ is called the index surface. It is of prime importance in the frame work of the present book, with a prominent role in both linear and nonlinear optics.

The difference between the two possible values, *i.e.* $n^+(\omega,\vec{u}) - n^-(\omega,\vec{u})$, is called the birefringence associated with the direction of propagation $\vec{u}$. This quantity is therefore defined as positive.

It may be useful depending on the circumstances to use the angles of spherical coordinates (θ,φ) of the direction of propagation instead of its Cartesian coordinates in Eq. (5.46). The two systems of coordinates are linked by the following relations according to Figure 5.1:

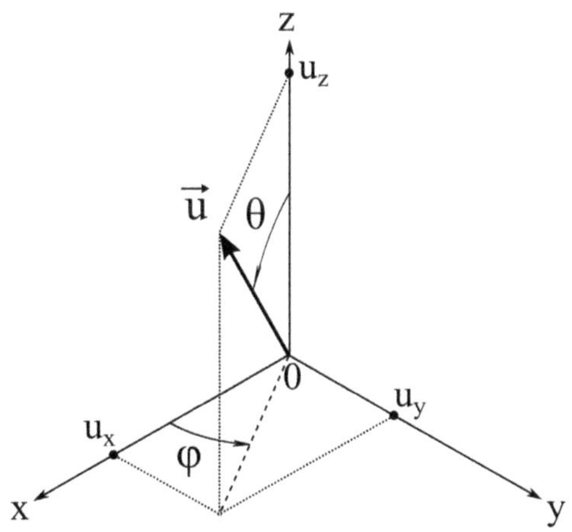

FIGURE 5.1 Cartesian (u_x, u_y, u_z) and angles of spherical coordinates (θ, φ) of the unit wave vector $\vec{u}$ in the dielectric frame $(\overrightarrow{Ox}, \overrightarrow{Oy}, \overrightarrow{Oz})$.

$$\begin{cases} u_x = \cos(\varphi)\sin(\theta) \\ u_y = \sin(\varphi)\sin(\theta) \\ u_z = \cos(\theta) \end{cases} \tag{5.47}$$

The index surfaces corresponding to the three optical classes are detailed in the three next parts.

A remark about the refractive index solutions depicted in Eq. (5.46) is welcome here. Indeed, the derived method to express Eq. (5.46), based on the resolution of the determinant nullity of the linear equation of light propagation, is fully general and suffers no approximation, which makes it relevant for any of the three optical classes and any direction of propagation. Historically, there exist other forms of angular-dependent refractive index solutions, labeled the Fresnel solutions of the so-called Fresnel equation that is not detailed here but can be seen in reference (Nye 1985) [1]. Indeed, because light propagation is deterministic and its refractive index solutions require unicity whatever the chosen approach, both Fresnel solutions and those developed here in Eq. (5.46) describe exactly the same behavior. Still, although historical, the Fresnel approach relies on assumption, making it not fully complete since it is not valid along the dielectric axes and does not extend to the isotropic class.

5.2.1.2 Isotropic Optical Class

In that case, the three principal refractive indices are equal, *i.e.* $n_x(\omega) = n_y(\omega) = n_z(\omega) \equiv n(\omega)$, so that it comes from Eq. (5.43) according to Eq. (5.24):

$$\begin{cases} A = n^{-4}(\omega)\left[u_x^2 + u_y^2 + u_z^2\right] = n^{-4}(\omega) \\ B = -2n^4(\omega)\left[u_x^2 + u_y^2 + u_z^2\right] = -2n^{-2}(\omega) \\ C = 1 \end{cases} \tag{5.48}$$

For the isotropic optical class, it results thus that $\Delta = 0$ according to Eq. (5.44) and Eq. (5.48). As a consequence, Eq. (5.46) corresponds to the degenerated double-root of the second-degree Eq. (5.42), *i.e.*:

$$n^{-}(\omega,\vec{u}) = n^{+}(\omega,\vec{u}) = \sqrt{\frac{-B}{2A}} \equiv n(\omega) \tag{5.49}$$

Equation (5.49) puts in emphasis two features of isotropic media: there is no birefringence at any direction of propagation on the one hand, and the refractive index does not depend on the direction of propagation on the other hand. The index surface is then a sphere. It is the case of crystals belonging to the cubic crystal classes as well as of glasses, liquids, or gas. This situation is well described in Figure 5.2.

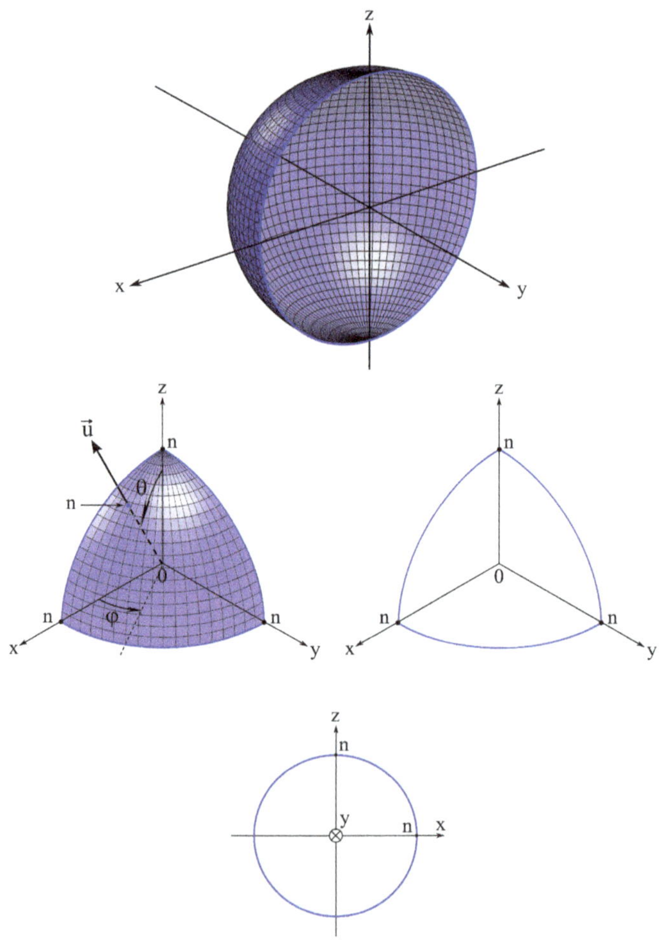

FIGURE 5.2 Index surface of the isotropic optical class at a given circular frequency ω where the indices (x,y,z) refer to the dielectric frame $(\overrightarrow{Ox},\overrightarrow{Oy},\overrightarrow{Oz})$. Top: half-space. Middle: space octant where (θ,φ) are the angles of spherical coordinates of the unit wave vector $\vec{u}$ and n is the single refractive index. Bottom: section of the index surface in the $(\overrightarrow{Ox},\overrightarrow{Oz})$ plane.

Figure 5.2 middle-left shows that the refractive index n is given by the distance from the origin to the sheet of the index surface in the direction of propagation $\vec{u}$ that is considered. It also appears that the index surface in the case of the isotropic optical class belongs to the static sphere group $\frac{\infty}{m}\frac{\infty}{m}$.

5.2.1.3 Uniaxial Optical Class

In that case $n_x(\omega) = n_y(\omega) \equiv n_o(\omega) \neq n_z(\omega) \equiv n_e(\omega)$. Thus Eq. (5.43) combined with $u_x^2 + u_y^2 = 1 - u_z^2$ coming from Eq. (5.24) gives:

$$\begin{cases} A = n_o^{-2}(\omega) n_e^{-2}(\omega)\left[1 - u_z^2\right] + n_o^{-4}(\omega) u_z^2 \\ B = -\left(n_o^{-2}(\omega) + n_e^{-2}(\omega)\right)\left[1 - u_z^2\right] - 2n_o^{-4}(\omega) u_z^2 \\ C = 1 \end{cases} \tag{5.50}$$

Equation (5.50) can be also written as a function of the angles of spherical coordinates according to Eq. (5.47), *i.e.*:

$$\begin{cases} A = n_o^{-2}(\omega) n_e^{-2}(\omega)\sin^2(\theta) + n_o^{-4}(\omega)\cos^2(\theta) \\ B = -\left(n_o^{-2}(\omega) + n_e^{-2}(\omega)\right)\sin^2(\theta) - 2n_o^{-4}(\omega)\cos^2(\theta) \\ C = 1 \end{cases} \tag{5.51}$$

It is interesting to proceed step by step regarding the value of θ, *i.e.*: $\theta = 0$, $\theta = \pi/2$ and $0 < \theta < \pi/2$.

For $\theta = 0$, which corresponds to a propagation along $\overrightarrow{Oz}$, Eq. (5.51) becomes:

$$\begin{cases} A = n_o^{-4}(\omega) \\ B = -2n_o^{-2}(\omega) \\ C = 1 \end{cases} \tag{5.52}$$

Then according to Eq. (5.44) and Eq. (5.52), it results that $\Delta = 0$, and the corresponding double-root of Eq. (5.42) locally appears as degenerated, with:

$$n^-(\omega, \theta = 0) = n^+(\omega, \theta = 0) = \sqrt{\frac{-B}{2A}} \equiv n_o(\omega) \tag{5.53}$$

This situation along $\overrightarrow{Oz}$ of a uniaxial crystal is exactly the same as in any direction of propagation in an isotropic medium since the birefringence is nil. $\overrightarrow{Oz}$ is called the optical axis (OA). There is only one single optical axis in such media, hence the name of "uniaxial" crystal.

For $\theta = \pi / 2$, which corresponds to a propagation at any angle φ in the $(\overrightarrow{Ox}, \overrightarrow{Oy})$ plane, Eq. (5.51) writes:

$$\begin{cases} A = n_o^{-2}(\omega) n_e^{-2}(\omega) \\ B = -\left(n_o^{-2}(\omega) + n_e^{-2}(\omega)\right) \\ C = 1 \end{cases} \tag{5.54}$$

Then according to Eq. (5.44), it comes:

$$\Delta = \left[n_o^{-2}(\omega) - n_e^{-2}(\omega)\right]^2 \tag{5.55}$$

This discriminant is different from 0, so that there are two different values for the refractive index, which are:

$$n^{\pm}(\omega, \theta = \pi / 2) = \sqrt{\frac{\left[n_o^{-2}(\omega) + n_e^{-2}(\omega)\right] \pm \sqrt{\left[n_o^{-2}(\omega) - n_e^{-2}(\omega)\right]^2}}{2 n_o^{-2}(\omega) n_e^{-2}(\omega)}} \tag{5.56}$$

Thus, the final calculation of $n^{\pm}(\omega, \theta = \pi / 2)$ requires knowledge of the sign of $n_e^{-2}(\omega) - n_o^{-2}(\omega)$, that is to say the sign of the uniaxial optical class that is considered.

- In the case of the positive uniaxial class, *i.e.* $n_o(\omega) < n_e(\omega)$, then $\sqrt{n_o^{-2}(\omega) - n_e^{-2}(\omega)} = n_o^{-2}(\omega) - n_e^{-2}(\omega)$ so that Eq. (5.56) gives:

$$\begin{cases} n^{+}(\omega, \theta = \pi / 2) = n_e(\omega) \\ n^{-}(\omega, \theta = \pi / 2) = n_0(\omega) \end{cases} \tag{5.57}$$

- In the case of the negative uniaxial class, *i.e.* $n_o(\omega) > n_e(\omega)$, then $\sqrt{n_o^{-2}(\omega) - n_e^{-2}(\omega)} = -\left|n_o^{-2}(\omega) - n_e^{-2}(\omega)\right|$ so that Eq. (5.56) gives:

$$\begin{cases} n^{+}(\omega, \theta = \pi / 2) = n_o(\omega) \\ n^{-}(\omega, \theta = \pi / 2) = n_e(\omega) \end{cases} \tag{5.58}$$

Equation (5.57) and Eq. (5.58) show that the section of the index surface in the $(\overrightarrow{Ox}, \overrightarrow{Oy})$ plane of a positive or negative uniaxial crystal is described by two circles corresponding to the ordinary and extraordinary principal refractive indices, respectively.

For $0<\theta<\pi/2$, which corresponds to any propagation out of the $\overrightarrow{Ox}$ axis or out of the $(\overrightarrow{Ox},\overrightarrow{Oy})$ plane, the discriminant given by Eq. (5.44) has to be directly calculated using Eq. (5.51), which gives:

$$\Delta=\sin^4(\theta)\left[n_o^{-2}(\omega)-n_e^{-2}(\omega)\right]^2 \tag{5.59}$$

Then Eq. (5.46) is expressed as:

$$n^{\pm}(\omega,\theta)=\sqrt{\frac{\left[\left(n_o^{-2}(\omega)+n_e^{-2}(\omega)\right)\sin^2(\theta)+2n_o^{-4}(\omega)\cos^2(\theta)\right]\pm\sin^2(\theta)\left|n_o^{-2}(\omega)-n_e^{-2}(\omega)\right|}{2n_o^{-2}(\omega)n_e^{-2}(\omega)\sin^2(\theta)+2n_o^{-4}(\omega)\cos^2(\theta)}} \tag{5.60}$$

with $+\sin^2(\theta)\left|n_o^{-2}(\omega)-n_e^{-2}(\omega)\right|$ for a positive uniaxial crystal, *i.e.* $n_o(\omega)<n_e(\omega)$, and $-\sin^2(\theta)\left|n_o^{-2}(\omega)-n_e^{-2}(\omega)\right|$ for a negative uniaxial crystal, *i.e.* $n_o(\omega)>n_e(\omega)$.

- Then in the case of the positive uniaxial class, Eq. (5.60) reduces to:

$$\begin{cases} n^{+}(\omega,\theta)=\dfrac{1}{\sqrt{n_o^{-2}(\omega)\cos^2(\theta)+n_e^{-2}(\omega)\sin^2(\theta)}} \\ n^{-}(\omega,\theta)=n_o(\omega) \end{cases} \tag{5.61}$$

- And in the case of the negative uniaxial class, Eq. (5.60) writes:

$$\begin{cases} n^{+}(\omega,\theta)=n_o(\omega) \\ n^{-}(\omega,\theta)=\dfrac{1}{\sqrt{n_o^{-2}(\omega)\cos^2(\theta)+n_e^{-2}(\omega)\sin^2(\theta)}} \end{cases} \tag{5.62}$$

Note that Eq. (5.61) and Eq. (5.62) are general and can reduce to Eq. (5.53) when $\theta=0$, and respectively to Eq. (5.57) and Eq. (5.58) when $\theta=\pi/2$. Actually, the birefringence $n^{+}(\omega,\theta,\varphi)-n^{-}(\omega,\theta,\varphi)$ varies continuously from $\overrightarrow{Oz}$, where it is nil, to the $(\overrightarrow{Ox},\overrightarrow{Oy})$ plane where it reaches a maximal value, *i.e.* $n_e(\omega)-n_o(\omega)$ for the positive uniaxial class and $n_o(\omega)-n_e(\omega)$ for the negative uniaxial class. It is also customary to define the birefringence as the quantity $n_e(\omega)-n_o(\omega)$, so that it directly informs about the optical sign of the uniaxial crystal by being equal to $n^{+}(\omega,\theta,\varphi)-n^{-}(\omega,\theta,\varphi)>0$ or $n^{-}(\omega,\theta)-n^{+}(\omega,\theta)<0$. But this second and historical convention cannot be used in general for biaxial crystals for which the notion of ordinary and extraordinary refractive indices cannot be applied out of the principal planes of the dielectric frame, as will be explained in Part 5.3.

Figure 5.3 and Figure 5.4 allow us to represent the geometry of the index surface of positive and negative uniaxial crystals, respectively.

From a geometrical point of view, Eq. (5.61) and Eq. (5.63), as well as Figure 5.3 and Figure 5.4, indicate that any section of the index surface containing $\overrightarrow{Oz}$, and so any section perpendicular to the $(\overrightarrow{Ox},\overrightarrow{Oy})$ plane, is described by a circle and an ellipse: each of them corresponds to the ordinary or extraordinary refractive index according to the sign of the uniaxial class. Considering the full space, the index surface is a double-sheet surface: the external surface is an ellipsoid and the internal one is a sphere in the case of a positive uniaxial crystal, while it is the reverse for a negative uniaxial crystal. These two sheets are in contact only along the optical axis, *i.e.* in $\overrightarrow{Oz}$, which defines the two umbilical points of the index surface. Such contacts along the optical axis are tangential for uniaxial crystals. The ellipsoid sheet is often named the extraordinary sheet, written $n_e(\omega,\theta,\varphi)$, while the spherical one is called the ordinary sheet, written $n_o(\omega,\theta,\varphi) = n_o(\omega)$.

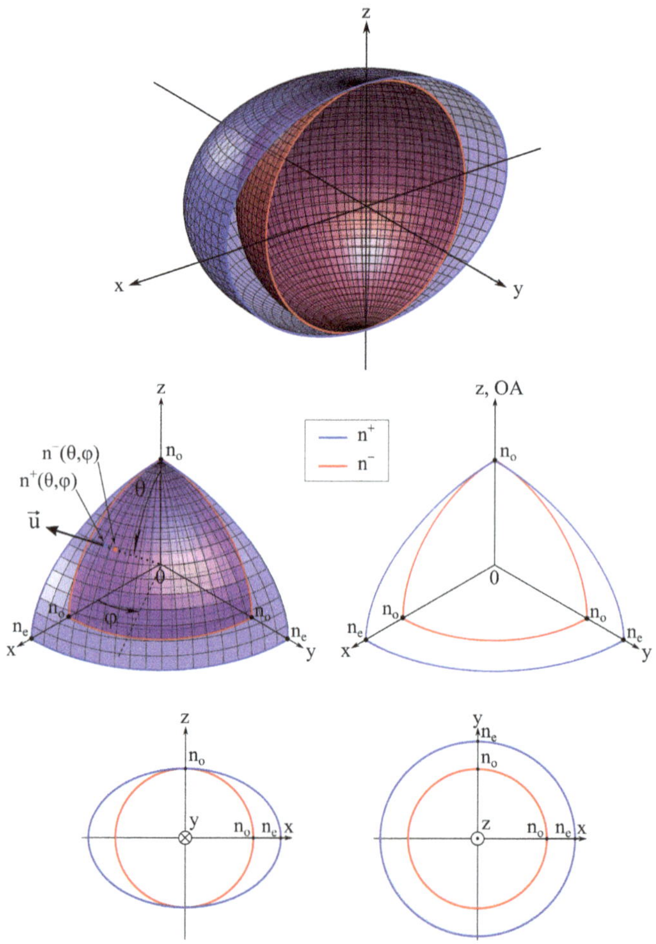

FIGURE 5.3 Index surface of the positive uniaxial optical class at a given circular frequency ω in which the indices (x,y,z) refer to the dielectric frame $(\overrightarrow{Ox},\overrightarrow{Oy},\overrightarrow{Oz})$. Top: half-space. Middle: space octant in which (θ,φ) are the angles of spherical coordinates of the unit wave vector $\vec{u}$ and $n^+(\theta,\varphi)$ and $n^-(\theta,\varphi)$ the two corresponding possible refractive indices; n_o and n_e are the ordinary and extraordinary principal refractive indices, and OA denotes the optical axis. Bottom: section of the index surface in the $(\overrightarrow{Ox},\overrightarrow{Oz})$ plane (left), which can be also any other planes containing $\overrightarrow{Oz}$, and in the $(\overrightarrow{Ox},\overrightarrow{Oy})$ plane (right).

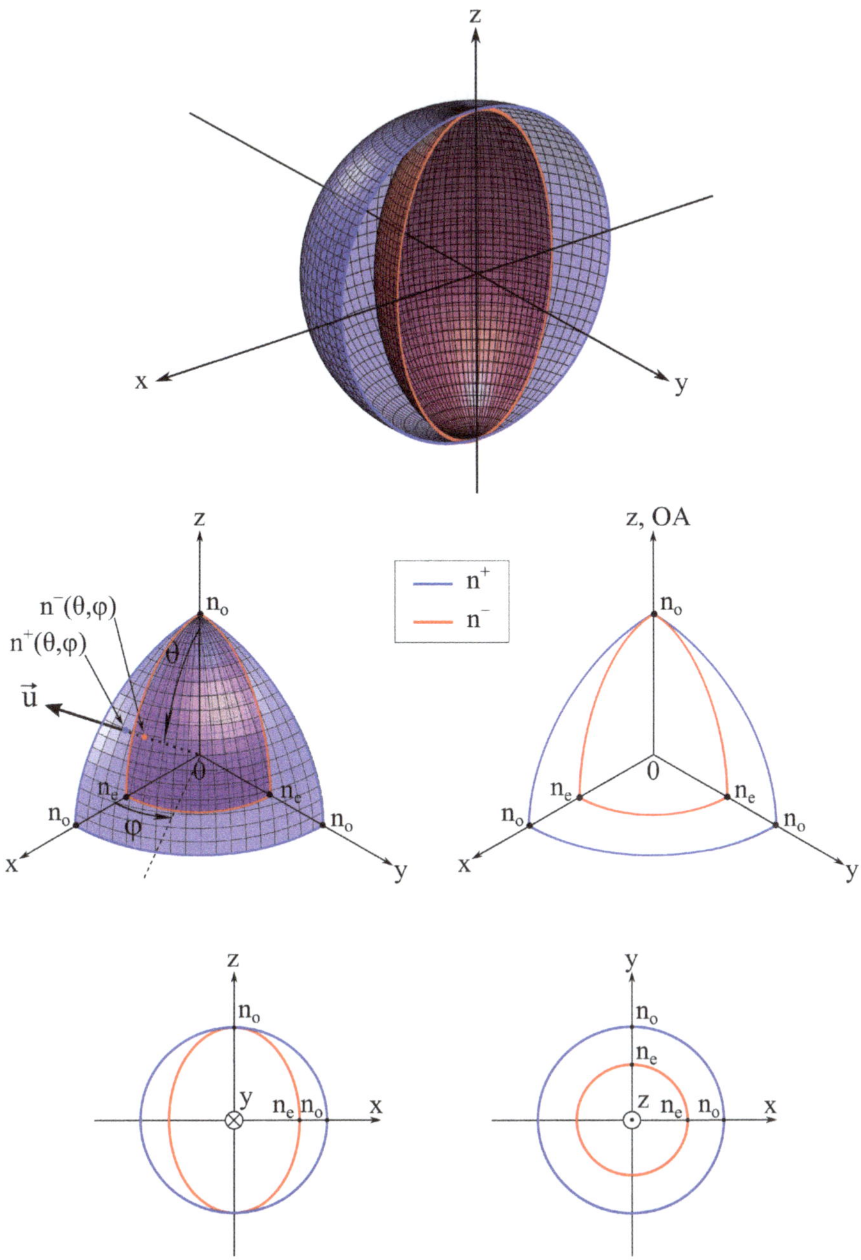

FIGURE 5.4 Index surface of the negative uniaxial optical class at a given circular frequency ω in which the indices (x, y, z) refer to the dielectric frame $(\overrightarrow{Ox},\overrightarrow{Oy},\overrightarrow{Oz})$. Top: half-space. Middle: space octant in which (θ,φ) are the angles of spherical coordinates of the unit wave vector $\vec{u}$ and $n^{+}(\theta,\varphi)$ and $n^{-}(\theta,\varphi)$ the two corresponding possible refractive indices; n_o and n_e are the ordinary and extraordinary principal refractive indices, and OA denotes the optical axis. Bottom: section of the index surface in the $(\overrightarrow{Ox},\overrightarrow{Oz})$ plane (left), which can be also any other planes containing $\overrightarrow{Oz}$, and in the $(\overrightarrow{Ox},\overrightarrow{Oy})$ plane (right).

Figure 5.3 and Figure 5.4 also well show that the index surface have a symmetry of revolution around $\overrightarrow{Oz}$, which also appears in Eq. (5.59) and Eq. (5.60) since they do not depend on the φ angle. As a consequence, the index surface of a uniaxial crystal belongs to the symmetry group equivalent to a static cylinder, *i.e.* $\left(\frac{\infty}{m}m\right)$, as defined in Part 2.1.2, $\overrightarrow{Oz}$ being the rotation axis of infinite order.

5.2.1.4 Biaxial Optical Class

In that case, the three principal refractive indices are completely independent, *i.e.* $n_x(\omega) \neq n_y(\omega) \neq n_z(\omega)$. The use of Eq. (5.46) allows us to calculate the angular distribution of the refractive index.

The analytical expressions of $n^+(\theta,\varphi)$ and $n^-(\theta,\varphi)$ are simple in the principal planes $(\overrightarrow{Ox},\overrightarrow{Oy})$ and $(\overrightarrow{Oy},\overrightarrow{Oz})$. They are described by circles and ellipses, as in the case of the uniaxial optical class.

In the $(\overrightarrow{Ox},\overrightarrow{Oy})$ plane, *i.e.* $\theta = \pi/2$, the two solutions are the following according to the optical sign:

- for $n_x(\omega) < n_y(\omega) < n_z(\omega)$,

$$\begin{cases} n^+(\omega,\varphi) = n_z(\omega) \\ n^-(\omega,\varphi) = \dfrac{1}{\sqrt{n_y^{-2}(\omega)\cos^2(\varphi) + n_x^{-2}(\omega)\sin^2(\varphi)}} \end{cases} \tag{5.63}$$

- for $n_x(\omega) > n_y(\omega) > n_z(\omega)$,

$$\begin{cases} n^+(\omega,\varphi) = \dfrac{1}{\sqrt{n_y^{-2}(\omega)\cos^2(\varphi) + n_x^{-2}(\omega)\sin^2(\varphi)}} \\ n^-(\omega,\varphi) = n_z(\omega) \end{cases} \tag{5.64}$$

In the $(\overrightarrow{Oy},\overrightarrow{Oz})$ plane, *i.e.* $\varphi = \pi/2$, the two solutions are the following according to the optical sign:

- for $n_x(\omega) < n_y(\omega) < n_z(\omega)$,

$$\begin{cases} n^+(\omega,\theta) = \dfrac{1}{\sqrt{n_y^{-2}(\omega)\cos^2(\theta) + n_z^{-2}(\omega)\sin^2(\theta)}} \\ n^-(\omega,\theta) = n_x(\omega) \end{cases} \tag{5.65}$$

- for $n_x(\omega) > n_y(\omega) > n_z(\omega)$,

$$\begin{cases} n^+(\omega,\theta) = n_x(\omega) \\ n^-(\omega,\theta) = \dfrac{1}{\sqrt{n_y^{-2}(\omega)\cos^2(\theta) + n_z^{-2}(\omega)\sin^2(\theta)}} \end{cases} \tag{5.66}$$

The case of the $(\overrightarrow{Ox},\overrightarrow{Oz})$ plane, *i.e.* $\varphi = 0$, is more complicated because there exists an umbilical point at an angle, written V_z, such as $0 < V_z < \pi/2$. It is important to notice that it is different compared to the uniaxial optical class for which the umbilical point is necessarily along $\overrightarrow{Oz}$, *i.e.* $\theta = 0$, according to Eq. (5.53). Then for a biaxial crystal, the $(\overrightarrow{Ox},\overrightarrow{Oz})$ plane is divided into two parts, $0 \leq \theta \leq V_z$ and $V_z \leq \theta \leq \pi/2$, with a singularity, *i.e.* an umbilical point, at $\theta = V_z$. In these two portions of space, $n^+(\theta,\varphi)$ and $n^-(\theta,\varphi)$ are each described by a portion of circle and a portion of ellipse, so that V_z can be easily calculated by considering the intersection of a continuous circle by a continuous ellipse over $\theta = 0$ and $\theta = \pi/2$, namely $n^+(\omega,\theta = V_z) = n^-(\omega,\theta = V_z)$, *i.e.*:

$$n_y(\omega) = \frac{1}{\sqrt{n_x^{-2}(\omega)\cos^2[V_z(\omega)] + n_z^{-2}(\omega)\sin^2[V_z(\omega)]}} \tag{5.67}$$

which gives:

$$V_z(\omega) = \arcsin\sqrt{\frac{n_y^{-2}(\omega) - n_x^{-2}(\omega)}{n_z^{-2}(\omega) - n_x^{-2}(\omega)}} \tag{5.68}$$

Then the two solutions for the refractive index the $(\overrightarrow{Ox},\overrightarrow{Oz})$ plane are:

- for $n_x(\omega) < n_y(\omega) < n_z(\omega)$,

$0 \leq \theta \leq V_z$

$$\begin{cases} n^+(\omega,\theta) = n_y(\omega) \\ n^-(\omega,\theta) = \dfrac{1}{\sqrt{n_x^{-2}(\omega)\cos^2(\theta) + n_z^{-2}(\omega)\sin^2(\theta)}} \end{cases} \tag{5.69}$$

$V_z \leq \theta \leq \pi/2$

$$\begin{cases} n^+(\omega,\theta) = \dfrac{1}{\sqrt{n_x^{-2}(\omega)\cos^2(\theta) + n_z^{-2}(\omega)\sin^2(\theta)}} \\ n^-(\omega,\theta) = n_y(\omega) \end{cases} \tag{5.70}$$

- for $n_x(\omega) > n_y(\omega) > n_z(\omega)$,

$$0 \le \theta \le V_z$$

$$\begin{cases} n^{+}(\omega,\theta) = \dfrac{1}{\sqrt{n_x^{-2}(\omega)\cos^2(\theta) + n_z^{-2}(\omega)\sin^2(\theta)}} \\ n^{-}(\omega,\theta) = n_y(\omega) \end{cases} \tag{5.71}$$

$$V_z \le \theta \le \pi/2$$

$$\begin{cases} n^{+}(\omega,\theta) = n_y(\omega) \\ n^{-}(\omega,\theta) = \dfrac{1}{\sqrt{n_x^{-2}(\omega)\cos^2(\theta) + n_z^{-2}(\omega)\sin^2(\theta)}} \end{cases} \tag{5.72}$$

For any plane containing $\overrightarrow{Oz}$ and out of the $(\overrightarrow{Ox},\overrightarrow{Oz})$ and $(\overrightarrow{Oy},\overrightarrow{Oz})$, $n^{+}(\theta,\varphi)$ and $n^{-}(\theta,\varphi)$ cannot be described by circles and ellipses. So, in the case of the biaxial optical class, the index surface is not depicted by a sphere and an ellipsoid, which differs from the case of the uniaxial optical class, as shown in Figure 5.5 and Figure 5.6.

Figure 5.5 and Figure 5.6 well show that there are four umbilical points defined by pair along the two optical axes, hence the name of "biaxial" for that kind of media. The birefringence $n^{+}(\theta,\varphi) - n^{-}(\theta,\varphi)$ is thus nil for a propagation along these two specific directions located in $(\overrightarrow{Ox},\overrightarrow{Oz})$ at an angle V_z from $\overrightarrow{Oz}$.

Note that historically, the angle V_z has been used for defining the optical sign of a biaxial crystal: it is positive if $V_z > \pi/4$, and negative if $V_z < \pi/4$ (Nesse 1991) [2]. This convention is still often used but it is not satisfying since it does not ensure the continuity between the uniaxial and biaxial characters. Actually, it must be possible to switch from one to the other by lifting or not the degeneracy between n_x and n_y. Thus the only definition of the optical sign compatible with this continuity criterion is $n_x < n_y < n_z \Leftrightarrow n_o < n_e$ for the positive sign and $n_x > n_y > n_z \Leftrightarrow n_o > n_e$ for the negative sign.

Figure 5.5 and Figure 5.6 allow us to identify that the three principal planes $(\overrightarrow{Ox},\overrightarrow{Oy})$, $(\overrightarrow{Oy},\overrightarrow{Oz})$, and $(\overrightarrow{Ox},\overrightarrow{Oz})$ are symmetry mirrors. Thus the index surface of a biaxial crystal belongs to the symmetry group *mmm*, which is an orthorhombic crystal class, as defined in Part 2.1.1.2.

Since the external and internal sheets of the index surface of a biaxial crystal are not spheres or ellipsoids, it is not possible to speak of ordinary or extraordinary sheets as in the case of uniaxial crystals. Nevertheless, it will be explained in Part 5.3 how it is still possible to define ordinary and extraordinary refractive indices in the case of biaxial crystals only in the principal planes of the dielectric frame, but at the cost of a certain complexity in the plane containing the optical axes, *i.e.* $(\overrightarrow{Ox},\overrightarrow{Oz})$.

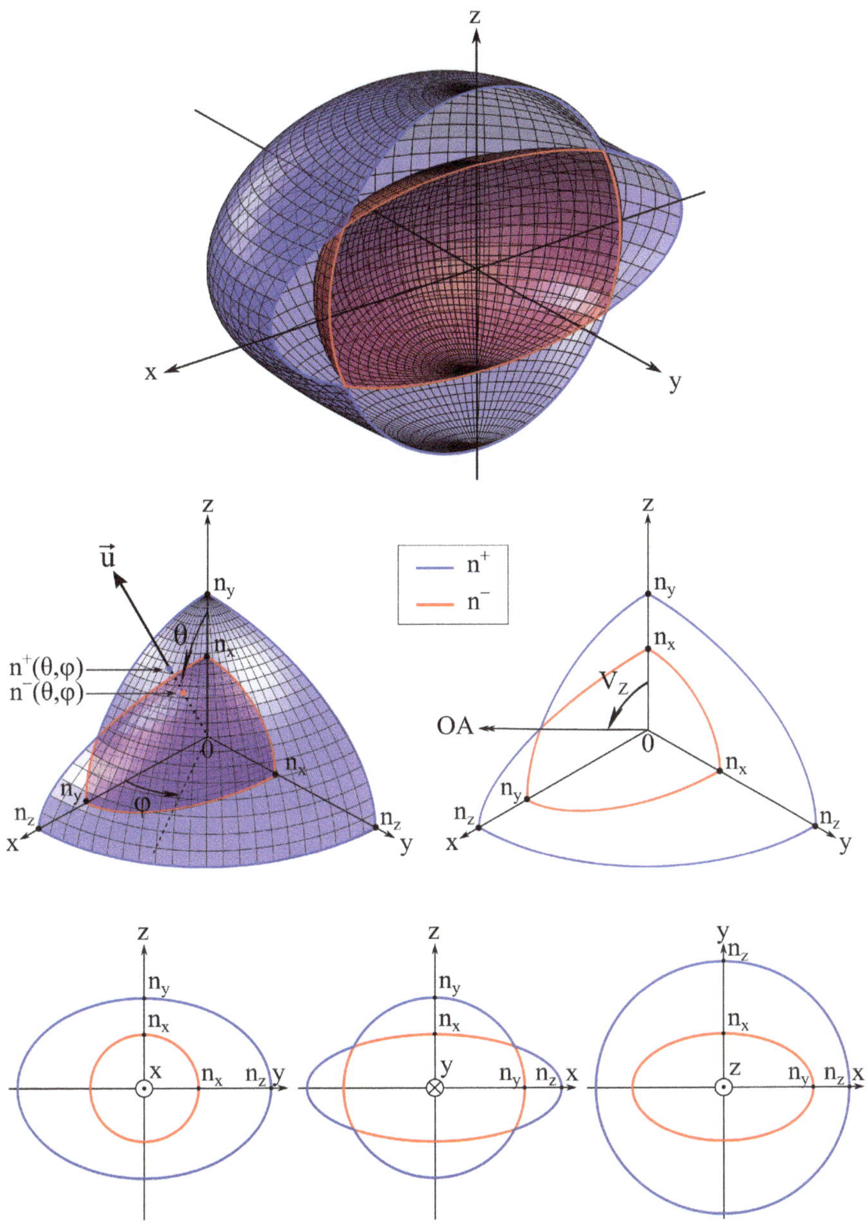

FIGURE 5.5 Index surface of the positive biaxial optical class at a given circular frequency ω in which the indices (x,y,z) refer to the dielectric frame $(\overrightarrow{Ox},\overrightarrow{Oy},\overrightarrow{Oz})$. Top: half-space. Middle: space octant in which (θ,φ) are the angles of spherical coordinates of the unit wave vector $\vec{u}$, and $n^{+}(\theta,\varphi)$ and $n^{-}(\theta,\varphi)$ the two corresponding possible refractive indices; n_x, n_y, and n_z are the principal refractive indices, and OA denotes the optical axes. Bottom: section of the index surface in the $(\overrightarrow{Oy},\overrightarrow{Oz})$ plane (left), $(\overrightarrow{Ox},\overrightarrow{Oz})$ plane (middle), and $(\overrightarrow{Ox},\overrightarrow{Oy})$ plane (right).

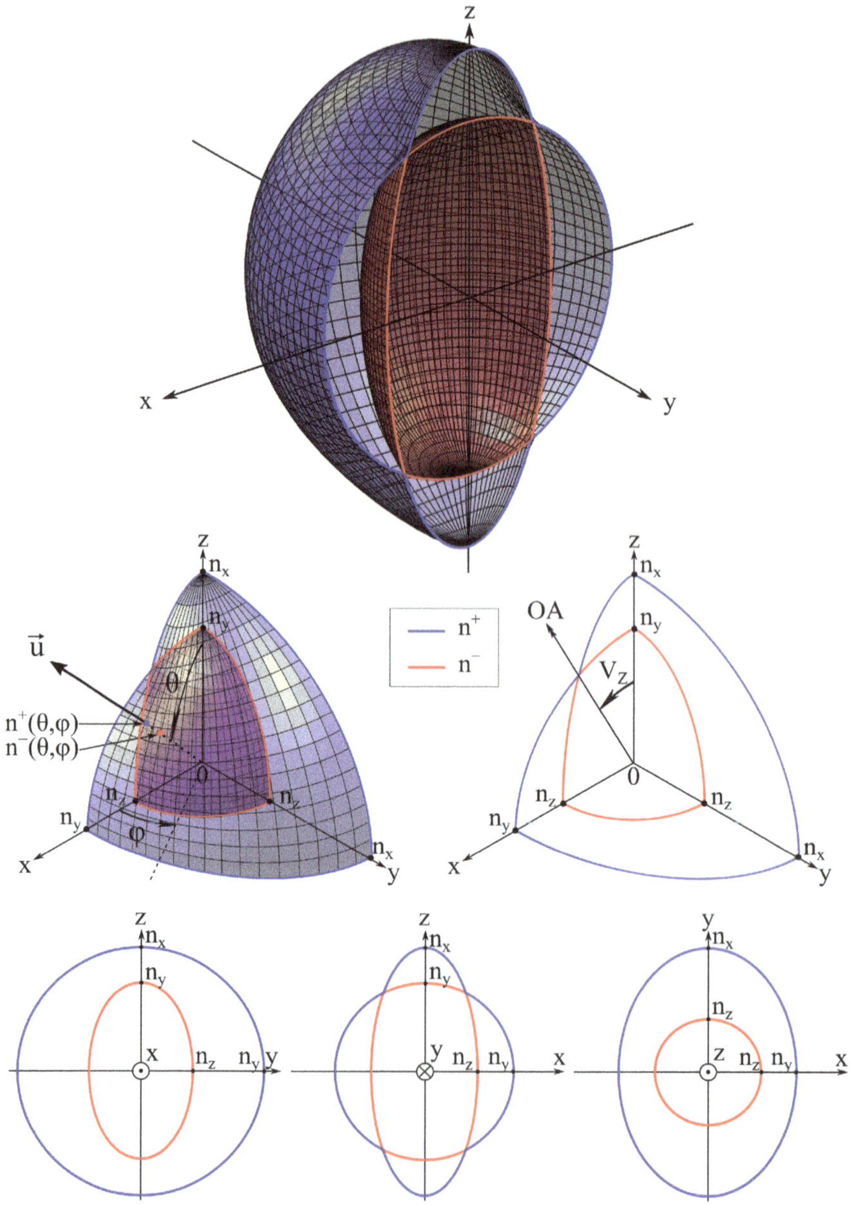

FIGURE 5.6 Index surface of the negative biaxial optical class at a given circular frequency ω in which the indices (x,y,z) refer to the dielectric frame $(\overrightarrow{Ox},\overrightarrow{Oy},\overrightarrow{Oz})$. Top: half-space. Middle: space octant in which (θ,φ) are the angles of spherical coordinates of the unit wave vector $\vec{u}$ and $n^+(\theta,\varphi)$ and $n^-(\theta,\varphi)$ the two corresponding possible refractive indices; n_x, n_y, and n_z are the principal refractive indices, and OA denotes the optical axes. Bottom: section of the index surface in the $(\overrightarrow{Oy},\overrightarrow{Oz})$ plane (left), $(\overrightarrow{Ox},\overrightarrow{Oz})$ plane (middle), and $(\overrightarrow{Ox},\overrightarrow{Oy})$ plane (right).

5.2.2 Phase Velocity Surface

The calculations performed in Part 5.2.1 allow us to directly access the angular distribution of the phase velocity of light whose general expression is given by Eq. (5.26).

Thus in the general case and in any direction of propagation $\vec{u}(\theta,\varphi)$, there exist two possible values for the phase velocity, written $\mathrm{v}^-(\omega,\vec{u})$ and $\mathrm{v}^+(\omega,\vec{u})$, expressed as:

$$\begin{cases} \mathrm{v}^+(\omega,\vec{u}) = \dfrac{c}{n^-(\omega,\vec{u})} \\ \mathrm{v}^-(\omega,\vec{u}) = \dfrac{c}{n^+(\omega,\vec{u})} \end{cases} \tag{5.73}$$

Note that $\mathrm{v}^+(\omega,\vec{u}) > \mathrm{v}^-(\omega,\vec{u})$ since $\mathrm{v}^+(\omega,\vec{u})$ corresponds to $n^-(\omega,\vec{u})$ and $\mathrm{v}^-(\omega,\vec{u})$ to $n^+(\omega,\vec{u})$.

As for the refractive index, there are principal values, called the principal phase velocities:

- $\mathrm{v}_x(\omega) = \mathrm{v}_y(\omega) = \mathrm{v}_z(\omega) \equiv \mathrm{v}(\omega)$ for the isotropic optical class;
- $\mathrm{v}_x(\omega) = \mathrm{v}_y(\omega) \equiv \mathrm{v}_o(\omega) > \mathrm{v}_z(\omega) \equiv \mathrm{v}_e(\omega)$ or $\mathrm{v}_x(\omega) = \mathrm{v}_y(\omega) \equiv \mathrm{v}_o(\omega) < \mathrm{v}_z(\omega) \equiv \mathrm{v}_e(\omega)$ for the positive and negative uniaxial optical classes, respectively, where $\mathrm{v}_o(\omega)$ and $\mathrm{v}_e(\omega)$ are the ordinary and extraordinary phase velocities, respectively;
- $\mathrm{v}_x(\omega) > \mathrm{v}_y(\omega) > \mathrm{v}_z(\omega)$ or $\mathrm{v}_x(\omega) < \mathrm{v}_y(\omega) < \mathrm{v}_z(\omega)$ for the positive and negative biaxial optical class, respectively.

The specific expressions of $\mathrm{v}^\pm(\omega,\vec{u}) \equiv \mathrm{v}^\pm(\omega,\theta,\varphi)$ regarding the optical classes can be obtained by using directly Eq. (5.49) in the case of the isotropic optical class, Eq. (5.53), Eq. (5.56), Eq. (5.57), Eq. (5.58), Eq. (5.60), Eq. (5.61), Eq. (5.62) in the case of the positive and negative uniaxial classes, and Eq. (5.63), Eq. (5.64), Eq. (5.65), Eq. (5.66), Eq. (5.69), Eq. (5.70), Eq. (5.71), and Eq. (5.72) for the positive and negative biaxial optical class. The corresponding angular distributions are given in Figure 5.7 in which a space octant is considered.

Due to the relation of inversion between the index surface and the phase velocity surface according to Eq. (5.73), there is an inversion of topology between the two kinds of surfaces for both the uniaxial and biaxial optical classes regarding their respective optical sign. This appears clearly by comparing the internal and external sheets in Figure 5.7 with those of Figure 5.3 to Figure 5.6. But the symmetry group remains the same, namely the group of the sphere $\left(\dfrac{\infty}{m}\dfrac{\infty}{m}\right)$, the cylinder group $\left(\dfrac{\infty}{m}m\right)$, and the group of the orthorhombic crystal class (mmm) for the isotropic, the uniaxial, and the biaxial optical classes, respectively.

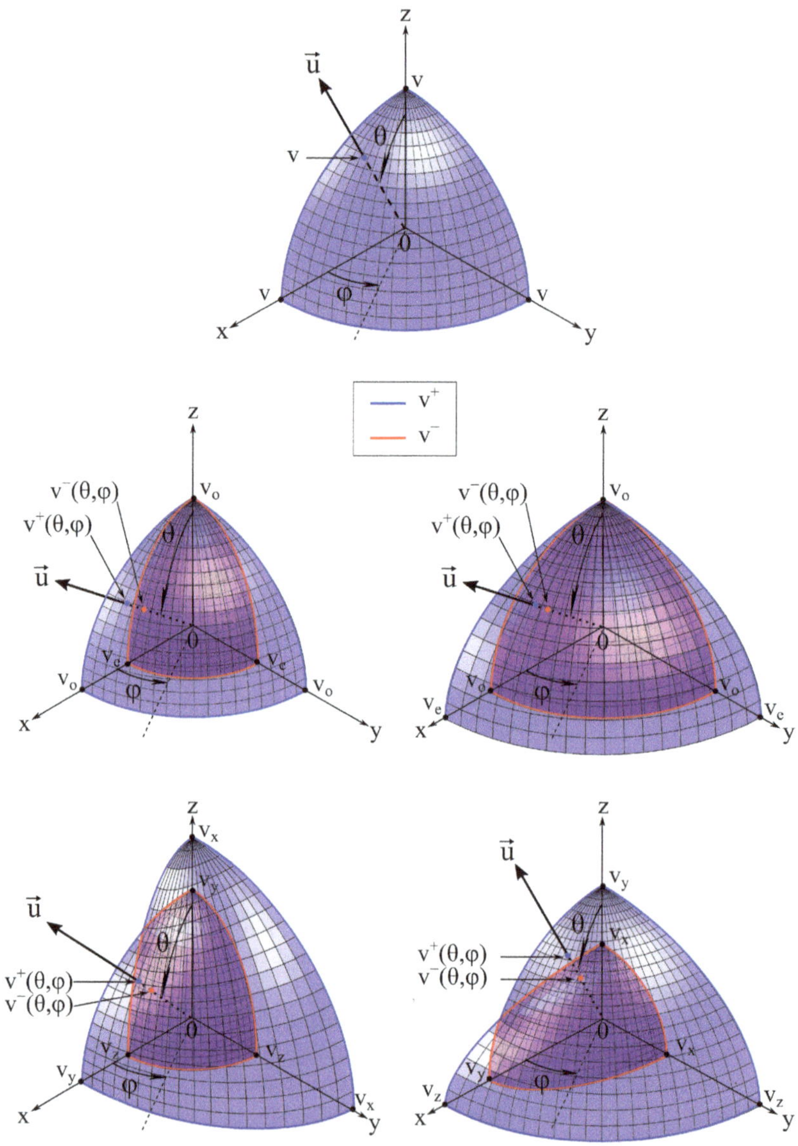

FIGURE 5.7 Phase velocity surface represented in a space octant at a given circular frequency ω in which the indices (x,y,z) refer to the dielectric frame $(\overrightarrow{Ox},\overrightarrow{Oy},\overrightarrow{Oz})$, and (θ,φ) are the angles of spherical coordinates of the unit wave vector $\vec{u}$. Top: isotropic optical class, in which v is the single phase velocity of light. Middle: positive (left) and negative (right) uniaxial optical classes, in which $\mathrm{v}^{+}(\theta,\varphi)$ and $\mathrm{v}^{-}(\theta,\varphi)$ are the two corresponding possible phase velocities of light; v_o and v_e are the ordinary and extraordinary principal phase velocities respectively, and OA denotes the optical axis. Bottom: positive (left) and negative (right) biaxial optical classes, in which $\mathrm{v}^{+}(\theta,\varphi)$ and $\mathrm{v}^{-}(\theta,\varphi)$ are the two corresponding possible phase velocities of light; v_x, v_y and v_z are the principal phase velocities, and OA denotes the optical axes.

5.3 POLARIZATION EIGENMODES

The calculation of the angular distribution of the refractive index having been done, it is now necessary to determine the corresponding angular distribution of the electric field that is solution of Eq. (5.30).

5.3.1 General Vector Configuration of an Optical Electromagnetic Wave

Before solving Eq. (5.30) regarding the two possible values of the refractive index n^- and n^+ in a given direction of propagation, it is important to have a general picture of the relative orientations of the different vector quantities that describe an electromagnetic wave propagating in matter, *i.e.*: the electric field $\vec{E}(\omega,\vec{r})$, the electric displacement $\vec{D}(\omega,\vec{r})$ defined in Eq. (1.97), the magnetic field $\vec{H}(\omega,\vec{r})$, and the magnetic induction $\vec{B}(\omega,\vec{r})$ defined in Eq. (1.98) associated with the wave vector $\vec{k}(\omega)$. As we did in Part 5.1 and Part 5.2, when the electric field was the only one to be considered and expressed by Eq. (5.9), the plane wave is used for describing all the fields, *i.e.*:

$$\vec{F}(\omega,\vec{r})=\vec{f}(\omega)F_0(\omega)e^{i\vec{k}(\omega)\cdot\vec{r}} \tag{5.74}$$

where $\vec{F}$ stands for $\vec{E}$, $\vec{D}$, $\vec{H}$, and $\vec{B}$, the unit vector $\vec{f}$ for $\vec{e}$, $\vec{d}$, $\vec{h}$, and $\vec{b}$, and the real amplitude F_0 stands for E_0, D_0, H_0, and B_0.

All the vectors are expressed in the dielectric frame $(\overrightarrow{Ox},\overrightarrow{Oy},\overrightarrow{Oz})$, so that the wave vector $\vec{k}(\omega)$ and the position $\vec{r}$ are given by Eq. (5.10) and $\vec{f}(\omega)$ by:

$$\vec{f}(\omega)=\begin{pmatrix} f_x \\ f_y \\ f_z \end{pmatrix} \tag{5.75}$$

It is then interesting to come back to the four Maxwell equations: Eq. (1.100), Eq. (1.101), Eq. (1.102), Eq. (1.103).

The calculation of the two Maxwell equations given by Eq. (1.100) and Eq. (1.102) according to Eq. (1.83), Eq. (5.74), Eq. (5.10), and Eq. (5.75) gives:

$$\begin{cases} div\left[\vec{D}(\omega,\vec{r})\right]=i\vec{k}(\omega).\vec{D}(\omega,\vec{r})=0 \\ div\left[\vec{B}(\omega,\vec{r})\right]=i\vec{k}(\omega).\vec{B}(\omega,\vec{r})=0 \end{cases} \tag{5.76}$$

These two equations show that the electric displacement and the magnetic induction are both perpendicular to the wave vector, *i.e.*:

$$\begin{cases} \vec{k}(\omega) \perp \vec{D}(\omega,\vec{r}) \\ \vec{k}(\omega) \perp \vec{B}(\omega,\vec{r}) \end{cases} \tag{5.77}$$

The calculation of the two other Maxwell equations Eq. (1.101) and Eq. (1.103) according to Eq. (1.84), Eq. (5.74), Eq. (5.10), and Eq. (5.75) gives:

$$\begin{cases} \overrightarrow{rot}\left[\vec{E}(\omega,\vec{r})\right] = i\vec{k}(\omega) \wedge \vec{E}(\omega,\vec{r}) = i\omega\vec{B} \\ \overrightarrow{rot}\left[\vec{H}(\omega,\vec{r})\right] = i\vec{k}(\omega) \wedge \vec{H}(\omega,\vec{r}) = -i\omega\vec{D} \end{cases} \tag{5.78}$$

Then it comes:

$$\begin{cases} \vec{B}(\omega,\vec{r}) \perp \vec{E}(\omega,\vec{r}) \\ \vec{D}(\omega,\vec{r}) \perp \vec{H}(\omega,\vec{r}) \end{cases} \tag{5.79}$$

The combination of Eq. (5.77) and Eq. (5.79) allows us to establish that $\vec{E}(\omega,\vec{r})$, $\vec{D}(\omega,\vec{r})$, and $\vec{k}(\omega)$ are located in the same plane on the one hand, and that $\vec{H}(\omega,\vec{r})$, $\vec{B}(\omega,\vec{r})$, and $\vec{k}(\omega)$ are also located in the same plane on the other hand, these two planes being perpendicular to each other, and $\vec{k}(\omega)$ belonging to their intersection. These two planes are called the electric vibration plane and the magnetic vibration plane, respectively.

The orientation of the magnetic field $\vec{H}(\omega,\vec{r})$ with respect to $\vec{B}(\omega,\vec{r})$ can be easily found for a non-magnetic medium, which is considered here. Actually, these two vectors are simply proportional to each other according to Eq. (1.98), so that it comes:

$$\vec{H}(\omega,\vec{r}) \,//\, \vec{B}(\omega,\vec{r}) \tag{5.80}$$

The situation is different for $\vec{E}(\omega,\vec{r})$ and $\vec{D}(\omega,\vec{r})$ that are not necessarily collinear. That can be easily seen by considering the combination of the electric constitutive equation Eq. (1.97) with Eq. (3.1) and Eq. (5.6) allowing the following tensor equation to be established:

$$\vec{D}(\omega,\vec{r}) = \varepsilon_o\left[Id + \chi^{(1)}(\omega)\right].\vec{E}(\omega,\vec{r}) = \varepsilon_o\varepsilon_r(\omega).\vec{E}(\omega,\vec{r}) \tag{5.81}$$

Note that this equation is another form of the electric constitutive equation initially defined by Eq. (1.97). It is reminded that the dot (.) in Eq. (5.81) corresponds to the contracted product, in the present case between a rank-2 tensor and a vector, leading thus to a vector quantity that is naturally compatible with the vector nature of the $\vec{D}(\omega,\vec{r})$ vector. The dielectric permittivity $\varepsilon_r(\omega)$ being a rank-2 tensor, the collinearity or lack thereof between $\vec{E}(\omega,\vec{r})$ and $\vec{D}(\omega,\vec{r})$ depends on the optical class: collinearity is always the rule for the isotropic class because the three diagonal tensor coefficients are equal in the dielectric frame, while non-collinearity occurs in the general case of the uniaxial and biaxial classes since two or three tensor coefficients are different. When there is no collinearity, the angle between $\vec{E}(\omega,\vec{r})$ and $\vec{D}(\omega,\vec{r})$ is called the spatial walk-off angle, written $\rho(\omega)$. This angle can be then expressed as:

$$\rho(\omega) = \cos^{-1}\left[\vec{d}(\omega)\cdot\vec{e}(\omega)\right] \tag{5.82}$$

It is relevant at this step to also take an interest in the direction of propagation of the energy that is carried out by the optical electromagnetic wave. It is given by the Umov-Poynting vector $\vec{S}(\omega,\vec{r})$ defined as:

$$\vec{S}(\omega,\vec{r}) = \vec{E}(\omega,\vec{r}) \wedge \vec{H}(\omega,\vec{r}) \tag{5.83}$$

Equation (5.83) indicates that $\vec{S}(\omega,\vec{r})$ is perpendicular to the plane $\left(\vec{E}(\omega,\vec{r}),\vec{H}(\omega,\vec{r})\right)$ so that it is located in the electric vibration plane since $\vec{H}(\omega,\vec{r}) \perp \vec{k}(\omega,\vec{r})$ according to Eq. (5.77) and Eq. (5.80). The other point is that when $\vec{D}(\omega,\vec{r})$ and $\vec{E}(\omega,\vec{r})$ are not collinear, then $\vec{k}(\omega,\vec{r})$ and $\vec{S}(\omega,\vec{r})$ are not collinear either. Furthermore, since $\vec{D}(\omega,\vec{r}) \perp \vec{k}(\omega,\vec{r})$ and $\vec{E}(\omega,\vec{r}) \perp \vec{S}(\omega,\vec{r})$ then $\vec{E}(\omega,\vec{r})$ and $\vec{S}(\omega,\vec{r})$ also make an angle equal to $\rho(\omega)$. It means that in the general case, the wave does not propagate in the same direction along which energy carries out. This non-intuitive situation has important repercussions on the nonlinear optical properties as will be seen in Chapter 7 and Chapter 8.

Note also that according to Eq. (5.83) and Eq. (5.74), the spatial evolution of the Umov-Poynting vector is given by:

$$\vec{S}(\omega,\vec{r}) = \vec{s}(\omega)S_0(\omega)e^{i\vec{k}(\omega)\cdot\vec{r}} \tag{5.84}$$

Figure 5.8 illustrates the general vector configuration of an optical electromagnetic wave according to the previous demonstrations.

This general configuration, *i.e.* coplanarity of $\left(\vec{D},\vec{E},\vec{k},\vec{S}\right)$, coplanarity of $\left(\vec{B},\vec{H},\vec{k}\right)$, and orthogonality between these two planes, remains the same whatever the considered propagation direction and the solicited refractive index Eigen values. However, it will be shown in the three following parts that there are strong specificities according to the optical class. Since non-magnetic media are considered here, the electric vibration plane $\left(\vec{D},\vec{E},\vec{k},\vec{S}\right)$ will be the plane of interest. It will be called the vibration plane in the following, written Π. For each optical class, Eq. (5.29) will be solved, which will lead to the determination of the unit electric field vector $\vec{e}(\omega,\vec{u})$ associated with the solicited refractive index $n(\omega,\vec{u})$ in the direction of propagation of unit vector $\vec{u}$. The spatial walk-off angle $\rho(\omega)$ will be also calculated.

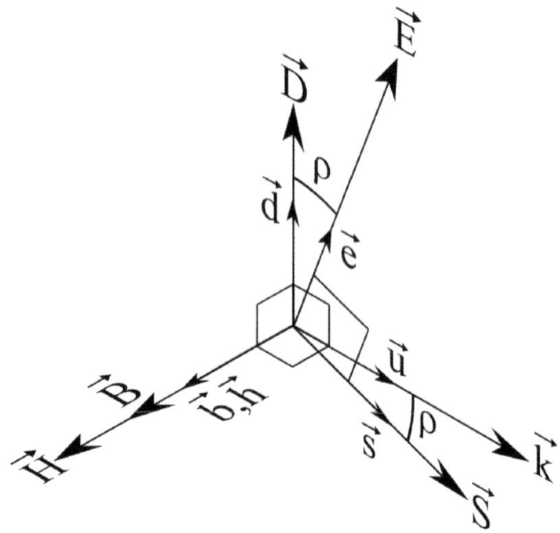

FIGURE 5.8 General vector configuration of an optical electromagnetic wave propagating in a non-magnetic medium: $\vec{E}$ is the electric field, $\vec{D}$ the electric displacement, $\vec{H}$ the magnetic field, $\vec{B}$ the magnetic induction, $\vec{S}$ the Umov-Poynting vector, $\vec{k}$ the wave vector, and ρ the spatial walk-off angle. $\left(\vec{D},\vec{E},\vec{k},\vec{S}\right)$ defines the electric vibration plane that is perpendicular to the $\left(\vec{B},\vec{H},\vec{k}\right)$ magnetic vibration plane.

5.3.2 Isotropic Optical Class

In this case, $\varepsilon_{r_{xx}}(\omega)=\varepsilon_{r_{yy}}(\omega)=\varepsilon_{r_{zz}}(\omega)\equiv\varepsilon_r(\omega)=n^2(\omega)$, and it was also demonstrated that $n(\omega,\vec{u})=n(\omega)$ for any direction of propagation $\vec{u}$, as shown by Figure 5.2, where $\vec{u}$ is the unit vector of the wave vector $\vec{k}$ as defined by Eq. (5.25). In these conditions Eq. (5.30) writes:

$$\begin{cases}\left[n^2(\omega)(1-u_x^2)-n^2(\omega)\right]e_x-n^2(\omega)u_xu_ye_y-n^2(\omega)u_xu_ze_z=0\\-n^2(\omega)u_yu_xe_x+\left[n^2(\omega)(1-u_y^2)-n^2(\omega)\right]e_y-n^2(\omega)u_yu_ze_z=0\\-n^2(\omega)u_zu_xe_x-n^2(\omega)u_zu_ye_y+\left[n^2(\omega)(1-u_z^2)-n^2(\omega)\right]e_z=0\end{cases}\tag{5.85}$$

Because $n(\omega,\vec{u})=n(\omega)$ is constant whatever the considered propagation direction, each of these three equations reduces to the following one:

$$u_xe_x+u_ye_y+u_ze_z=\vec{e}(\omega,\vec{u}).\vec{u}=0\tag{5.86}$$

Equation (5.86) means that the electric field remains perpendicular to the wave vector, *i.e.* $\vec{e}(\omega,\vec{u})\perp\vec{u}$, and that for any direction of the wave vector $\vec{u}$. This corresponds thus to a null walk-off angle, namely $\rho(\omega,\vec{u})=0$ in any case. As

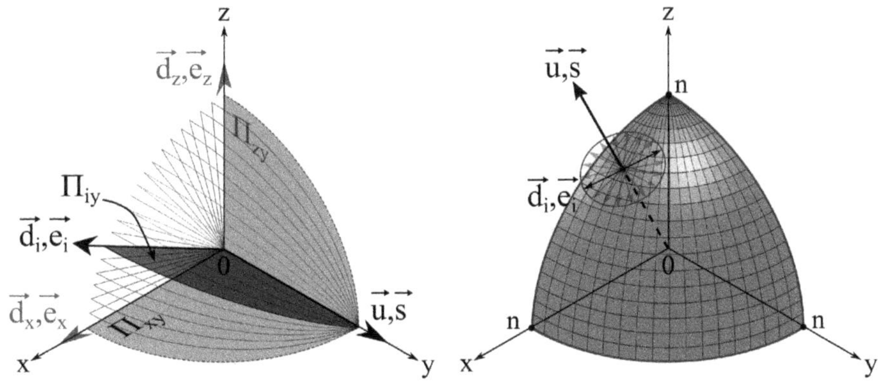

FIGURE 5.9 Possible vector configurations of an optical electromagnetic wave propagating in an isotropic medium and associated with a single wave vector of unit vector $\vec{u}$; $\vec{s}$ is the unit Umov-Poynting vector. Left: example of possible vibration planes Π_i, with $i = (1,2,3,...,n)$, containing the unit electric field and displacement vectors $(\vec{e_i},\vec{d_i})$. Right: link between the index surface and any possible direction of the electric field and displacement vectors $(\vec{e},\vec{d})$.

a consequence, the Umov-Poynting vector is collinear with the wave vector, *i.e.* $\vec{u}\,//\,\vec{s}(\omega,\vec{u})$.

The other important point is that Eq. (5.86) can be written for any direction of polarization $\vec{e}(\omega,\vec{u})$: this means that an isotropic medium allows any direction of polarization, *i.e.* any vibration plane Π, to exist. It is a major result that will make a strong difference with uniaxial and biaxial crystals as will be seen in the two following parts.

It is also interesting to notice that $\vec{e}(\omega,\vec{u})$ exactly corresponds to the tangent to the section of the index surface by the plane defined by Π: actually, any section containing the direction of propagation $\vec{u}$ is a circle according to Figure 5.2, and $\vec{u} \perp \vec{e}(\omega,\vec{u})$. As a consequence, $\vec{s}(\omega,\vec{u})$ is given by the direction normal to the index surface since $\vec{s}(\omega,\vec{u}) \perp \vec{e}(\omega,\vec{u})$ or by knowing that $\vec{s}(\omega,\vec{u})\,//\,\vec{u}$.

Figure 5.9 describes the corresponding vector configurations.

5.3.3 Uniaxial Optical Class

As a reminder in this case, $\varepsilon_{r_{xx}}(\omega) = \varepsilon_{r_{yy}}(\omega) \equiv \varepsilon_{r_o}(\omega) < \varepsilon_{r_{zz}}(\omega) \equiv \varepsilon_{r_e}(\omega)$ for the positive uniaxial class, or $\varepsilon_{r_{xx}}(\omega) = \varepsilon_{r_{yy}}(\omega) \equiv \varepsilon_{r_o}(\omega) > \varepsilon_{r_{zz}}(\omega) \equiv \varepsilon_{r_e}(\omega)$ for the negative uniaxial class. Thus there are only two principal refractive indices: $n_o(\omega) = \sqrt{\varepsilon_{r_o}(\omega)}$ and $n_e(\omega) = \sqrt{\varepsilon_{r_e}(\omega)}$, with $n_o(\omega) < n_e(\omega)$ for the positive optical sign and $n_o(\omega) > n_e(\omega)$ for the negative optical sign. The index surface is depicted in Figure 5.3 and Figure 5.4 for the two optical signs, showing that in a given direction of propagation $\vec{u}$, there are in general two possible values for the refractive index $n(\omega,\vec{u})$: $n^-(\omega,\vec{u}) = n_o(\omega)$

and $n^{+}(\omega,\vec{u})=n_e(\omega,\vec{u})$ for the positive optical sign, and $n^{-}(\omega,\vec{u})=n_e(\omega,\vec{u})$ and $n^{+}(\omega,\vec{u})=n_o(\omega)$ for the negative optical sign.

Then Eq. (5.30) may have to be split into two sets of equations in the general case: one relative to $n(\omega,\vec{u})=n^{+}(\omega,\vec{u})$ and the other one to $n(\omega,\vec{u})=n^{-}(\omega,\vec{u})$ respectively, which enables the determination of the corresponding Cartesian coordinates of the Eigenmodes of polarization. This is done hereafter by considering successively two locations for the direction of propagation, *i.e.*: $\theta=0$ and $0<\theta<\pi/2$.

For $\theta=0$, *i.e.* $(u_x=u_y=0,u_z=1)$, which corresponds to a propagation along $\overrightarrow{Oz}$ that is the optical axis (OA), it is the only case for which there is no splitting of Eq. (5.30) since there is degeneracy regarding the refractive index, *i.e.* $n^{+}(\omega,\theta=0)=n^{-}(\omega,\theta=0)=n_o(\omega)$. Then the three equalities of Eq. (5.30) simply write:

$$\begin{cases} 0=0 \\ 0=0 \\ -n_e^2(\omega)e_z=0 \end{cases} \tag{5.87}$$

Equation (5.87) thus reduces to $e_z=0$, which means that any direction of the unit electric field vector $\vec{e}(\omega,\vec{u})$ is allowed in the $(\overrightarrow{Ox},\overrightarrow{Oy})$ plane without any other restriction. Finally, along the optical axis of the uniaxial class, it means that the vector configuration of the corresponding electromagnetic wave is the same as that in an isotropic medium, *i.e.*: any vibration plane Π can exist along this specific propagation direction, $\vec{e}(\omega,\vec{u})\perp\vec{u}$, $\vec{u}//\vec{s}(\omega,\vec{u})$ and so $\rho(\omega)=0$.

For $\theta>0$, *i.e.* $(u_z\neq 1)$, which corresponds to any direction of propagation out of the optical axis, the two possible values of the refractive indices are $n_o(\omega,\vec{u})$ and $n_e(\omega,\vec{u})$, the optical sign being positive or negative as shown by Figure 5.3 and Figure 5.4. Then the resolution of Eq. (5.30) written with $n(\omega,\vec{u})=n_o(\omega,\vec{u})$ or $n(\omega,\vec{u})=n_e(\omega,\vec{u})$ allows us to define the corresponding unit electric field vectors $\vec{e}^{\,o}(\omega,\vec{u})$ and $\vec{e}^{\,e}(\omega,\vec{u})$, respectively.

In the case of the ordinary wave, *i.e.* $n(\omega,\vec{u})=n_o(\omega,\vec{u})$, Eq. (5.30) rewrites:

$$\begin{cases} \left[n_o^2(\omega)(1-u_x^2)-n_o^2(\omega)\right]e_x^o-n_o^2(\omega)u_xu_ye_y^o-n_o^2(\omega)u_xu_ze_z^o=0 \\ -n_o^2(\omega)u_xu_ye_x^o+\left[n_o^2(\omega)(1-u_y^2)-n_o^2(\omega)\right]e_y^o-n_o^2(\omega)u_yu_ze_z^o=0 \\ -n_o^2(\omega)u_xu_ze_x^o-n_o^2(\omega)u_yu_ze_y^o+\left[n_o^2(\omega)(1-u_z^2)-n_e^2(\omega)\right]e_z^o=0 \end{cases} \tag{5.88}$$

Identically to what happened for the isotropic class, the two first equalities of Eq. (5.88) are identical and rewrites:

$$u_xe_x^o+u_ye_y^o+u_ze_z^o=0 \tag{5.89}$$

Equation (5.89) corresponds to $\vec{e}^{\,o}(\omega,\vec{u})\cdot\vec{u}=0$, which means that $\vec{e}^{\,o}(\omega,\vec{u})\perp\vec{u}$ for any direction of propagation $\vec{u}$.

By inserting Eq. (5.89) in the third equality of Eq. (5.88), it comes:

$$\begin{cases} u_x e_x^o + u_y e_y^o + u_z e_z^o = 0 \\ n_o^2(\omega) u_z^2 e_z^o + \left[n_o^2(\omega)(1-u_z^2) - n_e^2(\omega)\right] e_z^o = \left[n_o^2(\omega) - n_e^2(\omega)\right] e_z^o = 0 \end{cases} \tag{5.90}$$

Since $n_o^2(\omega) \neq n_e^2(\omega)$ by definition, the third equality in Eq. (5.88) reduces to:

$$e_z^o = 0 \tag{5.91}$$

Thus, it means that the ordinary wave is always polarized in the $(\overrightarrow{Ox}, \overrightarrow{Oy})$ plane, whatever the considered propagation direction.

According to Eq. (5.91), Eq. (5.89) can also be written:

$$e_y^o = -\frac{u_x}{u_y} e_x^o \tag{5.92}$$

The fact that e_x^o and e_y^o are proportional one to each other on the one hand, u_x and u_y being real quantities, and that $e_z^o = 0$ on the other hand, indicates that the ordinary polarization Eigenmode is a linear polarization state of light, being oriented in the $(\overrightarrow{Ox}, \overrightarrow{Oy})$ plane.

Since $\vec{e}^o(\omega, \vec{u})$ is a unit vector, it comes:

$$\left[e_x^o\right]^2 + \left[e_y^o\right]^2 + \left[e_z^o\right]^2 = 1 \tag{5.93}$$

Equation (5.92) in Eq. (5.93) leads to:

$$\left[e_x^o\right]^2 + \left[e_y^o\right]^2 = \left[e_x^o\right]^2 \left[1 + \left(\frac{u_x}{u_y}\right)^2\right] = \left[e_x^o\right]^2 \left(\frac{1-u_z^2}{u_y^2}\right) = 1 \tag{5.94}$$

Equation (5.94), Eq. (5.92), and Eq. (5.91) allow us to get the three Cartesian coordinates of the ordinary unit electric field vector, *i.e.*:

$$\vec{e}^o(\omega, \vec{u}) = \begin{pmatrix} e_x^o \\ e_y^o \\ e_z^o \end{pmatrix} = \pm \begin{pmatrix} \dfrac{u_y}{\sqrt{1-u_z^2}} \\ -\dfrac{u_x}{\sqrt{1-u_z^2}} \\ 0 \end{pmatrix} \perp \overrightarrow{Oz} \tag{5.95}$$

The $\pm$ in Eq. (5.95) indicates that the two opposite directions are possible. These two solutions are relevant since an oscillating field is considered. However in the following, we will keep only the negative sign in order to lighten the writing. This choice is arbitrary. Equation (5.95) indicates that the unit electric field vector is orthogonal to $\overrightarrow{Oz}$.

It is useful to use the angles of spherical coordinates defined by Eq. (5.47) in order to express Eq. (5.95), which gives with the considered sign choice:

$$\vec{e}^{o}(\omega,\vec{u}) = \begin{pmatrix} e_x^o \\ e_y^o \\ e_z^o \end{pmatrix} = \begin{pmatrix} -\sin(\varphi) \\ +\cos(\varphi) \\ 0 \end{pmatrix} \perp \overrightarrow{Oz} \tag{5.96}$$

The calculation of $\vec{d}^{o}(\omega,\vec{u})$ can be easily performed at this step. Actually, knowing that $\vec{e}^{o}(\omega,\vec{u}) \perp \vec{u}$ according to Eq. (5.89), $\vec{d}^{o}(\omega,\vec{u}) \perp \vec{u}$ according to Eq. (5.77), and that $\left(\overrightarrow{d^o}, \overrightarrow{e^o}, \vec{u}\right)$ are coplanar, it comes that $\vec{d}^{o}(\omega,\vec{u}) // \vec{e}^{o}(\omega,\vec{u})$ so that:

$$\vec{d}^{o}(\omega,\vec{u}) = \begin{pmatrix} d_x^o \\ d_y^o \\ d_z^o \end{pmatrix} = \begin{pmatrix} e_x^o \\ e_y^o \\ e_z^o \end{pmatrix} = \begin{pmatrix} -\dfrac{u_y}{\sqrt{1-u_z^2}} \\ \dfrac{u_x}{\sqrt{1-u_z^2}} \\ 0 \end{pmatrix} = \begin{pmatrix} -\sin(\varphi) \\ \cos(\varphi) \\ 0 \end{pmatrix} \perp \overrightarrow{Oz} \tag{5.97}$$

The consequence is that:

$$\rho^{o}(\omega,\vec{u}) = Ar\cos\left[\vec{d}^{o}(\omega,\vec{u}) \cdot \vec{e}^{o}(\omega,\vec{u})\right] = 0 \tag{5.98}$$

So, the spatial walk-off angle of the ordinary wave is nil at any direction of propagation $\vec{u}$. The consequence of Eq. (5.98) is also that the Umov-Poynting vector is collinear with the wave vector, *i.e.* $\vec{s}^{o}(\omega,\vec{u}) = \vec{u}$, as shown in Figure 5.10. Thus the vibration plane, written Π^o, of an ordinary wave propagating in a positive or negative uniaxial crystal out of the optical axis is the same as that of any vibration plane associated to a propagation along the optical axis of the uniaxial crystal or in any direction of propagation in an isotropic medium.

Note also that Eq. (5.95), or equivalently Eq. (5.96), exactly corresponds to the tangent to a circle according to the properties of the conics. It means that $\vec{e}^{o}(\omega,\vec{u})$ can be given by the tangent to the section of the ordinary sheet of the index surface by a plane parallel to $(\overrightarrow{Ox},\overrightarrow{Oy})$: the tangent has to be taken at the point given by the intersection of $\vec{u}$ with the section of the index surface, such section of the ordinary index layer being a circle. This result could also have been obtained easily knowing that $\vec{e}^{o}(\omega,\vec{u})$ is perpendicular to $\vec{u}$.

Equation (5.96) and Eq. (5.97) allows us to specify the orientation of the electric field and displacement for a direction of propagation $\vec{u}$ located in one of the three principal planes, *i.e.*:

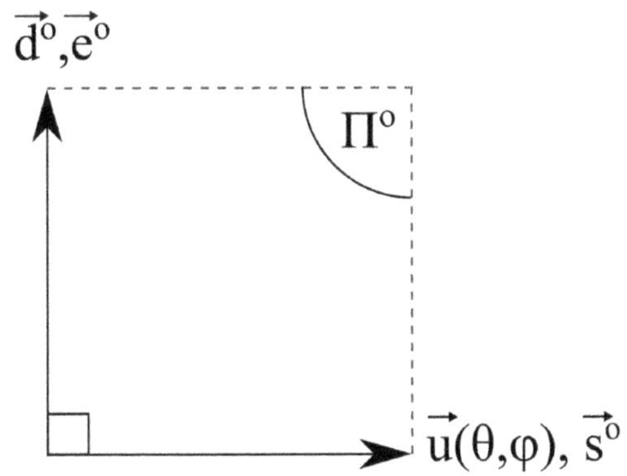

FIGURE 5.10 Ordinary vibration plane Π^o of an optical electromagnetic wave propagating in a positive or negative uniaxial crystal: the unit vector $\vec{u}(\theta,\varphi)$ stands for any direction of propagation out of the optical axis, $\vec{s}^o$ is the ordinary unit Umov-Poynting vector, $\vec{e}^o$ is the ordinary unit electric field vector, and $\vec{d}^o$ is the ordinary unit electric displacement vector.

- $\vec{e}^o(\omega,\vec{u}) = \vec{d}^o(\omega,\vec{u}) = \begin{pmatrix} -\sin(\varphi) \\ \cos(\varphi) \\ 0 \end{pmatrix}$ so that $\vec{u} \in (\overrightarrow{Ox},\overrightarrow{Oy})$,
- $\vec{e}^o(\omega,\vec{u}) = \vec{d}^o(\omega,\vec{u}) = \begin{pmatrix} 0 \\ 1 \\ 0 \end{pmatrix} = \overrightarrow{Oy}$ so that $\vec{u} \in (\overrightarrow{Ox},\overrightarrow{Oz})$,
- $\vec{e}^o(\omega,\vec{u}) = \vec{d}^o(\omega,\vec{u}) = \begin{pmatrix} -1 \\ 0 \\ 0 \end{pmatrix} = \overline{-Ox}$ so that $\vec{u} \in (\overrightarrow{Oy},\overrightarrow{Oz})$.

In the case of the extraordinary wave, *i.e.* $n(\omega,\vec{u}) = n_e(\omega,\vec{u})$, Eq. (5.30) now writes:

$$\begin{cases} \left\{[n_e(\omega,\vec{u})]^2(1-u_x^2) - n_o^2(\omega)\right\}e_x^e - [n_e(\omega,\vec{u})]^2 u_x u_y e_y^e - [n_e(\omega,\vec{u})]^2 u_x u_z e_z^e = 0 \\ -[n_e(\omega,\vec{u})]^2 u_x u_y e_x^e + \left\{[n_e(\omega,\vec{u})]^2(1-u_y^2) - n_o^2(\omega)\right\}e_y^e - [n_e(\omega,\vec{u})]^2 u_y u_z e_z^e = 0 \\ -[n_e(\omega,\vec{u})]^2 u_x u_z e_x^e - [n_e(\omega,\vec{u})]^2 u_y u_z e_y^e + \left\{[n_e(\omega,\vec{u})]^2(1-u_z^2) - n_e^2(\omega)\right\}e_z^e = 0 \end{cases} \quad (5.99)$$

After factorization, Eq. (5.99) can be expressed as:

$$e_x^e = \left[\frac{n_e(\omega,\vec{u})}{n_o(\omega)}\right]^2 \left\{e_x^e - u_x(u_x e_x^e + u_y e_y^e + u_z e_z^e)\right\} \quad (5.100a)$$

$$e_y^e = \left[\frac{n_e(\omega,\vec{u})}{n_o(\omega)}\right]^2 \left\{e_y^e - u_y(u_x e_x^e + u_y e_y^e + u_z e_z^e)\right\} \quad (5.100b)$$

$$e_z^e = \left[\frac{n_e(\omega,\vec{u})}{n_e(\omega)}\right]^2 \left\{e_z^e - u_z\left(u_x e_x^e + u_y e_y^e + u_z e_z^e\right)\right\} \tag{5.100c}$$

Equation (5.100a) in Eq. (5.100b) and Eq. (5.100c) gives:

$$e_y^e = \frac{u_y}{u_x} e_x^e \tag{5.101a}$$

$$e_z^e = -\frac{n_o^2(\omega)}{n_e^2(\omega)} \frac{(1-u_z^2)}{u_x u_z} e_x^e \tag{5.101b}$$

These equations show that the three Cartesian components are proportional one to each other, which means that $\vec{e}^e(\omega,\vec{u})$ is a linear polarization state, as it was found for $\vec{e}^o(\omega,\vec{u})$.

By inserting Eq. (5.101a) and Eq. (5.101b) into $\left[e_x^o\right]^2 + \left[e_y^o\right]^2 + \left[e_z^o\right]^2 = 1$, it comes:

$$\left[e_x^e\right]^2 \left\{1 + \frac{u_y^2}{u_x^2} + \frac{n_o^4(\omega)}{n_e^4(\omega)} \frac{(1-u_z^2)^2}{u_x^2 u_z^2}\right\} = 1 \tag{5.102}$$

Then by inserting $\left[u_x\right]^2 + \left[u_y\right]^2 + \left[u_z\right]^2 = 1$ into Eq. (5.102), it comes:

$$e_x^e = \pm \frac{n_e^2(\omega) u_x u_z}{\sqrt{1-u_z^2}\sqrt{n_0^4(\omega)(1-u_z^2) + n_e^4(\omega) u_z^2}} \tag{5.103a}$$

The insertion of Eq. (5.103a) into Eq. (5.101a) and Eq. (5.101b), by retaining only the negative sign in front of the ratio as was done previously for the ordinary wave, leads to:

$$e_y^e = -\frac{n_e^2(\omega) u_y u_z}{\sqrt{1-u_z^2}\sqrt{n_0^4(\omega)(1-u_z^2) + n_e^4(\omega) u_z^2}} \tag{5.103b}$$

$$e_z^e = \frac{n_o^2(\omega)(1-u_z^2)}{\sqrt{1-u_z^2}\sqrt{n_0^4(\omega)(1-u_z^2) + n_e^4(\omega) u_z^2}} \tag{5.103c}$$

Then $\vec{e}^e(\omega,\vec{u})$ writes:

$$\vec{e}^e(\omega,\vec{u}) = \begin{pmatrix} e_x^e \\ e_y^e \\ e_z^e \end{pmatrix} = \frac{1}{\sqrt{n_0^4(\omega)(1-u_z^2) + n_e^4(\omega) u_z^2}} \begin{pmatrix} -\dfrac{n_e^2(\omega) u_x u_z}{\sqrt{1-u_z^2}} \\ -\dfrac{n_e^2(\omega) u_y u_z}{\sqrt{1-u_z^2}} \\ \dfrac{n_0^2(\omega)\left(1-u_z^2\right)}{\sqrt{1-u_z^2}} \end{pmatrix} \tag{5.104}$$

Knowing $\vec{e}^{e}(\omega,\vec{u})$, it is easy to calculate the Cartesian coordinates of $\vec{d}^{e}(\omega,\vec{u})$. Actually, from $\vec{D}^{e}(\omega,\vec{u})=\vec{d}^{e}(\omega,\vec{u})\left\|\vec{D}^{e}(\omega,\vec{u})\right\|$, $\vec{E}^{e}(\omega,\vec{u})=\vec{e}^{e}(\omega,\vec{u})\left\|\vec{E}^{e}(\omega,\vec{u})\right\|$, and $\vec{D}^{e}(\omega,\vec{r})=\varepsilon_o\varepsilon_r(\omega).\vec{E}^{e}(\omega,\vec{r})$ arising from Eq. (5.54), it comes:

$$\vec{d}^{e}(\omega,\vec{u})=\varepsilon_o\frac{\left\|\vec{E}^{e}(\omega,\vec{u})\right\|}{\left\|\vec{D}^{e}(\omega,\vec{u})\right\|}\begin{pmatrix}\varepsilon_{r_o}^{(1)}e_x^e\\ \varepsilon_{r_o}^{(1)}e_y^e\\ \varepsilon_{r_e}^{(1)}e_z^e\end{pmatrix}=\varepsilon_o n_o^2(\omega)\frac{\left\|\vec{E}^{e}(\omega,\vec{u})\right\|}{\left\|\vec{D}^{e}(\omega,\vec{u})\right\|}\begin{pmatrix}e_x^e\\ e_y^e\\ \dfrac{n_e^2(\omega)}{n_o^2(\omega)}e_z^e\end{pmatrix} \tag{5.105}$$

Equation (5.104) in Eq. (5.105) gives:

$$\vec{d}^{e}(\omega,\vec{u})=\varepsilon_o\frac{\left\|\vec{E}^{e}(\omega,\vec{u})\right\|}{\left\|\vec{D}^{e}(\omega,\vec{u})\right\|}\frac{n_0^2(\omega)n_e^2(\omega)}{\sqrt{n_0^4(\omega)(1-u_z^2)+n_e^4(\omega)u_z^2}}\begin{pmatrix}-\dfrac{u_xu_z}{\sqrt{1-u_z^2}}\\ -\dfrac{u_yu_z}{\sqrt{1-u_z^2}}\\ \dfrac{\left(1-u_z^2\right)}{\sqrt{1-u_z^2}}\end{pmatrix} \tag{5.106}$$

Since $\vec{d}^{e}(\omega,\vec{u})$ is a unit vector, it comes:

$$\begin{aligned}\left\|\vec{d}^{e}(\omega,\vec{u})\right\|^2&=\left[\varepsilon_o\frac{\left\|\vec{E}^{e}(\omega,\vec{u})\right\|}{\left\|\vec{D}^{e}(\omega,\vec{u})\right\|}\frac{n_0^2(\omega)n_e^2(\omega)}{\sqrt{n_0^4(\omega)(1-u_z^2)+n_e^4(\omega)u_z^2}}\right]^2\\ &\times\left[\left(\frac{u_xu_z}{\sqrt{1-u_z^2}}\right)^2+\left(\frac{u_yu_z}{\sqrt{1-u_z^2}}\right)^2+\left(\frac{\left(1-u_z^2\right)}{\sqrt{1-u_z^2}}\right)^2\right]=1\end{aligned} \tag{5.107}$$

Because $\vec{u}$ is also a unit vector, the second bracket term in the right hand side of Eq. (5.107) simplifies and appears to equal unity, which writes:

$$\left(\frac{u_xu_z}{\sqrt{1-u_z^2}}\right)^2+\left(\frac{u_yu_z}{\sqrt{1-u_z^2}}\right)^2+\left(\frac{\left(1-u_z^2\right)}{\sqrt{1-u_z^2}}\right)^2=1 \tag{5.108}$$

As a direct consequence, the combination of Eq. (5.107) and Eq. (5.108) implies that:

$$\left\|\vec{D}^e(\omega,\vec{u})\right\| = \varepsilon_o \frac{n_0^2(\omega)n_e^2(\omega)}{\sqrt{n_0^4(\omega)(1-u_z^2)+n_e^4(\omega)u_z^2}}\left\|\vec{E}^e(\omega,\vec{u})\right\| \tag{5.109}$$

Thanks to this equation, Eq. (5.106) turns into a simplified expression that writes in Cartesian coordinates or in spherical coordinates defined by Eq. (5.47) and Figure 5.1:

$$\vec{d}^e(\omega,\vec{u}) = \begin{pmatrix} -\dfrac{u_x u_z}{\sqrt{1-u_z^2}} \\ -\dfrac{u_y u_z}{\sqrt{1-u_z^2}} \\ \dfrac{\left(1-u_z^2\right)}{\sqrt{1-u_z^2}} \end{pmatrix} = \begin{pmatrix} -\cos(\varphi)\cos(\theta) \\ -\sin(\varphi)\cos(\theta) \\ \sin(\theta) \end{pmatrix} \tag{5.110a}$$

Thanks to Eq. (5.110a), Eq. (5.104) rewrites:

$$\vec{e}^e(\omega,\vec{u}) = \begin{pmatrix} e_x^e \\ e_y^e \\ e_z^e \end{pmatrix} = \frac{1}{\sqrt{n_0^4(\omega)\sin^2(\theta)+n_e^4(\omega)\cos^2(\theta)}}\begin{pmatrix} -n_e^2(\omega)\cos(\varphi)\cos(\theta) \\ -n_e^2(\omega)\sin(\varphi)\cos(\theta) \\ n_0^2(\omega)\sin(\theta) \end{pmatrix} \tag{5.110b}$$

The comparison between Eq. (5.104) and Eq. (5.110b) immediately shows that $\vec{d}^e(\omega,\vec{u}) = \vec{e}^e(\omega,\vec{u})$ only for $u_z = 1$, namely for $\theta = 0$ that corresponds to a propagation along $\overrightarrow{Oz}$ as previously established, and for $u_z = 0$, namely for $\theta = \pi/2$ that corresponds to a propagation at any angle φ in the $(\overrightarrow{Ox},\overrightarrow{Oy})$. Then in these two particular sets of propagation directions, the spatial walk-off angle is nil, *i.e.* $\rho_{xy}^e(\omega,\vec{u}) = 0$. At the opposite, this angle is not nil for any other direction of propagation. This angle can be calculated from Eq. (5.104) and Eq. (5.110) as follows:

$$\rho^e(\omega,\vec{u}) = Ar\cos\left[\vec{d}^e(\omega,\vec{u})\cdot\vec{e}^e(\omega,\vec{u})\right] = Ar\cos\left[\frac{n_o^{-2}(\omega)u_z^2 + n_e^{-2}(\omega)\left(1-u_z^2\right)}{\sqrt{n_o^{-4}(\omega)u_z^2 + n_e^{-4}(\omega)(1-u_z^2)}}\right] \tag{5.111}$$

Equation (5.111) can also be expressed using the angles of spherical coordinates defined by Eq. (5.47) and Figure 5.1, which gives:

$$\rho^e(\omega,\vec{u}) = Ar\cos\left[\frac{n_o^{-2}(\omega)\cos^2(\theta)+n_e^{-2}(\omega)\sin^2(\theta)}{\sqrt{n_o^{-4}(\omega)\cos^2(\theta)+n_e^{-4}(\omega)\sin^2(\theta)}}\right] \tag{5.112}$$

This expression is valid for both positive and negative uniaxial crystals. However, there is a specificity linked to the optical sign. Thanks to Eq. (5.101a) and Eq. (5.101b), such specificity can be seen by calculating $\vec{e}^{e}(\omega,\vec{u}).\vec{u}$ that gives access to the angle γ^{e} between these two vectors:

$$\cos(\gamma^{e}) = \vec{e}^{e}(\omega,\vec{u}).\vec{u} = e_x^e u_x + e_y^e u_y + e_z^e u_z = e_x^e \left[u_x + \frac{u_y^2}{u_x} - \frac{n_o^2}{n_e^2}\frac{\left(1-u_z^2\right)}{u_x} \right] \quad (5.113)$$

By using Eq. (5.101a) and Eq. (5.101b), Eq. (5.113) writes:

$$\cos(\gamma^{e}) = \vec{e}^{e}(\omega,\vec{u}).\vec{u} = -\frac{\left(n_e^2 - n_o^2\right)}{n_o^2} e_z^e u_z \quad (5.114)$$

Equation (5.114) shows that:

- $\gamma^{e} > \frac{\pi}{2}$ for $\left(n_e^2 - n_o^2\right) > 0$, *i.e.* for a positive uniaxial crystal;
- $\gamma^{e} < \frac{\pi}{2}$ for $\left(n_e^2 - n_o^2\right) < 0$, *i.e.* for a negative uniaxial crystal.

This feature is well illustrated by Figure 5.11.

Actually this specificity of orientation according to the optical sign can also be established by considering the index surface as was previously done in the case of the ordinary wave. Indeed Eq. (5.104) that gives the Cartesian coordinates of $\vec{e}^{e}(\omega,\vec{u})$ also corresponds to the coordinates of the tangent vector $\vec{T}^{e}$ of the ellipse resulting

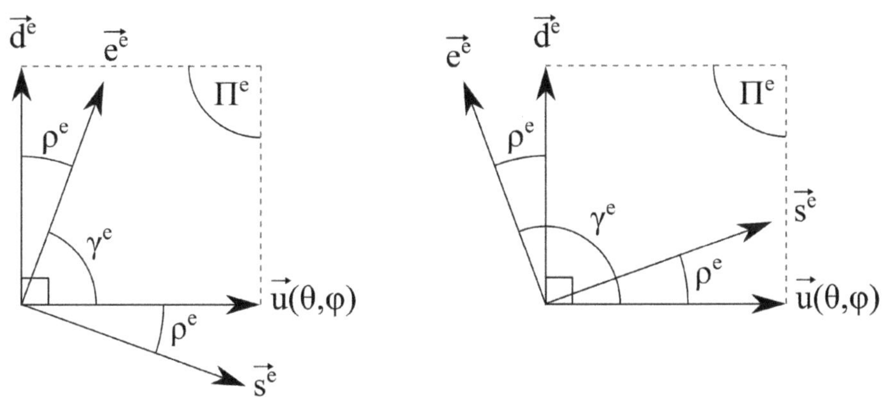

FIGURE 5.11 Extraordinary vibration plane Π^{e} of an optical electromagnetic wave propagating in a negative (left) or positive (right) uniaxial crystal: the unit vector $\vec{u}(\theta,\varphi)$ stands for any direction of propagation out of the optical axis, $\vec{s}^{e}$ is the extraordinary unit Umov-Poynting vector, $\vec{e}^{e}$ is the extraordinary unit electric field vector, $\vec{d}^{e}$ is the extraordinary unit electric displacement vector, ρ^{e} is the extraordinary spatial walk-off angle, and γ^{e} is the angle between $\vec{e}^{e}$ and $\vec{u}(\theta,\varphi)$.

from the section of the index surface by the vibration plane Π^e. The equation of the ellipse is given by $n^-(\omega,\theta)$ in Eq. (5.62) for a negative uniaxial crystal, and by $n^+(\omega,\theta)$ in Eq. (5.61) for a positive uniaxial crystal. The tangent has to be taken at the point given by the intersection of $\vec{u}$ with the ellipse, as shown in Figure (5.12). Note that the Umov-Poynting vector $\vec{s}^e(\omega,\vec{u})$ is necessarily orthogonal to the index surface at the same point since $\vec{s}^e(\omega,\vec{u}) \perp \vec{e}^e(\omega,\vec{u})$.

Figure 5.12 allows us to write the Cartesian coordinates of $\vec{e}^e$ using the angles of spherical coordinates (θ,φ), *i.e.*:

- for a negative uniaxial crystal

$$\vec{e}^e(\omega,\vec{u}) = \begin{pmatrix} -\cos(\varphi)\cos\left(\theta+\rho^e(\omega,\theta)\right) \\ -\sin(\varphi)\cos\left(\theta+\rho^e(\omega,\theta)\right) \\ \sin\left(\theta+\rho^e(\omega,\theta)\right) \end{pmatrix} \tag{5.115}$$

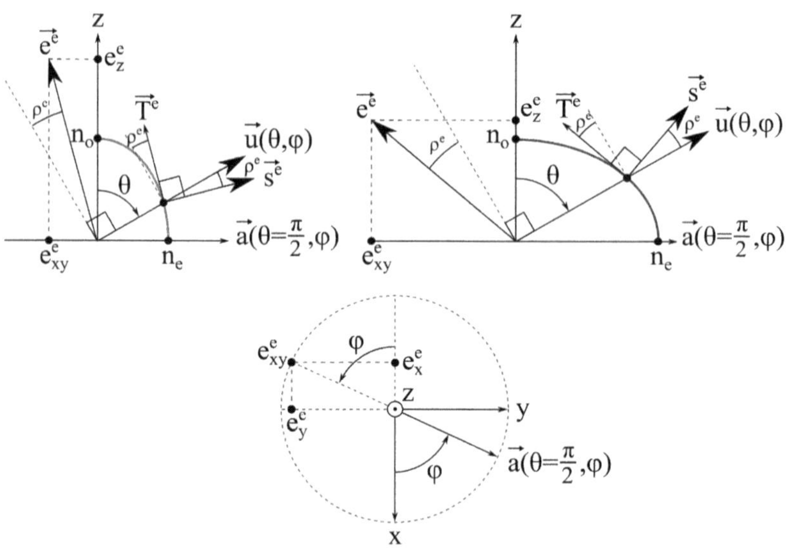

FIGURE 5.12 Relationships between the index surface and the extraordinary vibration plane Π^e of an optical electromagnetic wave propagating in a negative (top left) or positive (top right) uniaxial crystal: the index z refers to the $\overrightarrow{Oz}$ axis of the dielectric frame $(\overrightarrow{Ox},\overrightarrow{Oy},\overrightarrow{Oz})$, the unit vector $\vec{u}(\theta,\varphi)$ stands for any direction of propagation out of the optical axis, $\vec{s}^e$ is the extraordinary unit Umov-Poynting vector, $\vec{e}^e$ is the extraordinary unit electric field vector, ρ^e is the extraordinary spatial walk-off angle, $\vec{a}(\theta=\pi/2,\varphi)$ is any vector located in the $(\overrightarrow{Ox},\overrightarrow{Oy})$ plane where (θ,φ) are the angles of spherical coordinates in the dielectric frame, n_o and n_e are the ordinary and extraordinary principal refractive indices. Projection of $\vec{e}^e$ in the $(\overrightarrow{Ox},\overrightarrow{Oy})$ plane (center bottom): (x,y,z) refer to the dielectric frame, e_x^e, e_y^e, and e_{xy}^e are projections of $\vec{e}^e$ in the $(\overrightarrow{Ox},\overrightarrow{Oy})$ plane.

- for a positive uniaxial crystal

$$\vec{e}^{e}(\omega,\vec{u})=\begin{pmatrix}-\cos(\varphi)\cos\left(\theta-\rho^{e}(\omega,\theta)\right)\\-\sin(\varphi)\cos\left(\theta-\rho^{e}(\omega,\theta)\right)\\\sin\left(\theta-\rho^{e}(\omega,\theta)\right)\end{pmatrix}\tag{5.116}$$

where the spatial walk-off angle $\rho^{e}(\omega,\theta)$ is defined by Eq. (5.112).

Note that the walk-off angle $\rho^{e}(\omega,\theta)$ is thus defined as a positive angle, whatever the optical sign of the uniaxial crystal on the one hand, and of the considered direction of propagation on the other hand.

These equations directly correspond to Eq. (5.104) that is expressed as a function of the Cartesian coordinates.

Equation (5.112), Eq. (5.115), and Eq. (5.116) allow us to get the orientation of the electric field for any direction of propagation $\vec{u}$ located in one of the three principal planes of a negative or positive uniaxial crystal, *i.e.*:

- $$\vec{e}^{e}(\omega,\vec{u})=\begin{pmatrix}0\\0\\1\end{pmatrix}=\overrightarrow{Oz}\ \text{ for } \vec{u}\in(\overrightarrow{Ox},\overrightarrow{Oy})\tag{5.117a}$$

- $$\vec{e}^{e}(\omega,\vec{u})=\begin{pmatrix}-\cos\left(\theta\pm\rho^{e}(\omega,\theta)\right)\\0\\\sin\left(\theta\pm\rho^{e}(\omega,\theta)\right)\end{pmatrix}\text{ for } \vec{u}\in(\overrightarrow{Ox},\overrightarrow{Oz})\tag{5.117b}$$

- $$\vec{e}^{e}(\omega,\vec{u})=\begin{pmatrix}0\\-\cos\left(\theta\pm\rho^{e}(\omega,\theta)\right)\\\sin\left(\theta\pm\rho^{e}(\omega,\theta)\right)\end{pmatrix}\text{ for } \vec{u}\in(\overrightarrow{Oy},\overrightarrow{Oz})\tag{5.117c}$$

The sign is (+) for a negative uniaxial crystal, and (–) for a positive one.

It is possible at this step of the calculations to know the relative orientation between the ordinary and extraordinary vector configurations. Actually from Eq. (5.97) and Eq. (5.100) dealing with $\vec{d}^{o}(\omega,\vec{u})$ and $\vec{d}^{e}(\omega,\vec{u})$ respectively, and from Eq. (5.96) and Eq. (5.115) or Eq. (5.116) corresponding to $\vec{e}^{o}(\omega,\vec{u})$ and $\vec{e}^{e}(\omega,\vec{u})$ respectively, it comes:

$$\begin{cases}\vec{d}^{o}(\omega,\vec{u}).\vec{d}^{e}(\omega,\vec{u})=0\\\vec{e}^{o}(\omega,\vec{u}).\vec{e}^{e}(\omega,\vec{u})=0\end{cases}\tag{5.118}$$

Then $\vec{d}^{o}(\omega,\vec{u})\perp\vec{d}^{e}(\omega,\vec{u})$, which indicates that the ordinary and extraordinary vibration planes Π^{o} and Π^{e} are orthogonal to each other, and $\vec{e}^{o}(\omega,\vec{u})\perp\vec{e}^{e}(\omega,\vec{u})$, as shown in Figure 5.13.

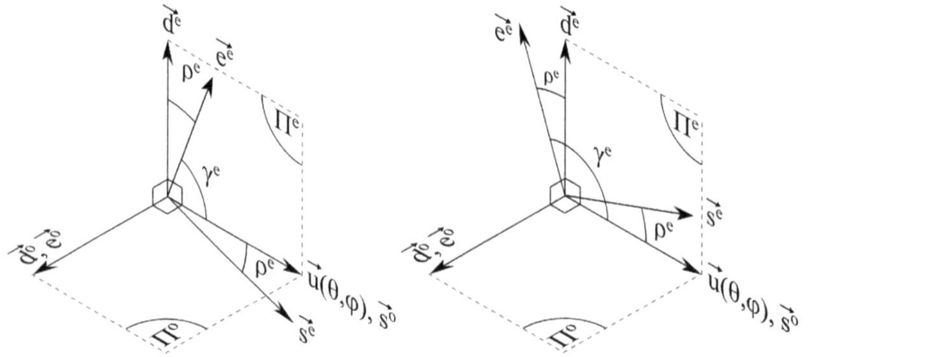

FIGURE 5.13 Ordinary and extraordinary vibration planes $\left(\Pi^o, \Pi^e\right)$ of an optical electromagnetic wave propagating in a negative (left) or positive (right) uniaxial crystal: the unit vector $\vec{u}(\theta,\varphi)$ stands for any direction of propagation out of the optical axis, $\left(\vec{s}^{\,o},\vec{s}^{\,e}\right)$ are the ordinary and extraordinary unit Umov-Poynting vectors, $\left(\vec{e}^{\,o},\vec{e}^{\,e}\right)$ are the ordinary and extraordinary unit electric field vectors, $\left(\vec{d}^{\,o},\vec{d}^{\,e}\right)$ are the ordinary and extraordinary unit electric displacement vectors, and ρ^e is the extraordinary spatial walk-off angle.

5.3.4 Biaxial Optical Class

As explained in Part 5.2.1.4, the optical biaxial class is defined by either $\varepsilon_{r_{xx}}(\omega) < \varepsilon_{r_{yy}}(\omega) < \varepsilon_{r_{zz}}(\omega)$ for the positive uniaxial class, or $\varepsilon_{r_{xx}}(\omega) > \varepsilon_{r_{yy}}(\omega) > \varepsilon_{r_{zz}}(\omega)$ for the negative uniaxial class. The three corresponding principal refractive indices are $n_x(\omega) = \sqrt{\varepsilon_{r_{xx}}(\omega)}$, $n_y(\omega) = \sqrt{\varepsilon_{r_{yy}}(\omega)}$, and $n_z(\omega) = \sqrt{\varepsilon_{r_{zz}}(\omega)}$ with $n_x(\omega) < n_y(\omega) < n_z(\omega)$ for the positive optical sign and $n_x(\omega) > n_y(\omega) > n_z(\omega)$ for the negative optical sign. Figure 5.3 and Figure 5.4 show the index surface for the two optical signs, showing that in a given direction of propagation $\vec{u}$, there are in general two possible values for the refractive index: $n^-(\omega,\vec{u})$ and $n^+(\omega,\vec{u})$ with $n^-(\omega,\vec{u}) < n^+(\omega,\vec{u})$.

Then, as in the case of the uniaxial optical class, Eq. (5.30) may have to be split into two sets of equations in the general case: one relative to $n(\omega,\vec{u}) = n^+(\omega,\vec{u})$ and the other one to $n(\omega,\vec{u}) = n^-(\omega,\vec{u})$ respectively, leading to the determination the corresponding Cartesian coordinates of the polarization Eigenmodes.

Much of the work has already been done in the previous part devoted to the uniaxial optical class, at least regarding the principal planes of the dielectric frame. Actually, as it was pointed out in Part 5.2.1.4, the sections of the biaxial index surface in $(\overrightarrow{Ox},\overrightarrow{Oy})$, $(\overrightarrow{Oy},\overrightarrow{Oz})$, and $(\overrightarrow{Ox},\overrightarrow{Oz})$ are described by circles and ellipses, which corresponds to the case of the uniaxial optical class. It means that the solutions of Eq. (5.30) will be of same kinds, so that the corresponding calculations will not be detailed in the following.

It is important to notice that the situation is different for a propagation out of the principal planes because no section of the index surface can be described by a circle or an ellipse.

Hereafter will be successively considered a direction of propagation $\vec{u}$ in $(\overrightarrow{Ox},\overrightarrow{Oy})$, $(\overrightarrow{Oy},\overrightarrow{Oz})$, $(\overrightarrow{Ox},\overrightarrow{Oz})$, and finally out of these planes.

For $\vec{u} \in (\overrightarrow{Ox},\overrightarrow{Oy})$ plane, *i.e.* at $\theta = \pi/2$, the two possible solutions $n^{+}(\omega,\vec{u})$ and $n^{-}(\omega,\vec{u})$ are given by Eq. (5.63) for $n_x(\omega) < n_y(\omega) < n_z(\omega)$ and Eq. (5.64) for $n_x(\omega) > n_y(\omega) > n_z(\omega)$. Solving Eq. (5.30) and subsequent calculations leads to:

- for $n_x(\omega) < n_y(\omega) < n_z(\omega)$,

$$\vec{e}^{+}(\omega,\vec{u}) = \vec{d}^{+}(\omega,\vec{u}) = \begin{pmatrix} 0 \\ 0 \\ 1 \end{pmatrix} = \overrightarrow{Oz} \tag{5.119}$$

$$\begin{cases} \vec{e}^{-}(\omega,\vec{u}) = \begin{pmatrix} -\sin\left(\varphi + \rho^{-}(\omega,\theta)\right) \\ \cos\left(\varphi + \rho^{-}(\omega,\theta)\right) \\ 0 \end{pmatrix} \perp \overrightarrow{Oz} \\ \vec{d}^{-}(\omega,\vec{u}) = \begin{pmatrix} -\sin(\varphi) \\ \cos(\varphi) \\ 0 \end{pmatrix} \perp \overrightarrow{Oz} \\ \rho^{-}(\omega,\vec{u}) = Ar\cos\left[\dfrac{n_y^{-2}(\omega)\cos^2(\varphi) + n_x^{-2}(\omega)\sin^2(\varphi)}{\sqrt{n_y^{-4}(\omega)\cos^2(\varphi) + n_x^{-4}(\omega)\sin^2(\varphi)}}\right] \end{cases} \tag{5.120}$$

- for $n_x(\omega) > n_y(\omega) > n_z(\omega)$,

$$\begin{cases} \vec{e}^{+}(\omega,\vec{u}) = \begin{pmatrix} -\sin\left(\varphi - \rho^{+}(\omega,\theta)\right) \\ \cos\left(\varphi - \rho^{+}(\omega,\theta)\right) \\ 0 \end{pmatrix} \perp \overrightarrow{Oz} \\ \vec{d}^{+}(\omega,\vec{u}) = \begin{pmatrix} -\sin(\varphi) \\ \cos(\varphi) \\ 0 \end{pmatrix} \perp \overrightarrow{Oz} \\ \rho^{+}(\omega,\vec{u}) = Ar\cos\left[\dfrac{n_y^{-2}(\omega)\cos^2(\varphi) + n_x^{-2}(\omega)\sin^2(\varphi)}{\sqrt{n_y^{-4}(\omega)\cos^2(\varphi) + n_x^{-4}(\omega)\sin^2(\varphi)}}\right] \end{cases} \tag{5.121}$$

$$\vec{e}^{-}(\omega,\vec{u}) = \vec{d}^{-}(\omega,\vec{u}) \begin{pmatrix} 0 \\ 0 \\ 1 \end{pmatrix} = \overrightarrow{Oz} \tag{5.122}$$

For $\vec{u} \in (\overrightarrow{Oy},\overrightarrow{Oz})$ plane, *i.e.* at $\varphi = \pi/2$, the two possible solutions $n^{+}(\omega,\vec{u})$ and $n^{-}(\omega,\vec{u})$ are given by Eq. (5.65) for $n_x(\omega) < n_y(\omega) < n_z(\omega)$ and Eq. (5.66) for $n_x(\omega) > n_y(\omega) > n_z(\omega)$. Then it comes:

- for $n_x(\omega) < n_y(\omega) < n_z(\omega)$,

$$\begin{cases} \vec{e}^{+}(\omega,\vec{u}) = \begin{pmatrix} 0 \\ -\cos\left(\theta - \rho^{+}(\omega,\theta)\right) \\ \sin\left(\theta - \rho^{+}(\omega,\theta)\right) \end{pmatrix} \\ \vec{d}^{+}(\omega,\vec{u}) = \begin{pmatrix} 0 \\ -\cos(\theta) \\ \sin(\theta) \end{pmatrix} \\ \rho^{+}(\omega,\vec{u}) = Ar\cos\left[\dfrac{n_y^{-2}(\omega)\cos^2(\theta) + n_z^{-2}(\omega)\sin^2(\theta)}{\sqrt{n_y^{-4}(\omega)\cos^2(\theta) + n_z^{-4}(\omega)\sin^2(\theta)}}\right] \end{cases} \tag{5.123}$$

$$\vec{e}^{-}(\omega,\vec{u}) = \vec{d}^{-}(\omega,\vec{u}) = \begin{pmatrix} -1 \\ 0 \\ 0 \end{pmatrix} = -\overrightarrow{Ox} \perp \overrightarrow{Oz} \tag{5.124}$$

- for $n_x(\omega) > n_y(\omega) > n_z(\omega)$,

$$\vec{e}^{+}(\omega,\vec{u}) = \vec{d}^{+}(\omega,\vec{u}) = \begin{pmatrix} -1 \\ 0 \\ 0 \end{pmatrix} = -\overrightarrow{Ox} \perp \overrightarrow{Oz} \tag{5.125}$$

$$\begin{cases} \vec{e}^{-}(\omega,\vec{u}) = \begin{pmatrix} 0 \\ -\sin(\varphi)\cos\left(\theta + \rho^{-}(\omega,\theta)\right) \\ \sin\left(\theta + \rho^{-}(\omega,\theta)\right) \end{pmatrix} \\ \vec{d}^{-}(\omega,\vec{u}) = \begin{pmatrix} 0 \\ -\cos(\theta) \\ \sin(\theta) \end{pmatrix} \\ \rho^{-}(\omega,\vec{u}) = \cos^{-1}\left[\dfrac{n_y^{-2}(\omega)\cos^2(\theta) + n_z^{-2}(\omega)\sin^2(\theta)}{\sqrt{n_y^{-4}(\omega)\cos^2(\theta) + n_z^{-4}(\omega)\sin^2(\theta)}}\right] \end{cases} \tag{5.126}$$

For $\vec{u} \in (\overrightarrow{Ox}, \overrightarrow{Oz})$, *i.e.* at $\varphi = 0$, the situation is more complicated due to the umbilical point of the index surface at $\theta = V_z$, which creates a discontinuity from either side, as shown by Figure 5.3 and Figure 5.4. That defines the two optical axes of the biaxial crystal.

For a propagation along these two directions, there is no birefringence, *i.e.* $n^+(\omega,\vec{u}) = n^-(\omega,\vec{u})$, so that the situation is the same as that along the optical axis of a uniaxial crystal, or along any direction of propagation in an isotropic medium: any direction is allowed for the polarization plane, with no spatial walk-off.

Out of the optical axes in the $(\overrightarrow{Ox}, \overrightarrow{Oz})$ plane, the two possible solutions $n^+(\omega,\vec{u})$ and $n^-(\omega,\vec{u})$ are given by Eq. (5.69) for $0 \le \theta < V_z$ and Eq. (5.70) for $V_z < \theta \le \pi/2$ in the case $n_x(\omega) < n_y(\omega) < n_z(\omega)$; and by Eq. (5.71) for $0 \le \theta < V_z$ and Eq. (5.72) for $V_z < \theta \le \pi/2$ in the case $n_x(\omega) > n_y(\omega) > n_z(\omega)$. The orientation of the electric field vector being directly linked to the topology of the two layers of the index surface as seen in Part 5.3.3, the vector configuration will be inverted from either side of an optical axis, which means that $\vec{e}^+(\omega,\vec{u})$ and $\vec{e}^-(\omega,\vec{u})$ at $0 \le \theta < V_z$ will correspond respectively to $\vec{e}^-(\omega,\vec{u})$ and $\vec{e}^+(\omega,\vec{u})$ at $V_z < \theta \le \pi/2$.

For $n_x(\omega) < n_y(\omega) < n_z(\omega)$

- at $0 \le \theta < V_z$

$$\vec{e}^+(\omega,\vec{u}) = \vec{d}^+(\omega,\vec{u}) = \begin{pmatrix} 0 \\ 1 \\ 0 \end{pmatrix} = \overrightarrow{Oy} \perp \overrightarrow{Oz} \tag{5.127}$$

$$\begin{cases} \vec{e}^-(\omega,\vec{u}) = \begin{pmatrix} -\cos(\varphi)\cos\left(\theta - \rho^-(\omega,\theta)\right) \\ 0 \\ \sin\left(\theta - \rho^-(\omega,\theta)\right) \end{pmatrix} \\ \vec{d}^-(\omega,\vec{u}) = \begin{pmatrix} -\cos(\theta) \\ 0 \\ \sin(\theta) \end{pmatrix} \\ \rho^-(\omega,\vec{u}) = \cos^{-1}\left[\dfrac{n_x^{-2}(\omega)\cos^2(\theta) + n_z^{-2}(\omega)\sin^2(\theta)}{\sqrt{n_x^{-4}(\omega)\cos^2(\theta) + n_z^{-4}(\omega)\sin^2(\theta)}} \right] \end{cases} \tag{5.128}$$

- at $V_z < \theta \le \pi/2$

$$\begin{cases} \vec{e}^{+}(\omega,\vec{u}) = \begin{pmatrix} -\cos(\varphi)\cos\left(\theta - \rho^{+}(\omega,\theta)\right) \\ 0 \\ \sin\left(\theta - \rho^{+}(\omega,\theta)\right) \end{pmatrix} \\ \vec{d}^{+}(\omega,\vec{u}) = \begin{pmatrix} -\cos(\theta) \\ 0 \\ \sin(\theta) \end{pmatrix} \\ \rho^{+}(\omega,\vec{u}) = \cos^{-1}\left[\dfrac{n_x^{-2}(\omega)\cos^2(\theta) + n_z^{-2}(\omega)\sin^2(\theta)}{\sqrt{n_x^{-4}(\omega)\cos^2(\theta) + n_z^{-4}(\omega)\sin^2(\theta)}}\right] \end{cases} \tag{5.129}$$

$$\vec{e}^{-}(\omega,\vec{u}) = \vec{d}^{-}(\omega,\vec{u}) = \begin{pmatrix} 0 \\ 1 \\ 0 \end{pmatrix} = \overrightarrow{Oy} \perp \overrightarrow{Oz} \tag{5.130}$$

For $n_x(\omega) > n_y(\omega) > n_z(\omega)$, the equations are the same as Eq. (5.127) to Eq. (5.130) by replacing (–,+) with (+,–) in the cos and sin functions.

From Eq. (5.119) to Eq. (5.130), it appears that $\vec{d}^{+}(\omega,\vec{u}).\vec{d}^{-}(\omega,\vec{u}) = 0$ in any direction of propagation $\vec{u}$ located in a principal plane of a biaxial crystal, $(\overrightarrow{Ox},\overrightarrow{Oy})$, $(\overrightarrow{Oy},\overrightarrow{Oz})$, or $(\overrightarrow{Ox},\overrightarrow{Oz})$, for both the negative and positive optical signs. It means that the Eigenmode polarization planes are orthogonal to each other, *i.e.* $\Pi^{+} \perp \Pi^{-}$. It appears also that $\vec{e}^{+}(\omega,\vec{u}).\vec{e}^{-}(\omega,\vec{u}) = 0$ so that $\vec{e}^{+}(\omega,\vec{u}) \perp \vec{e}^{-}(\omega,\vec{u})$ as shown in Figure 5.14. The general vector configuration is then the same as that in a uniaxial crystal.

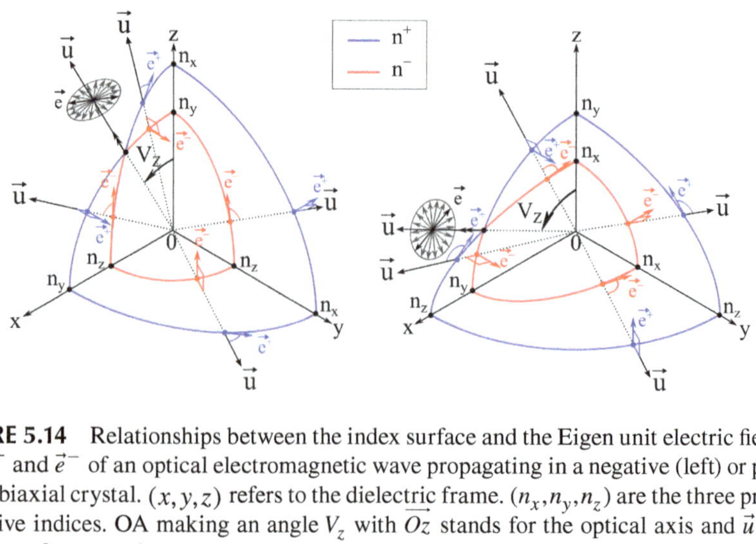

FIGURE 5.14 Relationships between the index surface and the Eigen unit electric field vectors $\vec{e}^{+}$ and $\vec{e}^{-}$ of an optical electromagnetic wave propagating in a negative (left) or positive (right) biaxial crystal. (x,y,z) refers to the dielectric frame. (n_x,n_y,n_z) are the three principal refractive indices. OA making an angle V_z with $\overrightarrow{Oz}$ stands for the optical axis and $\vec{u}$ for the direction of propagation.

The unit Umov-Poynting vector is not shown in Figure 5.14 for more visibility. Its relative orientation with the unit electric field vector is the same as that in the case of the uniaxial optical class: $\vec{s} \perp \vec{e}$ for a propagation along an optical axis, $\vec{s}^+ \perp \vec{e}^+$ and $\vec{s}^- \perp \vec{e}^-$ for a propagation out of an optical axis.

The situation corresponding to a propagation out of the principal planes cannot be simply deduced from the case of the uniaxial optical class. Thus, it is necessary to come back to Eq. (5.46) to access the equations of the two possible values of $n^+(\omega,\vec{u})$ and $n^-(\omega,\vec{u})$. The existence of these two solutions is indeed necessary for the resolution of Eq. (5.30) that writes:

$$\begin{cases} \left[\left(n^{\pm}(\omega,\vec{u})\right)^2 (1-u_x^2) - n_x^2(\omega)\right] e_x - \left(n^{\pm}(\omega,\vec{u})\right)^2 u_x u_y e_y - \left(n^{\pm}(\omega,\vec{u})\right)^2 u_x u_z e_z = 0 \\ -\left(n^{\pm}(\omega,\vec{u})\right)^2 u_y u_x e_x + \left[\left(n^{\pm}(\omega,\vec{u})\right)^2 (1-u_y^2) - n_y^2(\omega)\right] e_y - \left(n^{\pm}(\omega,\vec{u})\right)^2 u_y u_z e_z = 0 \\ -\left(n^{\pm}(\omega,\vec{u})\right)^2 u_z u_x e_x - \left(n^{\pm}(\omega,\vec{u})\right)^2 u_z u_y e_y + \left[\left(n^{\pm}(\omega,\vec{u})\right)^2 (1-u_z^2) - n_z^2(\omega)\right] e_z = 0 \end{cases} \tag{5.131}$$

Analytical solutions exist. Actually, they could be obtained by solving the nullity of the determinant of Eq. (5.131), which would lead to a polynomial of order 3 with respect to the variables $\left(n^{\pm}(\omega,\vec{u})\right)^2$ to determine. These solutions are so heavy to write that it is preferable to use a numerical integration of Eq. (5.131): one for $n^+(\omega,\vec{u})$ leading to $\vec{e}^+(\omega,\vec{u})$, and the other one for $n^-(\omega,\vec{u})$ leading to $\vec{e}^-(\omega,\vec{u})$. From these two solutions, it is possible to show that $\vec{e}^+(\omega,\vec{u}).\vec{e}^-(\omega,\vec{u}) \neq 0$ even if the corresponding vibration planes are orthogonal to each other, *i.e.* $\Pi^+ \perp \Pi^-$ as a consequence of $\vec{d}^+(\omega,\vec{u}).\vec{d}^-(\omega,\vec{u}) = 0$. It is also possible to establish that the two polarization Eigenmodes exhibit both a spatial walk-off $\rho^+(\omega,\vec{u})$ and $\rho^-(\omega,\vec{u})$, with $\rho^+(\omega,\vec{u}) \neq \rho^-(\omega,\vec{u}) \neq 0$ out of the dielectric planes, as shown in Figure 5.15 (Boulanger & Zyss 2013) [3].

Finally, it appears that neither $\vec{e}^+(\omega,\vec{u})$ nor $\vec{e}^-(\omega,\vec{u})$ is perpendicular to $\overrightarrow{Oz}$.

The frame (X,Y,Z) introduced in Figure 5.15 is a fixed frame, meaning that it does not move with the wave while propagating. It is so defined such because Z is along the direction of propagation $\vec{u}$, and the two other orthogonal axes $\overrightarrow{OX}$ and $\overrightarrow{OY}$ are defined so that they belong to the involved vibration planes for the considered propagation direction. There is no convention for the corresponding between the principal planes of the laboratory frame $\left(\overrightarrow{OX},\overrightarrow{OZ}\right)$ and $\left(\overrightarrow{OY},\overrightarrow{OZ}\right)$ and the Eigen vibration planes, Π^+ and Π^-. Then it can be chosen that Π^+ is in $\left(\overrightarrow{OX},\overrightarrow{OZ}\right)$, and Π^- in $\left(\overrightarrow{OY},\overrightarrow{OZ}\right)$ as in Figure 5.15, or the reverse, *i.e.* Π^- in $\left(\overrightarrow{OX},\overrightarrow{OZ}\right)$ and Π^+ in $\left(\overrightarrow{OY},\overrightarrow{OZ}\right)$. This laboratory frame will be extensively used in the following for describing both linear and nonlinear optical properties when the goal is to model the evolution of the electric field amplitude as a function of Z. This frame will then be the laboratory frame. It is different from the dielectric frame (x,y,z) that is linked to the crystal and in which the dielectric permittivity is expressed.

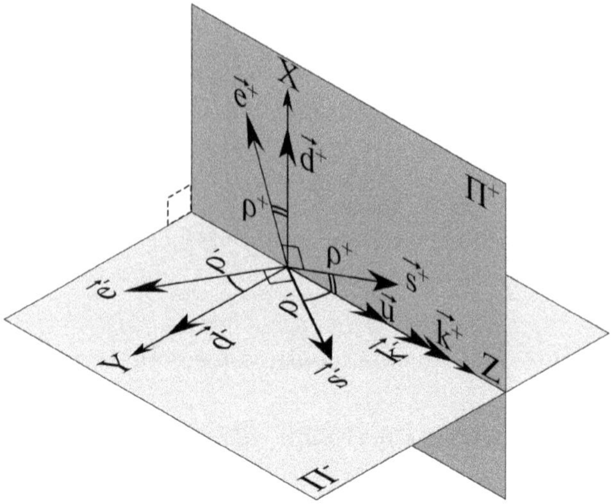

FIGURE 5.15 Eigen vibration planes $\left(\Pi^{+},\Pi^{-}\right)$ of an optical electromagnetic wave propagating in a negative or positive biaxial crystal: the unit vector $\vec{u}$ stands for any direction of propagation out of the optical axis, $\left(\vec{s}^{+},\vec{s}^{-}\right)$ are the Eigen unit Umov-Poynting vectors, $\left(\vec{e}^{+},\vec{e}^{-}\right)$ are the Eigen unit electric field vectors, $\left(\vec{d}^{+},\vec{d}^{-}\right)$ are the Eigen unit electric displacement vectors, and $\left(\rho^{+},\rho^{-}\right)$ are the associated spatial walk-off angles. $\left(X,Y,Z\right)$ is the laboratory frame, defined here by the propagation direction along the axis Z and by the orthogonal vibration planes, such a frame being different from the dielectric frame in general.

As pointed out at the end of Part 5.2.1, it is still possible to define ordinary and extraordinary waves in the case of biaxial crystals. But this analogy with uniaxial crystals can be done only in the principal planes of the dielectric frame of biaxial crystals because they are the only portions of space where the sections of the index surface are described by circles and ellipses. Nevertheless, there is often confusion about the circle being systematically assigned to the ordinary sheet and the ellipse to the extraordinary one. In other words, the ordinary wave is considered to be the only one with no spatial walk-off. It is true for a uniaxial crystal, except in the equatorial plane $(\overrightarrow{Ox},\overrightarrow{Oy})$ where the section of the index surface is described by two circles, for both the ordinary and extraordinary waves. So, in this plane, the extraordinary wave exhibits no spatial walk-off either.

At this step, it is important to notice that Eq. (5.96) or Eq. (5.97) shows that the unit ordinary electric field vector is orthogonal to $\overrightarrow{Oz}$, *i.e.*:

$$\vec{e}^{o}(\omega,\vec{u}) \perp \overrightarrow{Oz} \tag{5.132}$$

Equation (5.132) is thus the pure definition of an ordinary polarization. The extraordinary polarization can be then defined as being orthogonal to the ordinary one, *i.e.*:

$$\vec{e}^{e}(\omega,\vec{u}) \perp \vec{e}^{o}(\omega,\vec{u}) \tag{5.133}$$

TABLE 5.1
Definition of Ordinary and Extraordinary Polarization Eigenmodes in Positive Biaxial Crystals

Relation of order	$n_x(\omega) < n_y(\omega) < n_z(\omega)$			
Principal planes	$(\overrightarrow{Ox},\overrightarrow{Oy})$	$(\overrightarrow{Oy},\overrightarrow{Oz})$	$(\overrightarrow{Ox},\overrightarrow{Oz})$ $0 \le \theta < V_z$	$(\overrightarrow{Ox},\overrightarrow{Oz})$ $V_z < \theta \le \pi/2$
Eigenmodes $(\vec{e}^{\,+},\vec{e}^{\,-})$	$(\vec{e}^{\,e},\vec{e}^{\,o})$	$(\vec{e}^{\,e},\vec{e}^{\,o})$	$(\vec{e}^{\,o},\vec{e}^{\,e})$	$(\vec{e}^{\,e},\vec{e}^{\,o})$

TABLE 5.2
Definition of Ordinary and Extraordinary Polarization Eigenmodes in Negative Biaxial Crystals

Relation of order	$n_x(\omega) > n_y(\omega) > n_z(\omega)$			
Principal planes	$(\overrightarrow{Ox},\overrightarrow{Oy})$	$(\overrightarrow{Oy},\overrightarrow{Oz})$	$(\overrightarrow{Ox},\overrightarrow{Oz})$ $0 \le \theta < V_z$	$(\overrightarrow{Ox},\overrightarrow{Oz})$ $V_z < \theta \le \pi/2$
Eigenmodes $(\vec{e}^{\,+},\vec{e}^{\,-})$	$(\vec{e}^{\,o},\vec{e}^{\,e})$	$(\vec{e}^{\,o},\vec{e}^{\,e})$	$(\vec{e}^{\,e},\vec{e}^{\,o})$	$(\vec{e}^{\,o},\vec{e}^{\,e})$

As described in the dielectric planes of a biaxial crystal by Eq. (5.119) to Eq. (5.130), it appears that there is always a unit electric field vector, either $\vec{e}^{\,-}(\omega,\vec{u})$ or $\vec{e}^{\,+}(\omega,\vec{u})$, that is orthogonal to $\overrightarrow{Oz}$. As a consequence, this corresponding field in each of the three dielectric planes can be considered an ordinary mode, the other one being the extraordinary mode, as shown in Table 5.1 and Table 5.2.

Regarding the frequent confusion about the link between the topology of the index surface and the "ordinary" or "extraordinary" character of the wave, it is interesting to notice that the situation is also unusual in the $(\overrightarrow{Ox},\overrightarrow{Oy})$ plane of a biaxial crystal, as mentioned earlier in the case of a uniaxial crystal. Indeed for a biaxial crystal, it is the ordinary wave that exhibits a spatial walk-off, contrary to the extraordinary one, which can be seen in Figure 5.5, Figure 5.6, Figure 5.14, Table 5.1, and Table 5.2.

5.4 EFFECTS OF REFRACTION ON THE WAVE VECTOR AND UMOV-POYNTING VECTOR

The refraction is an optical phenomenon occurring when light propagates through two different media. One is typically the air, called here "Medium 1" that is an isotropic medium, and the other one, named "Medium 2", is a crystal that can belong to the isotropic, uniaxial, or biaxial optical classes.

Refraction, which is usually accompanied by deflection of the incident beam, is governed by the well-known Snell-Descartes law that is derived from the continuity of the tangential components of the electric field at the boundary of the two media (Born & Wolf 1980) [4].

5.4.1 Refraction between Air and an Isotropic Crystal

Figure 5.16 illustrates in a simple way the Snell-Descartes law when Medium 2, with refractive index $n_2(\omega)$, is an isotropic medium, the refractive index of Medium 1 being $n_1(\omega)$. The boundary of Medium 1 and Medium 2, also called the interface between the two media, is supposed to be planar.

As shown in Figure 5.16, the conservation of the tangential components of the wave vectors at the boundary of the two media imposes the following equalities:

$$\left\|\vec{k}_1(\omega,\vec{u}_1)\right\|\sin\left[i_1\right]=\left\|\vec{k}_2(\omega,\vec{u}_2)\right\|\sin\left[i_2(\omega)\right](=\overrightarrow{OM}) \tag{5.134}$$

where $\vec{u}_1$ and $\vec{u}_2$ are the unit vectors of the incident and refracted wave vectors, respectively.

The combination of Eq. (5.25) and Eq. (5.26) established in Part 5.2 leads in the present case to:

$$\left\|\vec{k}_i(\omega,\vec{u}_i)\right\|=\frac{\omega}{c}n_i(\omega) \tag{5.135}$$

with $i=(1,2)$.

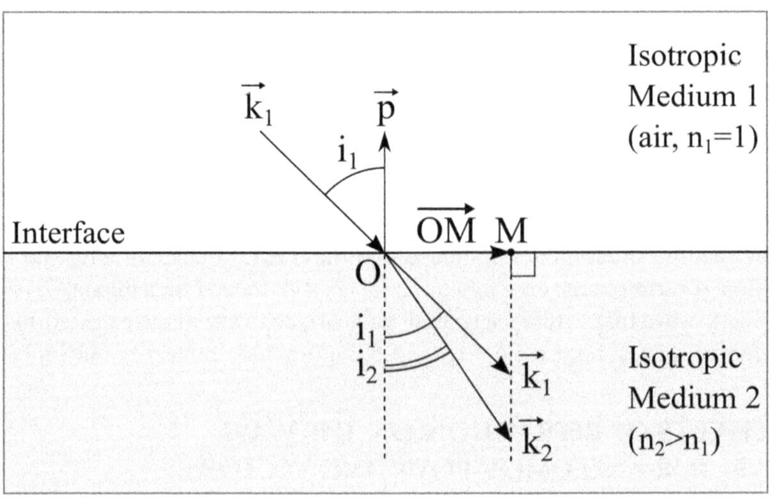

FIGURE 5.16 Scheme of refraction between air (Medium 1) and another isotropic medium (Medium 2). Unit vector $\vec{p}$ is perpendicular to the surface of the planar interface between the two media. The straight line corresponds to the section of the interface between Medium 1 and Medium 2 in the plane of incidence $(\vec{k}_1,\vec{p})$ where $\vec{k}_1$ is the wave vector of the incident beam. n_1 and n_2 are the refractive indices of the two media, with $n_1=1<n_2$. The wave vector of the refracted wave is $\vec{k}_2$. i_1 and i_2 are the incident and refracted angles, respectively. $\overrightarrow{OM}$ is the projection of both $\vec{k}_1$ and $\vec{k}_2$ on the surface of the boundary.

Then, Eq. (5.135) inserted into Eq. (5.134) immediately leads to the Snell-Descartes law as commonly stated, *i.e.*:

$$n_1(\omega)\sin[i_1] = n_2(\omega)\sin[i_2(\omega)] \tag{5.136}$$

Equation (5.136) shows that if $i_1 \neq 0$ then $i_2(\omega) \neq i_1$ when $n_2(\omega) \neq n_1(\omega)$, which means that refraction leads to a deflection of the incident beam. Furthermore, $i_2(\omega) < i_1$ when $n_2(\omega) > n_1(\omega)$, which is the case here since Medium 1 stands for the air. At the opposite, if $i_1 = 0$ then $i_2(\omega) = i_1 = 0$, and that even if $n_2(\omega) \neq n_1(\omega)$: in this trivial case, there is no deflection of the incident beam.

Angle $i_2(\omega)$ can be easily calculated using Eq. (5.136), but there exists also a geometrical way based on the use of the index surface defined in Part 5.2. The method, commonly called the Huygens construction, is described by Figure 5.17.

The different steps detailed hereafter for constructing Figure 5.17 allow us to geometrically determine the direction of the wave vector $\vec{k}_2(\omega)$ of the refracted beam. This vector is in the plane of incidence, such plane being perpendicular to the interface and containing the wave vector $\vec{k}_1(\omega)$ of the incident beam.

- In a first step, the section of the index surface of Medium 1 in the plane of incidence, namely the plane that is perpendicular to the interface and that contains the wave vector $\vec{k}_1(\omega, \vec{u}_1)$, is drawn at the level of Medium 1.

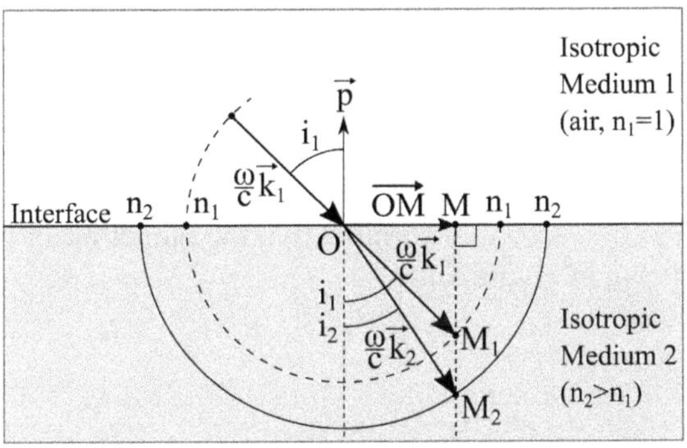

FIGURE 5.17 Use of the index surface for determining the direction of the wave vector refracted from the air (Medium 1) to another isotropic medium (Medium 2). Unit vector $\vec{p}$ is perpendicular to the surface of the planar interface between the two media. The straight line corresponds to the section of the interface between Medium 1 and Medium 2 in the plane of incidence $(\vec{k}_1, \vec{p})$ where $\vec{k}_1$ is the wave vector of the incident beam. The wave vector of the refracted wave is $\vec{k}_2$. The dashed circle stands for the section in the plane $(\vec{k}_1, \vec{p})$ of the index surface of Medium 1, that is artificially moved to Medium 2. The continuous circle corresponds to the section of the index surface of Medium 2 in the same plane as the previous one, *i.e.* $(\vec{k}_1, \vec{p})$. n_1 and n_2 are the refractive indices of the two media, with $n_1 = 1 < n_2$. $\overrightarrow{OM_1} = \frac{\omega}{c}\vec{k}_1$ with $\left\|\overrightarrow{OM_1}\right\| = n_1$, and $\overrightarrow{OM_2} = \frac{\omega}{c}\vec{k}_2$ with $\left\|\overrightarrow{OM_2}\right\| = n_2$ correspond to the incident and refracted waves, respectively, with ω the circular frequency and c the velocity of light in vacuum. $\overrightarrow{OM}$ is the projection of $\overrightarrow{OM_1}$ and $\overrightarrow{OM_2}$ on the surface of the boundary.

- In a second step, the wave vector $\vec{k}_1(\omega,\vec{u_1})$ of the incident wave is extended into Medium 2 in order to define the intersection point M_1 between $\vec{k}_1(\omega)$ and the index surface of Medium 1. Then it comes $\left\|\overrightarrow{OM_1}\right\| = n_1(\omega) = \frac{\omega}{c}\vec{k}_1(\omega,\vec{u_1})$.
- In a third step, $\overrightarrow{OM_1}$ is projected on the surface of the interface, which defines vector $\overrightarrow{OM}$.
- In a fourth step is drawn the section of the index surface of Medium 2 in the same plane as that of the index surface of Medium 1.
- In a fifth step, point M is lowered perpendicularly to the interface to the index surface of Medium 2, which defines point M_2.
- The sixth step allows us to finally determine the direction of the wave vector $\vec{k}_2(\omega,\vec{u_2})$ of the refracted beam. It is given by vector $\overrightarrow{OM_2}$ such as $\left\|\overrightarrow{OM_2}\right\| = n_2(\omega) = \frac{\omega}{c}\vec{k}_2(\omega,\vec{u_2})$.

It is important to notice that the sections of the index surfaces of Medium 1 and Medium 2 are the same in any plane containing the origin of the index surfaces since they are described by spheres in the case of isotropic media, as previously demonstrated in Part 5.2.1.2. These sections of the index surfaces of Medium 1 and Medium 2 are thus the same whatever the considered plane of incidence, for any position at the interface.

Equation (5.136) has been established by considering the wave vectors. It is also valid for the Umov-Poynting vectors since Medium 2 is an isotropic medium. Actually in this case, the wave vector is always collinear to the Umov-Poynting vector, as demonstrated in Part 5.3.2. Then i_1 and $i_2(\omega)$ correspond also to the incident and refracted Umov-Poynting vectors as shown in Figure 5.18.

The same approach can be used for modeling the refraction from Medium 2 to Medium 1, which occurs at the exit of Medium 2 once the wave has propagated through Medium 2. It is easy to determine that the output beam is collinear to the initial beam, as shown in Figure 5.18.

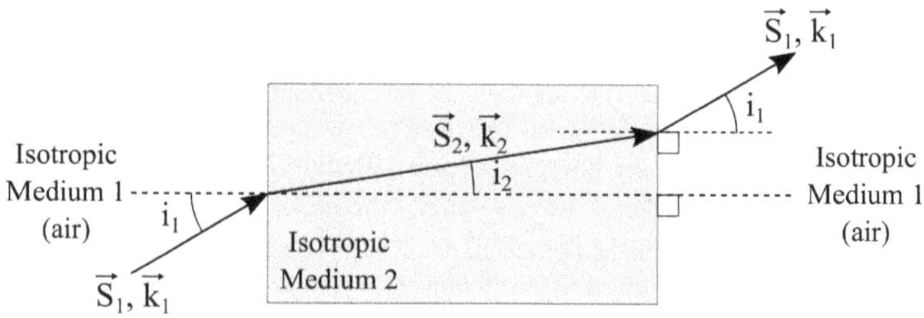

FIGURE 5.18 Configuration of wave vectors and Umov-Poynting vectors corresponding to the refraction between two isotropic media (Medium 1 and Medium 2). $\left(\vec{k}_i,\vec{S}_i\right)$ with $i=(1,2)$ are respectively the wave vector and Umov-Poynting vector in Medium 1 and Medium 2. i_1 and i_2 are the angles of propagation in Medium 1 and Medium 2, respectively.

Note that all the phenomenology described is the same regardless of the state of polarization of the incident beam since any direction of polarization is allowed in an isotropic medium.

5.4.2 Refraction between Air and Uniaxial and Biaxial Crystals

5.4.2.1 Double Refraction

The transition to the case in which Medium 2 is anisotropic, belonging to the uniaxial or biaxial optical class, can be done easily using the geometrical approach based on the index surface as was done previously for isotropic crystals. But here, the two layers of the index surface do not coincide in the general case, which leads to two possible directions for the wave vector of the refracted beam. This phenomenon is called double refraction. Figure 5.19 depicts the corresponding situation in the case in which the incident beam is oblique with respect to the boundary surface of the two media.

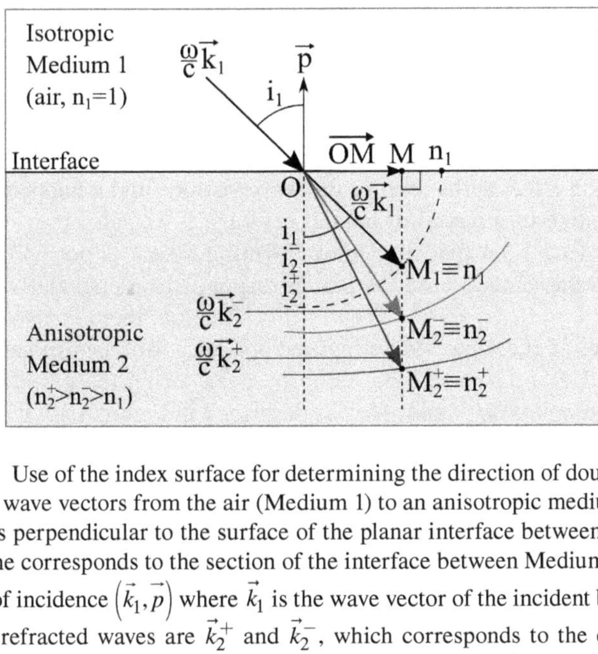

FIGURE 5.19 Use of the index surface for determining the direction of double refraction at the level of the wave vectors from the air (Medium 1) to an anisotropic medium (Medium 2). Unit vector $\vec{p}$ is perpendicular to the surface of the planar interface between the two media. The straight line corresponds to the section of the interface between Medium 1 and Medium 2 in the plane of incidence $\left(\vec{k}_1, \vec{p}\right)$ where $\vec{k}_1$ is the wave vector of the incident beam. The wave vectors of the refracted waves are $\vec{k}_2^+$ and $\vec{k}_2^-$, which corresponds to the effect of double refraction. The dashed circle stands for the section by plane $\left(\vec{k}_1, \vec{p}\right)$ of the index surface of Medium 1, that is artificially moved to Medium 2. The red and blue continuous curves correspond respectively to pieces of sections in plane $\left(\vec{k}_1, \vec{p}\right)$ of the internal and external layers of the index surface of Medium 2. n_1 is the refraction index of the air. n_2^+ and n_2^- are the refractive indices of Medium 2 ($n_2^{\pm} > n_1$) associated with $\vec{k}_2^+$ and $\vec{k}_2^-$. $\overrightarrow{OM_1} = \frac{\omega}{c}\vec{k}_1$ with $\left\|\overrightarrow{OM_1}\right\| = n_1$, and $\overrightarrow{OM_2}^{\pm} = \frac{\omega}{c}\vec{k}_2^{\pm}$ with $\left\|\overrightarrow{OM_2}^{\pm}\right\| = n_2^{\pm}$, ω being the circular frequency and c the velocity of light in vacuum. $\overrightarrow{OM}$ is the projection of $\overrightarrow{OM_1}$ and $\overrightarrow{OM_2}^{\pm}$ on the surface of the boundary. i_1 and $i_2^{\pm}$ are the incident and refracted angles, respectively.

Figure 5.19 describes the conservation of the tangential components of the wave vectors at the boundary of the two media, which writes:

$$\frac{c}{\omega}\left\|\vec{k}_1(\omega,\overrightarrow{u_1})\right\|\sin\left[i_1\right]=\frac{c}{\omega}\left\|\vec{k}_2^+(\omega,\vec{u}_2^+)\right\|\sin\left[i_2^+(\omega)\right]=\frac{c}{\omega}\left\|\vec{k}_2^-(\omega,\vec{u}_2^-)\right\|\sin\left[i_2^-(\omega)\right] \quad (5.137)$$

where $\vec{u}_1$ and $\vec{u}_2^\pm$ are the unit vectors of the wave vectors of the incident and refracted beams.

By combining Eq. (5.135) and Eq. (5.137) comes the Snell-Descartes law relative to an anisotropic medium:

$$n_1(\omega,\vec{u}_1)\sin\left[i_1\right]=n_2^+(\omega,\vec{u}_2^+)\sin\left[i_2^+(\omega)\right]=n_2^-(\omega,\vec{u}_2^-)\sin\left[i_2^-(\omega)\right] \quad (5.138)$$

It is interesting to distinguish the two possible cases regarding the incident angle: $i_1 \neq 0$ as shown Figure 5.19 (oblique incidence), and $i_1 = 0$ (normal incidence).

Equation (5.138) and Figure 5.19 show that if $i_1 \neq 0$, then $i_2^+(\omega) \neq i_2^-(\omega) \neq i_1$ when $n_2^+(\omega,\vec{u}_2^+) \neq n_2^-(\omega,\vec{u}_2^-) \neq n_1(\omega,\vec{u}_1)$, which means that the refraction leads to a deflection with respect to the incident wave vector $\vec{u}_1$, this deflection occurring in two distinct directions, *i.e.* $\vec{u}_2^+$ and $\vec{u}_2^-$. This phenomenon corresponds to the double refraction effect at the level of the wave vectors, and it happens in any direction of propagation in uniaxial or biaxial crystals. It was also previously shown in Part 5.3.3 and Part 5.3.4 that the Umov-Poynting vector is not collinear with the wave vector in the general case, which corresponds to the spatial walk-off effect, *i.e.* $\rho_2^+(\omega,\vec{u}_2^+) \neq \rho_2^-(\omega,\vec{u}_2^-) \neq 0$. Then in this situation, there is also double refraction at the level of the Umov-Poynting vectors. The two corresponding directions $\vec{S}_2^+(\omega,\vec{u}_2^+)$ and $\vec{S}_2^-(\omega,\vec{u}_2^-)$ can be found by considering the perpendicular to the index surface at points M_2^+ and M_2^-, as described in Figure 5.20. It is important to notice that such double refraction behavior starts as soon as light propagates in the anisotropic medium, thus directly from the interface.

Figure 5.20 well shows that the deflection of the Umov-Poynting vectors with respect to the incident beam ($i_1 - i_2^\pm - \rho_2^\pm$) is not the same as that of the wave vectors ($i_1 - i_2^\pm$) due to the spatial walk-off effect ($\rho_2^\pm \neq 0$). Furthermore, the deflection is not the same for the two Umov-Poynting vectors: $(i_1 - i_2^+ - \rho_2^+) \neq (i_1 - i_2^- - \rho_2^-)$ due to $(i_2^+ \neq i_2^-)$ and $(\rho_2^+ \neq \rho_2^-)$.

Two examples of calculation of i_2^+ and i_2^- are given here in the case of a positive uniaxial crystal ($n_o(\equiv n_2^-) < n_e(\equiv n_2^+)$).

- In the first example, the optical axis (OA) of Medium 2 is perpendicular to the plane of incidence of light. Thus the section by this plane of the index surface of Medium 2 gives two circles: one of radius $n_o(\equiv n_2^-)$ corresponding to the internal layer, and the other one of radius $n_e(\equiv n_2^+)$ corresponding to the external layer, as shown in Figure 5.21.

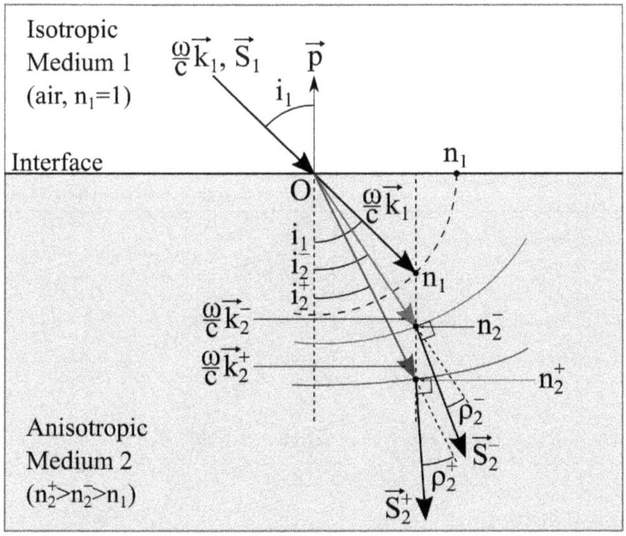

FIGURE 5.20 Use of the index surface for determining the direction of double refraction in oblique incidence at the level of the Umov-Poynting vectors from the air (Medium 1) to an anisotropic medium (Medium 2). Unit vector $\vec{p}$ is perpendicular to the surface of the plane interface between the two media. The straight line corresponds to the section of the interface between Medium 1 and Medium 2 in the plane of incidence $\left(\vec{k}_1, \vec{p}\right)$ where $\vec{k}_1$ is the wave vector of the incident beam. $\vec{S}_1(//\vec{k}_1)$ is the corresponding Umov-Poynting vector. The wave vectors of the refracted waves are $\vec{k}_2^+$ and $\vec{k}_2^-$. The dashed circle stands for the section in the plane $\left(\vec{k}_1, \vec{p}\right)$ of the index surface of Medium 1, that is artificially moved to Medium 2. The continuous curves correspond to pieces of sections in plane $\left(\vec{k}_1, \vec{p}\right)$ of the internal and external layers of the index surface of Medium 2. n_1 is the refraction index of the air. n_2^+ and n_2^- are the refractive indices of Medium 2 associated with $\vec{k}_2^+$ and $\vec{k}_2^-$. i_1 and $i_2^\pm$ are the angles of the incident and refracted wave vectors, respectively. $\vec{S}_2^+$ and $\vec{S}_2^-$ are the Umov-Poynting vectors of the refracted beam making the spatial walk-off angles ρ_2^+ and ρ_2^- with $\vec{k}_2^+$ and $\vec{k}_2^-$, respectively.

Equation (5.138) writes in the present case:

$$n_1(\omega)\sin\left[i_1\right] = n_e(\omega)\sin\left[i_e(\omega)\right] = n_o(\omega)\sin\left[i_o(\omega)\right] \tag{5.139}$$

It is then easy to calculate the refraction angles $i_o(\omega)$ and $i_e(\omega)$, *i.e.*:

$$\begin{cases} i_o(\omega) = arc\sin\left[\dfrac{n_1(\omega)}{n_o(\omega)}\sin(i_1)\right] \\ i_e(\omega) = arc\sin\left[\dfrac{n_1(\omega)}{n_e(\omega)}\sin(i_1)\right] \end{cases} \tag{5.140}$$

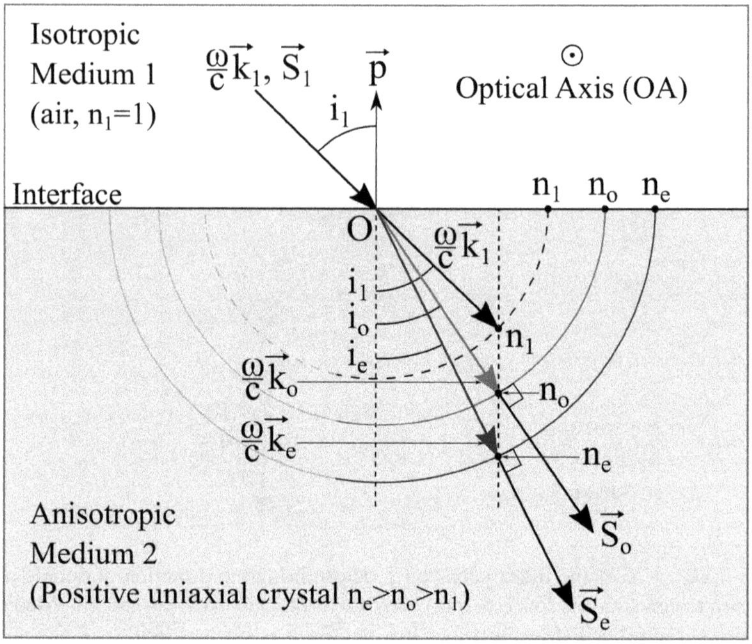

FIGURE 5.21 Double refraction in oblique incidence in the case of a positive uniaxial crystal for which the optical axis (OA) is along the interface and perpendicular to the plane of incidence $(\vec{k}_1, \vec{p})$ where $\vec{k}_1$ is the wave vector of the incident beam. n_1 is the refractive index of Medium 1, n_o and n_e are respectively the ordinary and extraordinary principal refractive indices of Medium 2. $\vec{k}_o$ and $\vec{k}_e$ are the ordinary and extraordinary wave vectors in Medium 2, respectively. $\vec{S}_o$ and $\vec{S}_e$ are the ordinary and extraordinary wave vectors in Medium 2, respectively. i_1 is the incidence angle, i_o and i_e are the ordinary and extraordinary refraction angles, respectively.

These two angles are also relative to the Umov-Poynting vectors $\vec{S}_o$ and $\vec{S}_e$ since $\vec{S}_o // \vec{k}_o$ and $\vec{S}_e // \vec{k}_e$.

- In the second example, the optical axis (OA) of Medium 2 is contained in the plane of incidence and the plane of interface between the two media. Thus the section of the index surface of Medium 2 by the considered plane of incidence gives a circle corresponding to the internal layer, and an ellipse corresponding to the external layer, as shown in Figure 5.22.

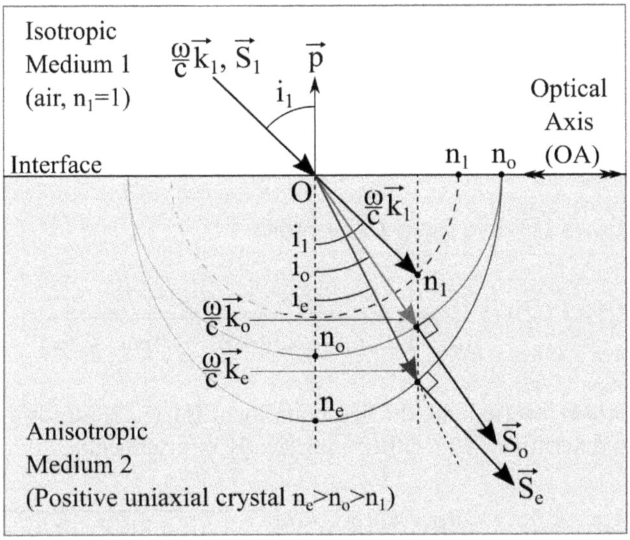

FIGURE 5.22 Double refraction in oblique incidence in the case of a positive uniaxial crystal for which the optical axis (OA) is contained in the plane of interface between Medium 1 and Medium 2 and in the plane of incidence $\left(\vec{k}_1,\vec{p}\right)$ where $\vec{k}_1$ is the wave vector of the incident beam. n_1 is the refractive index of Medium 1, n_o and n_e are respectively the ordinary and extraordinary principal refractive indices of Medium 2. $\vec{k}_o$ and $\vec{k}_e$ are the ordinary and extraordinary wave vectors in Medium 2, respectively. $\vec{S}_o$ and $\vec{S}_e$ are the ordinary and extraordinary wave vectors in Medium 2, respectively. i_1 is the incidence angle; i_o and i_e are the ordinary and extraordinary refraction angles, respectively. $n_e(i_e)$ is the extraordinary refractive index at angle i_e.

Equation (5.138) writes now:

$$n_1(\omega)\sin\left[i_1\right]=n_e\left(\omega,i_e(\omega)\right)\sin\left[i_e(\omega)\right]=n_o(\omega)\sin\left[i_o(\omega)\right] \tag{5.141}$$

The calculation of $i_o(\omega)$ is as easy as in the first example, *i.e.* from Eq. (5.141):

$$i_o(\omega)=arc\sin\left[\frac{n_1(\omega)}{n_o(\omega)}\sin(i_1)\right] \tag{5.142}$$

It is more complicated for $i_e(\omega)$ because it requires the knowledge of the corresponding refractive index, *i.e.* $n_e\left(\omega,i_e(\omega)\right)$, which can be obtained from the equation of the index surface. Equation (5.61) has to be used where $n^+(\omega,\theta)$ corresponds to

$n_e\left(\omega, i_e(\omega)\right)$. It is then necessary to change the variable, from the angle of spherical coordinates θ to the refraction angle $i_e(\omega)$. From Figure 5.1 and Figure 5.21, it is easy to show that these two angles verify:

$$i_e(\omega) + \theta = \frac{\pi}{2} \tag{5.143}$$

By inserting Eq. (5.143) into Eq. (5.61), it comes:

$$n_e(\omega, i_e(\omega)) = \frac{1}{\sqrt{n_o^{-2}(\omega)\sin^2\left[i_e(\omega)\right] + n_e^{-2}(\omega)\cos^2\left[i_e(\omega)\right]}} \tag{5.144}$$

Finally, Eq. (5.144) inserted into the first equality of Eq. (5.138) leads to the equation allowing us to determine $i_e(\omega)$, either numerically or graphically, *i.e.*:

$$\operatorname{cotan}\left[i_e(\omega)\right] = \frac{\cos\left[i_e(\omega)\right]}{\sin\left[i_e(\omega)\right]} = n_e(\omega)\sqrt{n_1^{-2}(\omega)\sin^{-2}\left[i_1\right] - n_o^{-2}(\omega)} \tag{5.145}$$

Beyond those two examples, that kind of calculation can be performed for any other situation of orientation of both a positive and a negative uniaxial crystal with respect to the plane of incidence of light, and also for any negative or positive biaxial crystal.

The refraction from Medium 2 to Medium 1 can be done using the same method, as described by Figure 5.23 corresponding to the vector configuration of Figure 5.20.

As shown in Figure 5.23, the angle of propagation is the same for the incident and output beams propagating in Medium 1. However, due to double refraction, there are two exit beams that propagate parallel to each other and are distant from each other by the quantity:

$$\delta(\omega) = L\left\{\tan\left[i_2^-(\omega) - \rho_2^-(\omega)\right] - \tan\left[i_2^+(\omega) - \rho_2^+(\omega)\right]\right\} \tag{5.146}$$

It is important to notice that the Snell-Descartes law, described by Eq. (5.138), is relative to the wave vectors, not to the Umov-Poynting vectors. However, it has to be applied where the light energy is present. It is why, in Figure 5.23, the origins of vectors $\vec{S}_1^+$ and $\vec{S}_1^-$ are located where the directions of vectors $\vec{S}_2^+$ and $\vec{S}_2^-$ intersect the exit surface.

The case of normal incidence, *i.e.* $i_1 = 0$, corresponds also to many situations in optics. In this situation, Eq. (5.138) leads to $i_2^+(\omega) = i_2^-(\omega) = i_1 = 0$, and that even if $n_2^+(\omega) \neq n_2^-(\omega) \neq n_1(\omega)$. That means that there is no double refraction at the level of the wave vectors. However, double refraction at the level of the Umov-Poynting vectors exists when spatial walk-off occurs, which is the case for a propagation out of the principal axes of the dielectric frame $(\underline{Ox}, \underline{Oy}, \underline{Oz})$ for both uniaxial and biaxial crystals. This situation is described by Figure 5.24.

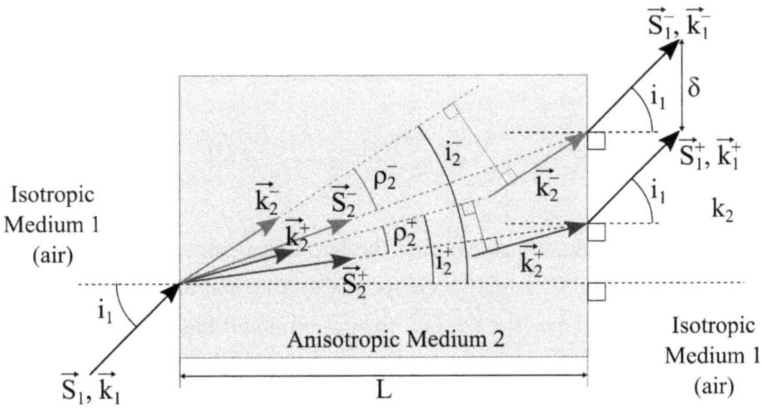

FIGURE 5.23 Configuration of wave vectors and Umov-Poynting vectors corresponding to the refraction in oblique incidence between two isotropic media (Medium 1 and Medium 2). The $\vec{k}$ and $\vec{S}$ stand for the different wave vectors and Umov-Poynting vectors: $(\vec{k}_1,\vec{S}_1)$ for the incident wave in Medium 1, $(\vec{k}_2^{\,+},\vec{S}_2^{\,+})$ and $(\vec{k}_2^{\,-},\vec{S}_2^{\,-})$ for the refracted wave from Medium 1 to Medium 2, $(\vec{k}_1^{\,+},\vec{S}_1^{\,+})$ and $(\vec{k}_1^{\,-},\vec{S}_1^{\,-})$ for the refracted wave from Medium 2 to Medium 1. i_1 is the angle of propagation in Medium 1; i_2^+ and i_2^- are the angles of propagation of the two possible Eigen polarization modes in Medium 2, ρ_2^+ and ρ_2^- are the corresponding angles of spatial walk-off. δ is the distance between the two parallel Umov-Poynting vectors $\vec{S}_1^{\,+}$ and $\vec{S}_1^{\,-}$. L is the geometrical length of Medium 2.

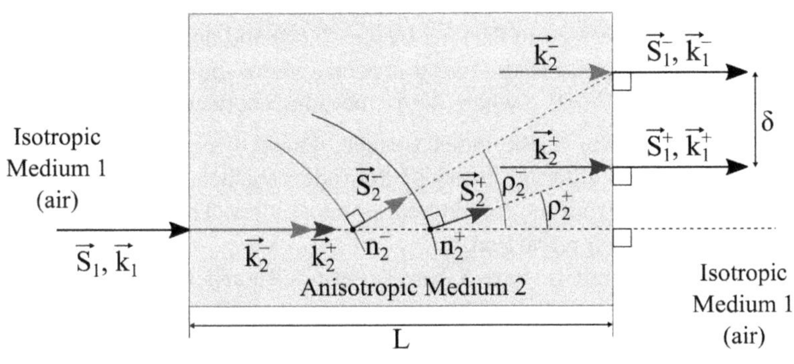

FIGURE 5.24 Configuration of wave vectors and Umov-Poynting vectors corresponding to the refraction in normal incidence between an isotropic medium (Medium 1) and a uniaxial or biaxial crystal (Medium 2). The $\vec{k}$ and $\vec{S}$ stand for the different wave vectors and Umov-Poynting vectors: $(\vec{k}_1,\vec{S}_1)$ for the incident wave in Medium 1, $(\vec{k}_2^{\,+},\vec{S}_2^{\,+})$ and $(\vec{k}_2^{\,-},\vec{S}_2^{\,-})$ for the refracted wave from Medium 1 to Medium 2, $(\vec{k}_1^{\,+},\vec{S}_1^{\,+})$ and $(\vec{k}_1^{\,-},\vec{S}_1^{\,-})$ for the refracted wave from Medium 2 to Medium 1. The continuous curves correspond to pieces of sections of the internal and external layers of the index surface of Medium 2. n_2^+ and n_2^- are the refractive indices of Medium 2 associated with $\vec{k}_2^{\,+}$ and $\vec{k}_2^{\,-}$. ρ_2^+ and ρ_2^- are the corresponding angles of spatial walk-off. δ is the distance between the two parallel Umov-Poynting vectors $\vec{S}_1^{\,+}$ and $\vec{S}_1^{\,-}$. L is the geometrical length of Medium 2.

As in the case of oblique incidence, the two refracted beams at the exit of Medium 2 propagate at the same angle, here $i_1 = 0$, and are separated by a distance $\delta(\omega)$ given by:

$$\delta(\omega) = L\left\{\tan\left[\rho_2^-(\omega)\right] - \tan\left[\rho_2^+(\omega)\right]\right\} \tag{5.147}$$

Equation (5.147) immediately shows that the two exit beams coincide, *i.e.* $\delta(\omega) = 0$, when $\rho_2^-(\omega) = \rho_2^+(\omega) = 0$, which corresponds to a propagation along the principal axes of the dielectric frame $(\overrightarrow{Ox}, \overrightarrow{Oy}, \overrightarrow{Oz})$ for uniaxial and biaxial crystals, and in the $(\overrightarrow{Ox}, \overrightarrow{Oy})$ plane for uniaxial crystal.

Independently of the considerations developed earlier, it is obvious that double refraction can only occur if both modes of propagation (+) and (–) are excited simultaneously. That can be effective by using an unpolarized incident wave or a wave linearly polarized showing non-zero projection in each of the two vibration planes, Π^+ or Π^- as defined in Figure 5.13 and Figure 5.15. This polarization feature will be addressed again in Part 5.5.

5.4.2.2 Conical Refraction

Conical refraction is the only phenomenon of linear crystal optics that is specific to the biaxial optical class. We are interested here in the so-called "internal conical refraction" that occurs when the light propagating in Medium 2 is along one of the two optical axes (OA) of any positive or negative biaxial crystal. The optical axes have been previously defined in Part 5.2 by Eq. (5.68) and depicted in Figure 5.5 and Figure 5.6: they correspond to the two directions where the birefringence is nil, *i.e.* $n^-(\omega, \theta = V_z) = n^+(\omega, \theta = V_z)$ where V_z is the angle between the wave vector and OA in the plane $(\overrightarrow{Ox}, \overrightarrow{Oz})$ of the index surface. These directions are also those for which the internal and external layers of the index surface exhibit a discontinuity, which is a singularity from the topological point of view. This later feature is at the origin of internal conical refraction.

As explained in detail in Part 5.3 and reminded earlier in the case of double refraction, the directions of the Umov-Poynting vectors associated with a given direction of propagation can be easily obtained by considering the perpendicular to the index surface at the corresponding points of the index surface. But contrary to the case of double refraction, in which there are only two directions for the refracted Umov-Poynting vectors as shown in Figure 5.24, here there are an infinity of Umov-Poynting vectors that are continuously distributed on a hollow cone to which the optical axis belongs, this hollow cone giving birth to a hollow cylinder at the exit of the biaxial crystal, as described by Figure 5.25.

Note that the exponents (+) and (–) no longer appear in the writing of $\vec{k}_2$ and $\vec{S}_2$ in Figure 5.25 because they belong to both layers (+) and (–) at the same time, these layers being in contact along the optical axis.

The angle of aperture of the internal cone, *i.e.* ρ_2 in Figure 5.25, is directly given by the angle of spatial walk-off in the direction of the optical axis. It can be

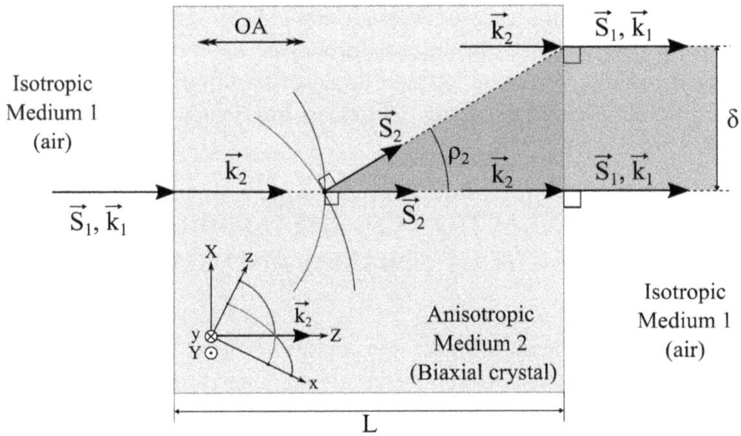

FIGURE 5.25 Configuration of wave vectors and Umov-Poynting vectors corresponding to internal conical refraction. Medium 1 stands for the air and Medium 2 for a positive or negative biaxial crystal. OA is the optical axis. The $\vec{k}$ and $\vec{S}$ stand for the different wave vectors and Umov-Poynting vectors: $(\vec{k}_1,\vec{S}_1)$ for the incident or exit wave in Medium 1, $(\vec{k}_2,\vec{S}_2)$ for the refracted waves from Medium 1 to Medium 2. (x,y,z) is the dielectric frame while (X,Y,Z) is the laboratory frame. The continuous curves correspond respectively to pieces of sections of the internal and external layers of the index surface of Medium 2. n_2^+ and n_2^-, with $n_2^+ = n_2^-$, are the refractive indices of Medium 2 corresponding to $\vec{k}_2$ along OA, *i.e.* at the umbilic of the index surface. ρ_2 is the angle of spatial walk-off. The parts delimited by the doted lines correspond to the internal cone (in Medium 2) and exit external cylinder (in Medium 1) generated by the Umov-Poynting vectors. δ is the diameter of the exit external beam with a cylindrical geometry. L is the geometrical length of Medium 2.

calculated using Eq. (5.128) and Eq. (5.129) at $\theta = V_z$ where V_z is given by Eq. (5.68). Then it comes:

$$\rho_2(\omega,V_z) = Ar\cos\left[\frac{n_x^{-2}(\omega)\cos^2(V_z)+n_z^{-2}(\omega)\sin^2(V_z)}{\sqrt{n_x^{-4}(\omega)\cos^2(V_z)+n_z^{-4}(\omega)\sin^2(V_z)}}\right] \tag{5.148}$$

The diameter of the exit hollow cone can be then calculated from Eq. (5.148) and Figure 5.25, which gives:

$$\delta(\omega) = L\tan\left[\rho_2(\omega,V_z)\right] \tag{5.149}$$

The polarization state of the refracted light will be discussed in Part 5.5.

There exists another kind of conical refraction that is called "external conical refraction". It occurs when the incident light is a convergent focused beam onto

Medium 2, contrary to the case of internal conical refraction for which it is necessary to use a parallel beam in order to propagate all the light along the optical axis. External conical refraction will not be described in the present book. But the reader can find detailed explanations in articles previously published, for example (Fève 1994, Turpin 2016) [5,6].

5.5 EFFECTS OF REFRACTION ON THE POLARIZATION STATES AT THE EXIT OF UNIAXIAL AND BIAXIAL CRYSTALS

These effects can be deduced from the considerations developed in Part 5.3 and Part 5.4. When refraction occurs between the air and another isotropic medium in normal incidence, there is no modification of the polarization state as already stipulated in Part 5.3.2. It means that light that is refracted in the air at the exit of the isotropic medium exhibits the same polarization state as the one that is incident on the isotropic medium. Thus, in this case there is no influence of refraction on polarization. The situation is even more complex when an anisotropic medium is involved, with a great diversity of situations. Indeed, there may or may not be modification of polarization depending on the direction of propagation, on the situation with respect to double refraction, and on the orientation of the polarization direction of the incident light. There are two scenarios: no change in polarization, and change. In the following, the incident wave is considered to be linearly polarized and propagating in normal incidence with the crystal entrance, which corresponds to a lot of practical situations in linear and nonlinear optics.

5.5.1 Propagation along the Optical Axis of Uniaxial Crystals

This situation is totally equivalent to the case of propagation along any direction in an isotropic medium as explained in Part 5.3.3: with normal incidence at the interface, it means no change in polarization of the exit beam compared with the incident beam, as shown in Figure 5.26.

5.5.2 Propagation out of the Optical Axes of Uniaxial or Biaxial Crystals and Incident Light Polarized in an Eigen Vibration Plane

This situation deals with a direction of propagation exhibiting or not a spatial walk-off effect at the level of the Umov-Poynting vectors as depicted in Figure 5.24 in Part 5.4.2. Concerning the case in which there is spatial walk-off, the following situations of propagation are involved: out of a principal axis of the dielectric frame $(\overrightarrow{Ox},\overrightarrow{Oy},\overrightarrow{Oz})$ in a biaxial crystal, or out of the equatorial plane $(\overrightarrow{Ox},\overrightarrow{Oy})$ in a uniaxial crystal. The complementary situations involve no spatial walk-off, namely: propagation along a principal axis of the dielectric frame in a biaxial crystal or in the equatorial plane in a uniaxial crystal.

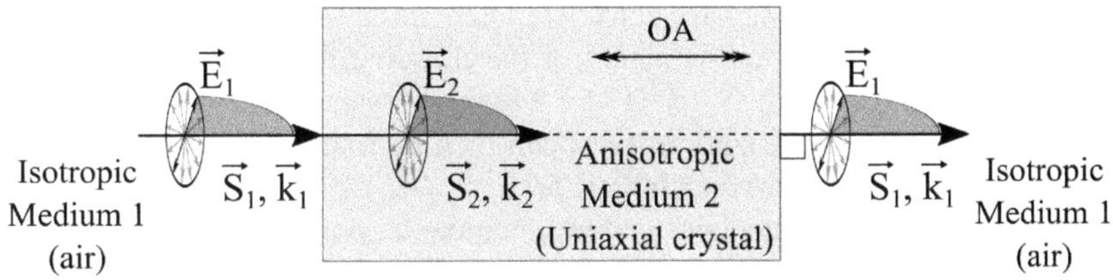

FIGURE 5.26 Polarization states corresponding to the refraction of a linearly polarized wave in normal incidence between an isotropic medium (Medium 1) and a uniaxial crystal (Medium 2) along its optical axis (OA). The $\vec{E}$, $\vec{k}$, and $\vec{S}$ stand for the different electric field vectors, wave vectors, and Umov-Poynting vectors: $(\vec{E}_1, \vec{k}_1, \vec{S}_1)$ for the incident and exit waves in Medium 1, $(\vec{E}_2, \vec{k}_2, \vec{S}_2)$ for the refracted wave from Medium 1 to Medium 2. The vibration plane in grey corresponds to any possible vibration plane of the linearly polarized wave.

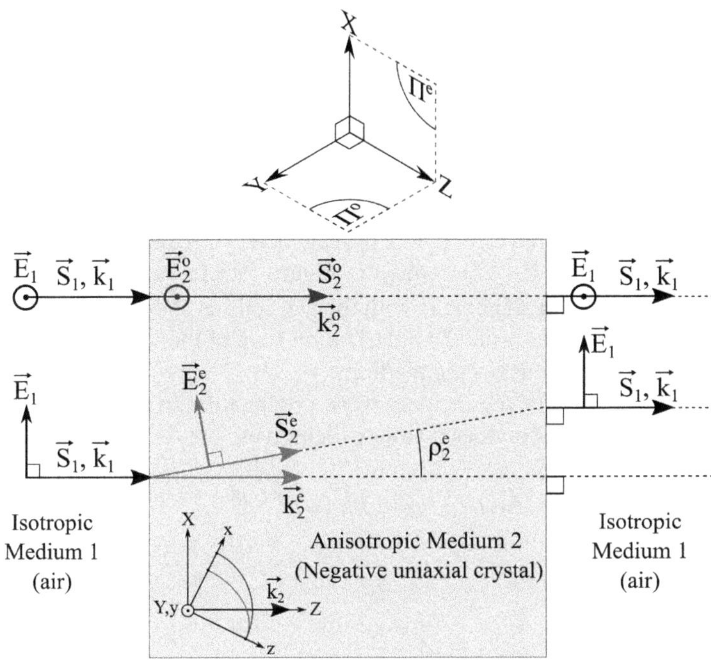

FIGURE 5.27 Polarization states corresponding to the refraction of a linearly polarized wave in normal incidence between an isotropic medium (Medium 1) and a positive uniaxial crystal (Medium 2) out of its optical axis (OA) and out of the principal plane of the dielectric frame. The $\vec{E}$, $\vec{k}$, and $\vec{S}$ stand for the different electric field vectors, wave vectors, and Umov-Poynting vectors: $\left(\vec{E}_1, \vec{k}_1, \vec{S}_1\right)$ for the incident and exit waves in Medium 1, $\left(\vec{E}_2^o, \vec{k}_2^o, \vec{S}_2^o\right)$ for the ordinary wave, and $\left(\vec{E}_2^e, \vec{k}_2^e, \vec{S}_2^e\right)$ for the extraordinary wave refracted wave from Medium 1 to Medium 2. ρ_2^e is the angle of spatial walk-off. $\left(\Pi^o, \Pi^e\right)$ are the two Eigen vibration planes associated with the considered direction of propagation. (x, y, z) is the dielectric frame while (X, Y, Z) is the laboratory frame.

Furthermore, the incident wave considered here has a linear polarization oriented in one of the two Eigen vibration planes of the chosen direction of propagation: Π^o or Π^e for a uniaxial crystal; Π^+ or Π^- for a biaxial crystal.

The corresponding polarization states can then be found by considering respectively Figure 5.13 and Figure 5.15 in Part 5.3.

In this configuration there is no modification of polarization since only one Eigenmode of polarization is excited within the crystal. Figure 5.27 gives the example of propagation out of the equatorial plane $(\overrightarrow{Ox},\overrightarrow{Oy})$ in a positive uniaxial crystal.

Figure 5.27 shows that the only effect of the output beam that is refracted in the air is a translation of the wave and Umov-Poynting vectors of the extraordinary wave in the walk-off angle plane.

5.5.3 Propagation out of the Optical Axis of Uniaxial or Biaxial Crystals, Incident Light Not Polarized in an Eigen Vibration Plane and No Double Refraction on the Umov-Poynting Vectors

Here are considered situations for which the propagation occurs along an axis of the dielectric frame (x,y,z), *i.e.* $\overrightarrow{Ox}$, $\overrightarrow{Oy}$, or $\overrightarrow{Oz}$ in the case of a biaxial crystal, or in the equatorial plane $(\overrightarrow{Ox},\overrightarrow{Oy})$ of a uniaxial crystal. Thus, there is no spatial walk-off. The incident beam has a linear polarization that is not oriented in one of the Eigen vibration planes of the considered direction of propagation. The calculations are performed in the laboratory frame (X,Y,Z) defined in Figure 5.15 in Part 5.3, and the incident polarization is taken with an angle α with the $\overrightarrow{OX}$ axis as shown in Figure 5.28.

The goal here is to determine the polarization state of the wave that is refracted in the air at the exit of the anisotropic medium.

The spatial evolution of the incident wave propagating in the air is described by Eq. (5.9) written in the laboratory frame, which gives:

$$\vec{E}(\omega,Z) = \vec{e}(\omega)E_0(\omega)e^{ik(\omega)Z} \tag{5.150}$$

with

$$k(\omega) = \frac{\omega}{c}n_{air}(\omega) \tag{5.151}$$

$n_{air}(\omega)$ is the refractive index of the air at the considered circular frequency ω.

As explained in Part 5.3, this wave is refracted in the anisotropic medium in the form of the two Eigen waves linearly polarized, these two waves corresponding to the projections of the incident polarized light onto the two polarization Eigenmodes, *i.e.*:

$$\begin{cases} \vec{E}^+(\omega,Z) = \vec{e}^+(\omega)E_0^+(\omega)e^{ik^+(\omega)Z} \\ \vec{E}^-(\omega,Z) = \vec{e}^-(\omega)E_0^-(\omega)e^{ik^-(\omega)Z} \end{cases} \tag{5.152}$$

where, according to the orientation convention taken in Figure 5.28,

$$\begin{cases} E_0^+(\omega) = E_0(\omega)\cos[\alpha] \\ E_0^-(\omega) = E_0(\omega)\sin[\alpha] \end{cases} \tag{5.153}$$

and

$$\begin{cases} k^+(\omega) = \dfrac{\omega}{c} n^+(\omega) \\ k^-(\omega) = \dfrac{\omega}{c} n^-(\omega) \end{cases} \tag{5.154}$$

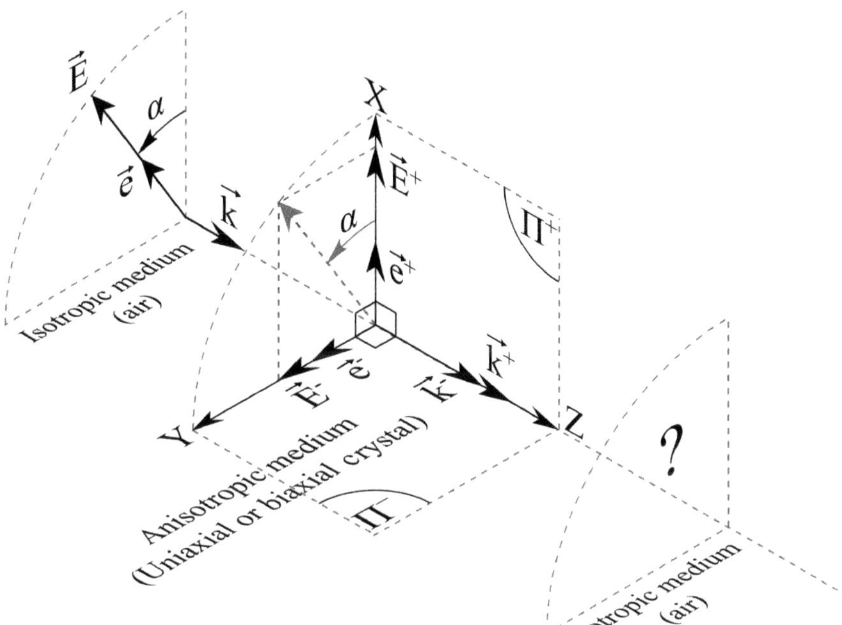

FIGURE 5.28 Polarization states corresponding to the refraction of a linearly polarized wave in normal incidence between an isotropic medium (air) and an anisotropic medium (uniaxial or biaxial crystals). (X,Y,Z) is the laboratory frame. The $\vec{E}$, $\vec{e}$, and $\vec{k}$ stand for the different electric field vectors, unit electric field vectors, and wave vectors: $(\vec{E},\vec{e},\vec{k})$ for the incident wave in the air, $(\vec{E}^{+,-},\vec{e}^{+,-},\vec{k}^{+,-})$ for the refracted waves in the Eigen vibration planes $\Pi^{+,-}$ of the anisotropic medium. All the wave vectors are along a principal axis of the dielectric frame for a biaxial crystal or in the equatorial plane for a uniaxial crystal. α is the angle between the incident linear polarization and the $\overrightarrow{OX}$ direction. The question mark means that the polarization state of the wave refracted at the exit of the anisotropic medium is not known at this stage of the study.

Thus, at the exit of the anisotropic medium, the two Eigen waves have accumulated different spatial phases, *i.e.* by taking L as the length of the anisotropic medium along the direction of propagation:

$$\begin{cases} \varphi^{+}(\omega,L)=\dfrac{\omega}{c}n^{+}(\omega)L \\ \varphi^{-}(\omega,L)=\dfrac{\omega}{c}n^{-}(\omega)L \end{cases} \tag{5.155}$$

The wave refracted in the air just at the exit of the anisotropic medium is then the linear combination of the two Eigen waves at this place. It is reminded here the absence of Eigen polarization mode in an isotropic medium, such as air. It is then interesting at this step to consider the time evolution of the wave propagating in the air. The corresponding electric field vector is described by the following expression in the laboratory frame (X,Y,Z):

$$\vec{E}(\omega,t,Z)=\begin{pmatrix} E_X \\ E_Y \\ E_Z \end{pmatrix}=\begin{pmatrix} E_{0X}\exp i[k(\omega)Z-\omega t+\varphi_X(\omega,L)] \\ E_{0Y}\exp i[k(\omega)Z-\omega t+\varphi_Y(\omega,L)] \\ 0 \end{pmatrix} \tag{5.156}$$

L is the length of the anisotropic medium, $k(\omega)$ the wave vector in air given by Eq. (5.151), and the initial phases $\varphi_X(\omega,L)$ corresponding to $\varphi^{+}(\omega,L)$ or $\varphi^{-}(\omega)$ and $\varphi_Y(\omega)$ to $\varphi^{-}(\omega,L)$ or $\varphi^{+}(\omega)$, respectively, according to the correspondence convention between the planes $(\overrightarrow{OX},\overrightarrow{OZ})$ and $(\overrightarrow{OY},\overrightarrow{OZ})$ of the laboratory frame on the one hand, and the Eigen vibration planes Π^{+} and Π^{-} on the other hand. It is important to notice that there exists another convention for writing the phase term in Eq. (5.155) which is $[\omega t-k(\omega)Z+\varphi_{X,Y}(\omega,L)]$ instead of $[k(\omega)Z-\omega t+\varphi_{X,Y}(\omega,L)]$ that is used in this book.

It is interesting for the following calculation to show the spatial dephasing between E_X and E_Y in the expression of $\vec{E}(\omega,t,Z)$. By properly resetting the origin of time, it is classically defined as $\Delta\varphi(\omega,L)=\varphi_Y(\omega,L)-\varphi_X(\omega,L)$ so that Eq. (5.156) can also write:

$$\vec{E}(\omega,t,Z)=\begin{pmatrix} E_X \\ E_Y \\ E_Z \end{pmatrix}=\begin{pmatrix} E_{0X}\exp i[k(\omega)Z-\omega t] \\ E_{0Y}\exp i[k(\omega)Z-\omega t+\Delta\varphi(\omega,L)] \\ 0 \end{pmatrix} \tag{5.157}$$

The following calculations are devoted to the study of the time evolution of the polarization state of the wave propagating in the air. It is not necessary for that purpose to use complex formalism, so that the real field, *i.e.* the real part of Eq. (5.156), can be considered, which writes:

$$\vec{E}(\omega,t,Z)=\begin{pmatrix} E_X \\ E_Y \\ E_Z \end{pmatrix}=\begin{pmatrix} E_{0X}\cos[k(\omega)Z-\omega t] \\ E_{0Y}\cos[k(\omega)Z-\omega t+\Delta\varphi(\omega,L)] \\ 0 \end{pmatrix} \tag{5.158}$$

Since α is the angle between the incident linear polarization and the $\overrightarrow{OX}$ direction, then Eq. (5.158) leads to the following parametric equation relative to the transverse coordinates of the refracted wave propagating in the air at the exit of the anisotropic medium:

$$\begin{cases} E_X = E_0(\omega)\cos\left[\alpha\right]\cos[k(\omega)Z - \omega t] \\ E_Y = E_0(\omega)\sin\left[\alpha\right]\cos[k(\omega)Z - \omega t + \Delta\varphi(\omega, L)] \end{cases} \tag{5.159}$$

where $E_0(\omega)$ is the amplitude of the incident wave, *i.e.* at the entrance of the anisotropic medium, and α the angle of its polarization with $\overrightarrow{OX}$ as shown in Figure 5.29.

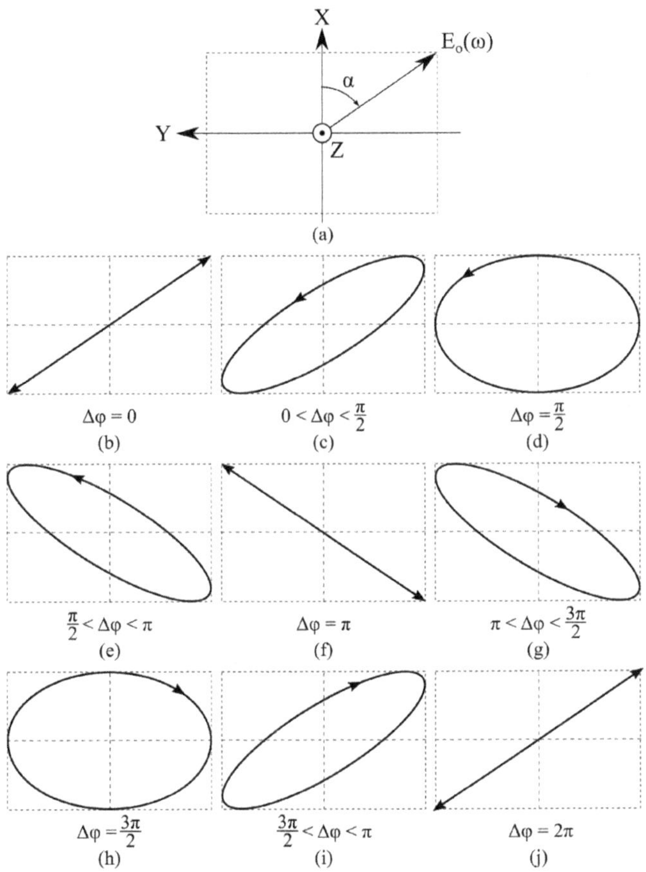

FIGURE 5.29 Polarization states of the wave at the exit of a uniaxial or biaxial crystal. (X,Y,Z) is the laboratory frame. (a) $\vec{E}(\omega)$ is the incident wave with $E_0(\omega)$ its amplitude and α the angle of polarization. Schemes (b) to (j) show the trajectory as a function of time of the electric field vector at the exit of the anisotropic medium: (b), (f), and (j) correspond to linear state of polarization while (c), (d), (e), (g), (h), and (i) depict an elliptical state of polarization. The arrows on the ellipses give the sense of rotation defining a left-handed polarization for (c), (d), and (e), and a right-handed polarization for (g), (h), and (i), the light propagating towards the reader, who is then in the fixed position of the receiver of light.

From Eq. (5.159), the following equation can be established without any difficulty:

$$\left(\frac{E_X}{\cos[\alpha]}\right)^2+\left(\frac{E_Y}{\sin[\alpha]}\right)^2-2\left(\frac{E_X}{\cos[\alpha]}\right)\left(\frac{E_Y}{\sin[\alpha]}\right)\cos[\Delta\varphi(\omega,L)]=$$
$$[E_0(\omega)]^2\sin^2[\Delta\varphi(\omega,L)] \quad (5.160)$$

Equation (5.160) is the general equation of an ellipse confined in a rectangle whose dimensions are defined by the incident wave: one side parallel to $\overrightarrow{OX}$ with a size of $2E_0(\omega)\cos[\alpha]$, and the other side parallel to $\overrightarrow{OY}$ with a size of $2E_0(\omega)\sin[\alpha]$. This means that in the general case, the end point of the electric field vector at the exit of an anisotropic medium describes an ellipse as a function of time, which corresponds to an elliptic polarization state. Equation (5.160) also shows that the form and orientation of the ellipse depends on the dephasing $\Delta\varphi(\omega,L)$ accumulated in the anisotropic medium, as shown in Figure 5.29.

There are two possible senses of travel of the end point of the electric vector describing the ellipse as a function of time. It is classically determined by the sign of $\frac{\partial E_Y}{\partial t}$ just at the exit of the anisotropic medium, *i.e.* at $Z=0$ and $t=0$. In these conditions and according to Eq. (5.159) it comes:

$$\left.\frac{\partial E_Y}{\partial t}\right|_{z=0,t=0}=\omega E_{0Y}\sin[\alpha]\sin[\Delta\varphi(\omega,L)] \quad (5.161)$$

Then $\left.\frac{\partial E_Y}{\partial t}\right|_{z=0,t=0}>0$ for $0<\Delta\varphi(\omega,L)<\pi$, which means a description of the ellipse in the anticlockwise sense leading to the so called "left-hand polarization" as depicted in Figure 5.29(c), Figure 5.29(d), and Figure 5.29(e). At the opposite, $\left.\frac{\partial E_Y}{\partial t}\right|_{z=0,t=0}<0$ for $\pi<\Delta\varphi(\omega,L)<2\pi$, which means a description of the ellipse in the clockwise sense leading to a "right-hand polarization" as shown in Figure 5.29(f), Figure 5.29(h), and Figure 5.29(i). The sense, right or left, is defined from the receiver's point of view located at a fixed position and looking at the electric field evolution in time, as shown in Figure 5.29 in which light propagates toward the reader, *i.e.* the receiver. Note that the axes of the ellipse are $\overrightarrow{OX}$ and $\overrightarrow{OY}$ when $\Delta\varphi(\omega,L)=(2p+1)\frac{\pi}{2}$ as shown in Figure 5.29(d) and Figure 5.29(h).

With adjusted length and solicited birefringence so that $\Delta\varphi(\omega,L)=(2p+1)\frac{\pi}{2}$, if the incident wave is polarized at $\alpha=\frac{\pi}{4}$ then the parallelepiped in which is inscribed the trajectory is a square of side equal to $\frac{\sqrt{2}}{2}E_0(\omega)$, which leads to a left-hand or right-hand circular polarization state according to $\Delta\varphi(\omega,L)$. The anisotropic medium is then called a "quarter wave plate" at the working wavelength λ_ω. This denomination comes from the fact that the phase shift comes from an optical path

δ which is a multiple of a quarter of the wavelength, *i.e.* $\Delta\varphi(\omega,L)=2\pi\dfrac{\delta}{\lambda_\omega}$ with $\delta=(2p+1)\dfrac{\lambda_\omega}{4}$.

Figure 5.29 also shows that there exists two situations for which the ellipse reduces to a line, which corresponds to a linear polarization state: it happens when $\Delta\varphi(\omega,L)=2p\pi$, as shown in Figure 5.29(b) and Figure 5.29(j), and $\Delta\varphi(\omega,L)=(2p+1)\pi$, as shown in Figure 5.29(f). In the first case, the wave at the exit of the anisotropic medium exhibits the same polarization as that of the wave at the entrance, which is called a "wave plate" configuration because the optical path is $\delta=p\lambda_\omega$. In the second case, the linear polarization has been rotated by an angle 2α with respect to the polarization of the incident wave; the anisotropic medium is then called a "half wave plate" since $\delta=(2p+1)\dfrac{\lambda_\omega}{2}$.

These interesting features of "wave plate", "half wave plate", and "quarter wave plate", or any other phase plates, are used in many optical experiments. When there is combination of several of such components, it is very convenient to use the Jones formalism in which the polarization state is represented by a two-component vector, the so-called Jones vector, and the anisotropic medium by a (2 × 2) matrix, called the Jones matrix (Hecht 1987, Collett 2005) [7,8]. In this case, the complex formalism is used with the wave being described by Eq. (5.156) that is rewritten in the following form:

$$\vec{E}(\omega,t,Z)=\begin{pmatrix} E_{0X}\exp i[\varphi_X] \\ E_{0Y}\exp i[\varphi_Y] \\ 0 \end{pmatrix}\exp i[k(\omega)Z-\omega t] \tag{5.162}$$

The Jones vector is a unit vector that depicts the polarization state of a polarized light, and it is then defined as:

$$J=\frac{1}{\sqrt{E_{0X}^2+E_{0Y}^2}}\begin{pmatrix} E_{0X} \\ E_{0Y}\exp i[\varphi_Y-\varphi_X] \end{pmatrix} \tag{5.163}$$

Using Eq. (5.163), it comes for example $J=\begin{pmatrix}1\\0\end{pmatrix}$ for a linear polarization in the $\overrightarrow{OX}$ direction, $J=\begin{pmatrix}0\\1\end{pmatrix}$ for a linear polarization in the $\overrightarrow{OY}$ direction, $J=\begin{pmatrix}\cos(\alpha)\\ \sin(\alpha)\end{pmatrix}$ for a linear polarization showing an angle α with the $\overrightarrow{OX}$ direction, $J=\dfrac{1}{\sqrt{2}}\begin{pmatrix}1\\1\end{pmatrix}$ for a linear polarization making an angle $\alpha=\dfrac{\pi}{4}$, $J=\dfrac{1}{\sqrt{2}}\begin{pmatrix}1\\-i\end{pmatrix}$ for a right-hand circular polarization since $\varphi_Y-\varphi_X=\dfrac{3\pi}{2}$, and $J=\dfrac{1}{\sqrt{2}}\begin{pmatrix}1\\i\end{pmatrix}$ for a left-hand circular polarization since $\varphi_Y-\varphi_X=\dfrac{\pi}{2}$.

This formalism will not be further developed here. More information, in particular the action on the Jones matrices on the Jones vectors, can be found in Hecht (1987) and Collett (2005) [7,8].

5.5.4 Propagation out of the Optical Axis of Uniaxial or Biaxial Crystals, Incident Light Not Polarized in an Eigen Vibration Plane and Double Refraction on the Umov-Poynting Vectors

Here the propagation occurs out of the principal axes of the dielectric frame $\overrightarrow{Ox}$, $\overrightarrow{Oy}$, or $\overrightarrow{Oz}$ in the case of a biaxial crystal, or out of the equatorial plane $(\overrightarrow{Ox},\overrightarrow{Oy})$ of a uniaxial crystal, so that there is a spatial walk-off effect. The incident beam has a linear polarization that is not oriented in one of the Eigen vibration planes of the considered direction of propagation with normal incidence to the interface. Two cases have to be considered according to the distance $\delta(\omega)$ between the two beams refracted at the exit of the anisotropic medium, as defined by Eq. (5.147).

If $\delta(\omega) > D$, in which D is the diameter of the beam, then the refraction at the exit of the anisotropic medium gives rise to two collinear but well-separated beams, *i.e.* without transverse spatial overlap, and with respectively orthogonal polarizations in the Eigen vibration planes as shown in the top Figure 5.30 in which a uniaxial crystal is taken as an example.

- If $\delta(\omega) < D$, then the refraction at the exit of the anisotropic medium leads to partially overlapped beams, and thus to a mixing between two linear polarization states and one elliptic polarization state in the common overlapping volume of the two beams. Such a situation leads to a complex transverse polarization structure of the refracted beam as depicted in the bottom of Figure 5.30.

5.5.5 Propagation along the Optical Axes of Biaxial Crystals

This situation corresponds to internal conical refraction described in Part 5.4.2.2 for the Umov-Poynting vector point of view. As it was shown, the optical axis does not create a discontinuity of the angular distribution of the spatial walk-off in the $(\overrightarrow{Ox},\overrightarrow{Oz})$ plane of the dielectric frame, which also appears well in Figure 5.25. Due to this continuity, the two rays of the cone located in this plane necessarily exhibit polarization states which are identical to those from either side of the optical axis, *i.e.* for $0 \leq \theta < V_z$ and $V_z < \theta \leq 90°$ in which V_z is the optical axis angle defined by Eq. (5.68). It means that the ray along the optical axis is polarized along the $\overrightarrow{Oy}$ axis while the diametrically opposite ray is polarized in the $(\overrightarrow{Ox},\overrightarrow{Oz})$ plane, as shown in Figure 5.31. The fact that the cone is continuous between these two rays necessarily leads to a continuity at the level of the polarization states (Fève 1994) [5]. The consequence is a continuous distribution of linear polarizations ranging between the $\overrightarrow{Oy}$ axis and the $(\overrightarrow{Ox},\overrightarrow{Oz})$ plane as depicted in Figure 5.31. By locating each ray by the polar coordinate angle β in the laboratory frame $(\overrightarrow{OX},\overrightarrow{OY},\overrightarrow{OZ})$, it comes for the corresponding unit electric field vector:

$$\vec{e}(\beta) = \begin{pmatrix} e_X \\ e_Y \\ e_Z \end{pmatrix} = \begin{pmatrix} \sin\left[\dfrac{\beta}{2}\right] \\ \cos\left[\dfrac{\beta}{2}\right] \\ 0 \end{pmatrix} \tag{5.164}$$

with $0 \leq \beta \leq \pi$.

Thus, $\beta = 0$ and $\beta = \pi$ respectively depict the extraordinary and ordinary polarization states of the cone located in the $(\overrightarrow{Ox}, \overrightarrow{Oz})$ plane of the dielectric frame.

This intuitive topological demonstration can be validated by a full calculation based on the angular distribution of the unit electric field given by Eq. (5.119) to Eq. (5.131). A (4 × 4) matrix formalism helps the calculation (De Smet 1993) [9].

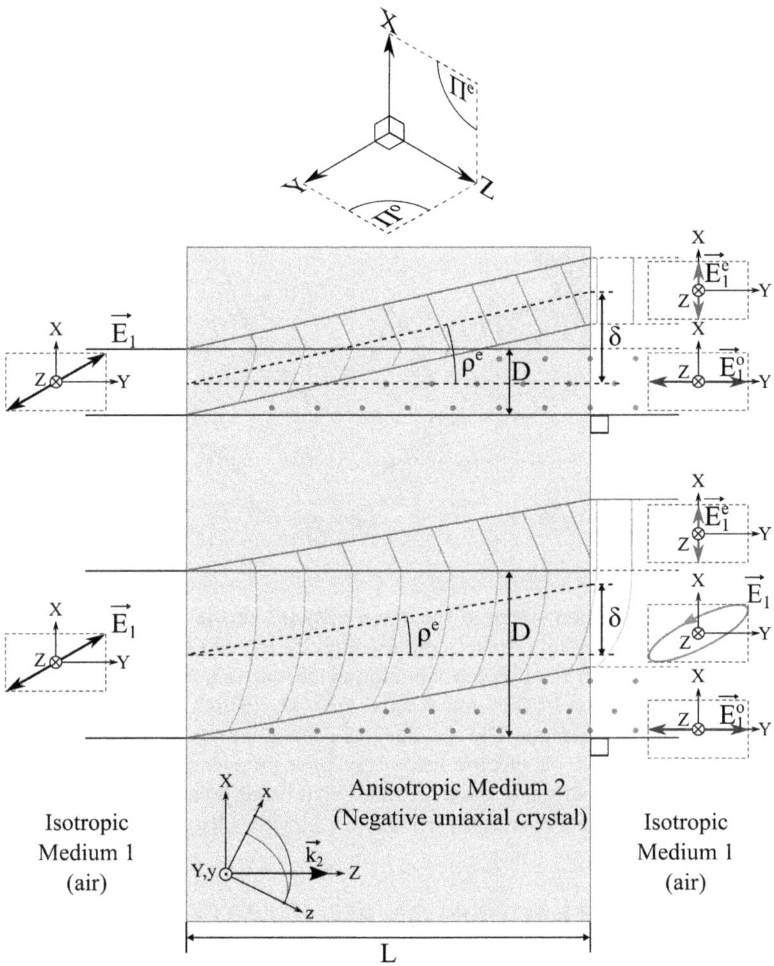

FIGURE 5.30 Polarization states of the wave refracted in the air (Isotropic Medium 1) at the exit of a uniaxial crystal (Anisotropic Medium 2), the entrance wave having a linear polarization not contained in an Eigen vibration plane and in normal incidence. (X,Y,Z) is the laboratory frame. $\vec{E}_1$ stands for the electric field of the wave propagating in the air: it is a linear polarization at the entrance, and an elliptic polarization in the common volume between the two refracted beams at the exit. $(\vec{E}_1^o, \vec{E}_1^e)$ correspond to the linearly polarized parts of the exit beam. ρ_1^e is the spatial walk-off angle. L is the crystal length, D is the beam diameter, and δ is the distance between the beam axes of the ordinary and extraordinary waves at the exit of Medium 2: $\delta > D$ (top) and $\delta < D$ (bottom).

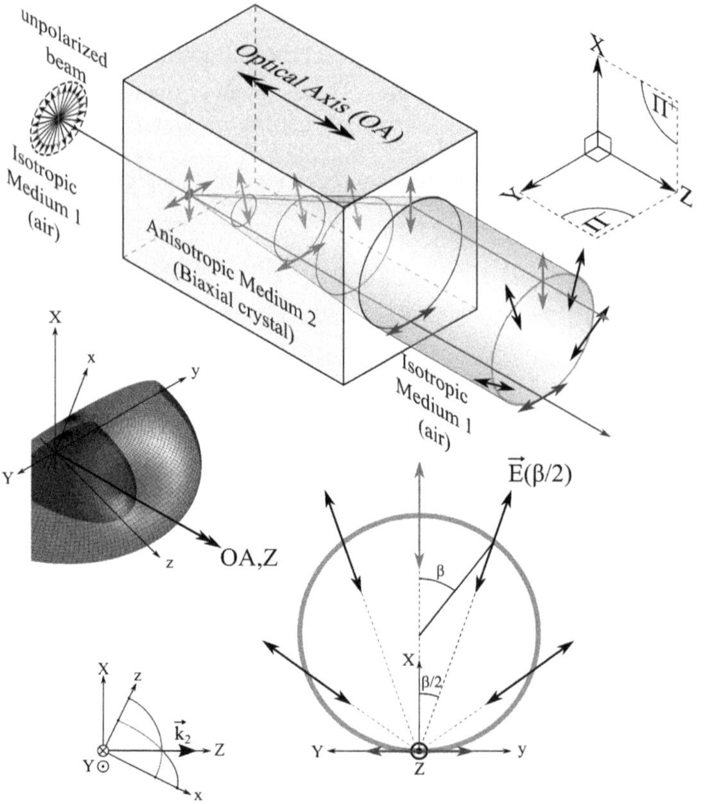

FIGURE 5.31 Polarization states of the wave refracted in the air (Isotropic Medium 1) at the exit of a biaxial crystal (Anisotropic Medium 2), the entrance wave having a linear polarization not contained in an Eigen vibration plane and in normal incidence. The wave propagates in the direction of the optical axis (OA). (x, y, z) is the dielectric frame while (X, Y, Z) is the laboratory frame. The circular line corresponds to the cone of internal conical refraction. $\vec{E}_1$ stands for the electric field of the wave propagating in the air: it is a single linear polarization at the entrance, and a distribution of linear polarizations around the cone. $\vec{E}_2$ stands for the electric field inside the biaxial crystal with a distribution of linear polarizations around the cone.

5.6 EFFECT OF REFRACTION ON REFLECTIVITY AND TRANSMISSION IN NORMAL INCIDENCE

Refraction between two media, namely Medium 1 and Medium 2, exhibiting different refractive indices also leads to the phenomenon of reflectivity, which is also called reflectance or reflection: it means that the light energy transmission from one medium to the other one is not equal to 100%. That can be established by writing the continuity of the tangential components of the electric and magnetic fields at the interface between the two media. This classical calculation is not detailed here since it is performed in numerous books, as for example in reference (Born & Wolf 1980) [4]. Different angles of incidence can be considered, and it appears that the reflectivity angular distribution, and hence the transmission angular distribution, depends on the light polarization orientation with respect to the incidence plane, *i.e.* the plane defined by the incident wave vector and the normal to the interface. This differentiated behavior no longer occurs under normal incidence, which is the configuration considered in the following chapters and in many practical situations. In this case, there may still be a polarization dependence, but this will only depend on the existence or not of birefringence in the direction that is considered. As in the previous parts, Medium 1 is air, and its refractive index is approximated to $n_1(\omega)=1$. Media 1 and 2 are considered to be lossless in the following.

When Medium 2 is isotropic, with refractive index $n_2(\omega)$, there is no birefringence for any direction of propagation. Thus, it can be shown that for any direction of polarization, the energy reflection coefficient $R(\omega)$ and the energy transmission coefficient $T(\omega)=1-R(\omega)$ are expressed as:

$$R(\omega)=\left[\frac{1-n_2(\omega)}{1+n_2(\omega)}\right]^2 \tag{5.165}$$

and

$$T(\omega)=\left[\frac{2}{1+n_2(\omega)}\right]^2 \tag{5.166}$$

The coefficients $R(\omega)$ and $T(\omega)$ are called the Fresnel energy reflection and transmission coefficients.

When Medium 2 is an anisotropic uniaxial or biaxial crystal, with Eigen refractive indices $n^+(\omega,\theta,\varphi)$ and $n^-(\omega,\theta,\varphi)$ in any direction of propagation with the angles of spherical coordinates (θ,φ), the corresponding Eigen polarization states $\vec{E}^+(\omega,\theta,\varphi)$ and $\vec{E}^-(\omega,\theta,\varphi)$ give rise to different energy reflection coefficients and different energy transmission coefficients, *i.e.*:

$$R^\pm(\omega,\theta,\varphi)=\left[\frac{1-n^\pm(\omega,\theta,\varphi)}{1+n^\pm(\omega,\theta,\varphi)}\right]^2 \tag{5.167}$$

and

$$T^\pm(\omega,\theta,\varphi)=\left[\frac{2}{1+n^\pm(\omega,\theta,\varphi)}\right]^2 \tag{5.168}$$

The refractive indices $n^+(\omega,\theta,\varphi)$ and $n^-(\omega,\theta,\varphi)$ are given in Part 5.2.1.3 in the case of uniaxial crystals and in Part 5.2.1.4 for biaxial crystals.

Note that according to Eq. (5.167) and Eq. (5.168), $R^+(\omega,\theta,\varphi)>R^-(\omega,\theta,\varphi)$ and $T^+(\omega,\theta,\varphi)<T^-(\omega,\theta,\varphi)$ since $n^+(\omega,\theta,\varphi)>n^-(\omega,\theta,\varphi)$ by definition.

Along an optical axis, *i.e.* $(\theta=0,\varphi=0)$ for a uniaxial crystal and $(\theta=0,\varphi=V_z)$ for a biaxial crystal, there is no birefringence, *i.e.* $n^+(\omega,\theta,\varphi)=n^-(\omega,\theta,\varphi)$, so that any direction of polarization is allowed and exhibits a unique energy reflection coefficient and a unique energy transmission coefficient.

- In the case of a uniaxial crystal, it comes:

$$\begin{cases} R(\omega,\theta=0,\varphi=0)=\left[\dfrac{1-n_o(\omega)}{1+n_o(\omega)}\right]^2 \\ T(\omega,\theta=0,\varphi=0)=\left[\dfrac{2}{1+n_o(\omega)}\right]^2 \end{cases} \tag{5.169}$$

where $n_o(\omega)$ is the ordinary principal refractive index.

- In the case of a biaxial crystal, it comes:

$$\begin{cases} R(\omega,\theta=0,\varphi=V_z)=\left[\dfrac{1-n_y(\omega)}{1+n_y(\omega)}\right]^2 \\ T(\omega,\theta=0,\varphi=V_z)=\left[\dfrac{2}{1+n_y(\omega)}\right]^2 \end{cases} \tag{5.170}$$

where $n_y(\omega)$ is the relevant principal refractive index since it is the fixed Eigen refractive index in the $(\overrightarrow{Ox},\overrightarrow{Oz})$ plane that contains the optical axis.

Then from the reflectivity and transmission point of views, an anisotropic medium behaves equivalently to an isotropic medium when the wave propagates along an optical axis.

It is important to notice that this phenomenon is exactly the same for the refraction from Medium 1 to Medium 2, *i.e.* at the entrance interface, and from Medium 2 to Medium 1, that is to say at the exit interface. As a consequence, the total transmission coefficient of Medium 2 is equal to $T(\omega)^2$ where $T(\omega)$ is given by Eq. (5.155) to Eq. (5.170) regarding the different possible situations.

5.7 SPECTRAL DISPERSION

The modeling in the framework of the Lorentz model performed in Chapter 1 allows us to access the frequency dispersion of the first-order electric susceptibility $\chi^{(1)}(\omega)$ given by Eq. (1.29a) and Eq. (1.29b) in a 1D picture. We are interested here in the transparency regime.

5.7.1 Refractive Index Dispersion—Sellmeier Equation

The medium is considered to be transparent at the angular frequency ω when $\omega<\omega_0$ is far from ω_0, where ω_0 is the resonance angular frequency of the harmonic oscillator that is considered. Actually, the damping term $-\Gamma\omega$ in Eq. (1.29a) or the driving term $+\Gamma\omega$ in Eq. (1.29b) are negligible compared to $(\omega_0^2-\omega^2)$, so that both Eq.

(1.29a) and Eq. (1.29b) tend toward the following real and positive quantity that is called the Sellmeier equation (Sellmeier 1871) [10].

$$\chi^{(1)}(\omega) = \frac{Nq^2}{m\,\varepsilon_0\left(\omega_0^2 - \omega^2\right)} \tag{5.171}$$

Any medium exhibits several oscillators, so that the angular frequency ω that is considered is systematically located between two given angular resonance frequencies. In the case of the visible range for example, one oscillator is usually in the ultraviolet (UV) with a resonance angular frequency ω_0^{UV}, while the other one is in the infrared (IR) at ω_0^{IR}. The modeling of $\chi^{(1)}(\omega)$ over the full transparency range, *i.e.* for $\omega_0^{UV} \ll \omega \ll \omega_0^{IR}$, is then performed by taking these two oscillators into account. When there is no coupling between the two oscillators, a good description can be obtained by summing two terms of the type of Eq. (5.171), which gives (Sellmeier 1871) [10]:

$$\chi^{(1)}(\omega) = \frac{N^{UV}q^2}{m\varepsilon_0\left[\left(\omega_0^{UV}\right)^2 - (\omega)^2\right]} + \frac{N^{IR}q^2}{m\varepsilon_0\left[\left(\omega_0^{IR}\right)^2 - (\omega)^2\right]} \tag{5.172}$$

where N^{UV} and N^{IR} are the related electron concentrations per volume unit of the UV and IR oscillators, respectively.

Equation (5.172) is called a two-term Sellmeier equation, while Eq. (5.171) is a one-term Sellmeier equation. Note that more than two terms can be considered regarding the number of different oscillators that can be involved.

Equation (5.172) corresponds to a 1D model, but it can be directly used for a 3D description, the relevant quantities being the diagonal matrix coefficients in Eq. (5.3), that is to say: $\chi_{xx}^{(1)}(\omega)$, $\chi_{yy}^{(1)}(\omega)$, and $\chi_{zz}^{(1)}(\omega)$. Then it comes from Eq. (5.172):

$$\chi_{aa}^{(1)}(\omega) = \frac{N^{UV}q^2}{m\varepsilon_0\left[\left(\omega_{0_a}^{UV}\right)^2 - (\omega)^2\right]} + \frac{N^{IR}q^2}{m\varepsilon_0\left[\left(\omega_{0_a}^{IR}\right)^2 - (\omega)^2\right]} \tag{5.173}$$

$a = (x, y, z)$ is a Cartesian index in the dielectric frame, and $\omega_{0_a}^{UV,IR}$ stands for the UV and IR resonance angular frequency that have to be taken into account for a light electric field collinear with the principal axes of the dielectric frame $\overrightarrow{Ox}$, $\overrightarrow{Oy}$, or $\overrightarrow{Oz}$.

Then according to Eq. (5.6), Eq. (5.8), Eq. (5.40), and Eq. (5.173), it comes for the square of the principal refractive indices:

$$n_a(\omega)^2 = 1 + \frac{N_a^{UV}q^2}{m\varepsilon_0\left[\left(\omega_{0_a}^{UV}\right)^2 - (\omega)^2\right]} + \frac{N_a^{IR}q^2}{m\varepsilon_0\left[\left(\omega_{0_a}^{IR}\right)^2 - (\omega)^2\right]} \tag{5.174}$$

Knowing that $\omega = 2\pi c / \lambda$, then Eq. (5.174) rewrites:

$$n_a(\lambda)^2 = 1 + \frac{\Omega_a^{UV}\lambda^2}{(\lambda)^2 - \left(\lambda_{0_a}^{UV}\right)^2} + \frac{\Omega_a^{IR}\lambda^2}{(\lambda)^2 - \left(\lambda_{0_a}^{IR}\right)^2} \tag{5.175}$$

with

$$\begin{cases} \Omega_a^{UV} = \dfrac{N_a^{UV} q^2 \left(\lambda_{0_a}^{UV}\right)^2}{4\pi^2 \varepsilon_0 mc^2} \\ \Omega_a^{IR} = \dfrac{N_a^{IR} q^2 \left(\lambda_{0_a}^{IR}\right)^2}{4\pi^2 \varepsilon_0 mc^2} \end{cases} \tag{5.176}$$

Equation (5.175) and Eq. (5.176) show that the refractive index decreases as a function of wavelength, *i.e.* $\partial n_a(\lambda)/\partial\lambda < 0$, as shown in Figure 5.32. This regime is called "normal dispersion". It corresponds to a propagation in the transparency range of the medium.

In practice, the equation that is used for fitting experimental data is Eq. (5.175) in which 1 is replaced by a fitting parameter, written A_a. Then, there are a total of five fitting parameters, $\{A_a, B_a, C_a, D_a, E_a\}$, the so-called Sellmeier coefficients, for each principal refractive index n_a that is conventionally expressed as (Boulanger 1998) [11]:

$$n_a(\lambda)^2 = A_a + \frac{B_a\lambda^2}{\lambda^2 - C_a} + \frac{D_a\lambda^2}{\lambda^2 - E_a} \tag{5.177}$$

where $\{B_a, D_a\} = \{\Omega_a^{UV}, \Omega_a^{IR}\}$ are given by Eq. (5.176), and $\{C_a, E_a\} = \{\lambda_{0_a}^{UV}, \lambda_{0_a}^{IR}\}$.

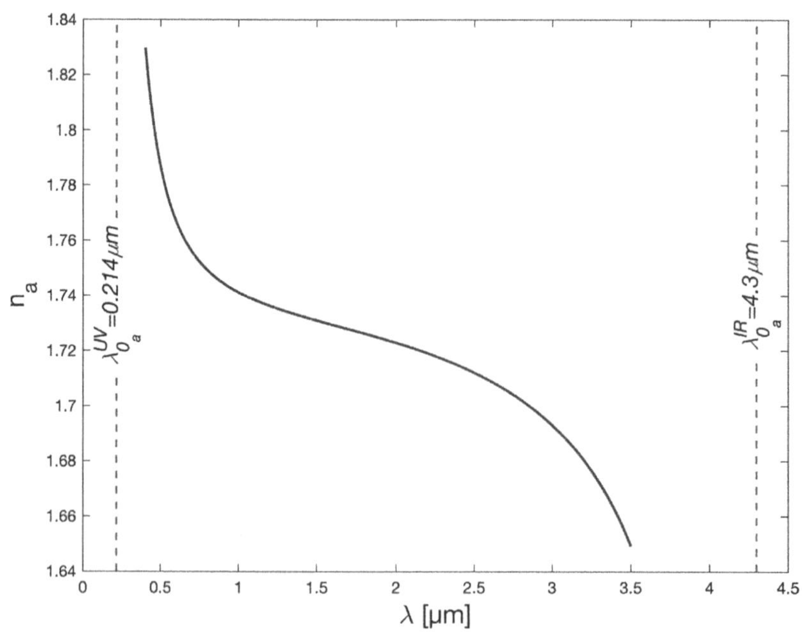

FIGURE 5.32 Typical wavelength dispersion of the refractive index in a transparency range: n_a is a principal refractive index, with $a = (x, y, z)$ relative to the dielectric frame, and λ is the wavelength. $\lambda_{0_a}^{UV}$ and $\lambda_{0_a}^{IR}$ are the resonance angular frequencies of the UV and IR oscillators delimiting the transparency range that is considered. The dashed lines correspond to asymptotes.

Equation (5.177) is also called a two-term Sellmeier equation.

For the isotropic optical class, the three principal refractive indices are equal, *i.e.* $n_x(\lambda) = n_y(\lambda) = n_z(\lambda)$. Thus, only one Sellmeier equation describes the dispersion equation of the refractive index, *i.e.*:

$$n(\lambda)^2 = A + \frac{B\lambda^2}{\lambda^2 - C} + \frac{D\lambda^2}{\lambda^2 - E} \tag{5.178}$$

For the uniaxial optical class, in which $n_x(\lambda) = n_y(\lambda) \equiv n_o(\lambda) \neq n_z(\lambda) \equiv n_e(\lambda)$, there are two Sellmeier equations to consider:

$$\begin{cases} n_o(\lambda)^2 = A_o + \dfrac{B_o\lambda^2}{\lambda^2 - C_o} + \dfrac{D_o\lambda^2}{\lambda^2 - E_o} \\[2ex] n_e(\lambda)^2 = A_e + \dfrac{B_e\lambda^2}{\lambda^2 - C_e} + \dfrac{D_e\lambda^2}{\lambda^2 - E_e} \end{cases} \tag{5.179}$$

And finally, for the optical biaxial class, where $n_x(\lambda) \neq n_y(\lambda) \neq n_z(\lambda)$, three Sellmeier equations are necessary for describing the refractive index:

$$\begin{cases} n_x(\lambda)^2 = A_x + \dfrac{B_x\lambda^2}{\lambda^2 - C_x} + \dfrac{D_x\lambda^2}{\lambda^2 - E_x} \\[2ex] n_y(\lambda)^2 = A_y + \dfrac{B_y\lambda^2}{\lambda^2 - C_y} + \dfrac{D_y\lambda^2}{\lambda^2 - E_y} \\[2ex] n_z(\lambda)^2 = A_z + \dfrac{B_z\lambda^2}{\lambda^2 - C_z} + \dfrac{D_z\lambda^2}{\lambda^2 - E_z} \end{cases} \tag{5.180}$$

Equation (5.178) through Eq. (5.180) allow us to calculate the wavelength dispersion of the Eigen refractive indices $n^+(\omega,\vec{u})$ and $n^-(\omega,\vec{u})$ at any direction of propagation $\vec{u}$ using the index surface equations established in Part 5.2.1.

Note that in certain circumstances, a one-term or several-term Sellmeier equation is not fully appropriate. Different kinds of polynomial expansions can be then used for fitting at best the experimental data, even if they have less immediate physical meaning than the straightforward Sellmeier equation like the following expression for example (Boursier 2014) [12]:

$$n_a(\lambda)^2 = A_a + \frac{B_a}{\lambda^2 - C_a} - D_a\lambda^2 \tag{5.181}$$

where $\{A_a, B_a, C_a, D_a\}$ are the new fitting parameters.

5.7.2 Phase and Group Velocities Dispersion

The wavelength dispersion of the phase velocity and group velocity can be calculated from the wavelength dispersion of the refractive index established previously in the transparency regime. The two-term Sellmeier equation is considered in the following calculations.

5.7.2.1 Phase Velocity Dispersion

The wavelength dispersion of a principal phase velocity $v_a(\lambda)$, with $a = (x, y, z)$, can be obtained from Eq. (5.26) and Eq. (5.177), which gives:

$$v_a(\lambda) = \frac{c}{n_a(\lambda)} = \frac{c}{\sqrt{A_a + \dfrac{B_a\lambda^2}{\lambda^2 - C_a} + \dfrac{D_a\lambda^2}{\lambda^2 - E_a}}} \tag{5.182}$$

Equation (5.182) indicates that the phase velocity increases as a function of wavelength, *i.e.* $\partial v_a(\lambda)/\partial\lambda > 0$, since $\partial n_a(\lambda)/\partial\lambda < 0$ when normal dispersion is considered. The typical behavior is shown in Figure 5.33.

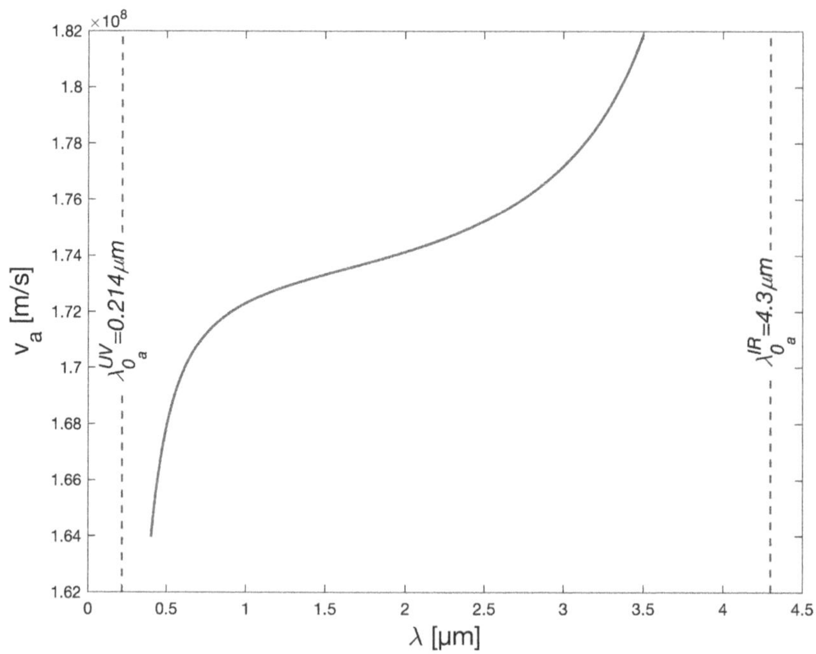

FIGURE 5.33 Typical wavelength dispersion of the phase velocity in a transparency range: v_a is a principal phase velocity, with $a = (x, y, z)$ relative to the dielectric frame, and λ is the wavelength. $\lambda_{0_a}^{UV}$ and $\lambda_{0_a}^{IR}$ are the resonance angular frequencies of the UV and IR oscillators delimiting the range that is considered. The dashed lines correspond to asymptotes.

Then the angular distribution of the phase velocity $v^{+}(\omega,\vec{u})$ and $v^{-}(\omega,\vec{u})$ performed in Part 5.2.2 can be calculated at any wavelength using Eq. (5.182).

5.7.2.2 Group Velocity Dispersion

The group velocity $\vec{v}_g$ is the velocity of the energy that is carried out by light. So, it is oriented in the direction of the Umov-Poynting vector defined in Part 5.3. The phase velocity $\vec{v}$ being the velocity of the wave and thus oriented in the wave vector direction, it comes that the angle between the wave and group velocities corresponds exactly to the spatial walk-off angle ρ as defined in Part 5.3 and shown in Figure 5.34.

The group velocity modulus of a wave propagating in direction $\vec{u}$ at the angular frequency ω is defined by the following relation (Yariv & Yeh 2002) [13]:

$$v_g(\omega,\vec{u}) = \left(\frac{\partial \left\| \vec{k}(\omega,\vec{u}) \right\|}{\partial \omega} \right)^{-1} \tag{5.183}$$

where $\left\| \vec{k}(\omega,\vec{u}) \right\|$ is the wave vector modulus.

From Eq. (5.25) and Eq. (5.26), it comes:

$$\left\| \vec{k}(\omega,\vec{u}) \right\| = \frac{\omega}{c} n(\omega,\vec{u}) \tag{5.184}$$

where $n(\omega,\vec{u})$ is the refractive index in direction $\vec{u}$, *i.e.* $n^{+}(\omega,\vec{u})$ or $n^{-}(\omega,\vec{u})$ according to the polarization that is considered.

Then the combination of Eq. (5.183) and Eq. (5.184) leads to:

$$v_g^{+/-}(\omega,\vec{u}) = \frac{c}{n^{+/-}(\omega,\vec{u}) + \omega \dfrac{\partial n^{+/-}(\omega,\vec{u})}{\partial \omega}} \tag{5.185}$$

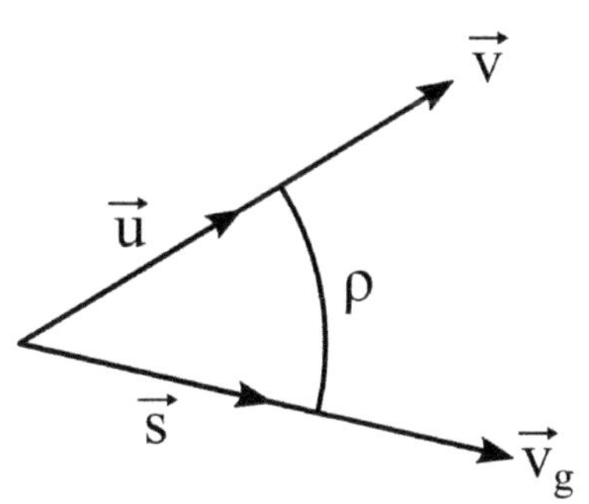

FIGURE 5.34 Vector configuration of the wave velocity $\vec{v}$ and group velocity $\vec{v}_g$ of light propagating in an anisotropic medium. $\vec{u}$ is the unit wave vector and $\vec{s}$ is the unit Umov-Poynting vector. ρ is the spatial walk-off angle.

It is convenient at this step to express v_g as a function of λ. It can easily be done knowing that $\omega = 2\pi c / \lambda$. Thus $\partial\omega / \partial\lambda = -2\pi c / \lambda^2$, which when inserted into Eq. (5.185) gives:

$$v_g^{+/-}(\lambda,\vec{u}) = \frac{c}{n^{+/-}(\lambda,\vec{u}) - \lambda \dfrac{\partial n^{+/-}(\lambda,\vec{u})}{\partial\lambda}} \tag{5.186}$$

The quantities $n^{+/-}(\lambda,\vec{u})$ and $\partial n^{+/-}(\lambda,\vec{u}) / \partial\lambda$ can be calculated according to the proper optical class by using the Sellmeier equations given by Eq. (5.178) through (5.180) and the refractive index equations given in Part 5.2.2.

There exist analytical expressions of $\partial n^{+/-}(\lambda,\vec{u}) / \partial\lambda$ and so of the group velocity dispersion. Thus, while restricting here to the simplest cases of the principal refractive indices, it corresponds to a principal group velocity $v_{g_a}(\lambda)$ expressed as:

$$v_{g_a}(\lambda) = \frac{c}{n_a(\lambda) - \lambda \dfrac{\partial n_a(\lambda)}{\partial\lambda}} \tag{5.187}$$

The principal refractive index $n_a(\lambda)$ is given by Eq. (5.177) from which it can also be calculated $\partial n_a(\lambda) / \partial\lambda$, *i.e.*:

$$\frac{\partial n_a(\lambda)}{\partial\lambda} = -\frac{\lambda}{n_a(\lambda)} \left[\frac{B_a C_a}{\left(\lambda^2 - C_a\right)^2} + \frac{D_a E_a}{\left(\lambda^2 - E_a\right)^2} \right] \tag{5.188}$$

Equation (5.187) and Eq. (5.188) immediately show that $\partial v_{g_a}(\lambda) / \partial\lambda > 0$, as shown in Figure 5.35.

It is interesting to notice that by multiplying Eq. (5.186) by $n^{+/-}(\lambda,\vec{u}) / n^{+/-}(\lambda,\vec{u})$ and using Eq. (5.73), it comes:

$$v_g^{+/-}(\lambda,\vec{u}) = \frac{v^{+/-}(\lambda,\vec{u})}{1 - \dfrac{\lambda}{n^{+/-}(\lambda,\vec{u})} \dfrac{\partial n^{+/-}(\lambda,\vec{u})}{\partial\lambda}} \tag{5.189}$$

Then it appears that the group velocity is equal to the phase velocity when the wavelength dispersion of the refractive index is nil.

Like the phase velocity that is defined from the refractive index, *i.e.* $v^{+/-}(\lambda,\vec{u}) = c / n^{+/-}(\lambda,\vec{u})$, it can be useful to define a group index $n_g^{+/-}(\lambda,\vec{u})$ linked with the group velocity, *i.e.*:

$$n_g^{+/-}(\lambda,\vec{u}) = \frac{c}{v_g^{+/-}(\lambda,\vec{u})} \tag{5.190}$$

Then the behavior of the wavelength dispersion of the group index is of the same kind as that of the refractive index.

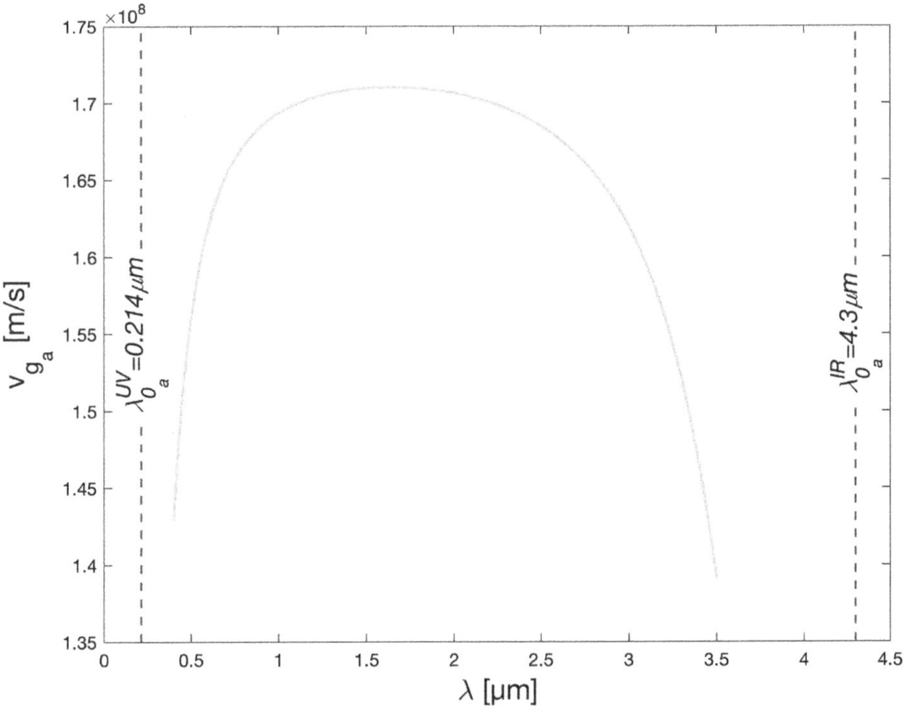

FIGURE 5.35 Typical wavelength dispersion of the group velocity in a transparency range: v_{g_a} is a principal group velocity, with $a=(x,y,z)$ relative to the dielectric frame, and λ is the wavelength. $\lambda_{0_a}^{UV}$ and $\lambda_{0_a}^{IR}$ are the resonance angular frequencies of the UV and IR oscillators delimiting the range that is considered. The dashed lines correspond to asymptotes.

5.8 SUMMARY

The combination of the linear induced polarization involving the real part of the first-order electric susceptibility tensor defined in Chapter 4 and the Maxwell equations allowed us to establish the propagation equation of light in a lossless medium. The projection of this equation on the three axes of the dielectric frame led to a system of three equations, *i.e.* Eq. (5.30), for which the consideration of the determinant has led to the angular distribution of the refractive index, the so-called index surface, and whose resolution allowed to determine the direction of the electric field of the propagated light. The isotropic, uniaxial, and biaxial optical classes have been then defined. These results are crucial tools for describing linear optical properties in a lossless or absorbing medium, as addressed in Chapter 6 for the latter, as well as for nonlinear optical properties as detailed in Chapter 7 and Chapter 8.

The difference between an isotropic medium and an anisotropic one clearly appears from these calculations. Actually, for the isotropic optical class, the refractive index does not vary as a function of the direction of propagation. Furthermore, any direction is allowed for the electric field of light, and that by keeping the same refractive index. At the opposite, in the anisotropic case, *i.e.* for the uniaxial and biaxial optical classes, there are in general two possible refractive index values for any direction of propagation, which defines the birefringence; there are also two associated and distinct directions for the electric fields. This difference in behavior also appears at the level of refraction with the phenomenon of double refraction in the case of uniaxial and biaxial crystals, or with conical refraction along the optical axes of biaxial crystals. Always regarding refraction, an anisotropic crystal can lead to a modification of the polarization state of light at the exit interface, contrary to an isotropic medium. All these features will have a strong impact on the nonlinear optical properties, in particular for their optimization.

Finally, the wavelength dispersion of the refractive index has been established from the Lorentz model detailed in Chapter 1. It will be shown in Chapter 7 and Chapter 8 that this phenomenon associated with birefringence is the key for achieving momentum conservation of photons in the corpuscular description, which stands for the so-called phase-matching in the wave description.

REFERENCES

[1] Nye J. F. 1957. Physical Properties of Crystals. Oxford: Clarendon Press.

[2] Nesse W. D. 1991. Introduction to Optical Mineralogy, 2nd ed., New York and Oxford: Oxford University Press.

[3] Boulanger B. and Zyss J. 2013. Non-linear optical properties—chapter 1.7. In International Tables for Crystallography, Vol. D: Physical Properties of Crystals, edited by A. Authier, pp. 181–222. Dordrecht, Netherlands: International Union of Crystallography, Kluwer Academic Publisher.

[4] Born M. and Wolf E. 1980. Principle of Optics, 6th ed., New York: Pergamon Press.

[5] Fève J. P., Boulanger B., and Marnier G. 1994. Experimental study of internal and external conical refractions in KTP. Optics Communications 105: 243–252.

[6] Turpin A., Loiko Y. V., Kalkandjiev T. K., and Mompart J. 2016. Laser and Photonics Reviews: 1–22. https://doi.org/10.1002/lpor.201600112.

[7] Hecht E. 1987. Optics, 2nd ed., San Francisco: Addison-Wesley.

[8] Collett E. 2005. Field Guide to Polarization. SPIE Field Guides vol. FG05. SPIE.

[9] De Smet D. J. 1993. 4 × 4 matrix formalism applied to internal conical refraction. Journal of the Optical Society of America A 10(1): 186–190.

[10] Sellmeier W. 1871. Zur Erkarung der abnormen Farbenfolge im Spectrum einiger. Substanzen Annalen der Physik und Chemie 210: 272–282.

[11] Boulanger B., Fève J. P., Marnier G., and Ménaert B. 1998. Methodology for nonlinear optical studies: Application to the isomorphs family $KTiOPO_4$, $KTiOAsO_4$, $RbTiOAsO_4$ and $CsTiOAsO_4$. Pure and Applied Optics: Journal of the European Optical Society Part A 7: 239–256.

[12] Boursier E., Segonds P., Boulanger B., Félix C., Debray J., Jegouso D., Ménaert B., Roshchupkin D., and Shoji I. 2014. Phase-matching directions, refined Sellmeier equations and second order nonlinear coefficient of the infrared Langatate crystal $La_3Ga_{5.5}Ta_{0.5}O_{14}$. Optics Letters 39(13): 4033–4036.
[13] Yariv A. and Yeh P. 2002. Optical Waves in Crystals. New York: Wiley.

6 Effects Based on the Imaginary Part of the First-Order Electric Susceptibility

6.1 LINEAR PROPAGATION EQUATION IN THE DIELECTRIC FRAME

When there is linear absorption or emission as described in Part 1.3.1, the equation of light propagation in the linear regime has the same form as in the case of a transparent medium, as described by Eq. (5.2), but in which the first-order electric susceptibility is to be considered a complex quantity. Then it writes at the position $\vec{r}$ and circular frequency ω:

$$\overrightarrow{rot}\left[\overrightarrow{rot}\left[\vec{E}(\omega,\vec{r})\right]\right]-\frac{\omega^2}{c^2}\vec{E}(\omega,\vec{r})=\omega^2\mu_0\varepsilon_0\hat{\chi}^{(1)}(\omega)\cdot\vec{E}(\omega,\vec{r}) \tag{6.1}$$

$\hat{\chi}^{(1)}(\omega)$ is the dimensionless complex first-order electric susceptibility that can be expressed as following (Born & Wolf 1980) [1]:

$$\hat{\chi}^{(1)}(\omega)=\chi^{(1)}(\omega)+i\chi^{(1)\prime}(\omega) \tag{6.2}$$

$\chi^{(1)}(\omega)$ is the real part already considered in the previous chapters, and $\chi^{(1)\prime}(\omega)$ the imaginary part introduced here to take into account linear absorption or emission effects.

It was fully demonstrated in Chapter 3 that $\chi^{(1)}(\omega)$ is a symmetric tensor in the case of a lossless medium. Without any demonstration, the same feature can be assumed for $\chi^{(1)\prime}(\omega)$ since both the real and imaginary parts describe the same property, *i.e.* the first-order electric susceptibility. On the other hand from a general point of view, there is no reason why the matrix of $\chi^{(1)\prime}(\omega)$ should be diagonal in the same principal frame as that of $\chi^{(1)}(\omega)$. Then it comes in the dielectric frame $(\overrightarrow{Ox},\overrightarrow{Oy},\overrightarrow{Oz})$:

$$\hat{\chi}^{(1)}(\omega)=\begin{pmatrix}\hat{\chi}^{(1)}_{xx}(\omega) & \hat{\chi}^{(1)}_{xy}(\omega) & \hat{\chi}^{(1)}_{xz}(\omega)\\ \hat{\chi}^{(1)}_{yx}(\omega) & \hat{\chi}^{(1)}_{yy}(\omega) & \hat{\chi}^{(1)}_{yz}(\omega)\\ \hat{\chi}^{(1)}_{zx}(\omega) & \hat{\chi}^{(1)}_{zy}(\omega) & \hat{\chi}^{(1)}_{zz}(\omega)\end{pmatrix}$$

$$=\begin{pmatrix}\chi^{(1)}_{xx}(\omega) & 0 & 0\\ 0 & \chi^{(1)}_{yy}(\omega) & 0\\ 0 & 0 & \chi^{(1)}_{zz}(\omega)\end{pmatrix}+i\begin{pmatrix}\chi^{(1)\prime}_{xx}(\omega) & \chi^{(1)\prime}_{xy}(\omega) & \chi^{(1)\prime}_{xz}(\omega)\\ \chi^{(1)\prime}_{yx}(\omega) & \chi^{(1)\prime}_{yy}(\omega) & \chi^{(1)\prime}_{yz}(\omega)\\ \chi^{(1)\prime}_{zx}(\omega) & \chi^{(1)\prime}_{zy}(\omega) & \chi^{(1)\prime}_{zz}(\omega)\end{pmatrix} \tag{6.3}$$

DOI: 10.1201/9780429091735-6

with

$$\begin{cases} \chi_{xy}^{(1)'}(\omega) = \chi_{yx}^{(1)'}(\omega) \\ \chi_{xz}^{(1)'}(\omega) = \chi_{zx}^{(1)'}(\omega) \\ \chi_{yz}^{(1)'}(\omega) = \chi_{zy}^{(1)'}(\omega) \end{cases} \tag{6.4}$$

Following the same way as in Chapter 5 for going from Eq. (5.2) to Eq. (5.7), Eq. (6.1) can be expressed as a function of the relative dielectric permittivity, which is a complex quantity here, written $\hat{\varepsilon}_r(\omega)$. Then the light equation of propagation writes:

$$\overrightarrow{rot}\left[\overrightarrow{rot}\left[\vec{E}(\omega,\vec{r})\right]\right] - \frac{\omega^2}{c^2}\hat{\varepsilon}_r(\omega)\cdot\vec{E}(\omega,\vec{r}) = \vec{0} \tag{6.5}$$

where

$$\hat{\varepsilon}_r(\omega) = \varepsilon_r(\omega) + i\varepsilon_r'(\omega) \tag{6.6}$$

with

$$\varepsilon_r'(\omega) = \chi^{(1)'}(\omega) \tag{6.7}$$

The real part $\varepsilon_r(\omega)$ is given by Eq. (5.6).

Then according to Eq. (6.3) and Eq. (6.4), it comes for the matrix of the complex dielectric permittivity expressed in the dielectric frame (Petit 2013) [2]:

$$\begin{aligned} \hat{\varepsilon}_r(\omega) &= \begin{pmatrix} \hat{\varepsilon}_{r_{xx}}(\omega) & \hat{\varepsilon}_{r_{xy}}(\omega) & \hat{\varepsilon}_{r_{xz}}(\omega) \\ \hat{\varepsilon}_{r_{yx}}(\omega) & \hat{\varepsilon}_{r_{yy}}(\omega) & \hat{\varepsilon}_{r_{yz}}(\omega) \\ \hat{\varepsilon}_{r_{zx}}(\omega) & \hat{\varepsilon}_{r_{zy}}(\omega) & \hat{\varepsilon}_{r_{zz}}(\omega) \end{pmatrix} \\ &= \begin{pmatrix} \varepsilon_{r_{xx}}(\omega) & 0 & 0 \\ 0 & \varepsilon_{r_{yy}}(\omega) & 0 \\ 0 & 0 & \varepsilon_{r_{zz}}(\omega) \end{pmatrix} + i\begin{pmatrix} \varepsilon'_{r_{xx}}(\omega) & \varepsilon''_{r_{xy}}(\omega) & \varepsilon'_{r_{xz}}(\omega) \\ \varepsilon'_{r_{yx}}(\omega) & \varepsilon'_{r_{yy}}(\omega) & \varepsilon'_{r_{yz}}(\omega) \\ \varepsilon'_{r_{zx}}(\omega) & \varepsilon'_{r_{zy}}(\omega) & \varepsilon'_{r_{zz}}(\omega) \end{pmatrix} \end{aligned} \tag{6.8}$$

with

$$\begin{cases} \varepsilon'_{r_{xy}}(\omega) = \varepsilon'_{r_{yx}}(\omega) \\ \varepsilon'_{r_{xz}}(\omega) = \varepsilon'_{r_{zx}}(\omega) \\ \varepsilon'_{r_{yz}}(\omega) = \varepsilon'_{r_{zy}}(\omega) \end{cases} \tag{6.9}$$

As for Eq. (5.7), the plane wave is still a possible solution of Eq. (6.5) provided that the electric field $\vec{E}(\omega,\vec{r})$ at the position $\vec{r}$ is given by:

$$\vec{E}(\omega,\vec{r}) = \vec{e}(\omega)E_0(\omega)e^{i\vec{k}(\omega)\cdot\vec{r}} \tag{6.10}$$

The Cartesian coordinates of the unit electric field vector $\vec{e}(\omega)$ and the position $\vec{r}$ are given by Eq. (5.10), and $\vec{\hat{k}}(\omega)$ is the complex wave vector expressed in the dielectric frame as:

$$\vec{\hat{k}}(\omega) = \begin{pmatrix} \hat{k}_x \\ \hat{k}_y \\ \hat{k}_z \end{pmatrix} \tag{6.11}$$

Using Eq. (6.10), the light propagation equation given by Eq. (6.5) becomes:

$$\overrightarrow{rot}\left[\overrightarrow{rot}\left[e^{i\vec{\hat{k}}(\omega)\cdot\vec{r}}\vec{e}(\omega)\right]\right] - \frac{\omega^2}{c^2} e^{i\vec{\hat{k}}(\omega)\cdot\vec{r}} \hat{\varepsilon}_r(\omega)\cdot\vec{e}(\omega) = \vec{0} \tag{6.12}$$

This equation has the same form as that in the transparent case, *i.e.* Eq. (5.12), except that the wave vector and the relative dielectric permittivity are now complex quantities. At this stage, the development of Eq. (6.12) is then similar to that followed from Eq. (5.12) to Eq. (5.21), which gives:

$$\left\{\vec{\hat{k}}(\omega) \wedge \left[\vec{\hat{k}}(\omega) \wedge \vec{e}(\omega)\right]\right\} + \frac{\omega^2}{c^2} \hat{\varepsilon}_r(\omega)\cdot\vec{e}(\omega) = \vec{0} \tag{6.13}$$

Taking into account the complex wave vector $\vec{\hat{k}}(\omega,\vec{u}) = \hat{k}(\omega,\vec{u})\vec{u} = \frac{\omega}{c}\hat{n}(\omega,\vec{u})\vec{u}$ along the direction of propagation of unit vector $\vec{u}$, with $\hat{n}(\omega,\vec{u})$ the dimensionless complex optical index, Eq. (6.13) rewrites:

$$\hat{n}^2(\omega,\vec{u})\left\{\vec{u} \wedge \left[\vec{u} \wedge \vec{e}(\omega)\right]\right\} + \hat{\varepsilon}_r(\omega)\cdot\vec{e}(\omega) = \vec{0} \tag{6.14}$$

The expansion of the double vectorial product in Eq. (6.14) leads to the following linear system of three equations:

$$\begin{cases} \left[\hat{n}^2(\omega,\vec{u})(1-u_x^2) - \hat{\varepsilon}_{r_{xx}}(\omega)\right]e_x - \left[\hat{n}^2(\omega,\vec{u})u_x u_y + \hat{\varepsilon}_{r_{xy}}(\omega)\right]e_y \\ -\left[\hat{n}^2(\omega,\vec{u})u_x u_z + \hat{\varepsilon}_{r_{xz}}(\omega)\right]e_z = 0 \\ \\ -\left[\hat{n}^2(\omega,\vec{u})u_y u_x + \hat{\varepsilon}_{r_{yx}}(\omega)\right]e_x + \left[\hat{n}^2(\omega,\vec{u})(1-u_y^2) - \hat{\varepsilon}_{r_{yy}}(\omega)\right]e_y \\ -\left[\hat{n}^2(\omega,\vec{u})u_y u_z + \hat{\varepsilon}_{r_{yz}}(\omega)\right]e_z = 0 \\ \\ -\left[\hat{n}^2(\omega,\vec{u})u_z u_x + \hat{\varepsilon}_{r_{zx}}(\omega)\right]e_x - \left[\hat{n}^2(\omega,\vec{u})u_z u_y + \hat{\varepsilon}_{r_{zy}}(\omega)\right]e_y \\ +\left[\hat{n}^2(\omega,\vec{u})(1-u_z^2) - \hat{\varepsilon}_{r_{zz}}(\omega)\right]e_z = 0 \end{cases} \tag{6.15}$$

Equation (6.15) has the same structure as Eq. (5.30), but here it deals with complex quantities and it involves also the extra-diagonal coefficients of the complex permittivity tensor, *i.e.* $\hat{\varepsilon}_{r_{ij}}(\omega)$ with $i \neq j$. For more clarity, it can be rewritten in a more compact form, *i.e.*:

$$\sum_j \left[\hat{n}^2(\omega,\vec{u})(\delta_{ij} - u_i u_j) - \hat{\varepsilon}_{r_{ij}}(\omega)\right] e_j = 0 \tag{6.16}$$

where *(i,j)* stands for the axis coordinates *(x,y,z)* in the dielectric frame.

The determination of the complex refractive optical index $\hat{n}$, for any direction propagation $\vec{u}$ in any absorbing or fluorescent medium, requires solving the linear equation system given by Eq. (6.15) or equivalently by Eq. (6.16). These forsook complex refractive optical indices correspond thus to the Eigen value solutions of the light propagation equation in matter, which will be addressed in the next part.

6.2 COMPLEX OPTICAL INDICES

Equation (6.15), or Eq. (6.16), can rewrite:

$$\hat{M}\begin{pmatrix} e_x \\ e_y \\ e_z \end{pmatrix} = 0 \tag{6.17}$$

The 3×3 matrix $\hat{M}$ is given by (Petit 2013) [2]:

$$\hat{M} = \begin{pmatrix} \hat{n}^2(\omega,\vec{u})(1-u_x^2) - \hat{\varepsilon}_{r_{xx}}(\omega) & -\hat{n}^2(\omega,\vec{u})u_x u_y - \hat{\varepsilon}_{r_{xy}}(\omega) & -\hat{n}^2(\omega,\vec{u})u_x u_z - \hat{\varepsilon}_{r_{xz}}(\omega) \\ -\hat{n}^2(\omega,\vec{u})u_y u_x - \hat{\varepsilon}_{r_{yx}}(\omega) & \hat{n}^2(\omega,\vec{u})(1-u_y^2) - \hat{\varepsilon}_{r_{yy}}(\omega) & -\hat{n}^2(\omega,\vec{u})u_y u_z - \hat{\varepsilon}_{r_{yz}}(\omega) \\ -\hat{n}^2(\omega,\vec{u})u_z u_x - \hat{\varepsilon}_{r_{zx}}(\omega) & -\hat{n}^2(\omega,\vec{u})u_z u_y - \hat{\varepsilon}_{r_{zy}}(\omega) & \hat{n}^2(\omega,\vec{u})(1-u_z^2) - \hat{\varepsilon}_{r_{zz}}(\omega) \end{pmatrix} \tag{6.18}$$

Equation (6.18) is a homogeneous linear equation system since its right-hand term is zero. Thus Eq. (6.18) has non-zero complex solutions, *i.e.* $\hat{n} \neq 0$, if the determinant of $\hat{M}$ is equal to zero:

$$Det(\hat{M}) = |\hat{M}| = 0 \tag{6.19}$$

The resolution of Eq. (6.19) is not complicated but heavy. The calculations are not detailed here, but it can be shown that the determinant of $\hat{M}$ is composed of six product terms that correspond each to a specific product of three components of $\hat{M}$. Each of these six terms has to be expanded, which provides a much larger amount of resulting individual terms. After the full expansion of each individual terms, it is expected to gather them as coefficients of a cubic polynomial expression of the $\hat{n}^2$, *i.e.*: $\left(\hat{n}^2\right)^3 = \hat{n}^6$, $\left(\hat{n}^2\right)^2 = \hat{n}^4$, $\left(\hat{n}^2\right)^1 = \hat{n}^2$, and $\left(\hat{n}^2\right)^0 = 1$. Nevertheless, the coefficient related to the $\left(\hat{n}^2\right)^3 = \hat{n}^6$ term is always identically null, whatever the considered direction of light propagation. As a consequence, the determinant always reduces to a quadratic polynomial expression of the $\hat{n}^2$ variable, which makes it much easier to further solve. The identical nullity of the term $\hat{n}^6$ may appear surprising. However, it bears much physical meaning since it directly deals with the dimensionality of our physical space, *i.e.* 3D, in which the Maxwell equations are written to depict light propagation on the one hand, and with the transverse behavior of the electric displacement on the other hand.

Another simplification is brought by the weak absorption hypothesis (or weak fluorescence hypothesis), which corresponds to the case in which the electric field amplitude shows an infinitesimal evolution over the wavelength scale during the propagation. It means that imaginary parts of the complex permittivity coefficients of Eq. (6.8) are orders of magnitude weaker than the non-zero real parts of the complex permittivity coefficients expressed in the dielectric frame, which writes:

$$\varepsilon'_{r_{jk}}(\omega) << \varepsilon_{r_{ii}}(\omega) \tag{6.20}$$

As a consequence, the coefficients of the quadratic polynomial expression of $\hat{n}^2$ gathers terms of different orders of magnitude with respect to Eq. (6.20). In addition to products that include diagonal components, Eq. (6.19) leads to zeroth-order terms with respect to purely imaginary extra-diagonal components $\hat{\varepsilon}_{r_{i\neq j}}$; first-order terms proportional to one purely imaginary extra-diagonal component $\hat{\varepsilon}_{r_{i\neq j}}$; second-order terms proportional to the product of two purely imaginary extra-diagonal components $\hat{\varepsilon}_{r_{i\neq j}}\hat{\varepsilon}_{r_{k\neq l}}$; and finally third-order terms proportional to the product of three purely imaginary extra-diagonal components $\hat{\varepsilon}_{r_{i\neq j}}\hat{\varepsilon}_{r_{k\neq l}}\hat{\varepsilon}_{r_{m\neq n}}$. Therefore, based on the weak absorption/amplification hypothesis, these second-order and third-order terms are systematically disregarded and neglected. In these conditions, solving Eq. (6.19) is equivalent to solving the following quadratic polynomial expression of $\hat{n}^2$:

$$\hat{A}(\omega,\vec{u})\hat{n}^4(\omega,\vec{u}) + \hat{B}(\omega,\vec{u})\hat{n}^2(\omega,\vec{u}) + \hat{C}(\omega,\vec{u}) = 0 \tag{6.21}$$

with:

$$\begin{cases} \hat{A}(\omega,\vec{u}) = -\left(u_x^2\hat{\varepsilon}_{r_{xx}}(\omega) + u_y^2\hat{\varepsilon}_{r_{yy}}(\omega) + u_z^2\hat{\varepsilon}_{r_{zz}}(\omega)\right) \\ \qquad -2\left(u_xu_y\hat{\varepsilon}_{r_{xy}}(\omega) + u_yu_z\hat{\varepsilon}_{r_{yz}}(\omega) + u_xu_z\hat{\varepsilon}_{r_{xz}}(\omega)\right) \\ \hat{B}(\omega,\vec{u}) = \left(1-u_x^2\right)\hat{\varepsilon}_{r_{yy}}(\omega)\hat{\varepsilon}_{r_{zz}}(\omega) + \left(1-u_y^2\right)\hat{\varepsilon}_{r_{zz}}(\omega)\hat{\varepsilon}_{r_{xx}}(\omega) + \left(1-u_z^2\right)\hat{\varepsilon}_{r_{xx}}(\omega)\hat{\varepsilon}_{r_{yy}} \\ \qquad (\omega) + 2\left(u_xu_y\hat{\varepsilon}_{r_{zz}}(\omega)\hat{\varepsilon}_{r_{xy}}(\omega) + u_yu_z\hat{\varepsilon}_{r_{xx}}(\omega)\hat{\varepsilon}_{r_{yz}}(\omega) + u_xu_z\hat{\varepsilon}_{r_{yy}}(\omega)\hat{\varepsilon}_{r_{xz}}(\omega)\right) \\ \hat{C}(\omega,\vec{u}) = -\hat{\varepsilon}_{r_{xx}}(\omega)\hat{\varepsilon}_{r_{yy}}(\omega)\hat{\varepsilon}_{r_{zz}}(\omega) \end{cases} \tag{6.22}$$

The two solutions $\hat{n}^2$ of Eq. (6.21), called $\hat{n}^+\big|^2$ and $\hat{n}^-\big|^2$, obviously write as following:

$$\hat{n}^{\pm}\Big|^2(\omega,\vec{u}) = \frac{-\hat{B}(\omega,\vec{u}) \pm \sqrt{\hat{B}(\omega,\vec{u})^2 - 4\hat{A}(\omega,\vec{u})\hat{C}(\omega,\vec{u})}}{2\hat{A}(\omega,\vec{u})} \tag{6.23}$$

To go further in the establishment of analytical expressions, it is necessary to express the coefficients $\hat{A}$, $\hat{B}$, and $\hat{C}$ in terms of their real and imaginary parts, *i.e.*:

$$\begin{cases} \hat{A}(\omega,\vec{u}) = A(\omega,\vec{u}) + jA'(\omega,\vec{u}) \\ \hat{B}(\omega,\vec{u}) = B(\omega,\vec{u}) + jB'(\omega,\vec{u}) \\ \hat{C}(\omega,\vec{u}) = C(\omega,\vec{u}) + jC'(\omega,\vec{u}) \end{cases} \tag{6.24}$$

Then, by expanding in Eq. (6.22) the real and imaginary parts of the complex permittivity tensor components, *i.e.* $\hat{\varepsilon}_{r_{ij}} = \varepsilon_{r_{jk}}(\omega) + j\varepsilon'_{r_{jk}}(\omega)$, the resulting polynomial coefficients are the following: zeroth-order terms that are free from any contribution from the imaginary part coefficients $\varepsilon'_{r_{jk}}(\omega)$; first-order terms proportional to one imaginary part coefficient $\varepsilon'_{r_{jk}}(\omega)$; second-order terms proportional to the product of two imaginary part coefficients $\varepsilon'_{r_{ij}}(\omega)\varepsilon'_{r_{kl}}(\omega)$; and finally third-order terms proportional to the product of three imaginary part coefficients $\varepsilon'_{r_{ij}}(\omega)\varepsilon'_{r_{kl}}(\omega)\varepsilon'_{r_{mn}}(\omega)$. As mentioned earlier, the remaining second-order and third-order terms can be neglected according to Eq. (6.20). In these conditions, Eq. (6.22) with Eq. (6.24) give the three following couples of equations:

$$\begin{cases} A(\omega,\vec{u}) = -\left(u_x^2\varepsilon_{r_{xx}}(\omega) + u_y^2\varepsilon_{r_{yy}}(\omega) + u_z^2\varepsilon_{r_{zz}}(\omega)\right) \\ A'(\omega,\vec{u}) = -\begin{pmatrix} u_x^2\varepsilon'_{r_{xx}}(\omega) + u_y^2\varepsilon'_{r_{yy}}(\omega) + u_z^2\varepsilon'_{r_{zz}}(\omega) + 2u_xu_y\hat{\varepsilon}'_{r_{xy}}(\omega) \\ +2u_yu_z\hat{\varepsilon}'_{r_{yz}}(\omega) + 2u_xu_z\hat{\varepsilon}'_{r_{xz}}(\omega) \end{pmatrix} \end{cases} \tag{6.25a}$$

$$\begin{cases} B(\omega,\vec{u}) = u_x^2\left(\varepsilon_{r_{xx}}(\omega)\varepsilon_{r_{zz}}(\omega) + \varepsilon_{r_{xx}}(\omega)\varepsilon_{r_{yy}}(\omega)\right) + u_y^2\left(\varepsilon_{r_{yy}}(\omega)\varepsilon_{r_{xx}}(\omega) + \varepsilon_{r_{yy}}(\omega)\varepsilon_{r_{zz}}(\omega)\right) \\ \qquad +u_z^2\left(\varepsilon_{r_{zz}}(\omega)\varepsilon_{r_{yy}}(\omega) + \varepsilon_{r_{zz}}(\omega)\varepsilon_{r_{xx}}(\omega)\right) \\ B'(\omega,\vec{u}) = u_x^2\left(\varepsilon'_{r_{xx}}(\omega)\varepsilon_{r_{zz}}(\omega) + \varepsilon_{r_{xx}}(\omega)\varepsilon'_{r_{zz}}(\omega) + \varepsilon'_{r_{xx}}(\omega)\varepsilon_{r_{yy}}(\omega) + \varepsilon_{r_{xx}}(\omega)\varepsilon'_{r_{yy}}(\omega)\right) \\ \qquad +u_y^2\left(\varepsilon'_{r_{yy}}(\omega)\varepsilon_{r_{zz}}(\omega) + \varepsilon_{r_{yy}}(\omega)\varepsilon'_{r_{zz}}(\omega) + \varepsilon'_{r_{xx}}(\omega)\varepsilon_{r_{yy}}(\omega) + \varepsilon_{r_{xx}}(\omega)\varepsilon'_{r_{yy}}(\omega)\right) \\ \qquad +u_z^2\left(\varepsilon'_{r_{yy}}(\omega)\varepsilon_{r_{zz}}(\omega) + \varepsilon_{r_{yy}}(\omega)\varepsilon'_{r_{zz}}(\omega) + \varepsilon'_{r_{xx}}(\omega)\varepsilon_{r_{zz}}(\omega) + \varepsilon_{r_{xx}}(\omega)\varepsilon'_{r_{zz}}(\omega)\right) \\ \qquad +2\left(u_xu_y\varepsilon_{r_{zz}}(\omega)\varepsilon'_{r_{xy}}(\omega) + u_yu_z\varepsilon_{r_{xx}}(\omega)\varepsilon'_{r_{yz}}(\omega) + u_xu_z\varepsilon_{r_{yy}}(\omega)\varepsilon'_{r_{xz}}(\omega)\right) \end{cases} \tag{6.25b}$$

$$\begin{cases} C(\omega,\vec{u}) = -\varepsilon_{r_{xx}}(\omega)\varepsilon_{r_{yy}}(\omega)\varepsilon_{r_{zz}}(\omega) \\ C'(\omega,\vec{u}) = -\left(\varepsilon'_{r_{xx}}(\omega)\varepsilon_{r_{yy}}(\omega)\varepsilon_{r_{zz}}(\omega) + \varepsilon_{r_{xx}}(\omega)\varepsilon'_{r_{yy}}(\omega)\varepsilon_{r_{zz}}(\omega) + \varepsilon_{r_{xx}}(\omega)\varepsilon_{r_{yy}}(\omega)\varepsilon'_{r_{zz}}(\omega)\right) \end{cases} \tag{6.25c}$$

Based on the weak absorption or emission of light hypothesis described by Eq. (6.20), these coefficients undergo the following relationships:

$$\begin{cases} A'(\omega,\vec{u}) << A(\omega,\vec{u}) \\ B'(\omega,\vec{u}) << B(\omega,\vec{u}) \\ C'(\omega,\vec{u}) << C(\omega,\vec{u}) \end{cases} \tag{6.26}$$

According to Eq. (6.26), the identification of the real part of $\hat{n}^{\pm}\Big|^2(\omega,\vec{u})$ leads to a zero-order expression with respect to $\varepsilon'_{r_{jk}}(\omega)$ terms, that corresponds to:

$$\mathrm{Re}\left(\hat{n}^{\pm}\Big|^2(\omega,\vec{u})\right) = \frac{-B(\omega,\vec{u}) \pm \sqrt{B(\omega,\vec{u})^2 - 4A(\omega,\vec{u})C(\omega,\vec{u})}}{2A(\omega,\vec{u})} \tag{6.27}$$

Note that Eq. (6.27), with coefficients $A(\omega,\vec{u})$, $B(\omega,\vec{u})$, and $C(\omega,\vec{u})$ given by Eq. (6.25a), Eq. (6.25b), and Eq. (6.25c) respectively, exactly corresponds to Eq. (5.45) with the coefficients $A(\omega,\vec{u})$, $B(\omega,\vec{u})$, and $C(\omega,\vec{u})$ given by Eq. (5.43) while considering purely transparent media as detailed in Chapter 5. Actually, these coefficients differ by a renormalization factor of $-1/\varepsilon_{r_{xx}}\varepsilon_{r_{yy}}\varepsilon_{r_{zz}}$, or $-1/n_x^2n_y^2n_z^2$ according to Eq. (5.40), which does not affect the physics of the obtained solutions both in Chapter 5 and here.

In the same way according to Eq. (6.26), the identification of the imaginary part of $n^{\pm}\Big|^2(\omega,\vec{u})$ leads to a first-order expression with respect to $\varepsilon'_{r_{jk}}(\omega)$ terms, *i.e.*:

$$\begin{aligned} \mathrm{Im}\left(\hat{n}^{\pm}\Big|^2(\omega,\vec{u})\right) &= -n^{\pm}\Big|^2(\omega,\vec{u})\frac{A'(\omega,\vec{u})}{A(\omega,\vec{u})} \\ &+ \frac{-B'(\omega,\vec{u}) \pm \left[2B(\omega,\vec{u})B'(\omega,\vec{u}) - 4\left(A'(\omega,\vec{u})C(\omega,\vec{u}) + A(\omega,\vec{u})C'(\omega,\vec{u})\right)\right]}{4A(\omega,\vec{u})\sqrt{B(\omega,\vec{u})^2 - 4A(\omega,\vec{u})C(\omega,\vec{u})}} \end{aligned} \tag{6.28}$$

with coefficients $\{A(\omega,\vec{u}),A'(\omega,\vec{u})\}$, $\{B(\omega,\vec{u}),B'(\omega,\vec{u})\}$, and $\{C(\omega,\vec{u}),C'(\omega,\vec{u})\}$ given by Eq. (6.25a), Eq. (6.25b), and Eq. (6.25c) respectively.

Based on Eq. (6.26) on the one hand, and on Eq. (6.27) and Eq. (6.28) on the other hand, it appears that the weak absorption/amplification hypothesis upon the complex permittivity tensor directly impacts the complex optical indices, *i.e.*:

$$\mathrm{Im}\left(\hat{n}^{\pm}\Big|^2(\omega,\vec{u})\right) << \mathrm{Re}\left(\hat{n}^{\pm}\Big|^2(\omega,\vec{u})\right) \tag{6.29}$$

Then by noting the complex indices $\hat{n}^{\pm} = n^{\pm}(\omega,\vec{u}) + jn'^{\pm}(\omega,\vec{u})$, it comes for $\hat{n}^{\pm}\Big|^2$:

$$\begin{aligned} \hat{n}^{\pm}\Big|^2(\omega,\vec{u}) &= \left[n^{\pm}(\omega,\vec{u}) + jn'^{\pm}(\omega,\vec{u})\right]^2 \\ &= n^{\pm}\Big|^2(\omega,\vec{u}) - n'^{\pm}\Big|^2(\omega,\vec{u}) + 2\,jn^{\pm}\Big|(\omega,\vec{u})n'^{\pm}(\omega,\vec{u}) \\ &\approx n^{\pm}\Big|^2(\omega,\vec{u}) + 2jn^{\pm}\Big|(\omega,\vec{u})n'^{\pm}(\omega,\vec{u}) \end{aligned} \tag{6.30}$$

By a term-by-term identification of both the real and imaginary parts of Eq. (6.27) and Eq. (6.28) with those of Eq. (6.30), it appears that $n'^{\pm}\big|^2(\omega,\vec{u}) << n^{\pm}\big|^2(\omega,\vec{u})$. Moreover, the real part of the complex index is to be considered a positive quantity, in agreement with the commonly observed positive refractive indices in classical dielectric media. Then it comes for $\hat{n}^{\pm}(\omega,\vec{u}) = \sqrt{\hat{n}^{\pm}\big|^2(\omega,\vec{u})}$:

$$n^{\pm}(\omega,\vec{u}) = \mathrm{Re}\left(\hat{n}^{\pm}(\omega,\vec{u})\right) = \sqrt{\frac{-B(\omega,\vec{u}) \pm \sqrt{B(\omega,\vec{u})^2 - 4A(\omega,\vec{u})C(\omega,\vec{u})}}{2A(\omega,\vec{u})}} \tag{6.31}$$

and

$$n'^{\pm}(\omega,\vec{u}) = \mathrm{Im}\left(\hat{n}^{\pm}(\omega,\vec{u})\right) = \frac{\mathrm{Im}\left(\hat{n}^{\pm}\big|^2(\omega,\vec{u})\right)}{2n^{\pm}(\omega,\vec{u})} \tag{6.32}$$

Equation (6.31) exactly corresponds to the refractive index solutions given by Eq. (5.46) in the case of transparent media. It is reminded, as seen in Chapter 5, that the Eigen values of the refractive index verify by definition:

$$n^{+}(\omega,\vec{u}) = \mathrm{Re}\left(\hat{n}^{+}(\omega,\vec{u})\right) \geq n^{-}(\omega,\vec{u}) = \mathrm{Re}\left(\hat{n}^{-}(\omega,\vec{u})\right) \tag{6.33}$$

However there is no *a priori* systematic relation of order between n'^{+} and n'^{-}. As a consequence, it is theoretically possible to have $n'^{+}(\omega,\vec{u}) \geq n'^{-}(\omega,\vec{u})$ or $n'^{+}(\omega,\vec{u}) \leq n'^{-}(\omega,\vec{u})$ from a mathematical point of view: it will further be illustrated in biaxial media.

Equation (6.31) shows that the refractive indices $n^{\pm}(\omega,\vec{u})$ depend only on the real part of the complex permittivity tensor in the weak absorption or emission regime. As a consequence, the Eigen vectors of Eq. (6.19), which are the polarization Eigenmodes related to the complex Eigen solutions depicted by Eq. (6.23), do remain the same as those defined in Part 5.3 for transparent media. Strictly speaking, the polarization Eigenmode shall be orthogonal elliptical polarization modes while considering the complex Eigen solutions. However, in the weak absorption/emission regime, these polarization modes are to be considered orthogonal linear polarization modes, such as in the case of transparent media (Born & Wolf 1980) [1]. Thus all the conclusions from Chapter 5 remain directly applicable in weakly absorbing/emitting media.

Equation (6.32) indicates that the imaginary indices $n'^{\pm}(\omega,\vec{u})$ are supported by both imaginary and real components of the complex permittivity tensor, and also directly by the refractive indices $n^{\pm}(\omega,\vec{u})$. The related interpretation is that absorption/emission properties behave as perturbative effects of light propagation at typical scales in the range of the wavelength. The angular distribution of $n'^{\pm}(\omega,\vec{u})$, namely the imaginary index surface, will be treated in Part 6.4 for the different optical classes. But before that, it is necessary to precise the form of the complex permittivity tensor in each case, which will be done in Part 6.3.

6.3 COMPLEX RELATIVE PERMITTIVITY TENSOR AS A FUNCTION OF THE OPTICAL CLASS

The matrix of $\chi^{(1)\prime}(\omega)$, or $\varepsilon_r'(\omega)$, have the same form as that of $\chi^{(1)}(\omega)$, or $\varepsilon_r(\omega)$, *i.e.* a symmetrical (3×3) matrix. Then the application of Neumann's principle leads exactly to the same result for both of these tensors. The work has been fully detailed in Chapter 3 for $\chi^{(1)}(\omega)$, which had allowed us to define the three optical classes characterized by three different forms of matrices according to the different crystal classes.

For the cubic, hexagonal, tetragonal, trigonal, and orthorhombic crystal classes, it directly comes out that the dielectric frame $(\overrightarrow{Ox},\overrightarrow{Oy},\overrightarrow{Oz})$, with the orientation conventions defined in Chapter 3, diagonalizes both $\chi^{(1)}(\omega)$ and $\chi^{(1)\prime}(\omega)$, and so both $\varepsilon_r(\omega)$ and $\varepsilon_r'(\omega)$. Then Eq. (6.8) becomes:

$$\hat{\varepsilon}_r(\omega)=\begin{pmatrix}\hat{\varepsilon}_{r_{xx}}(\omega) & 0 & 0\\ 0 & \hat{\varepsilon}_{r_{yy}}(\omega) & 0\\ 0 & 0 & \hat{\varepsilon}_{r_{zz}}(\omega)\end{pmatrix} = \begin{pmatrix}\varepsilon_{r_{xx}}(\omega) & 0 & 0\\ 0 & \varepsilon_{r_{yy}}(\omega) & 0\\ 0 & 0 & \varepsilon_{r_{zz}}(\omega)\end{pmatrix}+i\begin{pmatrix}\varepsilon'_{r_{xx}}(\omega) & 0 & 0\\ 0 & \varepsilon'_{r_{yy}}(\omega) & 0\\ 0 & 0 & \varepsilon'_{r_{zz}}(\omega)\end{pmatrix} \tag{6.34}$$

For the cubic crystal class, which corresponds to the isotropic optical class, the three diagonal coefficients are equal in each matrix, *i.e.*:

$$\begin{aligned}\hat{\varepsilon}_{r_{xx}}(\omega)&=\varepsilon_{r_{xx}}(\omega)+i\varepsilon'_{r_{xx}}(\omega)=\hat{\varepsilon}_{r_{yy}}(\omega)=\varepsilon_{r_{yy}}(\omega)+i\varepsilon'_{r_{yy}}(\omega)=\hat{\varepsilon}_{r_{zz}}(\omega)\\&=\varepsilon_{r_{zz}}(\omega)+i\varepsilon'_{r_{zz}}(\omega)\equiv\hat{\varepsilon}_r=\varepsilon_r(\omega)+i\varepsilon'_r(\omega)\end{aligned} \tag{6.35}$$

For the hexagonal, tetragonal, and trigonal crystal classes, which define the uniaxial optical class, only two coefficients of the diagonal are equal, which gives using the "ordinary (o)" and "extraordinary (e)" writing convention:

$$\begin{aligned}&\hat{\varepsilon}_{r_{xx}}(\omega)=\varepsilon_{r_{xx}}(\omega)+i\varepsilon'_{r_{xx}}(\omega)=\hat{\varepsilon}_{r_{yy}}(\omega)=\varepsilon_{r_{yy}}(\omega)+i\varepsilon'_{r_{yy}}(\omega)\equiv\hat{\varepsilon}_{r_o}=\varepsilon_{r_o}(\omega)+i\varepsilon'_{r_o}(\omega)\\&\neq\hat{\varepsilon}_{r_{zz}}(\omega)=\varepsilon_{r_{zz}}(\omega)+i\varepsilon'_{r_{zz}}(\omega)\equiv\hat{\varepsilon}_{r_e}=\varepsilon_{r_e}(\omega)+i\varepsilon'_{r_e}(\omega)\end{aligned} \tag{6.36}$$

We consider that the relation of order taken between the real part of the ordinary and extraordinary quantities is conserved at the level of the imaginary quantities. The reason is that both quantities derive from the same quantity, namely the macroscopic polarizability of matter under light propagation. As a consequence, it can be directly written that $\varepsilon_{r_o}(\omega)<\varepsilon_{r_e}(\omega)$ and $\varepsilon'_{r_o}(\omega)<\varepsilon'_{r_e}(\omega)$ which implies that $\hat{\varepsilon}_{r_o}(\omega)<\hat{\varepsilon}_{r_e}(\omega)$ in the case of the positive uniaxial class, while $\varepsilon_{r_o}(\omega)>\varepsilon_{r_e}(\omega)$ and $\varepsilon'_{r_o}(\omega)>\varepsilon'_{r_e}(\omega)$ which implies $\hat{\varepsilon}_{r_o}(\omega)>\hat{\varepsilon}_{r_e}(\omega)$ in the case of the negative uniaxial class. Indeed we argue that the same electrons are involved, whether the real part or the imaginary part is concerned.

For the three orthorhombic crystal classes, which belong to the biaxial optical class, the three coefficients of the diagonal are different, so that it comes:

$$\begin{aligned}\hat{\varepsilon}_{r_{xx}}(\omega) &= \varepsilon_{r_{xx}}(\omega) + i\varepsilon'_{r_{xx}}(\omega) \neq \hat{\varepsilon}_{r_{yy}}(\omega) = \varepsilon_{r_{yy}}(\omega) + i\varepsilon'_{r_{yy}}(\omega) \neq \hat{\varepsilon}_{r_{zz}}(\omega)\\ &= \varepsilon_{r_{zz}}(\omega) + i\varepsilon'_{r_{zz}}(\omega)\end{aligned} \tag{6.37}$$

In the case of the three monoclinic crystal classes, that also belong to the biaxial optical class, it appears that the dielectric frame does not diagonalize $\varepsilon'_r(\omega)$. However, it is necessary for the following to clarify the relative orientation of the so-called "special axis", as defined in Chapter 2, with respect to the dielectric frame. It is reminded here that the dielectric frame is the frame that diagonalizes $\varepsilon_r(\omega)$: the arbitrary convention that defines the relation of order among the three principal refractive indices, i.e. $n_x(\omega) < n_y(\omega) < n_z(\omega)$ or $n_x(\omega) > n_y(\omega) > n_z(\omega)$, imposes the terminology $\overrightarrow{Ox}$, $\overrightarrow{Oy}$, and $\overrightarrow{Oz}$. As a consequence, each principal axis is unambiguously oriented. Due to symmetry reasons, the special axis corresponds to one of the three dielectric axes, but there is no reason that imposes which one. Therefore, three configurations have to be considered for writing Eq. (6.8) as detailed here:

- special axis along $\overrightarrow{Ox}$

$$\begin{aligned}\hat{\varepsilon}_r(\omega) &= \begin{pmatrix} \hat{\varepsilon}_{r_{xx}}(\omega) & 0 & 0 \\ 0 & \hat{\varepsilon}_{r_{yy}}(\omega) & \hat{\varepsilon}_{r_{yz}}(\omega) \\ 0 & \hat{\varepsilon}_{r_{zy}}(\omega) & \hat{\varepsilon}_{r_{zz}}(\omega) \end{pmatrix}\\ &= \begin{pmatrix} \varepsilon_{r_{xx}}(\omega) & 0 & 0 \\ 0 & \varepsilon_{r_{yy}}(\omega) & 0 \\ 0 & 0 & \varepsilon_{r_{zz}}(\omega) \end{pmatrix} + i \begin{pmatrix} \varepsilon'_{r_{xx}}(\omega) & 0 & 0 \\ 0 & \varepsilon'_{r_{yy}}(\omega) & \varepsilon'_{r_{yz}}(\omega) \\ 0 & \varepsilon'_{r_{zy}}(\omega) & \varepsilon'_{r_{zz}}(\omega) \end{pmatrix}\end{aligned} \tag{6.38a}$$

with

$$\begin{cases} \hat{\varepsilon}_{r_{yz}}(\omega) = \hat{\varepsilon}_{r_{zy}}(\omega) \\ \varepsilon'_{r_{yz}}(\omega) = \varepsilon'_{r_{zy}}(\omega) \end{cases} \tag{6.38b}$$

- special axis along $\overrightarrow{Oy}$

$$\begin{aligned}\hat{\varepsilon}_r(\omega) &= \begin{pmatrix} \hat{\varepsilon}_{r_{xx}}(\omega) & 0 & \hat{\varepsilon}_{r_{xz}}(\omega) \\ 0 & \hat{\varepsilon}_{r_{yy}}(\omega) & 0 \\ \hat{\varepsilon}_{r_{zx}}(\omega) & 0 & \hat{\varepsilon}_{r_{zz}}(\omega) \end{pmatrix}\\ &= \begin{pmatrix} \varepsilon_{r_{xx}}(\omega) & 0 & 0 \\ 0 & \varepsilon_{r_{yy}}(\omega) & 0 \\ 0 & 0 & \varepsilon_{r_{zz}}(\omega) \end{pmatrix} + i \begin{pmatrix} \varepsilon'_{r_{xx}}(\omega) & 0 & \varepsilon'_{r_{xz}}(\omega) \\ 0 & \varepsilon'_{r_{yy}}(\omega) & 0 \\ \varepsilon'_{r_{zx}}(\omega) & 0 & \varepsilon'_{r_{zz}}(\omega) \end{pmatrix}\end{aligned} \tag{6.39a}$$

with

$$\begin{cases} \hat{\varepsilon}_{r_{xz}}(\omega) = \hat{\varepsilon}_{r_{zx}}(\omega) \\ \varepsilon'_{r_{xz}}(\omega) = \varepsilon'_{r_{zx}}(\omega) \end{cases} \tag{6.39b}$$

- special axis along $\overrightarrow{Oz}$

$$\hat{\varepsilon}_r(\omega) = \begin{pmatrix} \hat{\varepsilon}_{r_{xx}}(\omega) & \hat{\varepsilon}_{r_{xy}}(\omega) & 0 \\ \hat{\varepsilon}_{r_{yx}}(\omega) & \hat{\varepsilon}_{r_{yy}}(\omega) & 0 \\ 0 & 0 & \hat{\varepsilon}_{r_{zz}}(\omega) \end{pmatrix}$$
$$= \begin{pmatrix} \varepsilon_{r_{xx}}(\omega) & 0 & 0 \\ 0 & \varepsilon_{r_{yy}}(\omega) & 0 \\ 0 & 0 & \varepsilon_{r_{zz}}(\omega) \end{pmatrix} + i \begin{pmatrix} \varepsilon'_{r_{xx}}(\omega) & \varepsilon'_{r_{xy}}(\omega) & 0 \\ \varepsilon'_{r_{yx}}(\omega) & \varepsilon'_{r_{yy}}(\omega) & 0 \\ 0 & 0 & \varepsilon'_{r_{zz}}(\omega) \end{pmatrix} \tag{6.40a}$$

with

$$\begin{cases} \hat{\varepsilon}_{r_{xy}}(\omega) = \hat{\varepsilon}_{r_{yx}}(\omega) \\ \varepsilon'_{r_{xy}}(\omega) = \varepsilon'_{r_{yx}}(\omega) \end{cases} \tag{6.40b}$$

Finally, in the case of the triclinic crystals, it also appears that the dielectric frame does not diagonalize $\varepsilon'_r(\omega)$ and all the non-diagonal terms of $\varepsilon'_r(\omega)$ are non-zero so that Eqs. (6.8) and (6.9) directly apply.

6.4 IMAGINARY INDEX SURFACE

The imaginary index surface is the equivalent of the index surface studied in Chapter 5 for the real part of the optical index. The indices will be expressed only as a function of the unit vector $\vec{u}$ in the following. It will then be easy to convert the expressions established hereafter as a function of the angles of spherical coordinates (θ,φ) counted within the dielectric frame thanks to Eq. (5.47).

Equation (6.32) combined with Eq. (6.31), Eq. (6.28), Eq. (6.25a), Eq. (6.25b), and Eq. (6.25c) are general equations so that they can be directly used to access to n' as a function of $\vec{u}$ for any optical class using the proper tensor $\hat{\varepsilon}$ as detailed in Part 6.3. However in some cases, it is easy to directly start the calculation at the levels of Eq. (6.18) and Eq. (6.19), which is done hereafter for the isotropic and uniaxial optical classes along any direction of propagation, as well as for one crystal class of the biaxial optical class, *i.e.* the orthorhombic one, along the dielectric principal axes and planes.

6.4.1 Isotropic Optical Class

As detailed earlier, the isotropic optical class corresponds to $\hat{\varepsilon}_{r_{xx}} = \hat{\varepsilon}_{r_{yy}} = \hat{\varepsilon}_{r_{zz}} \equiv \hat{\varepsilon}_r$ and $\hat{\varepsilon}_{r_{i \neq j}} = 0$ in the dielectric frame. Then the matrix describing the tensor of the real and imaginary parts of $\hat{\varepsilon}_r$ is the same.

Along the dielectric axes $\overrightarrow{Ox}$, $\overrightarrow{Oy}$, and $\overrightarrow{Oz}$, *i.e.* for $\vec{u}$ with the coordinates $(1,0,0)$, $(0,1,0)$, and $(0,0,1)$ respectively, Eq. (6.18) and Eq. (6.19) reduce to the same equation, *i.e.*:

$$\left(\hat{n}^2(\omega,\vec{u}) - \hat{\varepsilon}_r(\omega)\right)^2 = 0 \tag{6.41}$$

The doubly-degenerated solutions of Eq. (6.41) are $\hat{n}^2(\omega,\vec{u}) = [n(\omega,\vec{u}) + jn'(\omega,\vec{u})]^2 = \hat{\varepsilon}_r$, which can be written $\hat{n}^2(\omega) = [n(\omega) + jn'(\omega)]^2$ since $\hat{\varepsilon}_r$ does not depend on $\vec{u}$. Then according to $n'(\omega) << n(\omega)$, the real and imaginary parts of the complex optical index solution for a propagation along any of the three dielectric axes are:

$$\begin{cases} n(\omega) = \sqrt{\varepsilon_r(\omega)} \\ n'(\omega) = \dfrac{\varepsilon_r'(\omega)}{2n(\omega)} = \dfrac{\varepsilon_r'(\omega)}{2\sqrt{\varepsilon_r(\omega)}} \end{cases} \tag{6.42}$$

Note the real part is the same as that given by Eq. (5.49).

In the plane $(\overrightarrow{Ox},\overrightarrow{Oy})$, the coordinates of the unit wave vector $\vec{u}$ are $(\vec{u}_x,\vec{u}_y,0)$, so that the determinant of Eq. (6.18) and Eq. (6.19) can easily be calculated as:

$$\begin{aligned} &\left[\left(\hat{n}^2(\omega,\vec{u})u_y^2 - \hat{\varepsilon}_r(\omega)\right)\left(\hat{n}^2(\omega,\vec{u})u_x^2 - \hat{\varepsilon}_r(\omega)\right) - \hat{n}^4(\omega,\vec{u})u_x^2u_y^2\right] \\ &\left(\hat{n}^2(\omega,\vec{u}) - \hat{\varepsilon}_r(\omega)\right) = 0 \end{aligned} \tag{6.43}$$

Since the unit vector verifies $u_x^2 + u_y^2 = 1$, Eq. (6.43) reduces to Eq. (6.41), with the solutions given by Eq. (6.42) for any direction of propagation of $(\overrightarrow{Ox},\overrightarrow{Oy})$.

It is straight forward to show that this couple of solutions is the same in $(\overrightarrow{Ox},\overrightarrow{Oz})$ and $(\overrightarrow{Oy},\overrightarrow{Oz})$.

Then the angular distribution of the imaginary optical index is given by the same circle in each of the three dielectric planes as shown in Figure 6.1. The geometry is the same as in the case of the index surface depicted in Figure 5.2.

Out of the dielectric planes, Eq. (6.18) and Eq. (6.19) also reduce to Eq. (6.41) so that the couple of solutions for the real and imaginary parts of the optical index are given by Eq. (6.42). Then the imaginary index surface is a sphere, belonging to the group $\frac{\infty}{m}\frac{\infty}{m}$, which is the same as for the index surface given by Figure 5.2.

The imaginary index n' is given by the distance from the origin to the sheet of the imaginary index surface in any considered direction of propagation $\vec{u}$.

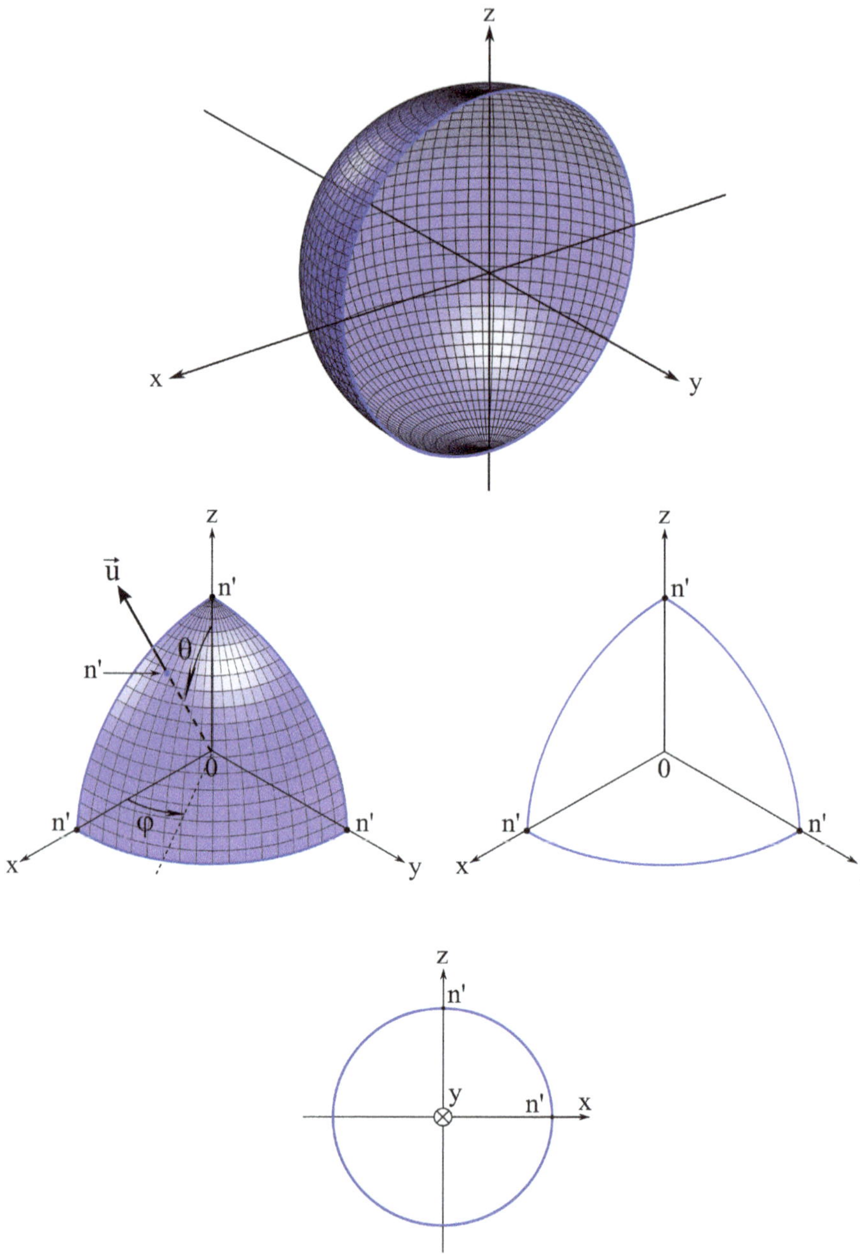

FIGURE 6.1 Imaginary index surface in the dielectric frame $(\overrightarrow{Ox},\overrightarrow{Oy},\overrightarrow{Oz})$ of a crystal belonging to the isotropic optical class. (θ,φ) are the angles of spherical coordinates of the unit vector $\vec{u}$ and n' is the single imaginary index.

6.4.2 Uniaxial Optical Class

It is reminded that the uniaxial optical class corresponds to $\hat{\varepsilon}_{r_{xx}}(\omega) = \hat{\varepsilon}_{r_{yy}}(\omega) \equiv \hat{\varepsilon}_{r_o}(\omega)$, $\hat{\varepsilon}_{r_{zz}}(\omega) \equiv \hat{\varepsilon}_{r_e}(\omega)$, and $\hat{\varepsilon}_{r_{i\neq j}}(\omega) = 0$ in the dielectric frame. We take the same relation of order for both the real and imaginary parts of the complex relative dielectric permittivity, *i.e.* $\varepsilon_{r_o}(\omega) < \varepsilon_{r_e}(\omega)$ and $\varepsilon'_{r_o}(\omega) < \varepsilon'_{r_e}(\omega)$ in the case of a positive uniaxial crystal; $\varepsilon_{r_o}(\omega) > \varepsilon_{r_e}(\omega)$ and $\varepsilon'_{r_o}(\omega) > \varepsilon'_{r_e}(\omega)$ for a negative uniaxial one. Such a hypothesis is physically reasonable because the property of the polarizability of an uniaxial medium deals simultaneously with both the real and imaginary parts of the dielectric permittivity. As for the isotropic optical class, the matrix for the real and imaginary parts of $\hat{\varepsilon}_r$ is the same.

Along the dielectric axes $\overrightarrow{Ox}$ and $\overrightarrow{Oy}$, Eq. (6.18) and Eq. (6.19) reduce to:

$$\left(-\hat{\varepsilon}_{r_o}(\omega)\right)\left(\hat{n}^2(\omega) - \hat{\varepsilon}_{r_o}(\omega)\right)\left(\hat{n}^2(\omega) - \hat{\varepsilon}_{r_e}(\omega)\right) = 0 \tag{6.44}$$

Since $\hat{\varepsilon}_{r_o}(\omega) \neq 0$, the two solutions of Eq. (6.44) are $\hat{n}_o^2(\omega) = \hat{\varepsilon}_{r_o}(\omega)$ and $\hat{n}_e^2(\omega) = \hat{\varepsilon}_{r_e}(\omega)$. Then, depending on the sign of the optical uniaxial class, there are two sets of complex optical index solutions, *i.e.*:

- for the positive uniaxial class

$$\begin{cases} n^+(\omega) = \sqrt{\varepsilon_{r_e}(\omega)} \equiv n_e(\omega) \\ n'^+(\omega) = \dfrac{\varepsilon'_{r_e}(\omega)}{2n_e(\omega)} = \varepsilon'_{r_e}(\omega)\Big/2\sqrt{\varepsilon_{r_e}(\omega)} \equiv n'_e(\omega) \end{cases} \tag{6.45}$$

$$\begin{cases} n^-(\omega) = \sqrt{\varepsilon_{r_o}(\omega)} \equiv n_o(\omega) \\ n'^-(\omega) = \dfrac{\varepsilon'_{r_o}(\omega)}{2n_o(\omega)} = \varepsilon'_{r_o}(\omega)\Big/2\sqrt{\varepsilon_{r_o}(\omega)} \equiv n'_o(\omega) \end{cases} \tag{6.46}$$

- for the negative uniaxial class

$$\begin{cases} n^+(\omega) = \sqrt{\varepsilon_{r_o}(\omega)} \equiv n_o(\omega) \\ n'^+(\omega) = \dfrac{\varepsilon'_{r_o}(\omega)}{2n_o(\omega)} = \varepsilon'_{r_o}(\omega)\Big/2\sqrt{\varepsilon_{r_o}(\omega)} \equiv n'_o(\omega) \end{cases} \tag{6.47}$$

$$\begin{cases} n^-(\omega) = \sqrt{\varepsilon_{r_e}(\omega)} \equiv n_e(\omega) \\ n'^-(\omega) = \dfrac{\varepsilon'_{r_e}(\omega)}{2n_e(\omega)} = \varepsilon'_{r_e}(\omega)\Big/2\sqrt{\varepsilon_{r_e}(\omega)} \equiv n'_e(\omega) \end{cases} \tag{6.48}$$

The quantities $n'_o(\omega)$ and $n'_e(\omega)$ are called the principal ordinary and extraordinary imaginary optical indices, respectively, $n_o(\omega)$ and $n_e(\omega)$ being the principal ordinary and extraordinary refractive indices.

Along $\overrightarrow{Oz}$, Eq. (6.18) and Eq. (6.19) reduce to:

$$\left(\hat{n}^2(\omega)-\hat{\varepsilon}_{r_o}(\omega)\right)^2\left(-\hat{\varepsilon}_{r_e}(\omega)\right)=0 \tag{6.49}$$

The two solutions of Eq. (6.49) are degenerate whatever the sign of the optical class, which leads to:

$$\begin{cases} n^+(\omega)=n^-(\omega)=\sqrt{\varepsilon_{r_o}(\omega)}\equiv n_o(\omega) \\ n'^+(\omega)=n'^-(\omega)=\varepsilon'_{r_o}(\omega)\Big/2\sqrt{\varepsilon_{r_o}(\omega)}\equiv n'_o(\omega) \end{cases} \tag{6.50}$$

In the plane $(\overrightarrow{Ox},\overrightarrow{Oy})$, the coordinates of $\vec{u}$ are $(\vec{u}_x,\vec{u}_y,0)$, so that the determinant of Eq. (6.18) and Eq. (6.19) can easily be calculated as:

$$\begin{aligned}&\left[\left(\hat{n}^2(\omega,\vec{u})u_y^2-\hat{\varepsilon}_{r_o}(\omega)\right)\left(\hat{n}^2(\omega,\vec{u})u_x^2-\hat{\varepsilon}_{r_o}(\omega)\right)-\hat{n}^4(\omega,\vec{u})u_x^2u_y^2\right]\\&\left(\hat{n}^2(\omega,\vec{u})-\hat{\varepsilon}_{r_e}(\omega)\right)=0\end{aligned} \tag{6.51}$$

Therefore, the two solutions of Eq. (6.51) are:

$$\begin{cases} \hat{n}_e|^2(\omega,\vec{u})=\hat{\varepsilon}_{r_e}(\omega) \\ \hat{n}_o|^2(\omega,\vec{u})=\dfrac{\hat{\varepsilon}_{r_o}(\omega)\hat{\varepsilon}_{r_o}(\omega)}{u_x^2\hat{\varepsilon}_{r_o}(\omega)+u_y^2\hat{\varepsilon}_{r_o}(\omega)}=\hat{\varepsilon}_{r_o}(\omega) \end{cases} \tag{6.52}$$

Under the weak absorption/emission hypothesis, and depending on the optical sign of the uniaxial class, the complex optical index solutions write:

- for the positive uniaxial class

$$\begin{aligned}&\begin{cases} n^+(\omega,\vec{u})=\sqrt{\varepsilon_{r_e}(\omega)}=n_e(\omega) \\ n'^+(\omega,\vec{u})=\dfrac{\varepsilon'_{r_e}(\omega)}{2n^+(\omega,\vec{u})}=\varepsilon'_{r_e}(\omega)\Big/2\sqrt{\varepsilon_{r_e}(\omega)} \end{cases}\\&\begin{cases} n^-(\omega,\vec{u})=\sqrt{\varepsilon_{r_o}(\omega)}=n_o(\omega) \\ n'^-(\omega,\vec{u})=\dfrac{\varepsilon'_{r_o}(\omega)}{2n^-(\omega,\vec{u})}=\varepsilon'_{r_o}(\omega)\Big/2\sqrt{\varepsilon_{r_o}(\omega)} \end{cases}\end{aligned} \tag{6.53}$$

- for the negative uniaxial class

$$\begin{cases} n^{+}(\omega,\vec{u}) = \sqrt{\varepsilon_{r_o}(\omega)} = n_o(\omega) \\ n'^{+}(\omega,\vec{u}) = \dfrac{\varepsilon'_{r_o}(\omega)}{2n^{+}(\omega,\vec{u})} = \varepsilon'_{r_o}(\omega)\Big/2\sqrt{\varepsilon_{r_o}(\omega)} \end{cases}$$
$$\begin{cases} n^{-}(\omega,\vec{u}) = \sqrt{\varepsilon_{r_e}(\omega)} = n_e(\omega) \\ n'^{-}(\omega,\vec{u}) = \dfrac{\varepsilon'_{r_e}(\omega)}{2n^{-}(\omega,\vec{u})} = \varepsilon'_{r_e}(\omega)\Big/2\sqrt{\varepsilon_{r_e}(\omega)} \end{cases} \tag{6.54}$$

Note that the real quantities are those obtained in Part 5.2.1.3.

In the plane $(\overrightarrow{Oy},\overrightarrow{Oz})$, the coordinates of $\vec{u}$ are $(0,\vec{u}_y,\vec{u}_z)$. Similarly to the approach detailed for the plane $(\overrightarrow{Ox},\overrightarrow{Oy})$, the determinant of Eq. (6.18) and Eq. (6.19) reduces to:

$$\begin{aligned} &\left[\left(\hat{n}^2(\omega,\vec{u})u_y^2 - \hat{\varepsilon}_{r_e}(\omega)\right)\left(\hat{n}^2(\omega,\vec{u})u_z^2 - \hat{\varepsilon}_{r_o}(\omega)\right) - \hat{n}^4(\omega,\vec{u})u_y^2u_z^2\right] \\ &\left(\hat{n}^2(\omega,\vec{u}) - \hat{\varepsilon}_{r_o}(\omega)\right) = 0 \end{aligned} \tag{6.55}$$

The two following solutions can then be calculated, *i.e.*:

$$\begin{cases} \hat{n}_o\big|^2(\omega,\vec{u}) = \hat{\varepsilon}_{r_o}(\omega) \\ \hat{n}_e\big|^2(\omega,\vec{u}) = \dfrac{\hat{\varepsilon}_{r_o}(\omega)\hat{\varepsilon}_{r_e}(\omega)}{u_y^2\hat{\varepsilon}_{r_e}(\omega) + u_z^2\hat{\varepsilon}_{r_o}(\omega)} \end{cases} \tag{6.56}$$

While the $\hat{n}_o\big|^2$ term is straightforward to handle, the second one $\hat{n}_e\big|^2$ requires additional linearization to show an explicit analytical expression, as detailed here:

$$\begin{aligned} \hat{n}_e\big|^2(\omega,\vec{u}) &= \frac{\varepsilon_{r_o}(\omega)\varepsilon_{r_e}(\omega)}{\left(u_y^2\varepsilon_{r_e}(\omega) + u_z^2\varepsilon_{r_o}(\omega)\right)} \frac{1 + j\left(\varepsilon'_{r_o}(\omega)\varepsilon_{r_e}(\omega) + \varepsilon_{r_o}(\omega)\varepsilon'_{r_e}(\omega)\right)\Big/\left(\varepsilon_{r_o}(\omega)\varepsilon_{r_e}(\omega)\right)}{1 + j\left(u_y^2\varepsilon'_{r_e}(\omega) + u_z^2\varepsilon'_{r_o}(\omega)\right)\Big/\left(u_y^2\varepsilon_{r_e}(\omega) + u_z^2\varepsilon_{r_o}(\omega)\right)} \\ &\approx n_e\big|^2(\omega,\vec{u})\left(1 + j\left[\frac{\varepsilon'_{r_o}(\omega)\varepsilon_{r_e}(\omega) + \varepsilon_{r_o}(\omega)\varepsilon'_{r_e}(\omega)}{\varepsilon_{r_o}(\omega)\varepsilon_{r_e}(\omega)} - \frac{u_y^2\varepsilon'_{r_e}(\omega) + u_z^2\varepsilon'_{r_o}(\omega)}{u_y^2\varepsilon_{r_e}(\omega) + u_z^2\varepsilon_{r_o}(\omega)}\right]\right) \end{aligned} \tag{6.57}$$

where $n_e\big|^2(\omega,\vec{u}) = \dfrac{\varepsilon_{r_o}(\omega)\varepsilon_{r_e}(\omega)}{\left(u_y^2\varepsilon_{r_e}(\omega) + u_z^2\varepsilon_{r_o}(\omega)\right)} = \left(u_y^2\varepsilon_{r_e}^{-1}(\omega) + u_z^2\varepsilon_{r_o}^{-1}(\omega)\right)^{-1}$, as detailed in Part 5.2.1.3.

Then it comes:

$$\hat{n}_e\big|^2(\omega,\vec{u}) = n_e\big|^2(\omega,\vec{u})\left(1 + j\frac{n_e^{\;2}(\omega,\vec{u})}{\varepsilon_{r_o}^2(\omega)}\left(\frac{\varepsilon'_{r_e}(\omega)}{\varepsilon_{r_e}(\omega)}u_y^2 + \frac{\varepsilon'_{r_o}(\omega)}{\varepsilon_{r_o}(\omega)}u_z^2\right)\right) \tag{6.58}$$

Therefore, by identifying the real and imaginary parts of the complex optical indices, and assuming again that $n'(\omega,\vec{u}) << n(\omega,\vec{u})$, the two complex index solutions write for each of the two optical signs of the uniaxial class:

- for the positive uniaxial class

$$\begin{cases} n^+(\omega,\vec{u}) = \left(u_y^2\varepsilon_{r_e}^{-1}(\omega) + u_z^2\varepsilon_{r_o}^{-1}(\omega)\right)^{-1/2} \\ n'^+(\omega,\vec{u}) = \dfrac{n^+\big|^3(\omega,\vec{u})}{2}\left(\dfrac{\varepsilon'_{r_e}(\omega)}{\varepsilon_{r_e}^2(\omega)}u_y^2 + \dfrac{\varepsilon'_{r_o}(\omega)}{\varepsilon_{r_o}^2(\omega)}u_z^2\right) \end{cases}$$
$$\begin{cases} n^-(\omega,\vec{u}) = \sqrt{\varepsilon_{r_o}(\omega)} = n_o(\omega) \\ n'^-(\omega,\vec{u}) = \dfrac{\varepsilon'_{r_o}(\omega)}{2n^-(\omega,\vec{u})} = \varepsilon'_{r_o}(\omega)\Big/2\sqrt{\varepsilon_{r_o}(\omega)} \end{cases} \tag{6.59}$$

- for the negative uniaxial class

$$\begin{cases} n^+(\omega,\vec{u}) = \sqrt{\varepsilon_{r_o}(\omega)} = n_o(\omega) \\ n'^+(\omega,\vec{u}) = \dfrac{\varepsilon'_{r_o}(\omega)}{2n^+(\omega,\vec{u})} = \varepsilon'_{r_o}(\omega)\Big/2\sqrt{\varepsilon_{r_o}(\omega)} \end{cases}$$
$$\begin{cases} n^-(\omega,\vec{u}) = \left(u_y^2\varepsilon_{r_e}^{-1}(\omega) + u_z^2\varepsilon_{r_o}^{-1}(\omega)\right)^{-1/2} \\ n'^-(\omega,\vec{u}) = \dfrac{n^-\big|^3(\omega,\vec{u})}{2}\left(\dfrac{\varepsilon'_{r_e}(\omega)}{\varepsilon_{r_e}^2(\omega)}u_y^2 + \dfrac{\varepsilon'_{r_o}(\omega)}{\varepsilon_{r_o}^2(\omega)}u_z^2\right) \end{cases} \tag{6.60}$$

In the plane $(\overrightarrow{Ox},\overrightarrow{Oz})$, the coordinates of the unit wave vector $\vec{u}$ are $(\vec{u}_x,0,\vec{u}_z)$. Similarly to the approach detailed in $(\overrightarrow{Oy},\overrightarrow{Oz})$, the determinant of Eq. (6.18) and Eq. (6.19) reduces to:

$$\begin{aligned} &\left[\left(\hat{n}^2(\omega,\vec{u})u_x^2 - \hat{\varepsilon}_{r_e}(\omega)\right)\left(\hat{n}^2(\omega,\vec{u})u_z^2 - \hat{\varepsilon}_{r_o}(\omega)\right) - \hat{n}^4(\omega,\vec{u})u_x^2u_z^2\right] \\ &\left(\hat{n}^2(\omega,\vec{u}) - \hat{\varepsilon}_{r_o}(\omega)\right) = 0 \end{aligned} \tag{6.61}$$

This expression is the same as Eq. (6.55) by simply replacing u_y with u_x. Then the corresponding solutions are given by Eq. (5.59) and Eq. (5.60) in the case of the positive and negative uniaxial class respectively after replacing u_y by u_x.

Out of the dielectric planes, the calculation of the determinant of Eq. (6.18) and Eq. (6.19) along the unit wave vector $\vec{u} = (\vec{u_x}, \vec{u_y}, \vec{u_z})$ for any direction of propagation has been solved using Eq. (6.23), Eq. (6.27), and Eq. (6.28). By invoking the uniaxial symmetry that corresponds to the invariance by continuous rotation around the optical axis, *i.e.* the z-axis, it can be shown that these general expressions reduce to the same expressions as those calculated in the planes and $(\overrightarrow{Oy}, \overrightarrow{Oz})$. As a consequence, it comes:

- for the positive uniaxial class

$$\begin{cases} n^{+}(\omega,\vec{u}) = \left(\left(u_x^2+u_y^2\right)\varepsilon_{r_e}^{-1}(\omega)+u_z^2\varepsilon_{r_o}^{-1}(\omega)\right)^{-1/2} \\ n'^{+}(\omega,\vec{u}) = \dfrac{n^{+}\Big|^{3}(\omega,\vec{u})}{2}\left(\dfrac{\varepsilon'_{r_e}(\omega)}{\varepsilon_{r_e}^{2}(\omega)}\left(u_x^2+u_y^2\right)+\dfrac{\varepsilon'_{r_o}(\omega)}{\varepsilon_{r_o}^{2}(\omega)}u_z^2\right) \end{cases}$$
$$\begin{cases} n^{-}(\omega,\vec{u}) = \sqrt{\varepsilon_{r_o}(\omega)} = n_o(\omega) \\ n'^{-}(\omega,\vec{u}) = \dfrac{\varepsilon'_{r_o}(\omega)}{2n^{-}(\omega,\vec{u})} = \varepsilon'_{r_o}(\omega)\Big/2\sqrt{\varepsilon_{r_o}(\omega)} \end{cases} \tag{6.62}$$

- for the negative uniaxial class

$$\begin{cases} n^{+}(\omega,\vec{u}) = \sqrt{\varepsilon_{r_o}(\omega)} = n_o(\omega) \\ n'^{+}(\omega,\vec{u}) = \dfrac{\varepsilon'_{r_o}(\omega)}{2n^{+}(\omega,\vec{u})} = \varepsilon'_{r_o}(\omega)\Big/2\sqrt{\varepsilon_{r_o}(\omega)} \end{cases}$$
$$\begin{cases} n^{-}(\omega,\vec{u}) = \left(\left(u_x^2+u_y^2\right)\varepsilon_{r_e}^{-1}(\omega)+u_z^2\varepsilon_{r_o}^{-1}(\omega)\right)^{-1/2} \\ n'^{-}(\omega,\vec{u}) = \dfrac{n^{-}\Big|^{3}(\omega,\vec{u})}{2}\left(\dfrac{\varepsilon'_{r_e}(\omega)}{\varepsilon_{r_e}^{2}(\omega)}\left(u_x^2+u_y^2\right)+\dfrac{\varepsilon'_{r_o}(\omega)}{\varepsilon_{r_o}^{2}(\omega)}u_z^2\right) \end{cases} \tag{6.63}$$

Such solutions show invariance with respect to the φ angle of the spherical coordinates of the unit vector $\vec{u}$, as illustrated in Figure 6.2 and Figure 6.3 in the case of the positive and negative uniaxial optical classes, respectively. The calculations are performed with the following fictive numerical values: $\varepsilon_{r_o}(\omega) = 2.6$, $\varepsilon_{r_e}(\omega) = 2.7$, $\varepsilon'_{r_o}(\omega) = 3.20\times10^{-5}$, and $\varepsilon'_{r_e}(\omega) = 8.25\times10^{-5}$ for Figure 6.2; $\varepsilon_{r_o}(\omega) = 2.7$, $\varepsilon_{r_e}(\omega) = 2.6$, $\varepsilon'_{r_o}(\omega) = 8.25\times10^{-5}$, and $\varepsilon'_{r_e}(\omega) = 3.30\times10^{-5}$ for Figure 6.3.

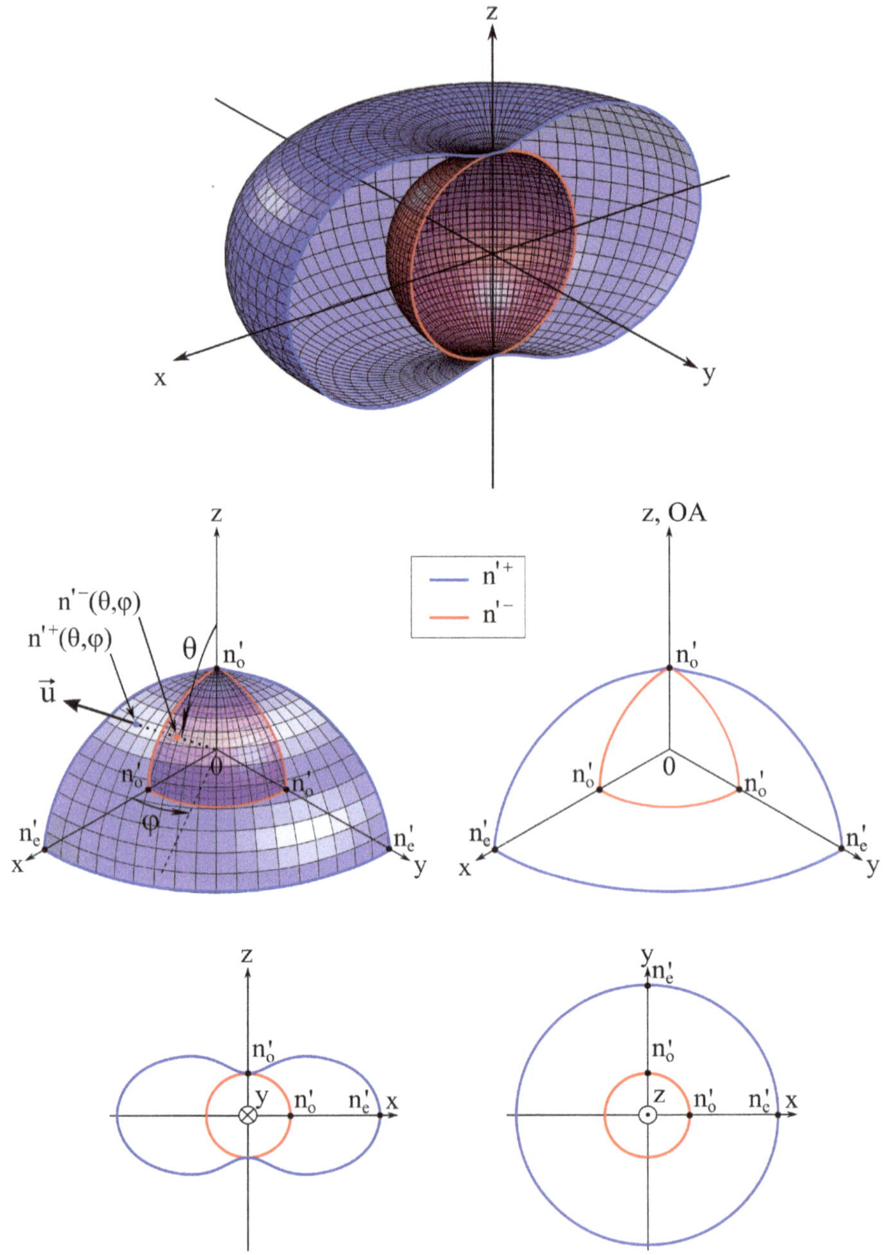

FIGURE 6.2 Imaginary index surface of a crystal belonging to the positive uniaxial optical class where (x,y,z) refers to the dielectric frame $(\overrightarrow{Ox},\overrightarrow{Oy},\overrightarrow{Oz})$. Top: half-space angular distribution showing rotation invariance with respect to the z-axis. Middle: octant space in which (θ,φ) are the angles of spherical coordinates of the unit vector $\vec{u}$, and $n'^{+}(\theta,\varphi)$ and $n'^{-}(\theta,\varphi)$ are the two corresponding imaginary indices; $n_o{}'$ and $n_e{}'$ refer to the ordinary and extraordinary principal imaginary indices, and OA denotes the optical axis that correspond to the z-axis. Bottom: section of the imaginary index in the $(\overrightarrow{Ox},\overrightarrow{Oz})$ plane (left), and in the $(\overrightarrow{Ox},\overrightarrow{Oy})$ plane (right).

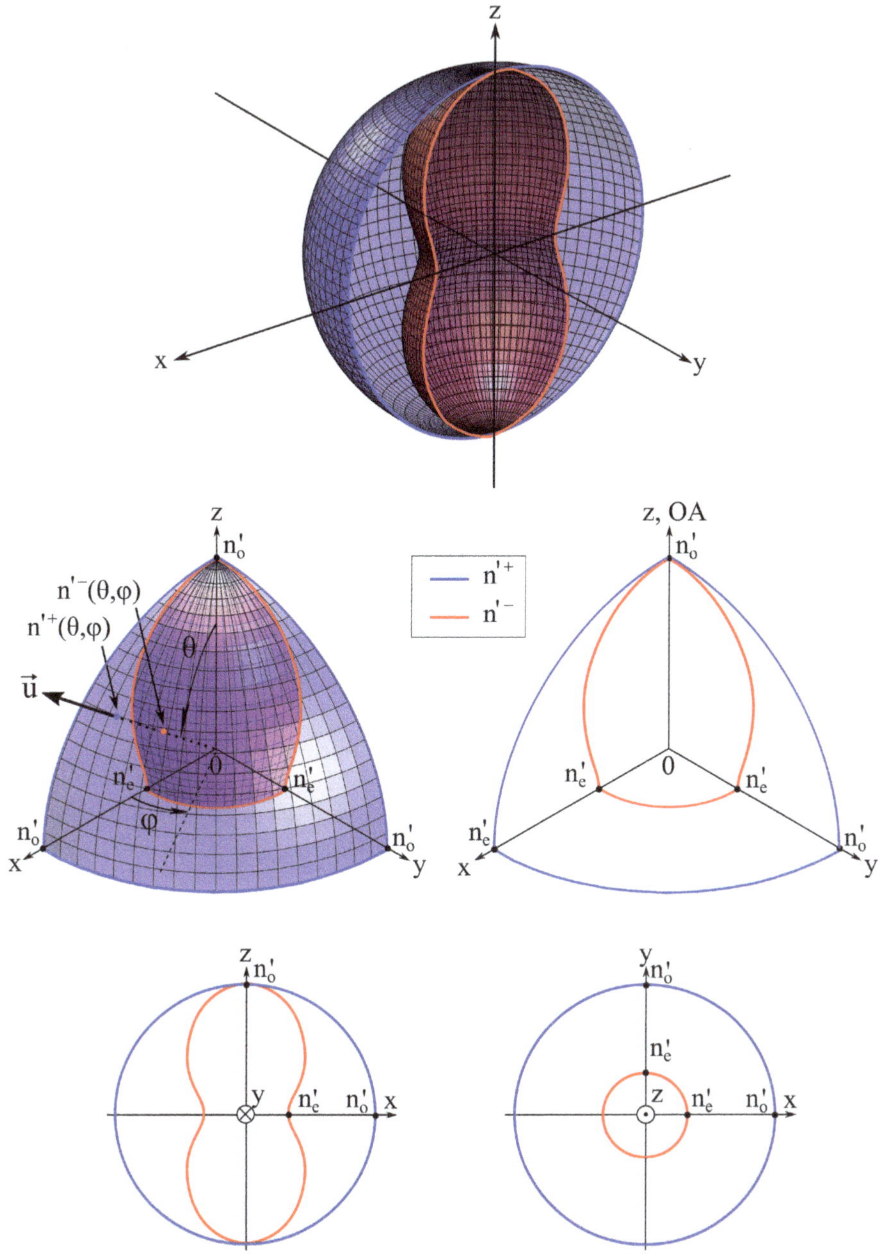

FIGURE 6.3 Imaginary index surface of a crystal belonging to the negative uniaxial optical class where (x,y,z) refers to the dielectric frame $(\overrightarrow{Ox},\overrightarrow{Oy},\overrightarrow{Oz})$. Top: half-space angular distribution showing rotation invariance with respect to the z-axis. Middle: octant space in which (θ,φ) are the angles of spherical coordinates of the unit vector $\vec{u}$, and $n'^{+}(\theta,\varphi)$ and $n'^{-}(\theta,\varphi)$ are the two corresponding imaginary indices; $n_o{}'$ and $n_e{}'$ refer to the ordinary and extraordinary principal imaginary indices, and OA denotes the optical axis that correspond to the z-axis. Bottom: section of the imaginary index in the $(\overrightarrow{Ox},\overrightarrow{Oz})$ plane (left), and in the $(\overrightarrow{Ox},\overrightarrow{Oy})$ plane (right).

Figure 6.2 and Figure 6.3 can be compared to Figure 5.3 and Figure 5.4, respectively. It then appears that the difference between the imaginary and real angular distributions of the optical index is only in the topology of the extraordinary layer around the optical axis $\overrightarrow{Oz}$: there is a bi-lobar shape compared with an elliptical one in the case of the index surface. In the case of the uniaxial optical class, the optical axis remains the same rotation axis of infinite order for both the real and imaginary optical distributions. Then the angular distribution of the imaginary index surface belongs to the static cylinder symmetry group $\frac{\infty}{m}m$, the same as for the index surface for such an optical class.

6.4.3 Biaxial Optical Class

In the dielectric frame, the tensor of the real part ε_r of $\hat{\varepsilon}_r$ is diagonal, with $\varepsilon_{r_{xx}} \neq \varepsilon_{r_{yy}} \neq \varepsilon_{r_{zz}}$. But there exist differences between the three corresponding crystal classes, *i.e.* the orthorhombic, monoclinic, and triclinic classes, at the level of the imaginary part ε_r' as mentioned in Part 6.3. Such differences do influence the imaginary index surface, as shown hereafter. Only the case of a positive optical sign, *i.e.* $\varepsilon_{r_{xx}} < \varepsilon_{r_{yy}} < \varepsilon_{r_{zz}}$, will be considered for the sake of brevity.

6.4.3.1 Orthorhombic Crystal Classes

In this case, the tensor ε_r' is also diagonal, with $\varepsilon'_{r_{xx}} \neq \varepsilon'_{r_{yy}} \neq \varepsilon'_{r_{zz}}$. Then, as for the isotropic and uniaxial optical class, the matrix of the real and imaginary parts of $\hat{\varepsilon}_r$ are identical.

Along $\overrightarrow{Ox}$, the coordinates of $\vec{u}$ are (1,0,0), so that the determinant of Eq. (6.18) and Eq. (6.19) can easily be calculated while considering Eq. (6.8) and Eq. (6.9). By keeping first-order terms only, the determinant reduces to:

$$\left(-\hat{\varepsilon}_{r_{xx}}\right)\left(\hat{n}^2(\omega,\vec{u}_x)-\hat{\varepsilon}_{r_{yy}}(\omega)\right)\left(\hat{n}^2(\omega,\vec{u}_x)-\hat{\varepsilon}_{r_{zz}}(\omega)\right)=0 \tag{6.64}$$

Since $\hat{\varepsilon}_{r_{xx}} \neq 0$, the two solutions of Eq. (6.64) are $\hat{n}^2(\omega,\vec{u}_x)=\hat{\varepsilon}_{r_{yy}}(\omega)$ and $\hat{n}^2(\omega,\vec{u}_x)=\hat{\varepsilon}_{r_{zz}}(\omega)$. Therefore, the real and imaginary solutions are:

$$\begin{cases} n^+(\omega)=\sqrt{\varepsilon_{r_{zz}}(\omega)} \equiv n_z(\omega) \\ n'^+(\omega)=\dfrac{\varepsilon'_{r_{zz}}(\omega)}{2n^+(\omega)}=\varepsilon'_{r_{zz}}(\omega)\Big/2\sqrt{\varepsilon_{r_{zz}}(\omega)} \equiv n'_z(\omega) \end{cases} \tag{6.65}$$

and

$$\begin{cases} n^-(\omega)=\sqrt{\varepsilon_{r_{yy}}(\omega)} \equiv n_y(\omega) \\ n'^-(\omega)=\dfrac{\varepsilon'_{r_{yy}}(\omega)}{2n^-(\omega)}=\varepsilon'_{r_{yy}}(\omega)\Big/2\sqrt{\varepsilon_{r_{yy}}(\omega)} \equiv n'_y(\omega) \end{cases} \tag{6.66}$$

In the same manner, along $\overrightarrow{Oy}$, for which the coordinates of $\vec{u}$ are (0,1,0), it comes:

$$\left(-\hat{\varepsilon}_{r_{yy}}\right)\left(\hat{n}^2(\omega,\vec{u}_y)-\hat{\varepsilon}_{r_{xx}}(\omega)\right)\left(\hat{n}^2(\omega,\vec{u}_y)-\hat{\varepsilon}_{r_{zz}}(\omega)\right)=0 \tag{6.67}$$

Since $\hat{\varepsilon}_{r_{yy}} \neq 0$, the solutions are:

$$\begin{cases} n^+(\omega)=\sqrt{\varepsilon_{r_{zz}}(\omega)}\equiv n_z(\omega) \\ n'^+(\omega)=\dfrac{\varepsilon'_{r_{zz}}(\omega)}{2n^+(\omega)}=\varepsilon'_{r_{zz}}(\omega)\Big/2\sqrt{\varepsilon_{r_{zz}}(\omega)}\equiv n'_z(\omega) \end{cases} \tag{6.68}$$

and

$$\begin{cases} n^-(\omega)=\sqrt{\varepsilon_{r_{xx}}(\omega)}\equiv n_x(\omega) \\ n'^-(\omega)=\dfrac{\varepsilon'_{r_{xx}}(\omega)}{2n^-(\omega)}=\varepsilon'_{r_{xx}}(\omega)\Big/2\sqrt{\varepsilon_{r_{xx}}(\omega)}\equiv n'_x(\omega) \end{cases} \tag{6.69}$$

Similarly, along $\overrightarrow{Oz}$, the coordinates of $\vec{u}$ are (0,0,1) so that the determinant of Eq. (6.18) and Eq. (6.19) reduces to:

$$\left(\hat{n}^2(\omega,\vec{u}_z)-\hat{\varepsilon}_{r_{xx}}\right)\left(\hat{n}^2(\omega,\vec{u}_z)-\hat{\varepsilon}_{r_{yy}}(\omega)\right)\left(-\hat{\varepsilon}_{r_{zz}}(\omega)\right)=0 \tag{6.70}$$

Since $\hat{\varepsilon}_{r_{zz}} \neq 0$, it comes:

$$\begin{cases} n^+(\omega)=\sqrt{\varepsilon_{r_{yy}}(\omega)}\equiv n_y(\omega) \\ n'^+(\omega)=\dfrac{\varepsilon'_{r_{yy}}(\omega)}{2n^+(\omega)}=\varepsilon'_{r_{yy}}(\omega)\Big/2\sqrt{\varepsilon_{r_{yy}}(\omega)}\equiv n'_y(\omega) \end{cases} \tag{6.71}$$

and

$$\begin{cases} n^-(\omega)=\sqrt{\varepsilon_{r_{xx}}(\omega)}\equiv n_x(\omega) \\ n'^-(\omega)=\dfrac{\varepsilon'_{r_{xx}}(\omega)}{2n^-(\omega)}=\varepsilon'_{r_{xx}}(\omega)\Big/2\sqrt{\varepsilon_{r_{xx}}(\omega)}\equiv n'_x(\omega) \end{cases} \tag{6.72}$$

The quantities $n'_x(\omega)$, $n'_y(\omega)$, and $n'_z(\omega)$ are the principal imaginary optical indices of the orthorhombic crystal, while $n_x(\omega)$, $n_y(\omega)$, and $n_z(\omega)$ are the principal refractive indices.

In $(\overrightarrow{Ox},\overrightarrow{Oy})$, the coordinates of the unit wave vector $\vec{u}$ are $(\vec{u}_x,\vec{u}_y,0)$, which gives for the determinant:

$$\begin{gathered}\left[\left(\hat{n}^2(\omega,\vec{u})u_y^2-\hat{\varepsilon}_{r_{xx}}(\omega)\right)\left(\hat{n}^2(\omega,\vec{u})u_x^2-\hat{\varepsilon}_{r_{yy}}(\omega)\right)-\left(\hat{n}^2(\omega,\vec{u})u_xu_y\right)^2\right] \\ \left(\hat{n}^2(\omega,\vec{u})-\hat{\varepsilon}_{r_{zz}}(\omega)\right)=0\end{gathered} \tag{6.73}$$

The two corresponding solutions are:

$$\begin{cases} \hat{n}^{+2}(\omega,\vec{u}) = \hat{\varepsilon}_{r_{zz}}(\omega) \\ \hat{n}^{-2}(\omega,\vec{u}) = \dfrac{\hat{\varepsilon}_{r_{xx}}(\omega)\hat{\varepsilon}_{r_{yy}}(\omega)}{u_x^2\hat{\varepsilon}_{r_{xx}}(\omega)+u_y^2\hat{\varepsilon}_{r_{yy}}(\omega)} \end{cases} \tag{6.74}$$

While the $\hat{n}^+|^2$ term is straightforward to handle, the second $\hat{n}^-|^2$ requires additional linearization to show an explicit analytical expression, *i.e.*:

$$\begin{aligned} \hat{n}^-|^2(\omega,\vec{u}) &= \frac{\varepsilon_{r_{xx}}(\omega)\varepsilon_{r_{yy}}(\omega)}{\left(u_x^2\varepsilon_{r_{xx}}(\omega)+u_y^2\varepsilon_{r_{yy}}(\omega)\right)} \frac{1+j\left(\varepsilon'_{r_{xx}}(\omega)\varepsilon_{r_{yy}}(\omega)+\varepsilon_{r_{xx}}(\omega)\varepsilon'_{r_{yy}}(\omega)\right)\Big/\left(\varepsilon_{r_{xx}}(\omega)\varepsilon_{r_{yy}}(\omega)\right)}{1+j\left(u_x^2\varepsilon'_{r_{xx}}(\omega)+u_y^2\varepsilon'_{r_{yy}}(\omega)\right)\Big/\left(u_x^2\varepsilon_{r_{xx}}(\omega)+u_y^2\varepsilon_{r_{yy}}(\omega)\right)} \\ &\approx n^-|^2(\omega,\vec{u})\left(1+j\left[\frac{\varepsilon'_{r_{xx}}(\omega)\varepsilon_{r_{yy}}(\omega)+\varepsilon_{r_{xx}}(\omega)\varepsilon'_{r_{yy}}(\omega)}{\varepsilon_{r_{xx}}(\omega)\varepsilon_{r_{yy}}(\omega)} - \frac{u_x^2\varepsilon'_{r_{xx}}(\omega)+u_y^2\varepsilon'_{r_{yy}}(\omega)}{u_x^2\varepsilon_{r_{xx}}(\omega)+u_y^2\varepsilon_{r_{yy}}(\omega)}\right]\right) \end{aligned} \tag{6.75}$$

where $n^-|^2(\omega,\vec{u}) = \dfrac{\varepsilon_{r_{xx}}(\omega)\varepsilon_{r_{yy}}(\omega)}{\left(u_x^2\varepsilon_{r_{xx}}(\omega)+u_y^2\varepsilon_{r_{yy}}(\omega)\right)} = \left(u_x^2\varepsilon_{r_{yy}}^{-1}(\omega)+u_y^2\varepsilon_{r_{xx}}^{-1}(\omega)\right)^{-1}$, which is the equivalent of Eq. (5.63).

Thus Eq. (6.75) rewrites:

$$\hat{n}^{-2}(\omega,\vec{u}) = n^{-2}(\omega,\vec{u})\left[1+jn^{-2}(\omega,\vec{u})\left(\frac{\varepsilon'_{r_{yy}}(\omega)}{\varepsilon^2_{r_{yy}}(\omega)}u_x^2+\frac{\varepsilon'_{r_{xx}}(\omega)}{\varepsilon^2_{r_{xx}}(\omega)}u_y^2\right)\right] \tag{6.76}$$

Therefore, by identifying the real and imaginary parts of the complex optical indices, and by assuming again $n'(\omega,\vec{u}) << n(\omega,\vec{u})$, the two complex index solutions write:

$$\begin{cases} n^+(\omega,\vec{u}) = \sqrt{\varepsilon_{r_{zz}}(\omega)} \\ n'^+(\omega,\vec{u}) = \dfrac{\varepsilon'_{r_{zz}}(\omega)}{2n^+(\omega,\vec{u})} = \varepsilon'_{r_{zz}}(\omega)\Big/2\sqrt{\varepsilon_{r_{zz}}(\omega)} \end{cases} \tag{6.77}$$

and

$$\begin{cases} n^-(\omega,\vec{u}) = \left(u_x^2\varepsilon_{r_{yy}}^{-1}(\omega)+u_y^2\varepsilon_{r_{xx}}^{-1}(\omega)\right)^{-1/2} \\ n'^-(\omega,\vec{u}) = \dfrac{n^-(\omega,\vec{u})^3}{2}\left(\dfrac{\varepsilon'_{r_{yy}}(\omega)}{\varepsilon^2_{r_{yy}}(\omega)}u_x^2+\dfrac{\varepsilon'_{r_{xx}}(\omega)}{\varepsilon^2_{r_{xx}}(\omega)}u_y^2\right) \end{cases} \tag{6.78}$$

In $(\overrightarrow{Oy},\overrightarrow{Oz})$, the coordinates of the unit wave vector $\vec{u}$ are $(0,\overrightarrow{u_y},\overrightarrow{u_z})$. Similarly to the approach previously detailed for $(\overrightarrow{Ox},\overrightarrow{Oy})$, it comes for the determinant of Eq. (6.18) and Eq. (6.19):

$$\left[\left(\hat{n}^2(\omega,\vec{u})u_z^2-\hat{\varepsilon}_{r_{yy}}(\omega)\right)\left(\hat{n}^2(\omega,\vec{u})u_y^2-\hat{\varepsilon}_{r_{zz}}(\omega)\right)-\left(\hat{n}^2(\omega,\vec{u})u_yu_z\right)^2\right]$$
$$\left(\hat{n}^2(\omega,\vec{u})-\hat{\varepsilon}_{r_{xx}}(\omega)\right)=0 \quad (6.79)$$

Then, the same approach leads to the identification of the real and imaginary parts of the two complex optical index solutions, *i.e.*:

$$\begin{cases} n^+(\omega,\vec{u})=\left(u_y^2\varepsilon_{r_{zz}}^{-1}(\omega)+u_z^2\varepsilon_{r_{yy}}^{-1}(\omega)\right)^{-1/2} \\ n'^+(\omega,\vec{u})=\dfrac{n^+(\omega,\vec{u})^3}{2}\left(\dfrac{\varepsilon'_{r_{zz}}(\omega)}{\varepsilon_{r_{zz}}^2(\omega)}u_y^2+\dfrac{\varepsilon'_{r_{yy}}(\omega)}{\varepsilon_{r_{yy}}^2(\omega)}u_z^2\right) \end{cases} \quad (6.80)$$

and

$$\begin{cases} n^-(\omega,\vec{u})=\sqrt{\varepsilon_{r_{xx}}(\omega)} \\ n'^-(\omega,\vec{u})=\dfrac{\varepsilon'_{r_{xx}}(\omega)}{2n^-(\omega,\vec{u})}=\varepsilon'_{r_{xx}}(\omega)\Big/2\sqrt{\varepsilon_{r_{xx}}(\omega)} \end{cases} \quad (6.81)$$

In the plane $(\overrightarrow{Ox},\overrightarrow{Oz})$, the coordinates of the unit wave vector $\vec{u}$ are $(\overrightarrow{u_x},0,\overrightarrow{u_z})$. Following the same approach than that for the planes $(\overrightarrow{Ox},\overrightarrow{Oy})$ and $(\overrightarrow{Oy},\overrightarrow{Oz})$, the determinant of Eq. (6.18) and Eq. (6.19) reduces to:

$$\left[\left(\hat{n}^2(\omega,\vec{u})u_z^2-\hat{\varepsilon}_{r_{xx}}(\omega)\right)\left(\hat{n}^2(\omega,\vec{u})u_{xx}^2-\hat{\varepsilon}_{r_{zz}}(\omega)\right)-\left(\hat{n}^2(\omega,\vec{u})u_xu_z\right)^2\right]$$
$$\left(\hat{n}^2(\omega,\vec{u})-\hat{\varepsilon}_{r_{yy}}(\omega)\right)=0 \quad (6.82)$$

Since this plane containing the optical axes, as explained in Chapter 5, the rest of the calculation depends on the location the propagation direction $\vec{u}(\theta,\varphi)$ with respect to the optical axis angle V_z, *i.e.* $0\le\theta\le V_z$ or $V_z\le\theta\le\pi/2$ where V_z is defined by Eq. (5.68). Then, assuming again $n'(\omega,\vec{u})<<n(\omega,\vec{u})$, it comes:

- for $0\le\theta\le V_z$

$$\begin{cases} n^+(\omega,\vec{u})=\sqrt{\varepsilon_{r_{yy}}(\omega)} \\ n'^+(\omega,\vec{u})=\dfrac{\varepsilon'_{r_{yy}}(\omega)}{2n^+(\omega,\vec{u})}=\varepsilon'_{r_{yy}}(\omega)\Big/2\sqrt{\varepsilon_{r_{yy}}(\omega)} \end{cases} \quad (6.83)$$

and

$$\begin{cases} n^{-}(\omega,\vec{u}) = \left(u_x^2 \varepsilon_{r_{zz}}^{-1}(\omega) + u_z^2 \varepsilon_{r_{xx}}^{-1}(\omega)\right)^{-1/2} \\ n'^{-}(\omega,\vec{u}) = \dfrac{n^{-}(\omega,\vec{u})^3}{2}\left(\dfrac{\varepsilon'_{r_{zz}}(\omega)}{\varepsilon^2_{r_{zz}}(\omega)} u_x^2 + \dfrac{\varepsilon'_{r_{xx}}(\omega)}{\varepsilon^2_{r_{xx}}(\omega)} u_z^2\right) \end{cases} \tag{6.84}$$

- for $V_z \le \theta \le \pi/2$

$$\begin{cases} n^{+}(\omega,\vec{u}) = \left(u_x^2 \varepsilon_{r_{zz}}^{-1}(\omega) + u_z^2 \varepsilon_{r_{xx}}^{-1}(\omega)\right)^{-1/2} \\ n'^{+}(\omega,\vec{u}) = \dfrac{n^{+}(\omega,\vec{u})^3}{2}\left(\dfrac{\varepsilon'_{r_{zz}}(\omega)}{\varepsilon^2_{r_{zz}}(\omega)} u_x^2 + \dfrac{\varepsilon'_{r_{xx}}(\omega)}{\varepsilon^2_{r_{xx}}(\omega)} u_z^2\right) \end{cases} \tag{6.85}$$

and

$$\begin{cases} n^{-}(\omega,\vec{u}) = \sqrt{\varepsilon_{r_{yy}}(\omega)} \\ n'^{-}(\omega,\vec{u}) = \dfrac{\varepsilon'_{r_{yy}}(\omega)}{2n^{-}(\omega,\vec{u})} = \varepsilon'_{r_{yy}}(\omega)\Big/2\sqrt{\varepsilon_{r_{yy}}(\omega)} \end{cases} \tag{6.86}$$

Out of the principal planes, it is more convenient to directly use Eq. (6.32) combined with Eq. (6.31), Eq. (6.28), Eq. (6.25a), Eq. (6.25b), and Eq. (6.25c) in order to access to n' as a function of $\vec{u}$ using the $\hat{\varepsilon}$ tensor given by Eq. (6.34) and Eq. (6.37).

Hereafter in Figure 6.4 is plotted the imaginary index surface based on the calculations given earlier. The calculations are performed with the following fictive numerical values: $\varepsilon_{r_{xx}}(\omega) = 2.71$, $\varepsilon_{r_{yy}}(\omega) = 2.89$, $\varepsilon_{r_{zz}}(\omega) = 2.93$, $\varepsilon'_{r_{xx}}(\omega) = 3.29 \times 10^{-5}$, $\varepsilon'_{r_{yy}}(\omega) = 6.80 \times 10^{-5}$, $\varepsilon'_{r_{zz}}(\omega) = 13.75 \times 10^{-5}$.

As already mentioned in Part 6.2 and as shown in Figure 6.4, there is no *a priori* systematic relation of order between n'^{+} and n'^{-} in biaxial crystals such as orthorhombic ones. As a consequence, one systematically observes angular domains of orientations in the $(\overrightarrow{Ox},\overrightarrow{Oz})$ where $n^{+}(\omega,\vec{u}) = \mathrm{Re}\left(\hat{n}^{+}(\omega,\vec{u})\right) \ge n^{-}(\omega,\vec{u}) = \mathrm{Re}\left(\hat{n}^{-}(\omega,\vec{u})\right)$ according to Eq. (6.33) and $n'^{+}(\omega,\vec{u}) \ge n'^{-}(\omega,\vec{u})$ while angular domains show $n^{+}(\omega,\vec{u}) = \mathrm{Re}\left(\hat{n}^{+}(\omega,\vec{u})\right) \ge n^{-}(\omega,\vec{u}) = \mathrm{Re}\left(\hat{n}^{-}(\omega,\vec{u})\right)$ and $n'^{+}(\omega,\vec{u}) \le n'^{-}(\omega,\vec{u})$.

It is also interesting to compare Figure 6.4 with the index surface shown in Figure 5.5. In both cases, the three principal planes $(\overrightarrow{Ox},\overrightarrow{Oy})$, $(\overrightarrow{Oy},\overrightarrow{Oz})$, and $(\overrightarrow{Ox},\overrightarrow{Oz})$ are symmetry mirrors. Thus the imaginary index surface of an orthorhombic crystal also belongs to the orthorhombic symmetry group, *i.e.* mmm, with respect to the same axes as for the refractive index. Regarding the shape, it is identical in $(\overrightarrow{Ox},\overrightarrow{Oy})$. In $(\overrightarrow{Oy},\overrightarrow{Oz})$ and $(\overrightarrow{Ox},\overrightarrow{Oz})$, there is a circle for one section in both cases, but the other section is bi-lobar instead of being elliptical as for the index surface. The difference with the index surface is also strong out of the principal planes since the two layers intersect over a continuum surrounding each optical axis. Such a continuum directly derives from the orientation boundaries of angular domains of inversion from $n'^{+}(\omega,\vec{u}) \geq n'^{-}(\omega,\vec{u})$ to $n'^{+}(\omega,\vec{u}) \leq n'^{-}(\omega,\vec{u})$.

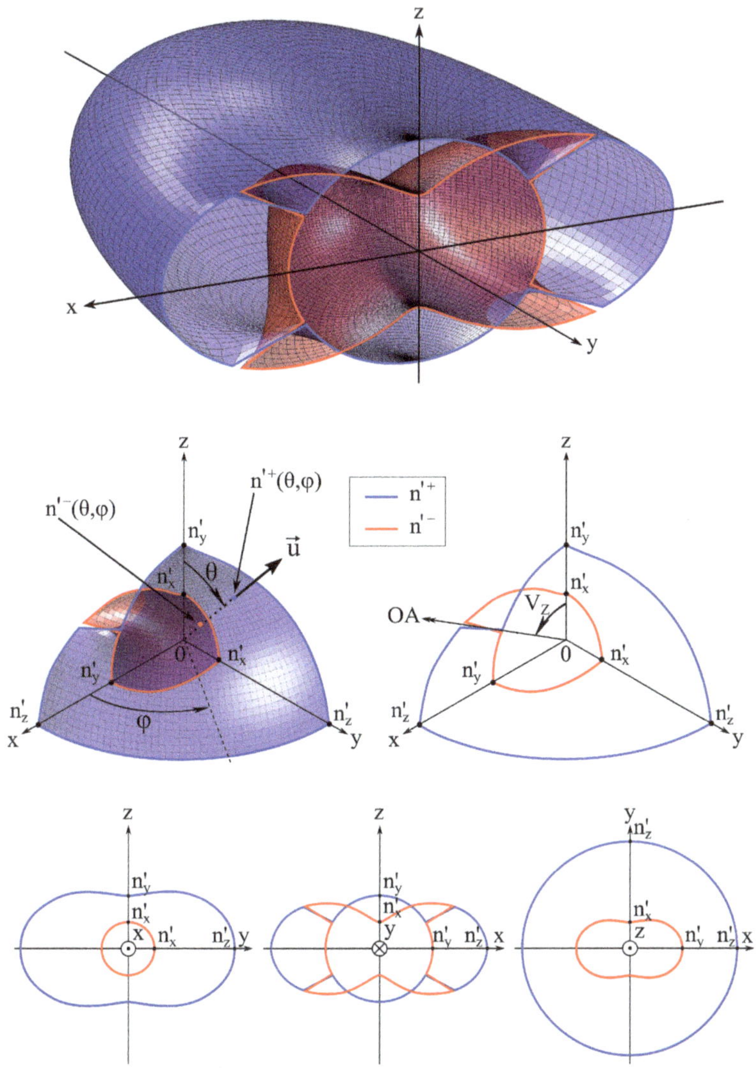

FIGURE 6.4 Imaginary index surface of an orthorhombic crystal belonging to the positive biaxial optical class where (x,y,z) refers to the dielectric frame $(\overrightarrow{Ox},\overrightarrow{Oy},\overrightarrow{Oz})$. Top: half-space angular distribution showing rotation invariance with respect to the z-axis. Middle: octant space in which (θ,φ) are the angles of spherical coordinates of the unit vector $\vec{u}$, and $n'^{+}(\theta,\varphi)$ and $n'^{-}(\theta,\varphi)$ are the two corresponding imaginary indices; n'_x, n'_y, and n'_z refer to the ordinary and extraordinary principal imaginary indices, and OA denotes the optical axes. Bottom: section of the imaginary index surface in the $(\overrightarrow{Oy},\overrightarrow{Oz})$ plane (left), $(\overrightarrow{Ox},\overrightarrow{Oz})$ plane (middle), and $(\overrightarrow{Ox},\overrightarrow{Oy})$ plane (right).

6.4.3.2 Monoclinic Crystal Classes

In this case, tensor ε_r' is not diagonal, with three different forms regarding the extra-diagonal coefficients according to the special axis as explained in Part 6.3: $\varepsilon_{r_{yz}}' = \varepsilon_{r_{zy}}' \neq 0$ when the special axis is along $\overrightarrow{Ox}$; $\varepsilon_{r_{xz}}' = \varepsilon_{r_{zx}}' \neq 0$ when it is $\overrightarrow{Oy}$; $\varepsilon_{r_{xy}}' = \varepsilon_{r_{yx}}' \neq 0$ when it is $\overrightarrow{Oz}$.

Here, we choose to directly use Eq. (6.32) combined with Eq. (6.31), Eq. (6.28), Eq. (6.25a), Eq. (6.25b), and Eq. (6.25c) to access n' as a function of $\vec{u}$, and that for any direction of propagation in the dielectric frame. The differentiation between the three possible orientations of the special axis is made at the levels of Eq. (6.25a), Eq. (6.25b), and Eq. (6.25c) according to Eq. (6.38), Eq. (6.39), and Eq. (6.40). Equation (6.25c) remains the same, while Eq. (6.25a) and Eq. (6.25b) become:

- with the special axis along $\overrightarrow{Ox}$

$$\begin{cases} A(\omega,\vec{u}) = -\left(u_x^2\varepsilon_{r_{xx}}(\omega) + u_y^2\varepsilon_{r_{yy}}(\omega) + u_z^2\varepsilon_{r_{zz}}(\omega)\right) \\ A'(\omega,\vec{u}) = -\left(u_x^2\varepsilon_{r_{xx}}'(\omega) + u_y^2\varepsilon_{r_{yy}}'(\omega) + u_z^2\varepsilon_{r_{zz}}'(\omega) + 2u_yu_z\hat{\varepsilon}_{r_{yz}}'(\omega)\right) \end{cases} \tag{6.87a}$$

$$\begin{cases} B(\omega,\vec{u}) = u_x^2\left(\varepsilon_{r_{xx}}(\omega)\varepsilon_{r_{zz}}(\omega) + \varepsilon_{r_{xx}}(\omega)\varepsilon_{r_{yy}}(\omega)\right) + u_y^2\left(\varepsilon_{r_{yy}}(\omega)\varepsilon_{r_{xx}}(\omega) + \varepsilon_{r_{yy}}(\omega)\varepsilon_{r_{zz}}(\omega)\right) \\ \quad + u_z^2\left(\varepsilon_{r_{zz}}(\omega)\varepsilon_{r_{yy}}(\omega) + \varepsilon_{r_{zz}}(\omega)\varepsilon_{r_{xx}}(\omega)\right) \\ B'(\omega,\vec{u}) = u_x^2\left(\varepsilon_{r_{xx}}'(\omega)\varepsilon_{r_{zz}}(\omega) + \varepsilon_{r_{xx}}(\omega)\varepsilon_{r_{zz}}'(\omega) + \varepsilon_{r_{xx}}'(\omega)\varepsilon_{r_{yy}}(\omega) + \varepsilon_{r_{xx}}(\omega)\varepsilon_{r_{yy}}'(\omega)\right) \\ \quad + u_y^2\left(\varepsilon_{r_{yy}}'(\omega)\varepsilon_{r_{zz}}(\omega) + \varepsilon_{r_{yy}}(\omega)\varepsilon_{r_{zz}}'(\omega) + \varepsilon_{r_{xx}}'(\omega)\varepsilon_{r_{yy}}(\omega) + \varepsilon_{r_{xx}}(\omega)\varepsilon_{r_{yy}}'(\omega)\right) \\ \quad + u_z^2\left(\varepsilon_{r_{yy}}'(\omega)\varepsilon_{r_{zz}}(\omega) + \varepsilon_{r_{yy}}(\omega)\varepsilon_{r_{zz}}'(\omega) + \varepsilon_{r_{xx}}'(\omega)\varepsilon_{r_{zz}}(\omega) + \varepsilon_{r_{xx}}(\omega)\varepsilon_{r_{zz}}'(\omega)\right) \\ \quad + 2\left(u_yu_z\varepsilon_{r_{xx}}(\omega)\varepsilon_{r_{yz}}'(\omega)\right) \end{cases} \tag{6.87b}$$

- with the special axis along $\overrightarrow{Oy}$

$$\begin{cases} A(\omega,\vec{u}) = -\left(u_x^2\varepsilon_{r_{xx}}(\omega) + u_y^2\varepsilon_{r_{yy}}(\omega) + u_z^2\varepsilon_{r_{zz}}(\omega)\right) \\ A'(\omega,\vec{u}) = -\left(u_x^2\varepsilon_{r_{xx}}'(\omega) + u_y^2\varepsilon_{r_{yy}}'(\omega) + u_z^2\varepsilon_{r_{zz}}'(\omega) + 2u_xu_z\varepsilon_{r_{xz}}'(\omega)\right) \end{cases} \tag{6.88a}$$

$$\begin{cases} B(\omega,\vec{u}) = u_x^2\left(\varepsilon_{r_{xx}}(\omega)\varepsilon_{r_{zz}}(\omega)+\varepsilon_{r_{xx}}(\omega)\varepsilon_{r_{yy}}(\omega)\right)+u_y^2\left(\varepsilon_{r_{yy}}(\omega)\varepsilon_{r_{xx}}(\omega)+\varepsilon_{r_{yy}}(\omega)\varepsilon_{r_{zz}}(\omega)\right) \\ \qquad +u_z^2\left(\varepsilon_{r_{zz}}(\omega)\varepsilon_{r_{yy}}(\omega)+\varepsilon_{r_{zz}}(\omega)\varepsilon_{r_{xx}}(\omega)\right) \\ B'(\omega,\vec{u}) = u_x^2\left(\varepsilon'_{r_{xx}}(\omega)\varepsilon_{r_{zz}}(\omega)+\varepsilon_{r_{xx}}(\omega)\varepsilon'_{r_{zz}}(\omega)+\varepsilon'_{r_{xx}}(\omega)\varepsilon_{r_{yy}}(\omega)+\varepsilon_{r_{xx}}(\omega)\varepsilon'_{r_{yy}}(\omega)\right) \\ \qquad +u_y^2\left(\varepsilon'_{r_{yy}}(\omega)\varepsilon_{r_{zz}}(\omega)+\varepsilon_{r_{yy}}(\omega)\varepsilon'_{r_{zz}}(\omega)+\varepsilon'_{r_{xx}}(\omega)\varepsilon_{r_{yy}}(\omega)+\varepsilon_{r_{xx}}(\omega)\varepsilon'_{r_{yy}}(\omega)\right) \\ \qquad +u_z^2\left(\varepsilon'_{r_{yy}}(\omega)\varepsilon_{r_{zz}}(\omega)+\varepsilon_{r_{yy}}(\omega)\varepsilon'_{r_{zz}}(\omega)+\varepsilon'_{r_{xx}}(\omega)\varepsilon_{r_{zz}}(\omega)+\varepsilon_{r_{xx}}(\omega)\varepsilon'_{r_{zz}}(\omega)\right) \\ \qquad +2\left(u_x u_z \varepsilon_{r_{yy}}(\omega)\varepsilon'_{r_{xz}}(\omega)\right) \end{cases} \tag{6.88b}$$

- with the special axis along $\overrightarrow{Oz}$

$$\begin{cases} A(\omega,\vec{u}) = -\left(u_x^2\varepsilon_{r_{xx}}(\omega)+u_y^2\varepsilon_{r_{yy}}(\omega)+u_z^2\varepsilon_{r_{zz}}(\omega)\right) \\ A'(\omega,\vec{u}) = -\left(u_x^2\varepsilon'_{r_{xx}}(\omega)+u_y^2\varepsilon'_{r_{yy}}(\omega)+u_z^2\varepsilon'_{r_{zz}}(\omega)+2u_x u_y\hat{\varepsilon}'_{r_{xy}}(\omega)\right) \end{cases} \tag{6.89a}$$

$$\begin{cases} B(\omega,\vec{u}) = u_x^2\left(\varepsilon_{r_{xx}}(\omega)\varepsilon_{r_{zz}}(\omega)+\varepsilon_{r_{xx}}(\omega)\varepsilon_{r_{yy}}(\omega)\right)+u_y^2\left(\varepsilon_{r_{yy}}(\omega)\varepsilon_{r_{xx}}(\omega)+\varepsilon_{r_{yy}}(\omega)\varepsilon_{r_{zz}}(\omega)\right) \\ \qquad +u_z^2\left(\varepsilon_{r_{zz}}(\omega)\varepsilon_{r_{yy}}(\omega)+\varepsilon_{r_{zz}}(\omega)\varepsilon_{r_{xx}}(\omega)\right) \\ B'(\omega,\vec{u}) = u_x^2\left(\varepsilon'_{r_{xx}}(\omega)\varepsilon_{r_{zz}}(\omega)+\varepsilon_{r_{xx}}(\omega)\varepsilon'_{r_{zz}}(\omega)+\varepsilon'_{r_{xx}}(\omega)\varepsilon_{r_{yy}}(\omega)+\varepsilon_{r_{xx}}(\omega)\varepsilon'_{r_{yy}}(\omega)\right) \\ \qquad +u_y^2\left(\varepsilon'_{r_{yy}}(\omega)\varepsilon_{r_{zz}}(\omega)+\varepsilon_{r_{yy}}(\omega)\varepsilon'_{r_{zz}}(\omega)+\varepsilon'_{r_{xx}}(\omega)\varepsilon_{r_{yy}}(\omega)+\varepsilon_{r_{xx}}(\omega)\varepsilon'_{r_{yy}}(\omega)\right) \\ \qquad +u_z^2\left(\varepsilon'_{r_{yy}}(\omega)\varepsilon_{r_{zz}}(\omega)+\varepsilon_{r_{yy}}(\omega)\varepsilon'_{r_{zz}}(\omega)+\varepsilon'_{r_{xx}}(\omega)\varepsilon_{r_{zz}}(\omega)+\varepsilon_{r_{xx}}(\omega)\varepsilon'_{r_{zz}}(\omega)\right) \\ \qquad +2\left(u_x u_y \varepsilon_{r_{zz}}(\omega)\varepsilon'_{r_{xy}}(\omega)\right) \end{cases} \tag{6.89b}$$

Note that the expressions of the real parts $A(\omega,\vec{u})$ and $B(\omega,\vec{u})$ remain the same in the three cases of orientation of the special axis.

6.4.3.2.1 Special Axis along $\overrightarrow{Oy}$

The case in which the special axis is along $\overrightarrow{Oy}$ is detailed hereafter from Eq. (6.88a) and Eq. (6.88b).

Figure 6.5 shows that the general shape of the imaginary index surface of a monoclinic crystal with the special axis along $\overrightarrow{Oy}$ is mostly the same as that of an orthorhombic crystal. The only difference is a distortion and rotation of the pattern in $(\overrightarrow{Ox},\overrightarrow{Oz})$ leading to the loss of $(\overrightarrow{Oy},\overrightarrow{Oz})$ and $(\overrightarrow{Ox},\overrightarrow{Oy})$ as symmetry mirrors. The remaining mirror is then $(\overrightarrow{Ox},\overrightarrow{Oz})$, that is to say the plane being perpendicular to the special axis. Then the resulting orientation symmetry group of the imaginary index surface is m (Petit 2013, Joly 2009) [2,3]. The calculations for Figure 6.5 are performed with the following fictive numerical values: $\varepsilon_{r_{xx}}(\omega) = 2.71$, $\varepsilon_{r_{yy}}(\omega) = 2.89$, $\varepsilon_{r_{zz}}(\omega) = 2.93$ and $\varepsilon'_{r_{xx}}(\omega) = 3.29 \times 10^{-5}$, $\varepsilon'_{r_{yy}}(\omega) = 6.80 \times 10^{-5}$, $\varepsilon'_{r_{zz}}(\omega) = 13.75 \times 10^{-5}$, and $\varepsilon'_{r_{xz}}(\omega) = -2.69 \times 10^{-5}$.

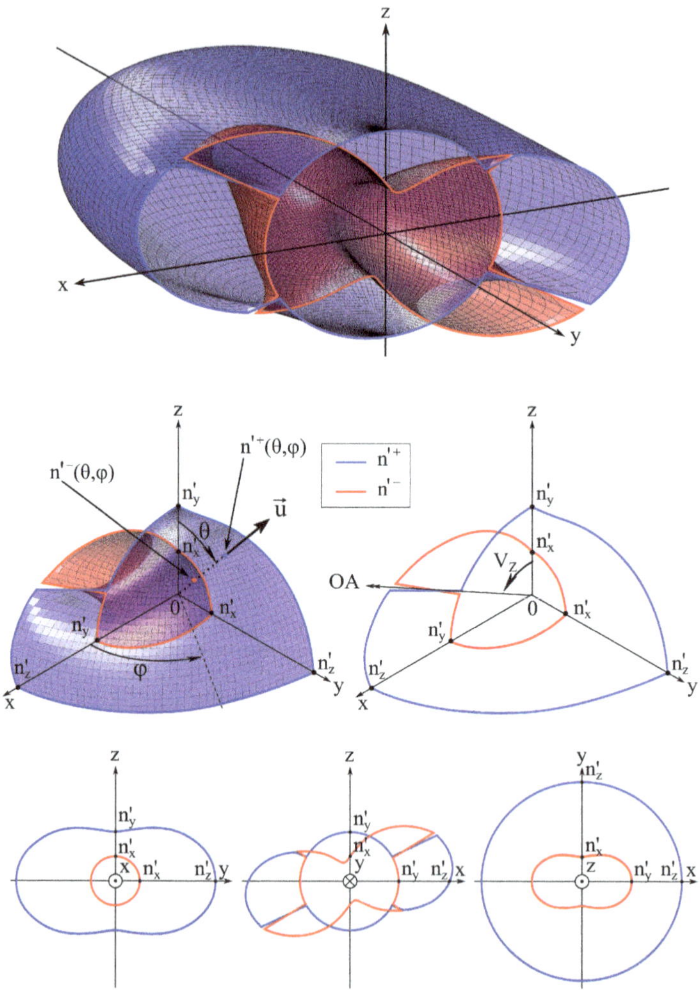

FIGURE 6.5 Imaginary index surface of a monoclinic crystal with the special axis along $\overrightarrow{Oy}$ and belonging to the positive biaxial optical class. (x,y,z) refers to the dielectric frame $(\overrightarrow{Ox},\overrightarrow{Oy},\overrightarrow{Oz})$. Top: half-space angular distribution showing rotation invariance with respect to the z-axis. Middle: octant space in which (θ,φ) are the angles of spherical coordinates of the unit vector $\vec{u}$, and $n'^{+}(\theta,\varphi)$ and $n'^{-}(\theta,\varphi)$ are the two corresponding imaginary indices; n'_x, n'_y, and n'_z refer to the ordinary and extraordinary principal imaginary indices, and OA denotes the optical axes. Bottom: section of the imaginary index surface in the $(\overrightarrow{Oy},\overrightarrow{Oz})$ plane (left), $(\overrightarrow{Ox},\overrightarrow{Oz})$ plane (middle), and $(\overrightarrow{Ox},\overrightarrow{Oy})$ plane (right).

In such a case of monoclinic crystal, the expressions of complex solutions remain the same as those of an orthorhombic crystal in the dielectric planes $(\overrightarrow{Ox},\overrightarrow{Oy})$ and $(\overrightarrow{Oy},\overrightarrow{Oz})$, as given by Eq. (6.77) and Eq. (6.78) on the one hand, and by Eq. (6.80) and Eq. (6.81) on the other hand. However, in the plane perpendicular to the special axis $\overrightarrow{Oy}$, namely plane $(\overrightarrow{Ox},\overrightarrow{Oz})$, the calculation of the determinant of Eq. (6.18) and Eq. (6.19) leads to the following expressions:

- for $0 \leq \theta \leq V_z$

$$\begin{cases} n^{+}(\omega,\vec{u}) = \sqrt{\varepsilon_{r_{yy}}(\omega)} \\ n'^{+}(\omega,\vec{u}) = \dfrac{\varepsilon'_{r_{yy}}(\omega)}{2n^{+}(\omega,\vec{u})} = \varepsilon'_{r_{yy}}(\omega)\Big/2\sqrt{\varepsilon_{r_{yy}}(\omega)} \end{cases} \tag{6.90a}$$

and

$$\begin{cases} n^{-}(\omega,\vec{u}) = \left[u_x^2\varepsilon_{r_{zz}}^{-1}(\omega) + u_z^2\varepsilon_{r_{xx}}^{-1}(\omega)\right]^{-1/2} \\ n'^{-}(\omega,\vec{u}) = \dfrac{n^{-}(\omega,\vec{u})^3}{2}\left[\dfrac{\varepsilon'_{r_{zz}}(\omega)}{\varepsilon^2_{r_{zz}}(\omega)}u_x^2 + \dfrac{\varepsilon'_{r_{xx}}(\omega)}{\varepsilon^2_{r_{xx}}(\omega)}u_z^2 - \dfrac{2\varepsilon'_{r_{xz}}(\omega)}{\varepsilon_{r_{xx}}(\omega)\varepsilon_{r_{zz}}(\omega)}u_x u_z\right] \end{cases} \tag{6.90b}$$

- for $V_z \leq \theta \leq \pi/2$

$$\begin{cases} n^{+}(\omega,\vec{u}) = \left[u_x^2\varepsilon_{r_{zz}}^{-1}(\omega) + u_z^2\varepsilon_{r_{xx}}^{-1}(\omega)\right]^{-1/2} \\ n'^{+}(\omega,\vec{u}) = \dfrac{n^{+}(\omega,\vec{u})^3}{2}\left[\dfrac{\varepsilon'_{r_{zz}}(\omega)}{\varepsilon^2_{r_{zz}}(\omega)}u_x^2 + \dfrac{\varepsilon'_{r_{xx}}(\omega)}{\varepsilon^2_{r_{xx}}(\omega)}u_z^2 - \dfrac{2\varepsilon'_{r_{xz}}(\omega)}{\varepsilon_{r_{xx}}(\omega)\varepsilon_{r_{zz}}(\omega)}u_x u_z\right] \end{cases} \tag{6.91a}$$

and

$$\begin{cases} n^{-}(\omega,\vec{u}) = \sqrt{\varepsilon_{r_{yy}}(\omega)} \\ n'^{-}(\omega,\vec{u}) = \dfrac{\varepsilon'_{r_{yy}}(\omega)}{2n^{-}(\omega,\vec{u})} = \varepsilon'_{r_{yy}}(\omega)\Big/2\sqrt{\varepsilon_{r_{yy}}(\omega)} \end{cases} \tag{6.91b}$$

By considering the $(\overrightarrow{Ox},\overrightarrow{Oz})$ plane of Figure 6.5 (bottom center), the angular distributions are characterized by Eq. (6.90a) to Eq. (6.91b), involving the four independent non-zero imaginary elements of the permittivity tensor. Note that when expressed in the dielectric frame, only the three diagonal imaginary elements of the permittivity tensor are involved in the two other dielectric planes $(\overrightarrow{Ox},\overrightarrow{Oy})$ and $(\overrightarrow{Oy},\overrightarrow{Oz})$. Thus only the $(\overrightarrow{Ox},\overrightarrow{Oz})$ plane bears the full information of the imaginary part of the permittivity tensor.

It is interesting at this step to determine the orientation of the imaginary dielectric frame $(\overrightarrow{Ox'},\overrightarrow{Oy'},\overrightarrow{Oz'})$ that is the frame in which tensor ε_r' is diagonal, *i.e.*:

$$\varepsilon_r'(\omega)=\begin{pmatrix}\varepsilon'_{r_{x'x'}}(\omega) & 0 & 0\\ 0 & \varepsilon'_{r_{y'y'}}(\omega) & 0\\ 0 & 0 & \varepsilon'_{r_{z'z'}}(\omega)\end{pmatrix} \tag{6.92}$$

It is possible to calculate the magnitude of $\varepsilon'_{r_{x'x'}}(\omega)$, $\varepsilon'_{r_{y'y'}}(\omega)$, and $\varepsilon'_{r_{z'z'}}(\omega)$, as well as of θ', as a function of the coefficients of tensor ε_r' in $(\overrightarrow{Ox},\overrightarrow{Oy},\overrightarrow{Oz})$ given by Eq. (6.39a), *i.e.*: $\varepsilon'_{r_{xx}}(\omega)$, $\varepsilon'_{r_{yy}}(\omega)$, $\varepsilon'_{r_{zz}}(\omega)$, and $\varepsilon'_{r_{xz}}(\omega)=\varepsilon'_{r_{zx}}(\omega)$. For that purpose, it is useful to use the transformation of a rank-2 polar tensor by a symmetry operation represented by its matrix (*a*), as performed in Chapter 2 for implementing the Neumann principle from Eq. (2.57) for example. In the present case, the transformation equation writes:

$$\varepsilon'_{r_{i'j'}}(\omega)=\sum_{mn} a_{im}a_{jn}\varepsilon'_{r_{mn}}(\omega) \tag{6.93}$$

with $(m,n)=(x,y,z)$, $(i',j')=(x',y',z')$, and matrix (*a*) corresponding to a rotation of an angle θ' around $\overrightarrow{Oy}$, *i.e.*:

$$(a)=\begin{pmatrix}\cos\theta' & 0 & -\sin\theta'\\ 0 & 1 & 0\\ \sin\theta' & 0 & \cos\theta'\end{pmatrix} \tag{6.94}$$

So, it comes:

$$\begin{cases}\varepsilon'_{r_{x'x'}}(\omega)=(\cos\theta')^2\varepsilon'_{r_{xx}}(\omega)+(\sin\theta')^2\varepsilon'_{r_{zz}}(\omega)-2\sin\theta'\cos\theta'\varepsilon'_{r_{xz}}(\omega)\\ \varepsilon'_{r_{y'y'}}(\omega)=1\\ \varepsilon'_{r_{z'z'}}(\omega)=(\sin\theta')^2\varepsilon'_{r_{xx}}(\omega)+(\cos\theta')^2\varepsilon'_{r_{zz}}(\omega)+2\sin\theta'\cos\theta'\varepsilon'_{r_{xz}}(\omega)\end{cases} \tag{6.95}$$

The angle θ' can be found by expressing the fact that the non-diagonal coefficients of ε_r' are nil in $(\overrightarrow{Ox'},\overrightarrow{Oy'},\overrightarrow{Oz'})$, *i.e.* $\varepsilon'_{r_{x'z'}}(\omega)=\varepsilon'_{r_{z'x'}}(\omega)=0$. Then according to Eq. (6.93) and Eq. (6.94), it comes:

$$\theta'=\frac{1}{2}\arctan\left(2\frac{\varepsilon'_{r_{xz}}(\omega)}{\varepsilon'_{r_{zz}}(\omega)-\varepsilon'_{r_{xx}}(\omega)}\right) \tag{6.96}$$

with $\varepsilon'_{r_{zz}}(\omega) > \varepsilon'_{r_{xx}}(\omega)$.

The angle θ' depends on the light circular frequency ω due to the dispersion of the coefficients of the relative dielectric permittivity tensor. Note that the ε_r tensor would no longer be diagonal in the $(\overrightarrow{Ox'},\overrightarrow{Oy'},\overrightarrow{Oz'})$ frame. It is also important to notice that $\overrightarrow{Ox'}$ and $\overrightarrow{Oz'}$ are not exactly along the long and short axes of the pattern, as shown in Figure 6.6 hereafter. Indeed, this orientation leads to the fact that the planes $(\overrightarrow{Ox'},\overrightarrow{Oy'})$ and $(\overrightarrow{Oy'},\overrightarrow{Oz'})$ are not symmetry mirrors of the index surface. The only symmetry element is a mirror m perpendicular to the special axis $\overrightarrow{Oy}//\overrightarrow{Oy'}$, which then corresponds to the plane $(\overrightarrow{Ox'},\overrightarrow{Oz'})$ that is collinear to $(\overrightarrow{Ox},\overrightarrow{Oz})$. The calculations for plotting Figure 6.6 are performed with the following fictive numerical values: $\varepsilon_{r_{xx}}(\omega) = 1.32$, $\varepsilon_{r_{yy}}(\omega) = 4.00$, $\varepsilon_{r_{zz}}(\omega) = 7.95$, $\varepsilon'_{r_{xx}}(\omega) = 2.3\times10^{-5}$, $\varepsilon'_{r_{yy}}(\omega) = 8.00\times10^{-5}$, $\varepsilon'_{r_{zz}}(\omega) = 22.56\times10^{-5}$, $\varepsilon'_{r_{xz}}(\omega) = -3.18\times10^{-5}$.

6.4.3.2.2 Special Axis along $\overrightarrow{Ox}$ or $\overrightarrow{Oz}$

In Figure 6.7 are given the sections of the imaginary index surface in the principal planes $(\overrightarrow{Oy},\overrightarrow{Oz})$, $(\overrightarrow{Ox},\overrightarrow{Oz})$, and $(\overrightarrow{Ox},\overrightarrow{Oy})$ in the case in which the special axis is along $\overrightarrow{Ox}$, on the one hand, and $\overrightarrow{Oz}$, on the other hand. The calculations are performed with the following fictive numerical values: $\varepsilon_{r_{xx}}(\omega) = 2.71$, $\varepsilon_{r_{yy}}(\omega) = 2.89$,

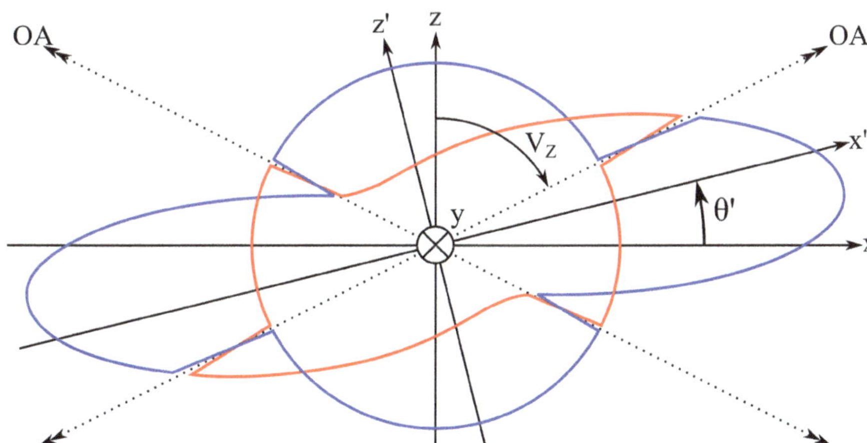

FIGURE 6.6 Section in the principal plane $(\overrightarrow{Ox},\overrightarrow{Oz})$ of the imaginary index surface of a monoclinic crystal with the special axis along $\overrightarrow{Oy}$ and belonging to the positive biaxial optical class. (x,y,z) refers to the dielectric frame $(\overrightarrow{Ox},\overrightarrow{Oy},\overrightarrow{Oz})$. (x',y',z') refers to the imaginary dielectric frame $(\overrightarrow{Ox'},\overrightarrow{Oy'},\overrightarrow{Oz'})$ that makes an angle θ' with the dielectric frame.

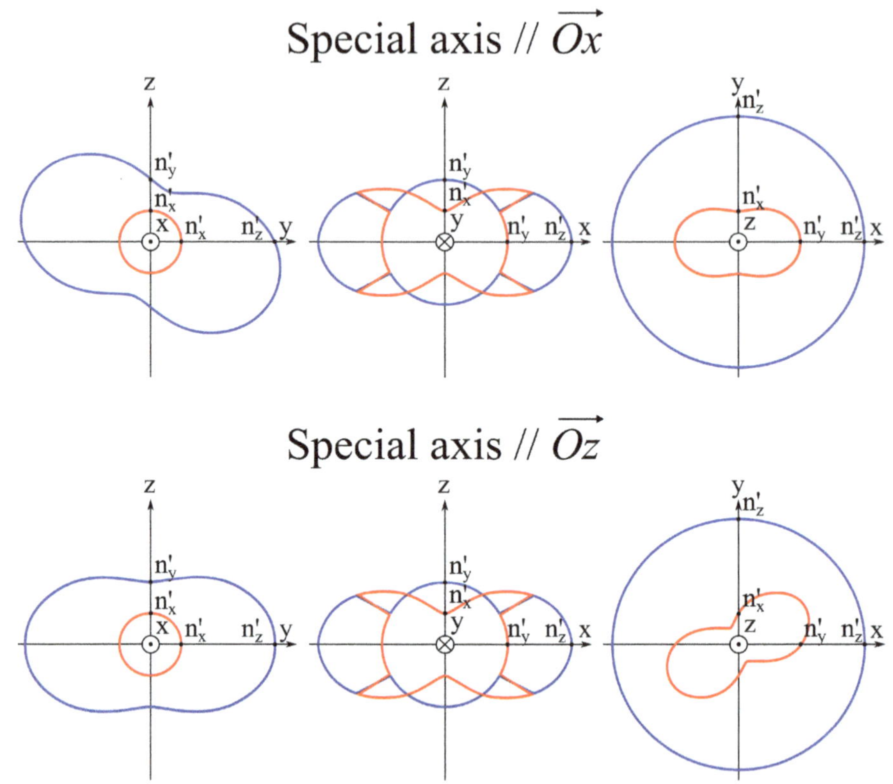

FIGURE 6.7 Imaginary index surface of a monoclinic crystal belonging to the positive biaxial optical class with the special axis along $\overrightarrow{Ox}$ (top) and $\overrightarrow{Oz}$ (bottom). (x,y,z) refers to the dielectric frame $(\overrightarrow{Ox},\overrightarrow{Oy},\overrightarrow{Oz})$: sections in the $(\overrightarrow{Oy},\overrightarrow{Oz})$ plane (left), $(\overrightarrow{Ox},\overrightarrow{Oz})$ plane (middle), and $(\overrightarrow{Ox},\overrightarrow{Oy})$ plane (right).

$\varepsilon_{r_{zz}}(\omega)=2.93$, $\varepsilon'_{r_{xx}}(\omega)=3.29\times10^{-5}$, $\varepsilon'_{r_{yy}}(\omega)=6.80\times10^{-5}$, $\varepsilon'_{r_{zz}}(\omega)=13.75\times10^{-5}$, and $\varepsilon'_{r_{yz}}(\omega)=-2.74\times10^{-5}$ for the special axis along $\overrightarrow{Ox}$ and $\varepsilon'_{r_{yz}}(\omega)=-2.68\times10^{-5}$ for the special axis along $\overrightarrow{Oz}$.

In the case in which the special axis is along $\overrightarrow{Ox}$, the expressions of the complex index keep being those detailed for the orthorhombic media in $(\overrightarrow{Ox},\overrightarrow{Oy})$, *i.e.* Eq. 6.77 and Eq. 6.78, and in $(\overrightarrow{Ox},\overrightarrow{Oz})$, *i.e.* Eq. 6.83 to Eq. 6.86. However, in $(\overrightarrow{Oy},\overrightarrow{Oz})$, Eq. (6.80) and Eq. (6.81) become:

$$\begin{cases} n^{+}(\omega,\vec{u}) = \left[u_y^2 \varepsilon_{r_{zz}}^{-1}(\omega) + u_z^2 \varepsilon_{r_{yy}}^{-1}(\omega)\right]^{-1/2} \\ n'^{+}(\omega,\vec{u}) = \dfrac{n^{+}(\omega,\vec{u})^3}{2}\left[\dfrac{\varepsilon'_{r_{zz}}(\omega)}{\varepsilon^2_{r_{zz}}(\omega)} u_y^2 + \dfrac{\varepsilon'_{r_{yy}}(\omega)}{\varepsilon^2_{r_{yy}}(\omega)} u_z^2 - \dfrac{2\varepsilon'_{r_{yz}}(\omega)}{\varepsilon_{r_{yy}}(\omega)\varepsilon_{r_{zz}}(\omega)} u_y u_z\right] \end{cases} \tag{6.97}$$

and

$$\begin{cases} n^{-}(\omega,\vec{u}) = \sqrt{\varepsilon_{r_{xx}}(\omega)} \\ n'^{-}(\omega,\vec{u}) = \dfrac{\varepsilon'_{r_{xx}}(\omega)}{2n^{-}(\omega,\vec{u})} = \varepsilon'_{r_{xx}}(\omega)\Big/2\sqrt{\varepsilon_{r_{xx}}(\omega)} \end{cases} \tag{6.98}$$

In the case where the special axis is along $\overrightarrow{Oz}$, the expressions of the complex index are those detailed for the orthorhombic media in $(\overrightarrow{Oy},\overrightarrow{Oz})$, *i.e.* Eq. (6.80) and Eq. (6.81), and in $(\overrightarrow{Ox},\overrightarrow{Oz})$, Eq. (6.83) to Eq. (6.86). However, in the plane perpendicular to the special axis $\overrightarrow{Oz}$, namely $(\overrightarrow{Ox},\overrightarrow{Oy})$, Eq. (6.77) and Eq. (6.78) become:

$$\begin{cases} n^{+}(\omega,\vec{u}) = \sqrt{\varepsilon_{r_{zz}}(\omega)} \\ n'^{+}(\omega,\vec{u}) = \dfrac{\varepsilon'_{r_{zz}}(\omega)}{2n^{+}(\omega,\vec{u})} = \varepsilon'_{r_{zz}}(\omega)\Big/2\sqrt{\varepsilon_{r_{zz}}(\omega)} \end{cases} \tag{6.99}$$

and

$$\begin{cases} n^{-}(\omega,\vec{u}) = \left[u_x^2 \varepsilon_{r_{yy}}^{-1}(\omega) + u_y^2 \varepsilon_{r_{xx}}^{-1}(\omega)\right]^{-1/2} \\ n'^{-}(\omega,\vec{u}) = \dfrac{n^{-}(\omega,\vec{u})^3}{2}\left[\dfrac{\varepsilon'_{r_{yy}}(\omega)}{\varepsilon^2_{r_{yy}}(\omega)} u_x^2 + \dfrac{\varepsilon'_{r_{xx}}(\omega)}{\varepsilon^2_{r_{xx}}(\omega)} u_y^2 - \dfrac{2\varepsilon'_{r_{xy}}(\omega)}{\varepsilon_{r_{xx}}(\omega)\varepsilon_{r_{yy}}(\omega)} u_x u_y\right] \end{cases} \tag{6.100}$$

As for the case in which the special axis is along $\overrightarrow{Oy}$, the resulting orientation symmetry group of the imaginary index surface is m: this mirror lies in the plane $(\overrightarrow{Oy},\overrightarrow{Oz})$ when the special axis is along $\overrightarrow{Ox}$, and in $(\overrightarrow{Ox},\overrightarrow{Oy})$ when the special axis is along $\overrightarrow{Oz}$. The imaginary axes that diagonalize the imaginary part of the complex permittivity can be determined by directly deriving the same procedure given from Eq. (6.93) to Eq. (6.96), while considering the proper rotation of the principal frame along the relevant special axis.

6.4.3.3 Triclinic Crystal Classes

In that case, all the non-diagonal terms of $\varepsilon'_r(\omega)$ are non-zero, with: $\varepsilon'_{r_{xy}} = \varepsilon'_{r_{yx}} \neq \varepsilon'_{r_{xz}} = \varepsilon'_{r_{zx}} \neq \varepsilon'_{r_{yz}} = \varepsilon'_{r_{zy}} \neq 0$. Here also, we directly use Eq. (6.32) combined with Eq. (6.31), Eq. (6.28), Eq. (6.25a), Eq. (6.25b), and Eq. (6.25c) to access n' as a function of $\vec{u}$, and that for any direction of propagation according to the dielectric frame. In such a triclinic crystal class, there is no more question of any special axis. It is possible to use the previous equations to plot the imaginary index surface, which is not done here. However, its topology and symmetry can be imagined on the basis of the monoclinic case which is close to the triclinic one. Indeed, the two sections of the imaginary layers in each of the principal planes of the dielectric frame $(\overrightarrow{Ox},\overrightarrow{Oy})$, $(\overrightarrow{Ox},\overrightarrow{Oz})$, and $(\overrightarrow{Oy},\overrightarrow{Oz})$ keep being a circular section and a bi-lobar section, respectively. Each of these three bi-lobar sections undergoes a rotation along the dielectric axis orthogonal to the associated principal plane, in direct relation with the associated non-diagonal imaginary element. In other words, in each dielectric plane, the bi-lobar distribution in a triclinic crystal behaves as that defined for the monoclinic crystal class with its special axis orthogonal to the considered dielectric plane: the analytical expressions to be considered correspond thus to Eq. (6.90a) to Eq. (6.91b), Eq. (6.97), and Eq. (6.98) or Eq. (6.99) and Eq. (6.100) according to the principal plane that is selected.

As for the monoclinic crystal classes, there are non-diagonal coefficients but the tensor ε'_r is symmetrical so that it can be diagonalized in the so-called imaginary dielectric frame $(\overrightarrow{Ox'},\overrightarrow{Oy'},\overrightarrow{Oz'})$ defined earlier. The only difference with the monoclinic case is that there is no special axis so that none of these three imaginary axes correspond to one of the principal axes of the dielectric frame. Because the dielectric frame and the imaginary frame do not present any common axis, the imaginary double-layer distribution does not show any more symmetry mirror. The only remaining symmetry is then the inversion center.

6.5 RELATIONSHIP WITH ABSORPTION AND EMISSION

The knowledge of the imaginary index surface $n'^{\pm}(\vec{u})$ is of prime importance for the description of any resonant linear optical property like linear absorption or emission as depicted by Figure 1.9 to Figure 1.11 in Chapter 1. Actually from Eq. (6.10), knowing that $\overrightarrow{\hat{k}^{\pm}}(\omega,\vec{u}) = \hat{k}^{\pm}(\omega,\vec{u})\vec{u} = \frac{\omega}{c}\hat{n}^{\pm}(\omega,\vec{u})\vec{u}$ with $\hat{n}^{\pm}(\omega,\vec{u}) = n^{\pm}(\omega,\vec{u}) + in'^{\pm}(\omega,\vec{u})$ and that the light intensity is given by $I(\omega) = \frac{n(\omega)}{2}\sqrt{\frac{\varepsilon_0}{\mu_0}}\left\|\vec{E}(\omega)\right\|^2$ (Yariv 1983) [6], it is easy to establish the following equation:

$$I^{\pm}(\omega,\vec{u},Z) = \frac{n^{\pm}(\omega,\vec{u})}{2}\sqrt{\frac{\varepsilon_0}{\mu_0}}\left\|E_0(\omega)\right\|^2 \exp[-\frac{2\omega}{c}n'^{\pm}(\omega,\vec{u})Z] \tag{6.101}$$

where Z is the space coordinate along the considered direction of propagation $\vec{u}$.

Contrary to the case of the refractive index whose sign is always positive since it is given by $n^{\pm}(\omega,\vec{u}) = c/\mathrm{v}^{\pm}(\omega,\vec{u})$, the sign of $n'^{\pm}(\omega,\vec{u})$ can be positive or negative according to the population of the matter energy levels that are considered during the light/matter interaction.

When $n'^{\pm}(\omega,\vec{u})>0$, there is an attenuation of the intensity during the propagation, *i.e.* $I^{\pm}(\omega,\vec{u},Z)\propto\exp[-\alpha^{\pm}(\omega,\vec{u})Z]$ with:

$$\alpha^{\pm}(\omega,\vec{u})=\frac{2\omega}{c}n'^{\pm}(\omega,\vec{u})>0 \tag{6.102}$$

$\alpha^{\pm}(\omega,\vec{u})$ is the so-called Beer-Lambert absorption coefficient. It corresponds for example to the case of a two-level system like that described in Figure 1.10 and Figure 1.11 of Chapter 1 below the intensity of saturation of the transition (Siegman 1986) [7].

When $n'^{\pm}(\omega,\vec{u})<0$, there is an amplification of the intensity during the propagation, *i.e.* $I^{\pm}(\omega,\vec{u},Z)\propto\exp[G^{\pm}(\omega,\vec{u})Z]$ with:

$$G^{\pm}(\omega,\vec{u})=-\frac{2\omega}{c}n'^{\pm}(\omega,\vec{u})>0 \tag{6.103}$$

$G^{\pm}(\omega,\vec{u})$ is the gain coefficient that exists in a three-level or four-level laser systems below the saturation intensity of the transition (Siegman 1986) [7].

The classical approach of Einstein's model using the population equations would allow us to obtain the expressions of $\alpha^{\pm}(\omega,\vec{u})$ and $G^{\pm}(\omega,\vec{u})$ as a function of the Einstein coefficients (Siegman 1986) [7]. Then the matching with Eq. (6.102) and Eq. (6.103) can lead to the link with $n'^{\pm}(\omega,\vec{u})$. Such calculations would bring the angular variation of the Einstein coefficients, which is usually not considered. This interesting extension of crystal optics is beyond the scope of this book in its present form.

6.6 SUMMARY

By introducing a complex first-order electric susceptibility tensor, and thus a complex relative dielectric permittivity tensor, at the level of the propagation equation of light and by considering the weak absorption/emission approximation, it has been possible to calculate the angular distribution of the imaginary optical index, also called the imaginary index surface. Regarding the shape, the main differences with the index surface detailed in Chapter 5 are as follows: in the case of the uniaxial optical class, a bi-lobar behavior; in the case of the biaxial optical class, a bi-lobar behavior, the existence of a continuum of contact points between the internal and external layers, and the fact that the orientation symmetry is lower for the monoclinic and triclinic crystal classes (Joly 2009, Petit 2010, Petit 2011) [3–5].

It has also been seen that the dielectric frame diagonalizes both the real and imaginary parts of the complex relative dielectric permittivity tensor for the isotropic and uniaxial optical classes as well as for one of the three crystal classes of the biaxial optical class, *i.e.* the orthorhombic ones. For the two others, *i.e.* the monoclinic and triclinic crystal classes, the complex dielectric permittivity tensor is diagonal in a specific frame called the imaginary dielectric frame. This frame is of prime importance from a practical point of view since it allows us to access the extremal values of the absorption or emission coefficient [2]. Furthermore, the orientation between the dielectric frame and the imaginary dielectric frame depends on the wavelength (Petit 2013, Joly 2010) [2,8]. Here appears the extreme complexity of these crystal classes since the orientation of the dielectric frame also depends on wavelength as pointed out in Chapter 3.

REFERENCES

[1] Born M. and Wolf E. 1980. Principle of Optics, 6th ed., New York: Pergamon Press.

[2] Petit Y., Joly S., Segonds P., and Boulanger B. 2013. Recent advances in monoclinic crystal optics. Laser and Photonics Reviews: 1–18. https://doi.org/10.1002/lpor.201200078.

[3] Joly S., Petit Y., Boulanger B., Segonds P., Félix C., Ménaert B., and Aka G. 2009. Absorption singular topology in biaxial crystals. Optics Express 17(22): 19869–19873.

[4] Petit Y., Segonds P., Joly S., and Boulanger B. 2010. Layer-inversion zones in angular distributions of linear spectroscopic properties in biaxial crystals. Luminescent Materials, Special issue of Materials, www.mdpi.com/journal/materials/special_issues/luminescent-materials/.

[5] Petit Y., Joly S., Segonds P., and Boulanger B. 2011. Refined modeling of angular distributions of linear absorption and fluorescence in biaxial crystals. Laser Physics 21(3–5).

[6] Yariv A. 1989. Quantum Electronics, 3rd ed., New York: John Wiley & Sons.

[7] Siegman A. E. 1986. Lasers. Mill Valley, CA: University Science Books.

[8] Joly S., Segonds P., Boulanger B., Petit Y., Félix C., and Ménaert B. 2010. Rotation of the absorption frame as a function of the electronic transition in the Nd^{3+}:$YCa_4O(BO_3)_3$ monoclinic crystal. Optics Express 18(18): 19169–19174.

7 Second-Order Interactions

7.1 COUPLED EQUATIONS

7.1.1 COMPLEX AMPLITUDES

According to Eq. (1.104) and Eq. (5.6), the equation of propagation of light at the circular frequency ω in the nonlinear regime writes:

$$\overrightarrow{rot}\left[\overrightarrow{rot}\left[\vec{E}(\omega,\vec{r})\right]\right]-\frac{\omega^2}{c^2}\varepsilon_r^{(1)}(\omega)\cdot\vec{E}(\omega,\vec{r})=\omega^2\mu_0\vec{P}^{NL}(\omega,\vec{r}) \tag{7.1}$$

As in the linear regime, the plane wave, which is given by Eq. (5.9), is a solution of Eq. (7.1). It is reminded here:

$$\vec{E}(\omega,\vec{r})=\vec{e}(\omega)E_0(\omega,\vec{r})\mathrm{e}^{i\vec{k}(\omega).\vec{r}} \tag{7.2}$$

The goal is to establish the spatial evolution of the complex amplitude $E_0(\omega,\vec{r})$ along the direction of propagation starting from Eq. (7.1).

By definition, $\overrightarrow{rot}\left[\vec{E}(\omega,\vec{r})\right]$ writes:

$$\overrightarrow{rot}\left[\vec{E}(\omega,\vec{r})\right]=\vec{\nabla}\wedge\left[E_0(\omega,\vec{r})\mathrm{e}^{i\vec{k}(\omega).\vec{r}}\,\vec{e}(\omega)\right] \tag{7.3}$$

with

$$\vec{\nabla}\wedge\left[f\vec{V}\right]=f\vec{\nabla}\wedge\vec{V}+\vec{\nabla}\left[f\right]\wedge\vec{V} \tag{7.4}$$

Thus at the position $\vec{r}$, it comes:

$$\overrightarrow{rot}\left[\vec{E}(\omega,\vec{r})\right]=E_0(\omega,\vec{r})\mathrm{e}^{i\vec{k}(\omega).\vec{r}}\,\vec{\nabla}\wedge\vec{e}(\omega)+\vec{\nabla}\left[E_0(\omega,\vec{r})\mathrm{e}^{i\vec{k}(\omega).\vec{r}}\right]\wedge\vec{e}(\omega) \tag{7.5}$$

The orientation of $\vec{e}(\omega)$ being independent of the position $\vec{r}$, the following simplification applies:

$$\vec{\nabla}\wedge\vec{e}(\omega)=\vec{0} \tag{7.6}$$

Furthermore, $\vec{\nabla}\left[E_0(\omega,\vec{r})\mathrm{e}^{i\vec{k}(\omega).\vec{r}}\right]$ develops as:

$$\vec{\nabla}\left[E_0(\omega,\vec{r})\mathrm{e}^{i\vec{k}(\omega).\vec{r}}\right]=E_0(\omega,\vec{r})\vec{\nabla}\left[\mathrm{e}^{i\vec{k}(\omega).\vec{r}}\right]+\mathrm{e}^{i\vec{k}(\omega).\vec{r}}\,\vec{\nabla}\left[E_0(\omega,\vec{r})\right] \tag{7.7}$$

DOI: 10.1201/9780429091735-7

with

$$\vec{\nabla}\left[e^{i\vec{k}(\omega).\vec{r}}\right] = e^{i\vec{k}(\omega).\vec{r}}\, i\vec{k}(\omega) \tag{7.8}$$

Then Eq. (7.7) further develops as:

$$\vec{\nabla}\left[E_0(\omega,\vec{r})e^{i\vec{k}(\omega).\vec{r}}\right] = iE_0(\omega,\vec{r})e^{i\vec{k}(\omega).\vec{r}}\,\vec{k}(\omega) + e^{i\vec{k}(\omega).\vec{r}}\,\vec{\nabla}\left[E_0(\omega,\vec{r})\right] \tag{7.9}$$

Thus Eq. (7.5) becomes:

$$\overrightarrow{rot}\left[\vec{E}(\omega,\vec{r})\right] = i\,e^{i\vec{k}(\omega).\vec{r}}\,E_0(\omega,\vec{r})\left[\vec{k}(\omega)\wedge\vec{e}(\omega)\right] + e^{i\vec{k}(\omega).\vec{r}}\left[\vec{\nabla}\left[E_0(\omega,\vec{r})\right]\wedge\vec{e}(\omega)\right] \tag{7.10}$$

Similarly, using the definition of Eq. (7.3) and considering also Eq. (7.9), $\overrightarrow{rot}\left[\overrightarrow{rot}\left[\vec{E}(\omega,\vec{r})\right]\right]$ develops as:

$$\begin{aligned}
&\overrightarrow{rot}\left[\overrightarrow{rot}\left[\vec{E}(\omega,\vec{r})\right]\right] = \vec{\nabla}\wedge\overrightarrow{rot}\left[\vec{E}(\omega,\vec{r})\right]\\
&= \vec{\nabla}\wedge\left\{iE_0(\omega,\vec{r})e^{i\vec{k}(\omega).\vec{r}}\left[\vec{k}(\omega)\wedge\vec{e}(\omega)\right] + e^{i\vec{k}(\omega).\vec{r}}\left[\vec{\nabla}\left[E_0(\omega,\vec{r})\right]\wedge\vec{e}(\omega)\right]\right\}\\
&= \vec{\nabla}\wedge\left\{iE_0(\omega,\vec{r})e^{i\vec{k}(\omega).\vec{r}}\left[\vec{k}(\omega)\wedge\vec{e}(\omega)\right]\right\} + \vec{\nabla}\wedge\left\{e^{i\vec{k}(\omega).\vec{r}}\left[\vec{\nabla}\left[E_0(\omega,\vec{r})\right]\wedge\vec{e}(\omega)\right]\right\}
\end{aligned} \tag{7.11}$$

By applying again Eq. (7.4), Eq. (7.11) becomes:

$$\begin{aligned}
&\overrightarrow{rot}\left[\overrightarrow{rot}\left[\vec{E}(\omega,\vec{r})\right]\right] = iE_0(\omega,\vec{r})e^{i\vec{k}(\omega).\vec{r}}\,\vec{\nabla}\wedge\left[\vec{k}(\omega)\wedge\vec{e}(\omega)\right]\\
&+i\vec{\nabla}\left[E_0(\omega,\vec{r})e^{i\vec{k}(\omega).\vec{r}}\right]\wedge\left[\vec{k}(\omega)\wedge\vec{e}(\omega)\right]\\
&+e^{i\vec{k}(\omega).\vec{r}}\,\vec{\nabla}\wedge\left\{\vec{\nabla}\left[E_0(\omega,\vec{r})\right]\wedge\vec{e}(\omega)\right\}\\
&+\vec{\nabla}\left[e^{i\vec{k}(\omega).\vec{r}}\right]\wedge\left\{\vec{\nabla}\left[E_0(\omega,\vec{r})\right]\wedge\vec{e}(\omega)\right\}
\end{aligned} \tag{7.12}$$

The second term of the right hand of Eq. (7.12) can be further developed using Eq. (7.8) and Eq. (7.9), which gives:

$$\begin{aligned}
&\overrightarrow{rot}\left[\overrightarrow{rot}\left[\vec{E}(\omega,\vec{r})\right]\right] = iE_0(\omega,\vec{r})e^{i\vec{k}(\omega).\vec{r}}\,\vec{\nabla}\wedge\left[\vec{k}(\omega)\wedge\vec{e}(\omega)\right]\\
&-e^{i\vec{k}(\omega).\vec{r}}\,E_0(\omega,\vec{r})\vec{k}(\omega)\wedge\left[\vec{k}(\omega)\wedge\vec{e}(\omega)\right]\\
&+i\,e^{i\vec{k}(\omega).\vec{r}}\,\vec{\nabla}\left[E_0(\omega,\vec{r})\right]\wedge\left[\vec{k}(\omega)\wedge\vec{e}(\omega)\right]\\
&+e^{i\vec{k}(\omega).\vec{r}}\,\vec{\nabla}\wedge\left\{\vec{\nabla}\left[E_0(\omega,\vec{r})\right]\wedge\vec{e}(\omega)\right\}\\
&+e^{i\vec{k}(\omega).\vec{r}}\,i\vec{k}(\omega)\wedge\left[\vec{\nabla}\left[E_0(\omega,\vec{r})\right]\wedge\vec{e}(\omega)\right]
\end{aligned} \tag{7.13}$$

Because the directions and modulus of $\vec{k}(\omega)$ and $\vec{e}(\omega)$ remain unchanged during the propagation, then $\left[\vec{\nabla}.\vec{k}(\omega)\right]=0$ and $\left[\vec{\nabla}.\vec{e}(\omega)\right]=0$ respectively, so that it comes:

$$\vec{\nabla}\wedge\left[\vec{k}(\omega)\wedge\vec{e}(\omega)\right]=\vec{k}(\omega)\left[\vec{\nabla}.\vec{e}(\omega)\right]-\vec{e}(\omega)\left[\vec{\nabla}.\vec{k}(\omega)\right]=0 \tag{7.14}$$

Thus Eq. (7.13) writes:

$$\begin{aligned}\overrightarrow{rot}\left[\overrightarrow{rot}\left[\vec{E}(\omega,\vec{r})\right]\right]&=-e^{i\vec{k}(\omega).\vec{r}}\,E_0(\omega,\vec{r})\vec{k}(\omega)\wedge\left[\vec{k}(\omega)\wedge\vec{e}(\omega)\right]\\&+i\,e^{i\vec{k}(\omega).\vec{r}}\,\vec{\nabla}\left[E_0(\omega,\vec{r})\right]\wedge\left[\vec{k}(\omega)\wedge\vec{e}(\omega)\right]\\&+e^{i\vec{k}(\omega).\vec{r}}\,\vec{\nabla}\wedge\left\{\vec{\nabla}\left[E_0(\omega,\vec{r})\right]\wedge\vec{e}(\omega)\right\}\\&+e^{i\vec{k}(\omega).\vec{r}}\,i\vec{k}(\omega)\wedge\left[\vec{\nabla}\left[E_0(\omega,\vec{r})\right]\wedge\vec{e}(\omega)\right]\end{aligned} \tag{7.15}$$

As used to expand Eq. (7.14), it is useful at this step to consider the following relation:

$$\vec{U}\wedge\left[\vec{V}\wedge\vec{W}\right]=\vec{V}\left[\vec{U}.\vec{W}\right]-\vec{W}\left[\vec{U}.\vec{V}\right] \tag{7.16}$$

Then Eq. (7.15) further develops as:

$$\begin{aligned}\overrightarrow{rot}\left[\overrightarrow{rot}\left[\vec{E}(\omega,\vec{r})\right]\right]&=-e^{i\vec{k}(\omega).\vec{r}}\,E_0(\omega,\vec{r})\vec{k}(\omega)\wedge\left[\vec{k}(\omega)\wedge\vec{e}(\omega)\right]\\&+i\,e^{i\vec{k}(\omega).\vec{r}}\,\vec{\nabla}\left[E_0(\omega,\vec{r})\right]\wedge\left[\vec{k}(\omega)\wedge\vec{e}(\omega)\right]\\&+e^{i\vec{k}(\omega).\vec{r}}\,\vec{\nabla}\left[E_0(\omega,\vec{r})\right]\left[\vec{\nabla}.\vec{e}(\omega)\right]-e^{i\vec{k}(\omega).\vec{r}}\,\vec{e}(\omega)\left\{\vec{\nabla}.\vec{\nabla}\left[E_0(\omega,\vec{r})\right]\right\}\\&+e^{i\vec{k}(\omega).\vec{r}}\,i\vec{k}(\omega)\wedge\left[\left[\vec{\nabla}E_0(\omega,\vec{r})\right]\wedge\vec{e}(\omega)\right]\end{aligned} \tag{7.17}$$

Since $\left[\vec{\nabla}.\vec{e}(\omega)\right]=0$ and $\vec{\nabla}.\vec{\nabla}\left[E_0(\omega,\vec{r})\right]=\Delta E_0(\omega,\vec{r})$, Eq. (7.17) finally writes:

$$\begin{aligned}\overrightarrow{rot}\left[\overrightarrow{rot}\left[\vec{E}(\omega,\vec{r})\right]\right]&=-e^{i\vec{k}(\omega).\vec{r}}\,E_0(\omega,\vec{r})\vec{k}(\omega)\wedge\left[\vec{k}(\omega)\wedge\vec{e}(\omega)\right]\\&+i\,e^{i\vec{k}(\omega).\vec{r}}\,\vec{\nabla}\left[E_0(\omega,\vec{r})\right]\wedge\left[\vec{k}(\omega)\wedge\vec{e}(\omega)\right]\\&-e^{i\vec{k}(\omega).\vec{r}}\,\vec{e}(\omega)\Delta E_0(\omega,\vec{r})+e^{i\vec{k}(\omega).\vec{r}}\,i\vec{k}(\omega)\wedge\left[\vec{\nabla}\left[E_0(\omega,\vec{r})\right]\wedge\vec{e}(\omega)\right]\end{aligned} \tag{7.18}$$

We consider that within the nonlinear interaction, the amplitude $E(\omega,\vec{r})$ does not vary significantly over a wavelength, which corresponds to the slowly varying envelop often verified, so that $\lambda\left|\vec{\nabla}E(\omega,\vec{r})\right|<<E(\omega,\vec{r})$. In this case, $\Delta E(\omega,\vec{r})$ can be neglected in Eq. (7.18) that becomes:

$$\begin{aligned}&\overrightarrow{rot}\left[\overrightarrow{rot}\left[\vec{E}(\omega,\vec{r})\right]\right]=-E_0(\omega,\vec{r})e^{i\vec{k}(\omega).\vec{r}}\,\vec{k}(\omega)\wedge\left[\vec{k}(\omega)\wedge\vec{e}(\omega)\right]\\&+ie^{i\vec{k}(\omega).\vec{r}}\,\vec{\nabla}\left[E_0(\omega,\vec{r})\right]\wedge\left[\vec{k}(\omega)\wedge\vec{e}(\omega)\right]\\&+ie^{i\vec{k}(\omega).\vec{r}}\,\vec{k}(\omega)\wedge\left[\vec{\nabla}\left[E_0(\omega,\vec{r})\right]\wedge\vec{e}(\omega)\right]\end{aligned}\tag{7.19}$$

Using Eq. (7.19) and after multiplying both parts of Eq. (7.1) by $e^{-i\vec{k}(\omega).\vec{r}}$, it comes:

$$\begin{aligned}&-E_0(\omega,\vec{r})\left\{\vec{k}(\omega)\wedge\left[\vec{k}(\omega)\wedge\vec{e}(\omega)\right]+\frac{\omega^2}{c^2}\varepsilon_r^{(1)}(\omega)\cdot\vec{e}(\omega)\right\}\\&+i\vec{\nabla}\left[E_0(\omega,\vec{r})\right]\wedge\left[\vec{k}(\omega)\wedge\vec{e}(\omega)\right]+i\vec{k}(\omega)\wedge\left[\vec{\nabla}\left[E_0(\omega,\vec{r})\right]\wedge\vec{e}(\omega)\right]\\&=\omega^2\mu_0\vec{P}^{NL}(\omega,\vec{r})e^{-i\vec{k}(\omega).\vec{r}}\end{aligned}\tag{7.20}$$

The next step consists of making the scalar product of the right and left sides of Eq. (7.20) with the electric field polarization unit vector $\vec{e}(\omega)$, *i.e.*:

$$\begin{aligned}&\vec{e}(\omega).\left\{\begin{aligned}&-E_0(\omega,\vec{r})\left\{\vec{k}(\omega)\wedge\left[\vec{k}(\omega)\wedge\vec{e}(\omega)\right]+\frac{\omega^2}{c^2}\varepsilon_r^{(1)}(\omega)\cdot\vec{e}(\omega)\right\}\\&+i\vec{\nabla}\left[E_0(\omega,\vec{r})\right]\wedge\left[\vec{k}(\omega)\wedge\vec{e}(\omega)\right]+i\vec{k}(\omega)\wedge\left[\vec{\nabla}\left[E_0(\omega,\vec{r})\right]\wedge\vec{e}(\omega)\right]\end{aligned}\right\}\\&=\omega^2\mu_0\vec{e}(\omega).\vec{P}^{NL}(\omega,\vec{r})e^{-i\vec{k}(\omega).\vec{r}}\end{aligned}\tag{7.21}$$

According to Eq. (5.21) of Chapter 5, Eq. (7.21) reduces to:

$$\begin{aligned}&i\vec{e}(\omega).\left\{\vec{\nabla}\left[E_0(\omega,\vec{r})\right]\wedge\left[\vec{k}(\omega)\wedge\vec{e}(\omega)\right]+\vec{k}(\omega)\wedge\left[\vec{\nabla}\left[E_0(\omega,\vec{r})\right]\wedge\vec{e}(\omega)\right]\right\}\\&=\omega^2\mu_0\vec{e}(\omega).\vec{P}^{NL}(\omega,\vec{r})e^{-i\vec{k}(\omega).\vec{r}}\end{aligned}\tag{7.22}$$

It is now useful to consider the following relations:

$$\left\{\begin{aligned}&\vec{U}\wedge\left[\vec{V}\wedge\vec{W}\right]+\vec{W}\wedge\left[\vec{U}\wedge\vec{V}\right]+\vec{V}\wedge\left[\vec{W}\wedge\vec{U}\right]=\vec{0}\\&\vec{U}\wedge\vec{V}=-\vec{V}\wedge\vec{U}\end{aligned}\right.\tag{7.23}$$

Then it comes for Eq. (7.22):

$$\begin{aligned}&\vec{\nabla}\left[E_0(\omega,\vec{r})\right]\wedge\left[\vec{k}(\omega)\wedge\vec{e}(\omega)\right]\\&+\vec{e}(\omega)\wedge\left[\vec{\nabla}\left[E_0(\omega,\vec{r})\right]\wedge\vec{k}(\omega)\right]\\&+\vec{k}(\omega)\wedge\left[\vec{e}(\omega)\wedge\vec{\nabla}\left[E_0(\omega,\vec{r})\right]\right]\\&=\vec{0}\end{aligned}\tag{7.24}$$

which also writes:

$$\vec{\nabla}\left[E_0(\omega,\vec{r})\right]\wedge\left[\vec{k}(\omega)\wedge\vec{e}(\omega)\right]-\vec{e}(\omega)\wedge\left[\vec{k}(\omega)\wedge\vec{\nabla}\left[E_0(\omega,\vec{r})\right]\right]$$
$$=\vec{k}(\omega)\wedge\left[\vec{\nabla}\left[E_0(\omega,\vec{r})\right]\wedge\vec{e}(\omega)\right] \tag{7.25}$$

Thus Eq. (7.21) becomes:

$$i\vec{e}(\omega).\left\{2\vec{\nabla}\left[E_0(\omega,\vec{r})\right]\wedge\left[\vec{k}(\omega)\wedge\vec{e}(\omega)\right]+\vec{e}(\omega)\wedge\left[\vec{\nabla}\left[E_0(\omega,\vec{r})\right]\wedge\vec{k}(\omega)\right]\right\}$$
$$=\omega^2\mu_0\vec{e}(\omega).\vec{P}^{NL}(\omega,\vec{r})\mathrm{e}^{-i\vec{k}(\omega).\vec{r}} \tag{7.26}$$

For the following, it is convenient to work in the laboratory frame (X, Y, Z) defined in Figure 5.15, this frame being not the dielectric frame (x, y, z). It is reminded in Figure 7.1 that the polarization is taken in the plane $\left(\overrightarrow{OX},\overrightarrow{OZ}\right)$ of the laboratory frame, such an arbitrary choice of orientation being general and carrying the full description of light polarization.

According to Figure 7.1, it comes:

$$\vec{k}(\omega).\vec{e}(\omega)=k(\omega)\sin(\rho(\omega)) \tag{7.27}$$

where $\rho(\omega)$ stands for the spatial walk-off angle defined in Part 5.4.

In the laboratory frame in which the Z-axis is chosen to be along the propagation direction of light, the relevant vectors write:

$$\vec{r}=\begin{pmatrix}X\\Y\\Z\end{pmatrix}\quad\vec{e}(\omega)=\begin{pmatrix}\sin(\rho(\omega))\\0\\\sin(\rho(\omega))\end{pmatrix}\quad\vec{k}(\omega)=\begin{pmatrix}0\\0\\k(\omega)\end{pmatrix}\quad\vec{\nabla}\left[E_0(\omega,\vec{r})\right]=\begin{pmatrix}0\\0\\\dfrac{\partial E_0(\omega,\vec{r})}{\partial Z}\end{pmatrix} \tag{7.28}$$

Equation (7.28) showing that $\vec{\nabla}\left[E_0(\omega,\vec{r})\right]$ and $\vec{k}(\omega)$ are collinear, so that $\vec{\nabla}\left[E_0(\omega,\vec{r})\right]\wedge\vec{k}(\omega)=\vec{0}$ in Eq. (7.26).

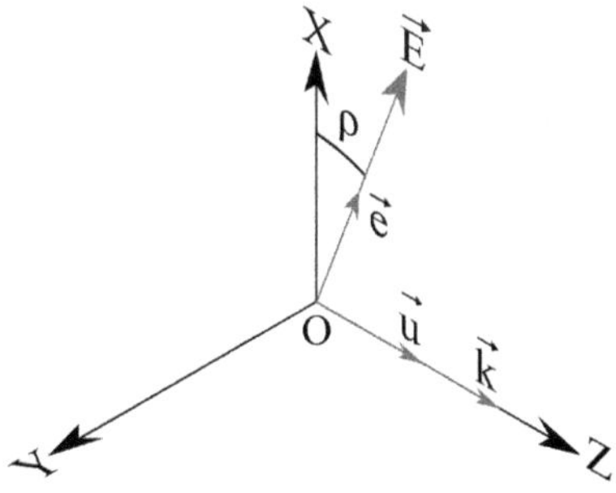

FIGURE 7.1 Definition of the laboratory frame (X, Y, Z).

Thus Eq. (7.26) becomes:

$$i\vec{e}(\omega).\left\{2\vec{\nabla}\left[E_0(\omega,\vec{r})\right]\wedge\left[\vec{k}(\omega)\wedge\vec{e}(\omega)\right]\right\}=\omega^2\mu_0\vec{e}(\omega).\vec{P}^{NL}(\omega,\vec{r})\mathrm{e}^{-ik(\omega)Z} \tag{7.29}$$

According to Eq. (7.23), it also comes:

$$\begin{aligned}
&\vec{e}(\omega).\left\{\vec{\nabla}\left[E_0(\omega,\vec{r})\right]\wedge\left[\vec{k}(\omega)\wedge\vec{e}(\omega)\right]\right\}\\
&=\vec{e}(\omega).\left\{\vec{k}(\omega)\wedge\left[\vec{\nabla}\left[E_0(\omega,\vec{r})\right]\wedge\vec{e}(\omega)\right]\right\}\\
&=-\vec{\nabla}\left[E_0(\omega,\vec{r})\right].\left\{\vec{e}(\omega)\wedge\left[\vec{k}(\omega)\wedge\vec{e}(\omega)\right]\right\}\\
&=-\vec{\nabla}\left[E_0(\omega,\vec{r})\right].\left\{\vec{k}(\omega)\left[\vec{e}(\omega).\vec{e}(\omega)\right]-\vec{e}(\omega)\left[\vec{e}(\omega).\vec{k}(\omega)\right]\right\}\\
&=-\vec{\nabla}\left[E_0(\omega,\vec{r})\right].\left\{\vec{k}(\omega)-\vec{e}(\omega)\left[\vec{e}(\omega).\vec{k}(\omega)\right]\right\}
\end{aligned} \tag{7.30}$$

Equation (7.30) develops as follows using Eq. (7.28):

$$\begin{aligned}
&\vec{e}(\omega).\left\{\vec{k}(\omega)\wedge\left[\vec{\nabla}\left[E_0(\omega,\vec{r})\right]\wedge\vec{e}(\omega)\right]\right\}\\
&=-\vec{\nabla}\left[E_0(\omega,\vec{r})\right].\left\{\vec{k}(\omega)-\vec{e}(\omega)\left[k(\omega)\sin(\rho(\omega))\right]\right\}\\
&=-\vec{\nabla}\left[E_0(\omega,\vec{r})\right].\vec{k}(\omega)+\vec{\nabla}\left[E_0(\omega,\vec{r})\right].\vec{e}(\omega)k(\omega)\sin(\rho(\omega))\\
&=-\frac{\partial E_0(\omega,\vec{r})}{\partial Z}k(\omega)+\frac{\partial E_0(\omega,\vec{r})}{\partial Z}e_Z(\omega)k(\omega)\sin(\rho(\omega))\\
&=-\frac{\partial E_0(\omega,\vec{r})}{\partial Z}k(\omega)+\frac{\partial E_0(\omega,\vec{r})}{\partial Z}\sin^2(\rho(\omega))k(\omega)\\
&=-\frac{\partial E_0(\omega,\vec{r})}{\partial Z}k(\omega)\left[1-\sin^2(\rho(\omega))\right]\\
&=-\frac{\partial E_0(\omega,\vec{r})}{\partial Z}k(\omega)\cos^2(\rho(\omega))
\end{aligned} \tag{7.31}$$

And then finally, Eq. (7.29) simplifies as:

$$\frac{\partial E_0(\omega,\vec{r})}{\partial Z}=i\frac{\omega^2\mu_0}{2k(\omega)\cos^2(\rho(\omega))}\vec{e}(\omega).\vec{P}^{NL}(\omega,\vec{r})\mathrm{e}^{-ik(\omega)Z} \tag{7.32}$$

with $\vec{P}^{NL}(\omega,\vec{r})=\vec{P}^{(2)}(\omega,\vec{r})$ here.

It is considered here that the three interacting waves propagate in the same direction, along the Z-axis. Then Eq. (7.32) can be applied to each wave, *i.e.* for $\omega\equiv(\omega_1,\omega_2,\omega_3)$, which gives:

$$\begin{cases} \dfrac{\partial E_1(Z)}{\partial Z} = i\dfrac{\omega_1^2 \mu_0}{2k_1 \cos^2 \rho_1} \vec{e}_1.\vec{P}_1^{(2)}(Z)\exp[-ik_1 Z] \\ \dfrac{\partial E_2(Z)}{\partial Z} = i\dfrac{\omega_2^2 \mu_0}{2k_2 \cos^2 \rho_2} \vec{e}_2.\vec{P}_2^{(2)}(Z)\exp[-ik_2 Z] \\ \dfrac{\partial E_3(Z)}{\partial Z} = i\dfrac{\omega_3^2 \mu_0}{2k_3 \cos^2 \rho_3} \vec{e}_3.\vec{P}_3^{(2)}(Z)\exp[-ik_3 Z] \end{cases} \tag{7.33}$$

where the writing is simplified, using $E_j(Z) \equiv E_0(\omega_j, \vec{r})$, $k_j \equiv k(\omega_j)$, $\rho_j \equiv \rho(\omega_j)$, $\vec{e}_j \equiv \vec{e}(\omega_j)$, and $\vec{P}_j^{(2)}(Z) \equiv \vec{P}^{(2)}(\omega_j, \vec{r})$. Note that, thanks to the judicious choice of the laboratory frame, each of the three polarization unit vectors related to the three interacting waves are either in the $\left(\overrightarrow{OX}, \overrightarrow{OZ}\right)$ or $\left(\overrightarrow{OY}, \overrightarrow{OZ}\right)$ planes, in agreement with Figure 7.1 and Figure 5.15.

The system given by Eq. (7.33) is coupled by the three second-order induced polarizations that are defined by Eq. (3.2). Combined with Eq. (7.2) and using the simplified notation given earlier, Eq. (3.2) leads to:

$$\begin{cases} \vec{P}_1^{(2)}(Z) = \varepsilon_0[\chi^{(2)}(\omega_1 = \omega_3 - \omega_2) : \vec{e}_3 \otimes \vec{e}_2] E_3(Z) E_2^*(Z) \exp[i(k_3 - k_2)Z] \\ \vec{P}_2^{(2)}(Z) = \varepsilon_0[\chi^{(2)}(\omega_2 = \omega_3 - \omega_1) : \vec{e}_3 \otimes \vec{e}_1] E_3(Z) E_1^*(Z) \exp[i(k_3 - k_1)Z] \\ \vec{P}_3^{(2)}(Z) = \varepsilon_0[\chi^{(2)}(\omega_3 = \omega_1 + \omega_2) : \vec{e}_1 \otimes \vec{e}_2] E_1(Z) E_2(Z) \exp[i(k_1 + k_2)Z] \end{cases} \tag{7.34}$$

Since Eq. (7.33) contains the projection of each second-order nonlinear polarization vector along its associated electric field polarization for each wave, it is interesting to develop the three contracted products in Eq. (7.34). These three scalar quantities write:

$$\begin{cases} \sum_{ijk} e_{1i} \chi_{ijk}^{(2)}(\omega_1 = \omega_3 - \omega_2) e_{3j} e_{2k} \\ \sum_{kji} e_{2k} \chi_{kji}^{(2)}(\omega_2 = \omega_3 - \omega_1) e_{3j} e_{1i} \\ \sum_{jik} e_{3j} \chi_{jik}^{(2)}(\omega_3 = \omega_1 + \omega_2) e_{1i} e_{2k} \end{cases} \tag{7.35}$$

where the Cartesian indices i, j, and k refer to the dielectric frame (x, y, z).

It clearly appears that the three quantities in Eq. (7.35) are equal assuming the ABDP symmetry described by Eq. (3.33) (Armstrong 1962) [1], *i.e.* in a lossless medium, and according to the commutativity of the products of unit electric field coordinates. Then a second-order effective coefficient $\chi_{eff}^{(2)}$ can be defined as:

$$\begin{aligned} \chi_{eff}^{(2)} &= \vec{e}_1.[\chi^{(2)}(\omega_1 = \omega_3 - \omega_2) : \vec{e}_3 \otimes \vec{e}_2] = \vec{e}_2.[\chi^{(2)}(\omega_2 = \omega_3 - \omega_1) : \vec{e}_3 \otimes \vec{e}_1] \\ &= \vec{e}_3.[\chi^{(2)}(\omega_3 = \omega_1 + \omega_2) : \vec{e}_1 \otimes \vec{e}_2] \end{aligned} \tag{7.36}$$

Then Eq. (7.33) writes (Armstrong 1962) [1]:

$$\begin{cases} \dfrac{\partial E_1(Z)}{\partial Z} = i\varepsilon_0 \kappa_1 \chi_{eff}^{(2)} E_3(Z) E_2^*(Z) \exp[i\Delta k Z] \\ \dfrac{\partial E_2(Z)}{\partial Z} = i\varepsilon_0 \kappa_2 \chi_{eff}^{(2)} E_3(Z) E_1^*(Z) \exp[i\Delta k Z] \\ \dfrac{\partial E_3(Z)}{\partial Z} = i\varepsilon_0 \kappa_3 \chi_{eff}^{(2)} E_1(Z) E_2(Z) \exp[-i\Delta k Z] \end{cases} \tag{7.37}$$

with

$$\kappa_i = \frac{\omega_i^2 \mu_0}{2k_i \cos^2 \rho_i} \qquad i = (1,2,3) \tag{7.38}$$

and

$$\Delta k = k_3 - k_1 - k_2 \tag{7.39}$$

The quantity $\Delta k Z$ is the spatial mismatch between the nonlinear induced polarization and the radiated field for each Fourier component, *i.e.* between $\vec{P}_1^{(2)}(Z)$ and $\vec{E}_1(Z)$, between $\vec{P}_2^{(2)}(Z)$ and $\vec{E}_2(Z)$, between $\vec{P}_3^{(2)}(Z)$ and $\vec{E}_3(Z)$. Then at this step, the nonlinear interaction appears as an interference in the classical sense of the term, *i.e.* between a matter wave (the induced polarization wave) and an optical wave, these waves having the same circular frequency.

It is also important to notice that Eq. (7.37) is valid for any kind of situation dealing with the incident beams: $E_1(0) = E_2(0) = 0$ and $E_3(0) \neq 0$ (parametric fluorescence); $E_1(0) \neq 0$, $E_2(0) \neq 0$, and $E_3(0) = 0$ (sum-frequency generation); $E_1(0) \neq 0$, $E_3(0) \neq 0$, and $E_2(0) = 0$ (difference-frequency generation at ω_2); $E_2(0) \neq 0$, $E_3(0) \neq 0$, and $E_1(0) = 0$ (difference-frequency generation at ω_1). The differentiation between these situations will be done at the integration step.

Finally, Eq. (3.37) well shows that the strength of the nonlinear coupling depends on two key parameters: $\chi_{eff}^{(2)}$, that depends on the second-order electric susceptibility but also on the first-order electric susceptibility via the electric fields, and Δk that depends only on the first-order electric susceptibility via the refractive indices. Thus, the importance of linear optics for nonlinear optics is well put in evidence here. It appears that the stronger the coupling, the larger the $\chi_{eff}^{(2)}$ and the smaller the Δk, the optimum being $\Delta k = 0$. The latter condition is called the "phase-matching": it corresponds to the "photon momentum conservation" in the corpuscular picture given by Eq. (1.104).

These two parameters will be the subject of specific studies in the following from Part 7.2 to Part 7.5.

7.1.2 Real Amplitudes

Replacing in Eq. (7.37) $E_j(Z)$ with its expression as a function of the real amplitude $A_j(Z)$ and the initial phase Φ_j, *i.e.* $E_j(Z) = A_j(Z)e^{i\Phi_j}$, the system of coupled equations can be expressed for the real amplitudes, which gives:

$$\begin{cases} \dfrac{\partial A_1(Z)}{\partial Z} = -\varepsilon_0 \kappa_1 \chi_{eff}^{(2)} A_3(Z) A_2(Z) \sin[\Delta k Z + \Delta\Phi] \\ \dfrac{\partial A_2(Z)}{\partial Z} = -\varepsilon_0 \kappa_2 \chi_{eff}^{(2)} A_3(Z) A_1(Z) \sin[\Delta k Z + \Delta\Phi] \\ \dfrac{\partial A_3(Z)}{\partial Z} = \varepsilon_0 \kappa_3 \chi_{eff}^{(2)} A_1(Z) A_2(Z) \sin[\Delta k Z + \Delta\Phi] \end{cases} \tag{7.40}$$

with

$$\Delta\Phi = \Phi_3 - \Phi_1 - \Phi_2 \tag{7.41}$$

It is of particular interest to write Eq. (7.40) when phase-matching is achieved, *i.e.* when $\Delta k = 0$, which gives:

$$\begin{cases} \dfrac{\partial A_1(Z)}{\partial Z} = -\varepsilon_0 \kappa_1 \chi_{eff}^{(2)} A_3(Z) A_2(Z) \sin[\Delta\Phi] \\ \dfrac{\partial A_2(Z)}{\partial Z} = -\varepsilon_0 \kappa_2 \chi_{eff}^{(2)} A_3(Z) A_1(Z) \sin[\Delta\Phi] \\ \dfrac{\partial A_3(Z)}{\partial Z} = \varepsilon_0 \kappa_3 \chi_{eff}^{(2)} A_1(Z) A_2(Z) \sin[\Delta\Phi] \end{cases} \tag{7.42}$$

This set of equations allows us to establish the relationships that exist between the initial phases in the different phase-matched nonlinear processes, knowing that there is a maximal energy exchange for $\sin[\Delta\Phi] = \pm 1$, *i.e.* for $\Delta\Phi = \pm\pi/2$ respectively.

$\Delta\Phi = +\pi/2$ corresponds to an optimal photon fusion $\{\omega_1 + \omega_2 \rightarrow \omega_3\}$ requiring $\left\{\dfrac{\partial A_1(Z)}{\partial Z} < 0, \dfrac{\partial A_2(Z)}{\partial Z} < 0, \dfrac{\partial A_3(Z)}{\partial Z} > 0\right\}$; then $\Phi_3 = \Phi_1 + \Phi_2 + \pi/2$ as shown in Figure 7.2.

$\Delta\Phi = -\pi/2$ corresponds to an optimal photon scission $\{\omega_3 \rightarrow \omega_1 + \omega_2\}$ requiring $\left\{\dfrac{\partial A_1(Z)}{\partial Z} > 0, \dfrac{\partial A_2(Z)}{\partial Z} > 0, \dfrac{\partial A_3(Z)}{\partial Z} < 0\right\}$; then $\Phi_1 + \Phi_2 = \Phi_3 + \pi/2$ for a spontaneous scission, $\Phi_2 = \Phi_3 - \Phi_1 + \pi/2$ for a scission stimulated at ω_1, and $\Phi_1 = \Phi_3 - \Phi_2 + \pi/2$ for a scission stimulated at ω_2, as shown in Figure 7.3.

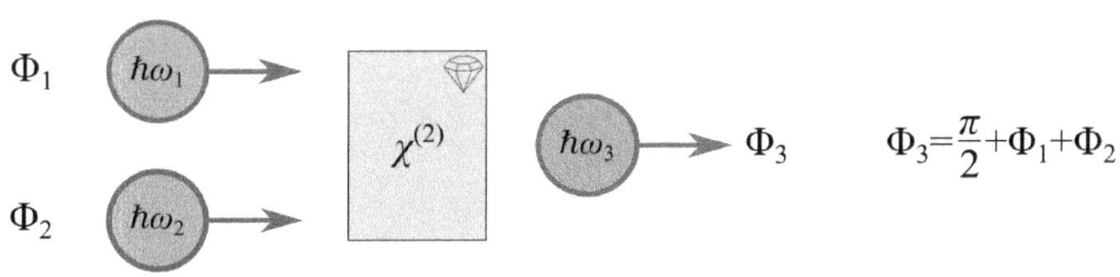

FIGURE 7.2 Initial phases configuration for a phase-matched second-order photon fusion process.

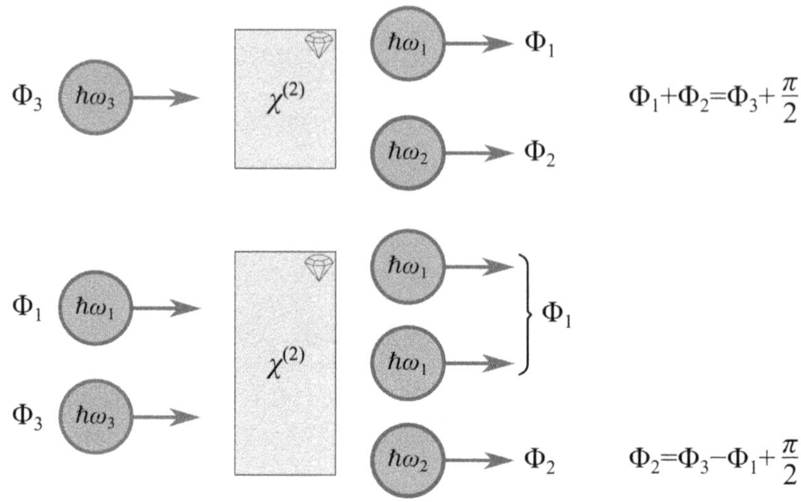

FIGURE 7.3 Initial phases configuration for second-order phase-matched photon spontaneous (top) and stimulated (bottom) scission processes.

It is important to notice that this optimal phase-mismatch of $\Delta\Phi = \pm\pi/2$ is not set by the experimenter but is selected "early", *i.e.* after several microns of propagation, thanks to the interference between the nonlinear polarization and the radiated field, which means that the best $\Delta\Phi$ "wins". Indeed, since the nonlinear interaction leads initially to an exponential growth of the newly radiated fields along the propagation direction, the configuration with the most efficient initial phase tends to dominate rapidly other possibilities of initial phase.

7.1.3 Intensities

It is easy to obtain the intensities coupled equations from Eq. (7.30) using the following relation:

$$I_j(Z) = \frac{1}{2}\frac{k_j}{\mu_0\omega_j}\left\|E_j(Z)\right\|^2 \tag{7.43}$$

Then it comes:

$$\frac{\partial I_j(Z)}{\partial Z} = \frac{1}{2}\frac{k_j}{\mu_o\omega_j}\left[E_j(Z)\frac{\partial E_j^*(Z)}{\partial Z} + E_j^*(Z)\frac{\partial E_j(Z)}{\partial Z}\right] \tag{7.44}$$

For any complex number C, it comes $[jC^* - jC] = 2\,\mathrm{Im}\{C\}$, so that it is possible to establish the following system:

$$\begin{cases} \dfrac{\partial I_1(Z)}{\partial Z} = -\dfrac{\omega_1}{2}\Psi(Z) \\ \dfrac{\partial I_2(Z)}{\partial Z} = -\dfrac{\omega_2}{2}\Psi(Z) \\ \dfrac{\partial I_3(Z)}{\partial Z} = +\dfrac{\omega_3}{2}\Psi(Z) \end{cases} \tag{7.45}$$

with

$$\Psi(Z) = \varepsilon_0 \chi_{eff}^{(2)} \operatorname{Im}\{E_3(Z)E_1^*(Z)E_2^*(Z)\exp[j\Delta kZ]\} \tag{7.46}$$

Equation (7.45) allows us to show the following equalities that are called the Manley-Rowe relations (Manley and Rowe 1959) [2]:

$$\frac{1}{\omega_1}\frac{\partial I_1(Z)}{\partial Z} = \frac{1}{\omega_2}\frac{\partial I_2(Z)}{\partial Z} = -\frac{1}{\omega_3}\frac{\partial I_3(Z)}{\partial Z} \tag{7.47}$$

These equalities govern the energy exchange between the interacting beams. They also indicate the conservation of the total intensity, *i.e.*:

$$\frac{\partial}{\partial Z}[I_1(Z)+I_2(Z)+I_3(Z)] = \frac{\Psi(Z)}{2}[\omega_3-\omega_1-\omega_2] = 0 \tag{7.48}$$

The quantity $[\omega_3-\omega_1-\omega_2]$ corresponds to the photon energy conservation given by Eq. (1.103). Such energy conservation is in direct relationship with the coherent behavior of these nonlinear processes, namely in which there is a defined phase relationship between the three interacting waves along the propagation, and in which the nonlinear medium participates to the photon fusion or fission processes without grabbing any energy amount. In this sense, since there is no irreversible energy transfer between the waves and matter, these nonlinear coherent processes appear intrinsically reversible.

7.1.4 Photon Flux Densities

The photon flux density N_i is the number of photons per unit of time and surface. It can be calculated from the intensity, *i.e.*:

$$N_i(Z) = \frac{I_i(Z)}{\hbar\omega_i} \tag{7.49}$$

Then from Eq. (7.45) and Eq. (7.49) it comes:

$$\begin{cases} \dfrac{\partial N_1(Z)}{\partial Z} = -\dfrac{\Psi(Z)}{2\hbar} \\ \dfrac{\partial N_2(Z)}{\partial Z} = -\dfrac{\Psi(Z)}{2\hbar} \\ \dfrac{\partial N_3(Z)}{\partial Z} = +\dfrac{\Psi(Z)}{2\hbar} \end{cases} \tag{7.50}$$

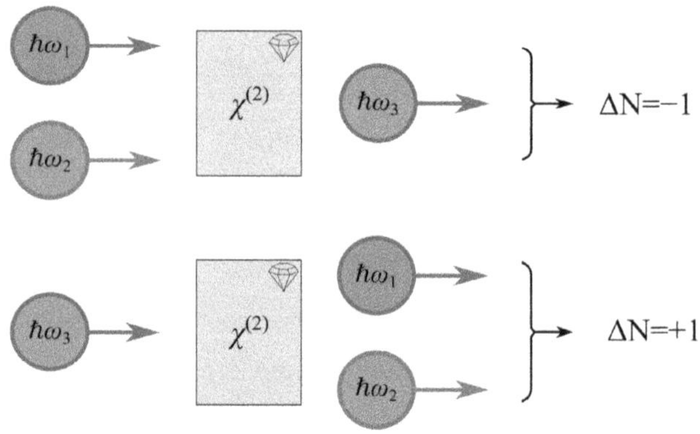

FIGURE 7.4 Schematic differential balance of the photon flux density ΔN due to photon fusion (top) and photon scission (down).

It is then possible to establish the following equalities:

$$\frac{\partial N_1(Z)}{\partial Z} = \frac{\partial N_2(Z)}{\partial Z} = -\frac{\partial N_3(Z)}{\partial Z} \tag{7.51}$$

These last equalities allow us to show that the total number of photons is not conserved during the nonlinear process:

$$\frac{\partial}{\partial Z}[N_1(Z)+N_2(Z)+N_3(Z)] = -\frac{\Psi(Z)}{2\hbar} \neq 0 \tag{7.52}$$

This feature can be well illustrated by Figure 7.4.

7.2 OUT OF PHASE-MATCHING

7.2.1 Generated Intensity in the Undepleted Pump Approximation

The out of phase-matching configuration corresponds to $\Delta k \neq 0$. In this case the efficiency of the nonlinear coupling is expected to be weak because the interference process is not fully constructive over the propagation length Z. It is then possible to consider that the intensities of the incident beams, called the "pump", do not vary as a function of Z, which corresponds to the so-called "undepleted pump approximation". Thus, in the case of sum-frequency generation for example, it comes $E_1(Z) \approx E_1(0)$ and $E_2(Z) \approx E_2(0)$ so that Eq. (7.30) reduces to:

$$\begin{cases} \dfrac{\partial E_1(Z)}{\partial Z} = 0 \\ \dfrac{\partial E_2(Z)}{\partial Z} = 0 \\ \dfrac{\partial E_3(Z)}{\partial Z} = i\varepsilon_0 \kappa_3 \chi_{eff}^{(2)} E_1(0) E_2(0) \exp[-i\Delta k Z] \end{cases} \tag{7.53}$$

It is easy to perform the integration, *i.e.*:

$$E_3(Z) = i\kappa_3\varepsilon_0\chi_{eff}^{(2)} E_1(0)E_2(0)\int_0^Z \exp[-i\Delta kZ]dZ \tag{7.54}$$

with

$$\int_0^Z \exp[-i\Delta kZ]dZ = \frac{2}{\Delta k}\exp[-i\frac{\Delta kZ}{2}]\sin[\frac{\Delta kZ}{2}] \tag{7.55}$$

Then it comes according to Eq. (7.43):

$$\|E_3(Z)\|^2 = 4\left[\varepsilon_0\kappa_3\right]^2[\chi_{eff}^{(2)}]^2\|E_1(0)\|^2\|E_2(0)\|^2\frac{1}{[\Delta k]^2}\sin^2[\frac{\Delta kZ}{2}] \tag{7.56}$$

Using Eq. (7.43) and Eq. (7.56), and knowing that $\mu_0\varepsilon_0c^2 = 1$, the generated intensity writes:

$$I_3(Z) = 8\varepsilon_0^2\sqrt{\frac{\mu_0}{\varepsilon_0}}\frac{n_3}{n_1n_2}[\kappa_3\chi_{eff}^{(2)}]^2 I_1(0)I_2(0)\frac{1}{[\Delta k]^2}\sin^2[\frac{\Delta kZ}{2}] \tag{7.57}$$

where n_j stands for the refractive index $n(\omega_j)$.

Finally, it comes according to Eq. (7.38) and one more time to $\mu_0\varepsilon_0c^2 = 1$:

$$I_3(Z) = 8\pi^2\sqrt{\frac{\mu_0}{\varepsilon_0}}\frac{1}{\lambda_3^2(\cos\rho_3)^4}\frac{[\chi_{eff}^{(2)}]^2}{n_1n_2n_3}I_1(0)I_2(0)\frac{1}{[\Delta k]^2}\sin^2[\frac{\Delta kZ}{2}] \tag{7.58}$$

Note that this equation is also valid for difference-frequency generation at ω_1, *i.e.* $I_2(0) \neq 0$, $I_3(0) \neq 0$, and $I_1(0) = 0$, by replacing indices 1, 2, 3 with 3, 2, 1 respectively, and also for difference-frequency generation at ω_2, *i.e.* $I_1(0) \neq 0$, $I_3(0) \neq 0$, and $I_2(0) = 0$, by replacing indices 1, 2, 3 with 1, 3, 2 respectively.

7.2.2 Parametric Coherence Length

Equation (7.58) shows that the intensity of the generated wave oscillates as a function of Z, as illustrated by Figure 7.5.

The periodicity of the generated intensity is $2l_c$ where l_c is the so-called parametric coherence length defined by:

$$l_c = \frac{\pi}{\Delta k} \tag{7.59}$$

Knowing that $k_j = (\omega_j / c)n_j$, Eq. (7.59) gives:

$$l_c = \frac{\pi c}{\omega_3n_3 - \omega_1n_1 - \omega_2n_2} \tag{7.60}$$

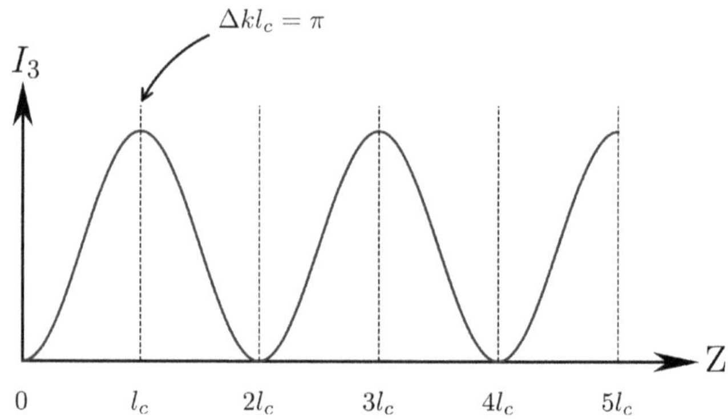

FIGURE 7.5 Evolution of the intensity of the generated wave along the direction of propagation in the case of an out of phase-matching nonlinear process. The half-period l_c is the parametric coherence length.

According to the typical values of the refractive indices, the parametric coherence length typically ranges between 1 µm and 100 µm, from ultraviolet to far infrared respectively. Then the interaction length over which is built the generated wave is very small, so that the corresponding intensity is weak, even at the optima, which validates the undepleted pump approximation. This feature had been previously anticipated before the integration step when considering Eq. (7.37) and noting that Δk governs the phase-mismatch between the nonlinear polarization and the radiated field. Equation (7.58) gives us access to this interference function, *i.e.* $\frac{1}{[\Delta k]^2}\sin^2[\frac{\Delta kZ}{2}]$. Then it appears that from 0 to l_c, the interference moves from fully constructive ($\Delta kZ = 0$) to fully destructive ($\Delta kZ = \pi$), and from l_c to $2l_c$ it moves from fully destructive ($\Delta kZ = \pi$) to fully constructive ($\Delta kZ = 2\pi$), and so on. The consequence is that the progressive building of the generated wave over the first coherence length is followed by its progressive destruction over the second one. This description corroborates the poor energy transfer efficiency towards the newly radiated field in the case of out-of-phase-matching nonlinear interactions.

7.2.3 Configurations of Polarization

Since three-wave processes are considered here, the only possible media have to be non-centrosymmetric in order to exhibit a non-zero second-order non-linearity, as seen in Chapter 3, and so a non-zero effective coefficient $\chi_{eff}^{(2)}$. Furthermore, it will be demonstrated hereafter that phase-matching can be reached in birefringent crystals. Thus, isotropic crystals are excluded here, in favor of keeping only the uniaxial or biaxial optical classes. For the latter ones, the possible directions of polarization ($\vec{e}_j$) and refractive indices (n_j) associated with a given direction of propagation are imposed, as demonstrated in Chapter 5. Actually, except along an optical axis, there are two possible Eigenmodes: (o) and (e) for uniaxial crystals, and (+) and (−) for biaxial crystals. Then, the three interacting waves are necessarily corresponding to one of these polarization modes, which gives $2^3 = 8$ possible combinations: (ooo), (eee), (oee), (eoo), (oeo), (ooe), (oee), (eoe), (eeo) for the uniaxial crystals, and (+++), (−−−), (+−−), (−++), (+−+), (++−), (+−−), (−+−), (−−+) for the biaxial ones, the modes being allocated to (ω_3, ω_1, ω_2) respectively.

In the case of a sum-frequency generation with a configuration (+−+) for example, it comes:

$$\begin{cases} \chi_{eff}^{(2)} = \vec{e_3}^{+} .[\chi^{(2)}(\omega_3 = \omega_1 + \omega_2) : \vec{e_1}^{-} \otimes \vec{e_2}^{+}] \\ I_3(Z) = 8\pi^2 \sqrt{\frac{\mu_0}{\varepsilon_0}} \frac{1}{\lambda_3^2 (\cos\rho_3)^4} \frac{[\chi_{eff}^{(2)}]^2}{n_1^- n_2^+ n_3^+} I_1(0) I_2(0) \frac{1}{[\Delta k]^2} \sin^2[\frac{\Delta k Z}{2}] \\ l_c = \frac{\pi c}{\omega_3 n_3^+ - \omega_1 n_1^- - \omega_2 n_2^+} \end{cases} \quad (7.61)$$

7.3 PHASE-MATCHING

7.3.1 Generated Intensity in the Undepleted Pump Approximation

A sum-frequency generation is also considered here for the following calculations in the undepleted pump approximation (UPA) and wherein phase-matching is assumed. UPA is easily accessible, by using short interaction length, weak non-linearity, or moderate incident intensities. Because $\Delta k = 0$, it is useful to multiply Eq. (7.58) by $\frac{Z^2/4}{Z^2/4}$, which gives:

$$I_3(Z) = 2\pi^2 \sqrt{\frac{\mu_0}{\varepsilon_0}} \frac{1}{\lambda_3^2 (\cos\rho_3)^4} \frac{[\chi_{eff}^{(2)}]^2}{n_1 n_2 n_3} I_1(0) I_2(0) Z^2 \,\mathrm{sinc}^2[\frac{\Delta k Z}{2}] \quad (7.62)$$

The corresponding curve is plotted hereafter in Figure 7.6.

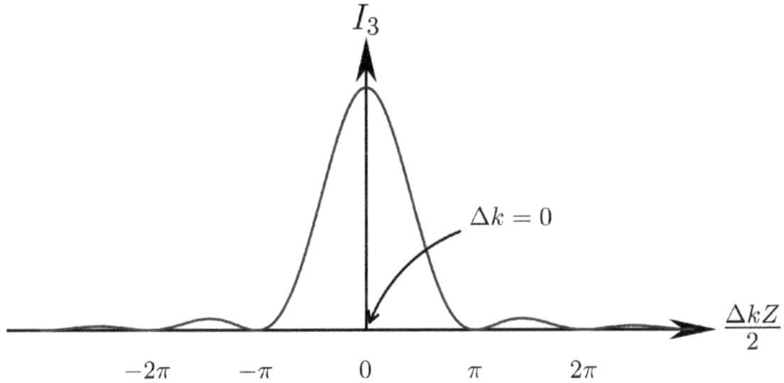

FIGURE 7.6 Evolution of the normalized intensity of the generated wave along the direction of propagation in the case of phase-matching at a given position Z in the nonlinear medium.

It appears that the optimum of generation well corresponds to $\Delta k = 0$, where the interference function is maximal, *i.e.* $\sin c^2[0] = 1$. In this case according to Eq. (7.62), I_3 depends on Z^2, such a dependence being relevant as long as the UPA remains valid. Then the interference is constructive over the entire interaction length leading to a continuous and unidirectional energy transfer from the incident waves to the generated wave. In that case $l_c \rightarrow \infty$ according to Eq. (7.59). These conclusions are obviously the same for difference-frequency generation because the interference function is the same as that of sum-frequency generation.

7.3.2 Birefringence Phase-Matching

Therefore, an important point now is to be able to determine the phase-matching conditions.

In the hypothetical situation of a non-dispersive medium, *i.e.* $n_3 = n_1 = n_2 \equiv n$, phase-matching would be always fulfilled because Eq. (7.39) reduces to the simple photon energy conversion as seen here:

$$\Delta k = \frac{1}{c}[\omega_3 n_3 - \omega_1 n_1 - \omega_2 n_2] = \frac{1}{c} n[\omega_3 - \omega_1 - \omega_2] = 0 \tag{7.63}$$

Such a situation is fantastic but it does not exist in reality, even in the middle of a transparency range wherein the dispersion is weak but not nil. The case of a non-centrosymmetric crystal of the cubic crystal system is now taken as an example. It belongs to the isotropic optical class, with a single refractive index as seen in Chapter 5. A situation of second-harmonic generation is also considered for the following. Then the phase-matching relation writes:

$$\Delta k = \frac{1}{c}[2\omega n_{2\omega} - \omega n_{\omega} - \omega n_{\omega}] = \frac{2\omega}{c}[n_{2\omega} - n_{\omega}] = 0 \tag{7.64}$$

Taking the same typical wavelength dispersion as that depicted in Figure 5.32, it clearly appears that it is impossible to fulfill Eq. (7.64), and that for any couple $(\lambda_\omega, \lambda_{2\omega})$. This situation is shown in Figure 7.7, with the example of the non-centrosymmetric cubic crystal GaAs (Skauli 2003) [3].

Finally, by considering an anisotropic crystal, being uniaxial or biaxial, it can be possible to fulfill Eq. (7.64) using birefringence as shown in Figure 7.8 with the example of the biaxial crystal $KTiOPO_4$ (KTP).

Figure 7.8 shows that by taking two different Eigenmodes at ω and 2ω, it can exist a direction for which the phase-matching is possible for a given couple $(\lambda_\omega, \lambda_{2\omega})$. It also appears that the configuration of modes is necessary (+) for the wave at ω and (−) for the wave at 2ω. Such a configuration can also be called “birefringence phase-matching” (BPM).

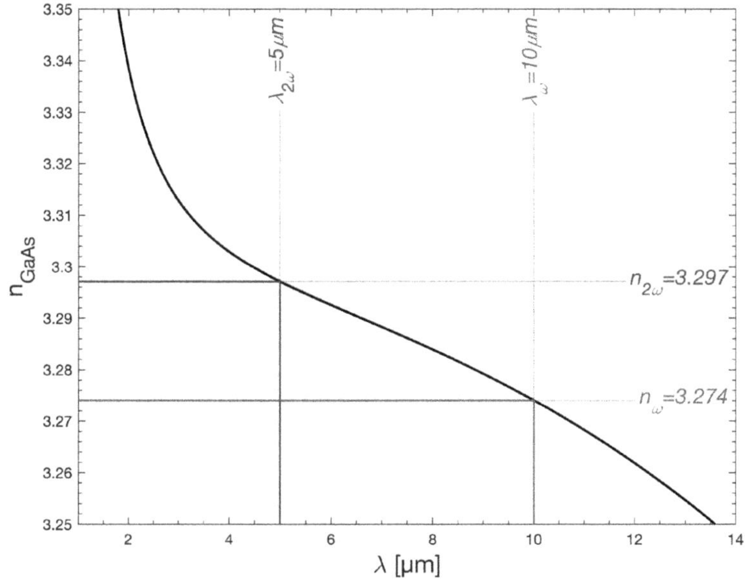

FIGURE 7.7 Illustration of the impossibility to reach phase-matching, *i.e.* $n_{2\omega} - n_{\omega} \neq 0$, at any wavelength in GaAs that is an isotropic crystal. An example is taken at $(\lambda_{\omega} = 10\mu\, m, \lambda_{2\omega} = 5\mu m)$.

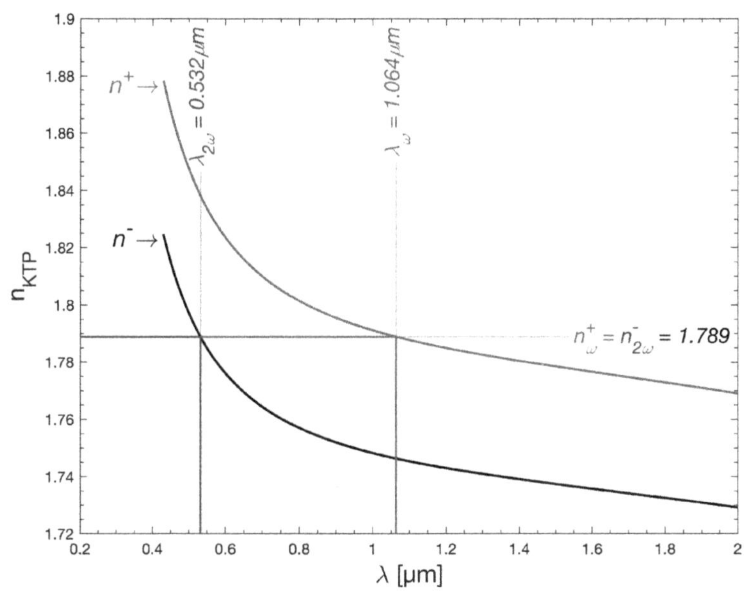

FIGURE 7.8 Illustration of the possibility to reach phase-matching, *i.e.* $n^{-}_{2\omega} - n^{+}_{\omega} = 0$, in the *yz*-plane at $\theta = 68°$ of the birefringent biaxial crystal KTP for Type I SHG $(\lambda_{\omega} = 1.064\ \mu m \rightarrow \lambda_{2\omega} = 0.532\ \mu m)$. n_{KTP} is the refractive index of KTP, n^{-} or n^{+}, and λ is the wavelength.

7.3.3 Configurations of Polarization

The graphical determination of the mode configuration for a birefringence phase-matched second-harmonic generation (SHG) performed previously can also be done by calculation (Boulanger 1991) [4]. Actually, the approach developed hereafter allows us to define all the possible configurations of BPM for any three-wave sum-frequency generation (SFG) or difference-frequency generation (DFG), *i.e.* SFG ($\omega_1 + \omega_2 \rightarrow \omega_3$), the degenerate case $\omega_1 = \omega_2$ corresponding to SHG, DFG ($\omega_3 - \omega_2 \rightarrow \omega_1$) and DFG ($\omega_3 - \omega_1 \rightarrow \omega_2$) with $\omega_1 \leq \omega_2 < \omega_3$. It is then necessary to consider both energy and momentum conservations reminded here:

$$\begin{cases} \omega_3 = \omega_1 + \omega_2 \\ \omega_3 n_3 = \omega_1 n_1 + \omega_2 n_2 \end{cases} \tag{7.65}$$

By combining these two equations, it immediately comes:

$$\frac{\omega_1}{\omega_2} = \frac{n_3 - n_2}{n_1 - n_3} > 0 \tag{7.66}$$

Equation (7.66) is compatible with the two following relations of order:

$$\begin{cases} n_1 < n_3 < n_2 \\ n_2 < n_3 < n_1 \end{cases} \tag{7.67}$$

Furthermore, in the case of normal dispersion and if $\omega_1 < \omega_2 < \omega_3$, the Eigen refractive indices verify the following relations of order:

$$\begin{cases} n_1^+ < n_2^+ < n_3^+ \\ n_1^- < n_2^- < n_3^- \end{cases} \tag{7.68}$$

The determination of the possible combinations of refractive indices can now be performed by testing the 2^3 possibilities regarding the two sets of relations of order given by Eq. (7.67) and Eq. (7.68). It is easy to establish that only three combinations fit: (n_3^-, n_1^+, n_2^+), (n_3^-, n_1^+, n_2^-), and (n_3^-, n_1^-, n_2^+) given in Table 7.1.

TABLE 7.1
Birefringence Phase-Matching Conditions for Three-Wave Sum-Frequency Generation (SFG) and Difference-Frequency Generation (DFG) of Types I, II, and III for the Biaxial and Uniaxial Optical Classes

Refractive indices	Polarizations	SFG ($\omega_3 = \omega_1 - \omega_2$)	DFG ($\omega_1 = \omega_3 - \omega_2$)	DFG ($\omega_2 = \omega_3 - \omega_1$)
(n_3^-, n_1^+, n_2^+)	$(\vec{e}_3^{\,-}, \vec{e}_1^{\,+}, \vec{e}_2^{\,+})$	I	II	III
(n_3^-, n_1^-, n_2^+)	$(\vec{e}_3^{\,-}, \vec{e}_1^{\,-}, \vec{e}_2^{\,+})$	II	III	I
(n_3^-, n_1^+, n_2^-)	$(\vec{e}_3^{\,-}, \vec{e}_1^{\,+}, \vec{e}_2^{\,-})$	III	I	II

The designation of the type of phase-matching, *i.e.* I, II, or III, is then defined according to the two refractive indices at the two circular frequencies that are added or subtracted: type I corresponds to interactions for which the two refractive indices belong to the same Eigenmode; the two Eigenmodes are different for types II and III, as shown in Table 7.1.

Note that each combination corresponds to one SFG and two DFG, and types II and III are equivalent in the case of SHG since $\omega_1 = \omega_2$.

7.3.4 Angular Distribution

The configurations of polarization under BPM condition having been determined, it is now possible to calculate the corresponding direction. We mainly consider here collinear phase-matching, $\vec{u}(\theta,\varphi)$ being the unit vector of the wave vectors of the three interacting waves in which (θ,φ) are the angles of spherical coordinates in the dielectric frame (x,y,z) as defined by Eq. (5.47) and Figure 5.1. Then the goal is to calculate these phase-matching coordinates, *i.e.* $(\theta_{PM},\varphi_{PM})$, verifying the following equation:

$$\omega_3 n_3(\theta_{PM},\varphi_{PM}) = \omega_1 n_1(\theta_{PM},\varphi_{PM}) + \omega_2 n_2(\theta_{PM},\varphi_{PM}) \tag{7.69}$$

where the combinations of refractive indices Eigenmodes $(n_3^{\pm}, n_1^{\pm}, n_2^{\pm})$ are given in Table 7.1.

A good tool for that purpose is undoubtedly the index surface, described in Part 5.2, for both uniaxial and biaxial crystals.

7.3.4.1 Uniaxial Crystals

The resolution of Eq. (7.69) can be performed analytically in a simple manner only if the solution corresponds geometrically to the intersection of a circle and an ellipse as shown hereafter with the example of type I SHG in a positive uniaxial crystal, *i.e.* $n_o < n_e$. According to Figure 7.8 and Table 7.1, the corresponding phase-matching relation writes:

$$n_{2\omega}^{-}(\theta_{PM},\varphi_{PM}) = n_{\omega}^{+}(\theta_{PM},\varphi_{PM}) \tag{7.70}$$

Then according to Eq. (5.61), Eq. (7.70) writes:

$$n_o(2\omega) = \frac{1}{\sqrt{n_o^{-2}(\omega)\cos^2(\theta_{PM}) + n_e^{-2}(\omega)\sin^2(\theta_{PM})}} \tag{7.71}$$

where n_o and n_e are the ordinary and extraordinary principal refractive indices, respectively.

It is important to notice that the angle φ_{PM} does not appear in Eq. (7.71). It is due to the fact that the ordinary layer of the index surface is a sphere and the extraordinary one is an ellipsoid of revolution around the optical axis, *i.e.* the z-axis, as shown in Figure 5.3. Then Eq. (7.71) is valid in any plane containing the z-axis, and so it is valid for any φ_{PM} angle.

Equation (7.70) corresponds to the intersection of a circle (left side) and an ellipse (right side), as also shown in Figure 7.9.

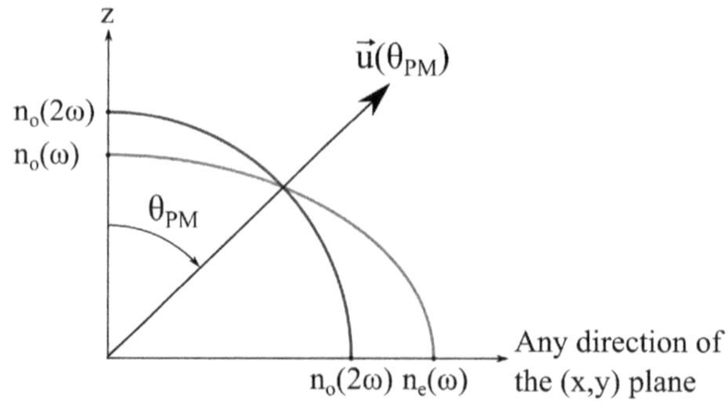

FIGURE 7.9 Geometrical determination of the BPM angle θ_{PM} of type I SHG in a positive uniaxial crystal. $\vec{u}(\theta_{PM})$ is the unit vector of the corresponding phase-matching direction. The circle and ellipse are the section of the index surface in any plane containing the z-axis. (x, y, z) is the dielectric frame.

Figure 7.9 well shows that there is an intersection between the layers at ω and 2ω only if the following relation of order is fulfilled:

$$(n_o(\omega) <) n_o(2\omega) \leq n_e(\omega) \tag{7.72}$$

Note that $n_o(\omega) < n_o(2\omega)$ is always fulfilled because of normal dispersion, and $n_o(2\omega) = n_e(\omega)$ would correspond to a phase-matching direction in the (x,y) plane, *i.e.* at $\theta_{PM} = \pi/2$.

By raising to the inverse square power each side of Eq. (7.71), it comes:

$$\theta_{PM} = \arcsin\left[\sqrt{\frac{n_o^{-2}(2\omega) - n_o^{-2}(\omega)}{n_e^{-2}(\omega) - n_o^{-2}(\omega)}}\right] \tag{7.73}$$

Equation (7.73) is valid for any φ_{PM} angle, which means that there is a continuum of phase-matching directions forming a cone of revolution around the z-axis with an aperture angle equal to θ_{PM}, as shown in Figure 7.10.

It can be convenient to plot the phase-matching cone of Figure 7.10 in a Wulff abacus, as shown by Figure 7.11.

The geometrical approach detailed earlier in the case of type I SHG in a positive uniaxial crystal can be generalized to any SFG or DFG processes of types I, II, or III in positive or negative uniaxial crystals. The corresponding angular solutions of Eq. (7.69) are also cones in both cases, the same ones as those described in Figure 7.10, or equivalently in Figure 7.11.

In any case, the phase-matching angle θ_{PM} varies as a function of wavelength due to the wavelength dispersion of the refractive indices. Next is given the real example of the negative uniaxial crystal $CdSiP_2$ shown in Figure 7.12. (Kemlin 2011) [5].

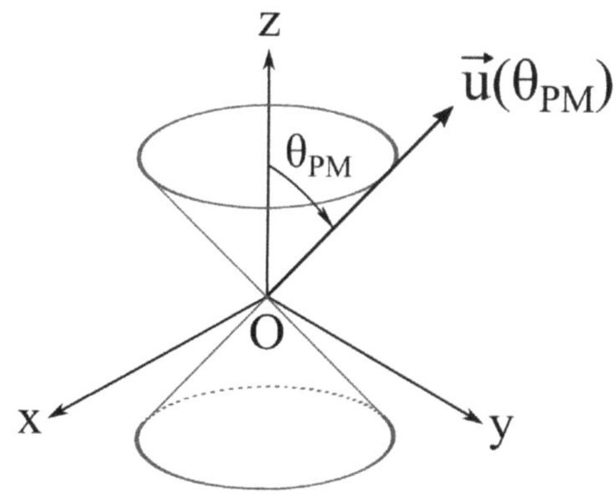

FIGURE 7.10 Birefringence phase-matching cone of a uniaxial crystal. The aperture angle is θ_{PM} and $\vec{u}(\theta_{PM})$ stands for the unit vector of any phase-matching direction. (x, y, z) is the dielectric frame.

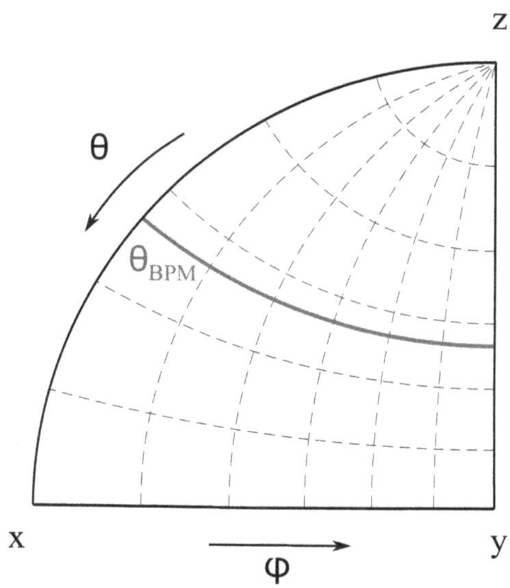

FIGURE 7.11 Birefringence phase-matching cone of a uniaxial crystal plotted in a Wulff abacus. The phase-matching angle is θ_{PM}, and (x, y, z) is the dielectric frame.

It is interesting to notice that the angular distribution depicted in Figure 7.11 belongs to the infinite point group $\left(\frac{\infty}{m}m\right)$ introduced in Chapter 2. It is the same symmetry grou[illegible] of the index surface of the uniaxial optical class, which is in fact that [illegible]linary layer, as shown in Chapter 5. This feature can be easily under[illegible] group theory. Actually, the symmetry group G_{BPM} of the birefrin[illegible]ching space of solutions $\vec{u}(\theta_{PM}, \varphi_{PM})$ is given by the intersection of the symmetry groups G_1, G_2, and G_3 of the three surfaces that compose Eq. (7.69), *i.e.* $\omega_1 n_1(\theta_{PM},\varphi_{PM})$, $\omega_2 n_2(\theta_{PM},\varphi_{PM})$, and $\omega_3 n_3(\theta_{PM},\varphi_{PM})$ respectively. Therefore, the symmetry group G_{BPM} writes:

$$G_{BPM} = G_3 \cap G_1 \cap G_2 \tag{7.74}$$

According to Table 7.1, (n_3, n_1, n_2) can be equal to (n_3^-, n_1^+, n_2^+), (n_3^-, n_1^+, n_2^-), and (n_3^-, n_1^-, n_2^+), with $G_i^- = \left(\frac{\infty}{m}\frac{\infty}{m}\right)$, *i.e.* the group of a static sphere, and $G_i^+ = \left(\frac{\infty}{m}m\right)$, *i.e.* the group of a static cylinder, in the case of a positive uniaxial crystal; it is the reverse in the case of a negative uniaxial crystal, *i.e.* $G_i^- = \left(\frac{\infty}{m}m\right)$ and $G_i^+ = \left(\frac{\infty}{m}\frac{\infty}{m}\right)$. Then for each of the three sets of refractive indices, Eq. (7.74) reduces to the intersection of $\left(\frac{\infty}{m}\frac{\infty}{m}\right)$ and $\left(\frac{\infty}{m}m\right)$, which is equal to $\left(\frac{\infty}{m}m\right)$ since $\left(\frac{\infty}{m}m\right) \subset \left(\frac{\infty}{m}\frac{\infty}{m}\right)$.

As demonstrated earlier for type I SHG, phase-matching solutions of SFG and DFG of types I, II, and III exist only if particular relations of order are fulfilled between the principal ordinary and extraordinary refractive indices at the circular frequencies of the different interacting waves (Fève 1993, Boulanger 2013) [6,7]. These relations of order are given by a generalization of Eq. (7.72) as shown in Table 7.2.

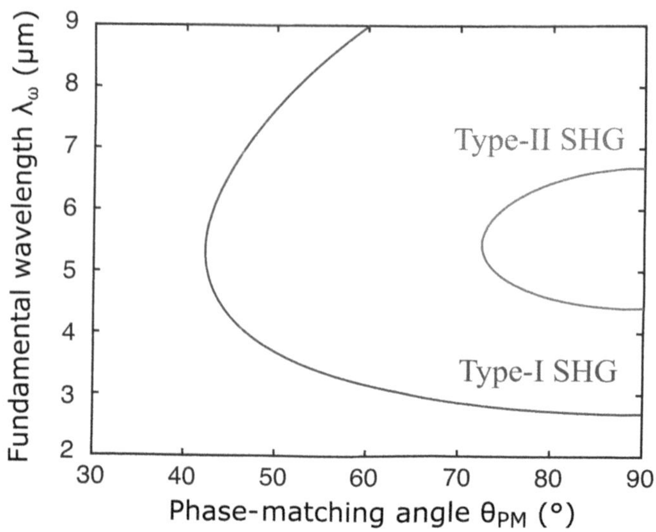

FIGURE 7.12 Variation of the birefringence phase-matching angle θ_{PM} with the fundame tal wavelength λ_ω calculated for types I and II SHG in the negative uniaxial crystal Cc

TABLE 7.2
Relations of Order between the Principal Ordinary and Extraordinary Refractive Indices at the Different Circular Frequencies Allowing Birefringence Phase-Matching

Optical sign	Polarizations	Relations of order allowing phase-matching
>0	$(\vec{e}_3^{\,-},\vec{e}_1^{\,+},\vec{e}_2^{\,+})$	$\omega_3 n_o(\omega_3) \le \omega_1 n_e(\omega_1) + \omega_2 n_e(\omega_2)$
	$(\vec{e}_3^{\,-},\vec{e}_1^{\,-},\vec{e}_2^{\,+})$	$\omega_3 n_o(\omega_3) \le \omega_1 n_o(\omega_1) + \omega_2 n_e(\omega_2)$
	$(\vec{e}_3^{\,-},\vec{e}_1^{\,+},\vec{e}_2^{\,-})$	$\omega_3 n_o(\omega_3) \le \omega_1 n_e(\omega_1) + \omega_2 n_o(\omega_2)$
<0	$(\vec{e}_3^{\,-},\vec{e}_1^{\,+},\vec{e}_2^{\,+})$	$\omega_3 n_e(\omega_3) \le \omega_1 n_o(\omega_1) + \omega_2 n_o(\omega_2)$
	$(\vec{e}_3^{\,-},\vec{e}_1^{\,-},\vec{e}_2^{\,+})$	$\omega_3 n_e(\omega_3) \le \omega_1 n_e(\omega_1) + \omega_2 n_o(\omega_2)$
	$(\vec{e}_3^{\,-},\vec{e}_1^{\,+},\vec{e}_2^{\,-})$	$\omega_3 n_e(\omega_3) \le \omega_1 n_o(\omega_1) + \omega_2 n_e(\omega_2)$

7.3.4.2 Biaxial Crystals

The case of the principal planes of biaxial crystals is in straight line with what has been previously seen for uniaxial crystals since ordinary and extraordinary waves can also be defined. The calculation of the phase-matching angles is then performed from the resolution of Eq. (7.69) with the refractive indices given by Eq. (5.63) to Eq. (5.72), the relevant equations to consider depending on the angular location in the optical frame, as detailed hereafter. Indeed, it can be directly seen that there are four possible locations according to the principal axes and optical axes as shown by the following Wulff abacus.

As previously, the example of type I SHG is taken, here in the case of a positive biaxial crystal, *i.e.* $n_x < n_y < n_z$. According to the corresponding index surface described in Chapter 5 and using the same graphical method as that illustrated Figure 7.9, there is an intersection between the layers at ω and 2ω in the different zones A, B, C, and D, as shown in Figure 7.13, only if the following relations of order are fulfilled:

- Zone A

$$n_x(2\omega) \le n_y(\omega) \tag{7.75}$$

- Zone B

$$n_y(2\omega) \le n_z(\omega) \tag{7.76}$$

- Zone C

$$\begin{cases} n_x(2\omega) < n_z(\omega) \le n_y(2\omega) \\ \text{or} \\ n_x(2\omega) \le n_z(\omega) < n_y(2\omega) \end{cases} \tag{7.77}$$

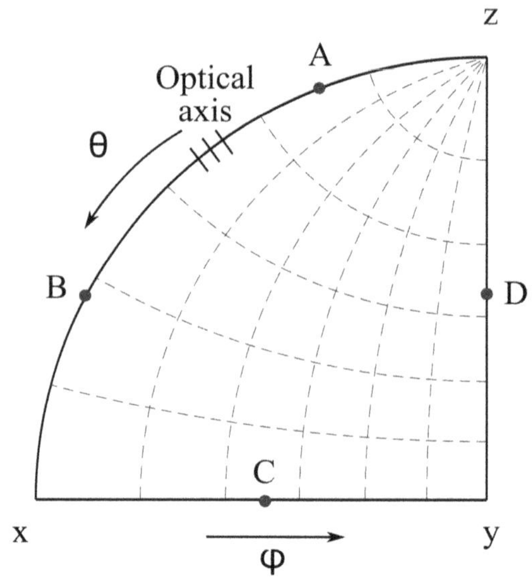

FIGURE 7.13 Possible birefringence phase-matching locations in the principal planes of a biaxial crystal. Dot A stands for any location between the z-axis and the optical axes of the three interacting waves, B between the optical axes and the x-axis, C between the x- and y-axis, D between the y- and z-axis, where (x, y, z) is the dielectric frame. The three lines crossing the xz-plane stand for the optical axes at the wavelengths of the three interacting waves.

- Zone D

$$\begin{cases} n_y(\omega) < n_x(2\omega) \leq n_z(\omega) \\ \text{or} \\ n_y(\omega) \leq n_x(2\omega) < n_z(\omega) \end{cases} \tag{7.78}$$

It is interesting at this step to compare the different zones. Regarding the equalities, $n_y(\omega) = n_x(2\omega)$ in Eq. (7.75) (zone A) and in Eq. (7.78) (zone D) corresponds to a phase-matching direction along the z-axis; $n_y(2\omega) = n_z(\omega)$ in Eq. (7.76) (zone B) and in Eq. (7.77) (zone C) corresponds to a phase-matching direction along the x-axis; $n_z(2\omega) = n_x(\omega)$ in Eq. (7.77) (zone C) and in Eq. (7.78) (zone D) corresponds to a phase-matching direction along the y-axis. As far as strict inequalities are concerned, *i.e.* out of the principal axes, it appears that there are incompatibilities between Eq. (7.76) and Eq. (7.77), which means that it is impossible to have a phase-matching cone joining zone B and zone C. It is the same situation between Eq. (7.75) and Eq. (7.78), so that a phase-matching cone joining zone A and zone D is not possible. On the other hand, there is compatibility between zones A and B, A and C, B and D, and C and D, which gives four possible topologies for the phase-matching cones as illustrated in Figure 7.14. The calculation of the phase-matching angles $(\theta_{PM}, \varphi_{PM})$ out of the principal planes is performed from the numerical resolution

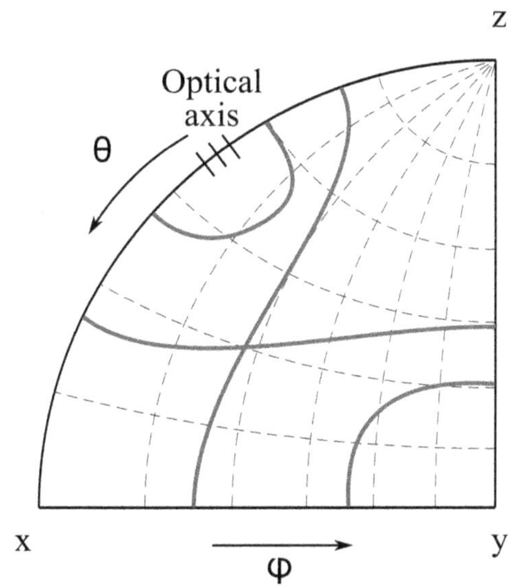

FIGURE 7.14 Possible birefringence phase-matching cones of a biaxial crystal plotted in a Wulff abacus. (x, y, z) is the dielectric frame.

of Eq. (7.69) and the refractive indices given by Eq. (5.46) for each of the allowed configurations of polarization under BPM condition.

These results can be generalized for any three-wave processes, *i.e.* types I, II and III SFG and DFG, in positive as well as negative biaxial crystals leading to the same four possible topologies, *i.e.* those given by Figure 7.14 (Fève 1993, Boulanger 2013) [6,7]. The corresponding inequalities can be found in these later references.

The comparison between Figure 7.11 and Figure 7.14 shows that the phase-matching cones can wrap around the z-axis, as for uniaxial crystals, but also around the y-axis and the optical axes. Another difference is that the perimeters of the phase-matching cones of biaxial crystal are not circular, contrary to the case of uniaxial crystals: this is associated with the loss of rotation invariance along the z-axis of the dielectric frame for the BPM solutions. Figure 7.15 well shows this feature with the example of types I and II SHG in the positive biaxial crystal $KTiOPO_4$ (KTP) (Boulanger 1994) [8].

In Figure 7.15, the phase-matching angular distribution of type I SHG corresponds to a cone joining zones B and D, while it is a cone joining zones C and D for type II SHG. The non-circularity of each cone well appears: $\theta_{PM} = 49°$ in zone B and $\theta_{PM} = 39°$ in zone D for type I SHG; $\varphi_{PM} = 23°$ in zone C and $\theta_{PM} = 72°$ in zone D for type II SHG.

As performed for uniaxial crystals, it is interesting to look at the symmetry group of the phase-matching angular distribution of biaxial crystals. If all the octans are considered, Figure 7.15 shows that all the cones have the same orientation symmetry elements, *i.e.* the three principal planes as mirrors m. It means that the corresponding point group is $G_{BPM} = (mmm)$, which is an orthorhombic crystal class.

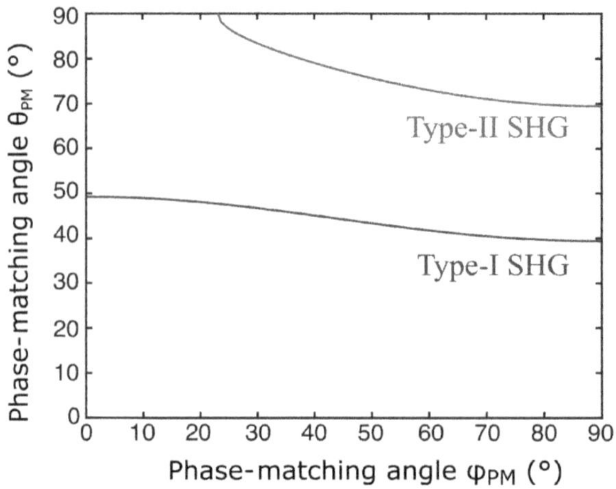

FIGURE 7.15 Birefringence phase-matching angles $(\theta_{PM},\varphi_{PM})$ calculated for types I and II SHG in the positive biaxial crystal $KTiOPO_4$ (KTP) pumped at the fundamental wavelength $\lambda_\omega = 1.064\ \mu m$. $\theta_{PM} = 90°$ corresponds to the xy-plane and $\varphi_{PM} = 90°$ to the yz-plane.

This result can be also found by considering the symmetry of the two layers of the index surface described in Chapter 5 and using Eq. (7.74): here, for each of the three interacting waves, $G_i^- = G_i^+ = (mmm)$ so that $G_1 = G_2 = G_3 = (mmm)$ leading to $G_{BPM} = G_3 \cap G_1 \cap G_2 = (mmm)$.

7.3.5 Collinear versus Non-Collinear Phase-Matching

The previous calculations concern the collinear phase-matching configuration, which is mainly used. However, the non-collinear phase-matching configuration can be interesting, in the case of multi-pump optical parametric amplification for example (Trophème 2012) [9]. In this case, the phase-matching relation has to take into account the angles between the wave vectors of the interacting waves as shown in Figure 7.16.

Then the vector phase-matching relation writes:

$$\vec{k}_3 - \vec{k}_1 - \vec{k}_2 = \vec{0} \tag{7.79}$$

According to Figure 7.16, the vector Eq. (7.79) leads to the two following scalar relations (Trophème 2013) [9]:

$$\begin{cases} \omega_3 n_3(\theta_3,\varphi_3)\cos\alpha_{31} = \omega_1 n_1(\theta_1,\varphi_1) + \omega_2 n_2(\theta_2,\varphi_2)\cos\alpha_{21} \\ \omega_3 n_3(\theta_3,\varphi_3)\sin\alpha_{31} = \omega_2 n_2(\theta_2,\varphi_2)\sin\alpha_{21} \end{cases} \tag{7.80}$$

with

$$\begin{cases} \cos\alpha_{31} = \sin\theta_1\cos\varphi_1\sin\theta_3\cos\varphi_3 + \sin\theta_1\sin\varphi_1\sin\theta_3\sin\varphi_3 + \cos\theta_1\cos\theta_3 \\ \cos\alpha_{21} = \sin\theta_1\cos\varphi_1\sin\theta_2\cos\varphi_2 + \sin\theta_1\sin\varphi_1\sin\theta_2\sin\varphi_2 + \cos\theta_1\cos\theta_2 \end{cases} \tag{7.81}$$

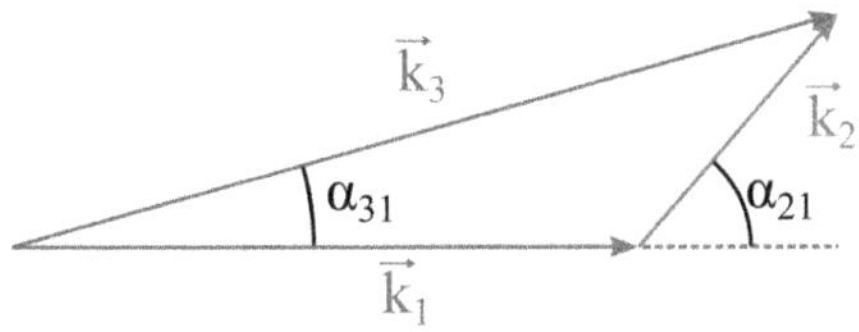

FIGURE 7.16 Angles between wave vectors to account for calculating the phase-matching relation of a three-wave non-collinear configuration.

Equation (7.80) and Eq. (7.81) relative to non-collinear phase-matching replace Eq. (7.69) relative to collinear phase-matching. A numerical calculation allows us to determine all the sets of possible phase-matching angles $\left[(\theta_1,\varphi_1),(\theta_2,\varphi_2),(\theta_3,\varphi_3),(\alpha_{21},\alpha_{31})\right]$ knowing the refractive indices $\left[n_1,n_2,n_3\right]$ at the circular frequencies $\left[\omega_1,\omega_2,\omega_3\right]$ in the directions $\left[(\theta_1,\varphi_1),(\theta_2,\varphi_2),(\theta_3,\varphi_3)\right]$ thanks to the index surface of the crystal that is considered.

Using the same way as for the collinear configuration, it is possible to show that non-collinear phase-matching is possible for only three combinations of polarization, *i.e.* the same as those of collinear phase-matching given in Table 7.1.

7.4 QUASI-PHASE-MATCHING

7.4.1 Reversal of the Sign of $\chi^{(2)}$

We saw previously in Part 7.2 according to Eq. (7.58) and Figure 7.5 that there is a periodical building and destruction of the generated wave during the propagation when the propagation occurs in a direction for which $\Delta k \neq 0$, namely corresponding to the "out of phase-matching" case. Such a direction of propagation is not interesting except if a phase reset of π between the induced polarization and the radiated field is introduced at each junction between two adjoining coherence lengths, as shown in Figure 7.17. In that case, the interference remains globally and locally constructive over the whole propagation length, as illustrated by the following figure. This scheme is called "quasi-phase-matching" (QPM) (Armstrong 1962) [1].

The behavior of Figure 7.17 can be described as a function by part, each part being the positive derivative path of $\sin^2[\frac{\Delta kZ}{2}]$ starting from the proper non-zero value of the generated intensity I_3 at the beginning of each l_c.

The best way to ensure QPM is to fabricate an artificially structured medium in which the sign of the nonlinear coefficient $\chi^{(2)}$ in the considered direction of propagation changes from a coherence length to another one as suggested by Figure 7.18.

The interest of such a configuration can be well understood by considering the second-order induced polarization at ω_3 over two consecutive parametric coherence lengths l_c and $2l_c$ for example. Then from Eq. (3.2) in the undepleted pump approximation, *i.e.* $E_1(Z)=E_1(0)$ and $E_2(Z)=E_2(0)$, it comes for the modulus of the induced polarization vector at ω_3:

- from 0 to l_c (domain $+\chi^{(2)}$)

$$P_3^{(2)}(Z)=\varepsilon_0[+\chi^{(2)}(\omega_3):\vec{e_1}\otimes\vec{e_2}]E_1(0)E_2(0)\exp j[(k_1+k_2)Z] \tag{7.82}$$

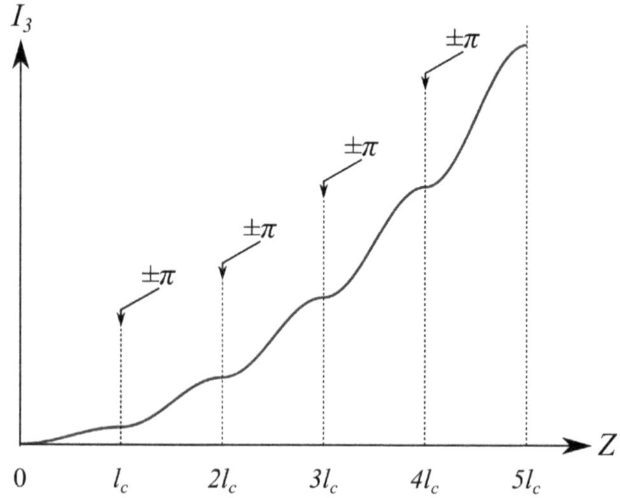

FIGURE 7.17 QPM principle illustrated by the schematic evolution of the generated intensity I_3 as a function of the interaction length Z in a direction for which $\Delta k \neq 0$. The quantity l_c is the parametric coherence length.

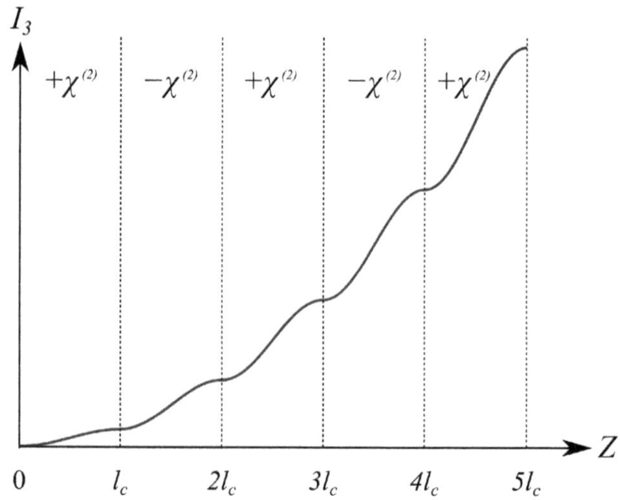

FIGURE 7.18 QPM performed in an artificial medium in which the sign of the nonlinear coefficient $\chi^{(2)}$ is reversed over a period equal to the parametric coherence length l_c.

- from l_c to $2l_c$ (domain $-\chi^{(2)}$)

$$P_3^{(2)}(Z) = \varepsilon_0[-\chi^{(2)}(\omega_3) : \overrightarrow{e_1} \otimes \overrightarrow{e_2}]E_1(0)E_2(0)\exp j[(k_1 + k_2)Z] \tag{7.83}$$

which is equivalent to

$$P_3^{(2)} = \varepsilon_0[+\chi^{(2)}(\omega_3) : \overrightarrow{e_1} \otimes \overrightarrow{e_2}]E_1(0)E_2(0)\exp j[(k_1 + k_2)Z \pm \pi] \tag{7.84}$$

Since

$$\exp j[\pm\pi] = -1 \tag{7.85}$$

The sign of the macroscopic second-order electric susceptibility tensor $\chi^{(2)}$ is directly linked to the crystal orientation, which can be understood according to Eq. (3.58), Eq. (3.59), and Eq. (3.60) in Chapter 3. These equations show that a $\chi^{(2)}$ medium devoted to QPM could be formed by a sequence of materials alternately rotated by an angle π: actually, any of the twenty-seven independent coefficients $\chi_{ijk}^{(2)}$ depend on the product $\cos\theta_i^b \cos\theta_j^b \cos\theta_k^b$, where $\theta_{i,j,k}^b$ is the angle between the axis of the bth bond of the crystal and the dielectric frame axis i (or j or k) where i, j, and k stand for x, y, or z, and $\cos(\theta_{i,j,k}^b \pm \pi) = -\cos(\theta_{i,j,k}^b)$. It is then easy to identify the coefficients for which the sign is going to be reversed thanks to a rotation of π around a given axis i, j, or k. These coefficients are given in the following figures for different configurations of sign inversions. The matrices that are shown correspond to that of the triclinic crystal class 1 for which all the coefficients are not nil, giving thus the full picture of the second-order case.

The configuration of Figure 7.19, *i.e.* rotation of π around the x-axis, is the one used in the case of the famous ferroelectric periodically poled (PP) crystals PPLiNbO$_3$ (PPLN) (Fejer 1992) [10] and PPKTiOPO$_4$ (PPKTP) (Karlson 1997) [11]. The point group of PPLN is *3m*. Then, according to the $\chi^{(2)}$ matrix shown in Figure 3.25 and Figure 7.19, it appears that all the five independent coefficients of PPLN change in sign. It is exactly the same for PPKTP that crystallizes in the point group *mm2* and for which all the seven independent coefficients given in Figure 3.23 can change in sign according to Figure 7.19.

Another famous crystal used for QPM is orientation-patterned GaAs (OP-GaAs) (Schunemann 2008) [12]. Its point group is $\bar{4}3m$ and the corresponding $\chi^{(2)}$ matrix is given in Figure 3.39. In that case there are six equal coefficients, *i.e.* $\chi_{xyz}^{(2)} = \chi_{xzy}^{(2)} = \chi_{yxz}^{(2)} = \chi_{yzx}^{(2)} = \chi_{zxy}^{(2)} = \chi_{zyx}^{(2)}$. These coefficients do not appear in Figure 7.19, Figure 7.20, and Figure 7.21, which means that a rotation of π around an axis of the dielectric frame cannot change the sign of the nonlinear coefficients of OP-GaAs. However, it is possible using two other configurations: a rotation of π around an axis oriented at $\pi/4$ from the x- and y-axis in the xy-plane, or a rotation of $\pi/2$ around the z-axis as shown in Figure 7.22 and Figure 7.23 in which all the possible coefficients are listed.

Note that other configurations could lead to a sign inversion of some coefficients.

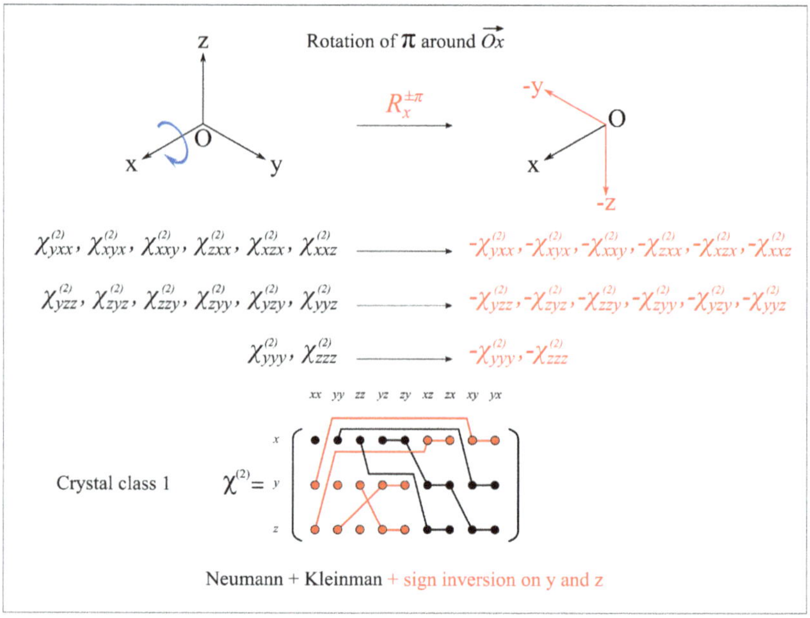

FIGURE 7.19 Second-order electric susceptibility tensor coefficients (red dots) for which the sign is reversed due to a rotation of π around the x-axis of the dielectric frame. The red dots in the tensor stand for the coefficients that can change in sign.

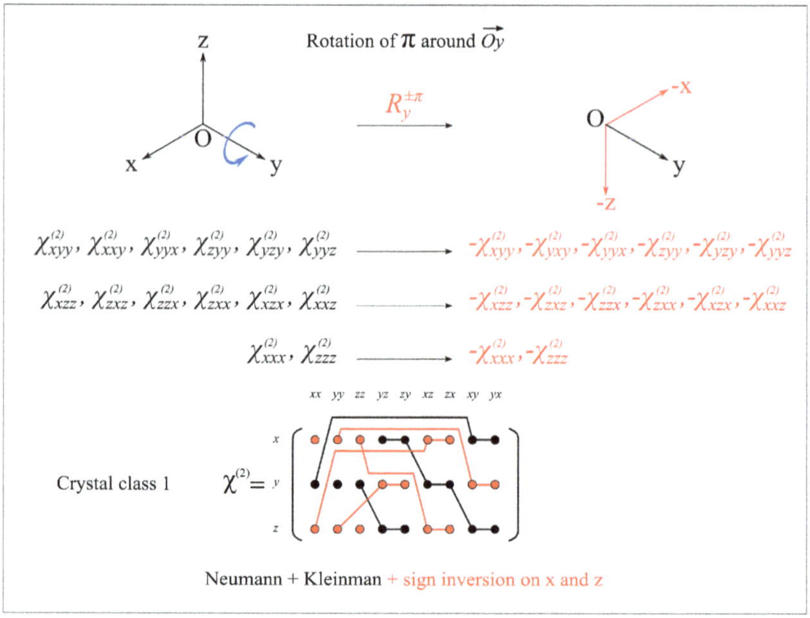

FIGURE 7.20 Second-order electric susceptibility tensor coefficients (red dots) for which the sign is reversed due to a rotation of π around the y-axis of the dielectric frame. The red dots in the tensor stand for the coefficients that can change in sign.

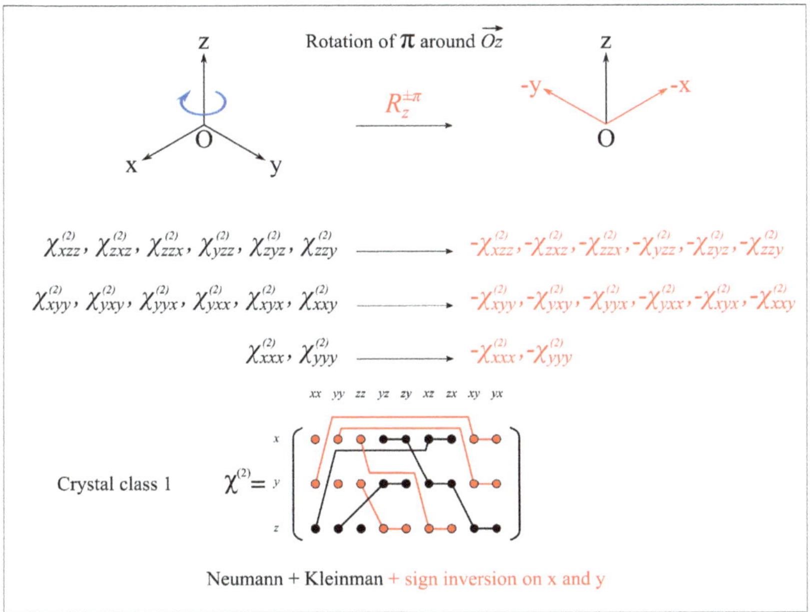

FIGURE 7.21 Second-order electric susceptibility tensor coefficients (red dots) for which the sign is reversed due to a rotation of π around the z-axis of the dielectric frame. The red dots in the tensor stand for the coefficients that can change in sign.

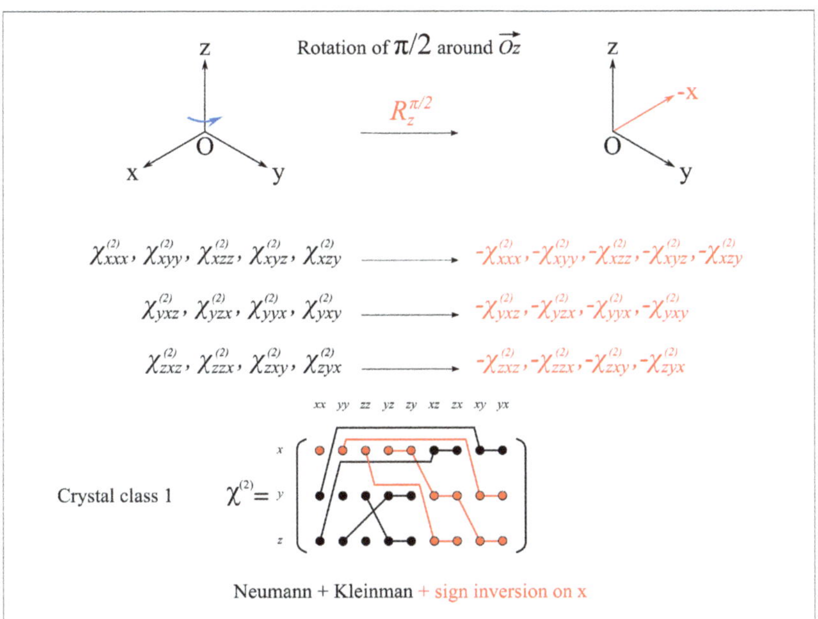

FIGURE 7.22 Second-order electric susceptibility tensor coefficients (red dots) for which the sign is reversed due to a rotation of $\pi/2$ around the z-axis of the dielectric frame. The red dots in the tensor stand for the coefficients that can change in sign.

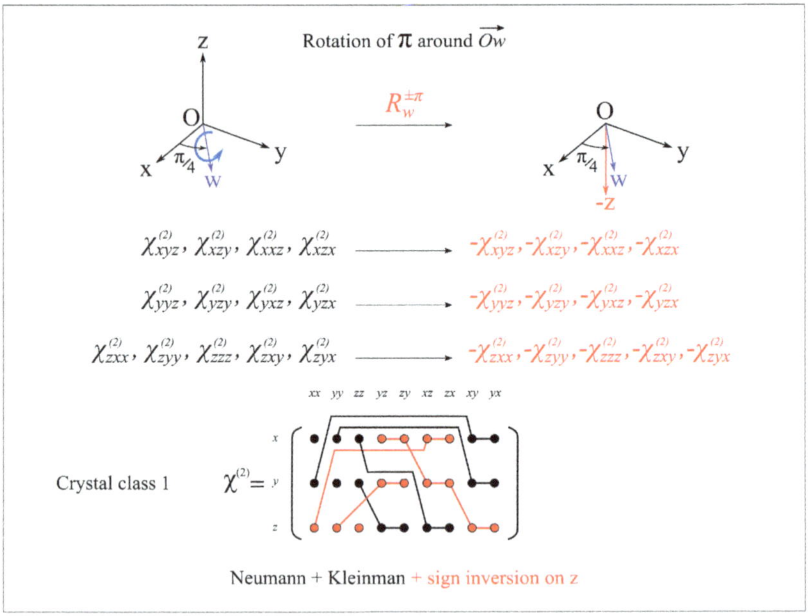

FIGURE 7.23 Second-order electric susceptibility tensor coefficients (red dots) for which the sign is reversed due to a rotation of π around an axis oriented at $\pi/4$ from the x- and y-axis in the xy-plane.

7.4.2 Fourier Approach—Quasi-Phase-Matching Relation

7.4.2.1 Positive First-Order Quasi-Phase-Matching

The intuitive approach described before shows that QPM should be achieved in a crystal in which the sign of $\chi^{(2)}$ is periodically reversed along Z that is the direction of propagation of the three interacting waves, which also leads to a spatial modulation of the effective coefficient $\chi_{eff}^{(2)}$ according to Eq. (7.36). Such a square periodic modulation of $\chi_{eff}^{(2)}(Z)$ is well described by the positive first term of a Fourier series, *i.e.*:

$$\chi_{eff}^{(2)}(Z) = \frac{2}{\pi}\chi_{eff}^{(2)} \exp[i\left|\vec{K}_G\right|Z] \tag{7.86}$$

with

$$\left|\vec{K}_G\right| \equiv K_G = \frac{2\pi}{\Lambda} \tag{7.87}$$

The effective coefficient $\chi_{eff}^{(2)}$ is defined by Eq. (7.36), K_G is the modulus of the so-called grating vector $\vec{K}_G$ along Z associated to the grating periodicity Λ such as $\frac{\Lambda}{2} = Z_+ = Z_-$ where Z_+ is the length where $\chi_{eff}^{(2)} > 0$ and Z_- the length where $\chi_{eff}^{(2)} < 0$.

According to Eq. (7.86), Eq. (7.54) becomes:

$$\begin{aligned} E_3(Z) &= i\kappa_3\varepsilon_0 E_1(0)E_2(0)\int_0^Z \chi_{eff}^{(2)}(Z)\exp[-i\Delta kZ]dZ \\ &= i\frac{2}{\pi}\kappa_3\chi_{eff}^{(2)}\varepsilon_0 E_1(0)E_2(0)\int_0^Z \exp[-i(\Delta k - K_G)Z]dZ \end{aligned} \tag{7.88}$$

Then the corresponding intensity writes:

$$\begin{aligned} I_3(Z) &= 2\pi^2\left(\frac{4}{\pi^2}\right)\sqrt{\frac{\mu_0}{\varepsilon_0}}\frac{1}{\lambda_3^2(\cos\rho_3)^4}\frac{[\chi_{eff}^{(2)}]^2}{n_1n_2n_3}I_1(0)I_2(0)Z^2\sin c^2[\frac{(\Delta k - K_G)Z}{2}] \\ &= 8\sqrt{\frac{\mu_0}{\varepsilon_0}}\frac{1}{\lambda_3^2(\cos\rho_3)^4}\frac{[\chi_{eff}^{(2)}]^2}{n_1n_2n_3}I_1(0)I_2(0)Z^2\sin c^2[\frac{(\Delta k - K_G)Z}{2}] \end{aligned} \tag{7.89}$$

Equation (7.89) replaces Eq. (7.62) relative to the case in which there is no spatial modulation of the effective coefficient. The interference function is also a $\sin c^2$ function, as shown by Figure 7.6, that is maximal when its argument is equal to 0, *i.e.*:

$$\Delta k - K_G = 0 \tag{7.90}$$

Equation (7.90) is the QPM relation. It expresses the fact that the modulus of the grating vector has to be equal to the phase-mismatch in the direction of propagation that is considered. It is then possible to define the suited grating periodicity. Actually, from Eq. (7.59), Eq. (7.87), and Eq. (7.90) it comes:

$$\Lambda = 2l_c \tag{7.91}$$

The development of Δk in Eq. (7.90) leads to:

$$k_3 - k_2 - k_1 - K_G = 0 \tag{7.92}$$

It is now possible to express Eq. (7.90) in terms of the refractive indices in the direction of the grating vector at the different wavelengths. Actually from Eq. (7.87) it comes:

$$\frac{n_3}{\lambda_3} - \frac{n_1}{\lambda_1} - \frac{n_2}{\lambda_2} - \frac{1}{\Lambda} = 0 \tag{7.93}$$

Equation (7.93) is the QPM relation that is used in practice. Note that this equation tends towards the BPM relation when the grating period tends toward infinity ($\Lambda \to +\infty$).

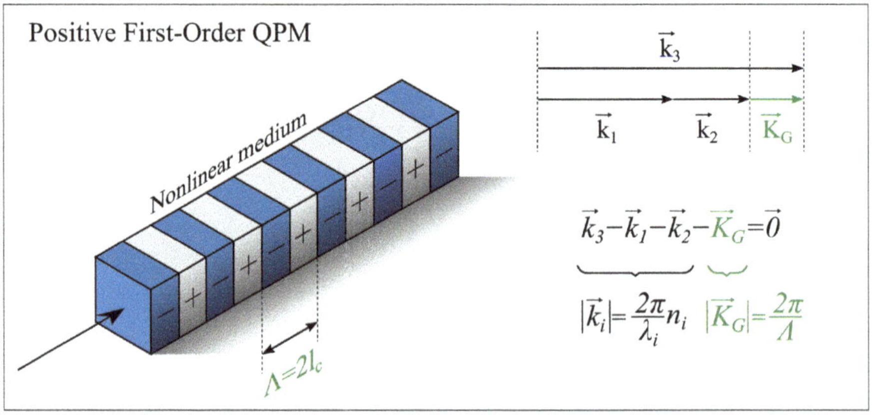

FIGURE 7.24 Conditions for achieving positive first-order quasi-phase-matching (QPM). The nonlinear medium exhibits a periodical spatial modulation of the sign (+/–) of the second-order electric susceptibility. $(\vec{k}_1,\vec{k}_2,\vec{k}_3)$ are the wave vectors of the three interacting waves at the wavelengths $(\lambda_1,\lambda_2,\lambda_3)$. $\vec{K}_G$ is the grating vector associated to the grating period Λ equal to twice the parametric coherence length l_c of the considered three-wave nonlinear interaction. The black arrow corresponds to the direction of propagation of the incident light.

The geometrical considerations established previously are summarized in Figure 7.24.

Comparing Eq. (7.62) and Eq. (7.89) with equal effective coefficients $\chi_{eff}^{(2)}$ and when BPM and positive first-order QPM are fulfilled, respectively, it comes that the only difference concerning the generated intensities as a function of the interacting length Z is that the positive first-order QPM intensity is $\frac{4}{\pi^2}$ smaller than the BPM intensity.

7.4.2.2 Positive and Negative mth-Order Quasi-Phase-Matching

The approach developed previously can be generalized to any other positive or negative order of the Fourier series. For the mth-order, the spatial modulation writes:

$$\chi_{eff}^{(2)}(Z)=\frac{2}{m\pi}\sin(m\pi D)\chi_{eff}^{(2)}\exp[iK_{Gm}Z] \tag{7.94}$$

with

$$K_{Gm}=\frac{2\pi m}{\Lambda}=mK_G \tag{7.95}$$

and

$$D=\frac{Z_+}{\Lambda} \tag{7.96}$$

The order m belongs to $\mathbb{Z}^*$, *i.e.* $m = \pm1, \pm2, \pm3, \pm4, ...$ and D is the so-called duty cycle.

The sine-function in Eq. (7.94) is maximal if the Fourier order is odd, *i.e.* $m = \pm1, \pm3, \pm5, ...$, and $D = \frac{1}{2}$, *i.e.* $Z_+ = Z_- = \frac{\Lambda}{2}$. In this case, Eq. (7.89) becomes:

$$I_3(Z) = 2\pi^2\left(\frac{4}{m^2\pi^2}\right)\sqrt{\frac{\mu_0}{\varepsilon_0}}\frac{1}{\lambda_3^2(\cos\rho_3)^4}\frac{[\chi_{eff}^{(2)}]^2}{n_1n_2n_3}I_1(0)I_2(0)Z^2\sin c^2[\frac{(\Delta k - K_{Gm})Z}{2}]$$
$$= \frac{8}{m^2}\sqrt{\frac{\mu_0}{\varepsilon_0}}\frac{1}{\lambda_3^2(\cos\rho_3)^4}\frac{[\chi_{eff}^{(2)}]^2}{n_1n_2n_3}I_1(0)I_2(0)Z^2\sin c^2[\frac{(\Delta k - K_{Gm})Z}{2}] \quad (7.97)$$

The mth order QPM is then fulfilled when $\Delta k - K_{Gm} = 0$, which leads to:

$$k_3 - k_2 - k_1 - K_{Gm} = 0 \quad (7.98)$$

and

$$\frac{n_3}{\lambda_3} - \frac{n_1}{\lambda_1} - \frac{n_2}{\lambda_2} - \frac{m}{\Lambda} = 0 \quad (7.99)$$

Equation (7.97) shows that the generated intensity is maximal for $m = \pm1$, *i.e.* for the first positive or negative Fourier order: the case $m = +1$ corresponds to Figure 7.24, and the case $m = -1$ is described in Figure 7.25.

Note that the positive first-order is the most considered.

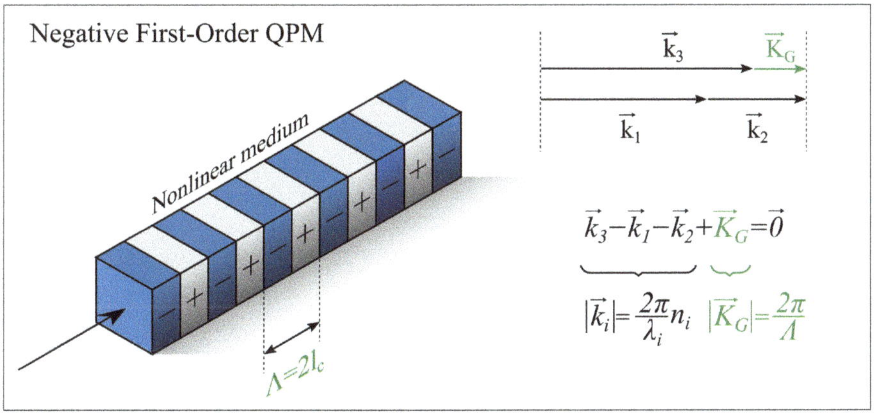

FIGURE 7.25 Conditions for achieving negative first-order quasi-phase-matching (QPM). The nonlinear medium exhibits a periodical spatial modulation of the sign (+/–) of the second-order electric susceptibility. $(\vec{k}_1, \vec{k}_2, \vec{k}_3)$ are the wave vectors of the three interacting waves at the wavelengths $(\lambda_1, \lambda_2, \lambda_3)$. $\vec{K}_G$ is the grating vector associated to the grating period Λ equal to twice the parametric coherence length l_c of the considered three-wave nonlinear interaction. The black arrow corresponds to the direction of propagation of the incident light.

7.4.3 Configurations of Polarization

Following the same way as in Part 7.3.3 for BPM, it is easy to establish the following relations, that are the equivalent of Eq. (7.65) and Eq. (7.66), from the energy conservation and the mth-order QPM relation, *i.e.*:

$$\begin{cases} \omega_3 = \omega_1 + \omega_2 \\ \dfrac{\omega_3}{c} n_3 = \dfrac{\omega_1}{c} n_1 + \dfrac{\omega_2}{c} n_2 + \dfrac{2\pi m}{\Lambda} \end{cases} \tag{7.100}$$

Then, by combining these two equations, it comes:

$$\frac{\omega_1}{\omega_2} = \frac{n_3 - n_2}{n_1 - n_3} - \frac{2\pi m}{\Lambda}\left[\frac{c}{\omega_2 (n_1 - n_3)}\right] = \frac{n_3 - n_2}{n_1 - n_3} - m\frac{\lambda_2}{\Lambda}\left[\frac{1}{(n_1 - n_3)}\right] > 0 \tag{7.101}$$

Equation (7.101) is the equivalent of Eq. (7.66) for BPM. For $m > 0$ using the inequalities approach used for BPM in Part 7.3.3, it can be shown that Eq. (7.101) can be verified for any of the 2^3 possible combinations of Eigen refractive indices $n_i^{\pm}$ if the proper value of Λ for a given m is found. These combinations are given in Table 7.3 according to SFG and DFG of types I, II, III, IV, V, VI, VII, and VIII for the isotropic, biaxial, and uniaxial optical classes. Type IV or type V, which corresponds to identical polarizations of the three interacting waves, are also called type 0.

TABLE 7.3
Positive-Order Quasi Phase-Matching (QPM)

Refractive indices	Polarizations	SFG ($\omega_3 = \omega_1 + \omega_2$)	DFG ($\omega_1 = \omega_3 - \omega_2$)	DFG ($\omega_2 = \omega_3 - \omega_1$)
(n_3^-, n_1^+, n_2^+)	$(\vec{e}_3^-, \vec{e}_1^+, \vec{e}_2^+)$	I	II	III
(n_3^-, n_1^-, n_2^+)	$(\vec{e}_3^-, \vec{e}_1^-, \vec{e}_2^+)$	II	III	I
(n_3^-, n_1^+, n_2^-)	$(\vec{e}_3^-, \vec{e}_1^+, \vec{e}_2^-)$	III	I	II
(n_3^-, n_1^-, n_2^-)	$(\vec{e}_3^-, \vec{e}_1^-, \vec{e}_2^-)$	IV	IV	IV
(n_3^+, n_1^+, n_2^+)	$(\vec{e}_3^+, \vec{e}_1^+, \vec{e}_2^+)$	V	V	V
(n_3^+, n_1^-, n_2^+)	$(\vec{e}_3^+, \vec{e}_1^-, \vec{e}_2^+)$	VI	VII	VIII
(n_3^+, n_1^+, n_2^-)	$(\vec{e}_3^+, \vec{e}_1^+, \vec{e}_2^-)$	VII	VIII	VI
(n_3^+, n_1^-, n_2^-)	$(\vec{e}_3^+, \vec{e}_1^-, \vec{e}_2^-)$	VIII	VI	VII

For $m < 0$, also using the approach described in Part 7.3.3, only types I, II, and III are compatible with Eq. (7.101). The corresponding configurations of polarization are thus the same as for BPM and are given in Table 7.1.

7.4.4 Angular Distribution

The first Fourier order, *i.e.* $m = \pm 1$, is considered in this part since it is the most favorable in term of amplitude of the associated Fourier component and also the most frequently encountered situation.

7.4.4.1 Angular Quasi-Phase-Matching–Grating Indicatrix

QPM is usually achieved in a configuration in which the wave vectors of the interacting waves $\vec{k}_i$ (i = 1, 2, 3) are collinear with the grating vector $\vec{K}_G$, which is considered here, knowing that any non-collinear scheme could be used. Furthermore, QPM can be realized for a propagation at any angle with respect to the grating vector, which is called angular quasi-phase-matching (AQPM) (Petit 2007) [13]. If the grating period is longer than the wavelengths of the interacting waves, *i.e.* $\Lambda > (\lambda_1, \lambda_2, \lambda_3)$, which is usually the case, then no diffraction by the grating occurs so that the collinear picture remains relevant, as shown in Figure 7.26.

In the case of collinear AQPM, it is then necessary to define an effective grating period $\Lambda_{eff}(\theta,\varphi)$ that is the projection of Λ on the considered direction $\vec{u}(\theta,\varphi)$, *i.e.* according to Eq. (5.47) and Figure 5.1:

- for $\vec{K}_G \,//\, \overrightarrow{Ox}$

$$\Lambda_{eff}(\theta,\varphi) = \frac{\Lambda}{|\sin\theta\cos\varphi|} \tag{7.102}$$

- for $\vec{K}_G \,//\, \overrightarrow{Oy}$

$$\Lambda_{eff}(\theta,\varphi) = \frac{\Lambda}{|\sin\theta\sin\varphi|} \tag{7.103}$$

- for $\vec{K}_G \,//\, \overrightarrow{Oz}$

$$\Lambda_{eff}(\theta,\varphi) = \frac{\Lambda}{|\cos\theta|} \tag{7.104}$$

That kind of calculation can be easily performed for any other orientation of $\vec{K}_G$ in the dielectric frame $(\overrightarrow{Ox}, \overrightarrow{Oy}, \overrightarrow{Oz})$.

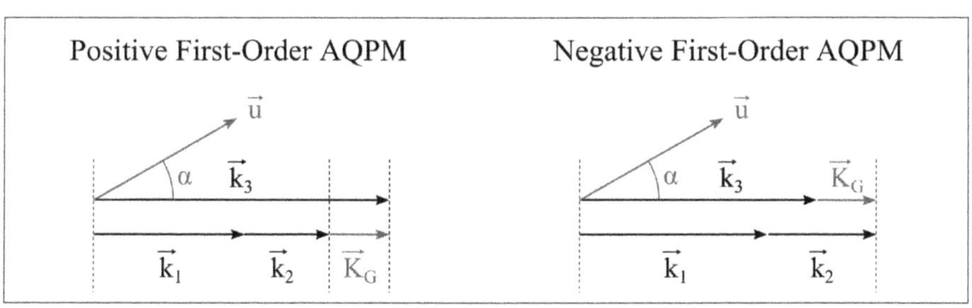

FIGURE 7.26 Collinear schemes of positive (left) and negative (right) first-order angular quasi-phase-matching (AQPM); $(\vec{k}_1, \vec{k}_2, \vec{k}_3)$ are the wave vectors of the three interacting waves; $\vec{K}_G$ is the grating period with $\left|\vec{K}_G\right| = \frac{2\pi}{\Lambda}$ where Λ is the grating period.

Then according to Eq. (7.99) with $m = \pm 1$, any AQPM direction, with $\left(\theta_{AQPM}, \varphi_{AQPM}\right)$ as angles of spherical coordinates, verifies:

$$\frac{n_3\left(\theta_{AQPM}, \varphi_{AQPM}\right)}{\lambda_3} - \frac{n_1\left(\theta_{AQPM}, \varphi_{AQPM}\right)}{\lambda_1} - \frac{n_2\left(\theta_{AQPM}, \varphi_{AQPM}\right)}{\lambda_2} \mp \frac{1}{\Lambda_{eff}\left(\theta_{AQPM}, \varphi_{AQPM}\right)} = 0 \tag{7.105}$$

with $-\frac{1}{\Lambda_{eff}\left(\theta_{AQPM}, \varphi_{AQPM}\right)}$ for $m = +1$ and $+\frac{1}{\Lambda_{eff}\left(\theta_{AQPM}, \varphi_{AQPM}\right)}$ for $m = -1$.

The function $\frac{1}{\Lambda_{eff}\left(\theta, \varphi\right)}$ is called the "grating indicatrix". It is given in Figure 7.27 for a grating vector along $\overrightarrow{Ox}$.

Figure 7.27 well shows that the grating indicatrix has a symmetry of revolution around $\vec{K}_G$ so that its symmetry group G_G is equivalent to a static cylinder, *i.e.* $G_G = \left(\frac{\infty}{m} m\right)$, as defined in Part 2.1.2, $\vec{K}_G$ being the rotation axis of infinite order.

The grating indicatrices in the case of $\vec{K}_G // \overrightarrow{Oy}$ and $\vec{K}_G // \overrightarrow{Oz}$ can be easily obtained from Figure 7.27 by a rotation of $\frac{\pi}{2}$ in the $\left(\overrightarrow{Ox}, \overrightarrow{Oy}\right)$ and $\left(\overrightarrow{Ox}, \overrightarrow{Oz}\right)$ planes, respectively.

7.4.4.2 Isotropic Crystals

Since the index surface of the isotropic optical class is a spherical single layer surface, which means only one value for the refractive index at the considered wavelength for any direction of propagation, Eq. (7.105) reduces to:

$$\frac{n_3}{\lambda_3} - \frac{n_1}{\lambda_1} - \frac{n_2}{\lambda_2} \mp \frac{1}{\Lambda_{eff}\left(\theta_{AQPM}, \varphi_{AQPM}\right)} = 0 \tag{7.106}$$

It is important to notice that the eight types of AQPM given in Table 7.3 lead to the same AQPM relation, *i.e.* Eq. (7.106), since any polarization corresponds to the same refractive index.

Furthermore, using the same approach based on inequalities between refractive indices for the definition of the types than that used in Part 7.3.3 for BPM and Part 7.4.3 for AQPM, it comes that Eq. (7.106) with the sign "+" is not possible (order $m = -1$). As a consequence, any $\chi^{(2)}$ process is forbidden in a cubic crystal for a negative first-order AQPM or any other negative order. Then, starting from the simple example of SHG, the AQPM relation can only correspond to the positive first-order, which writes:

$$\left(n_{2\omega} - n_{\omega}\right) - \frac{\lambda_{2\omega}}{\Lambda_{eff}\left(\theta_{AQPM}, \varphi_{AQPM}\right)} = 0 \tag{7.107}$$

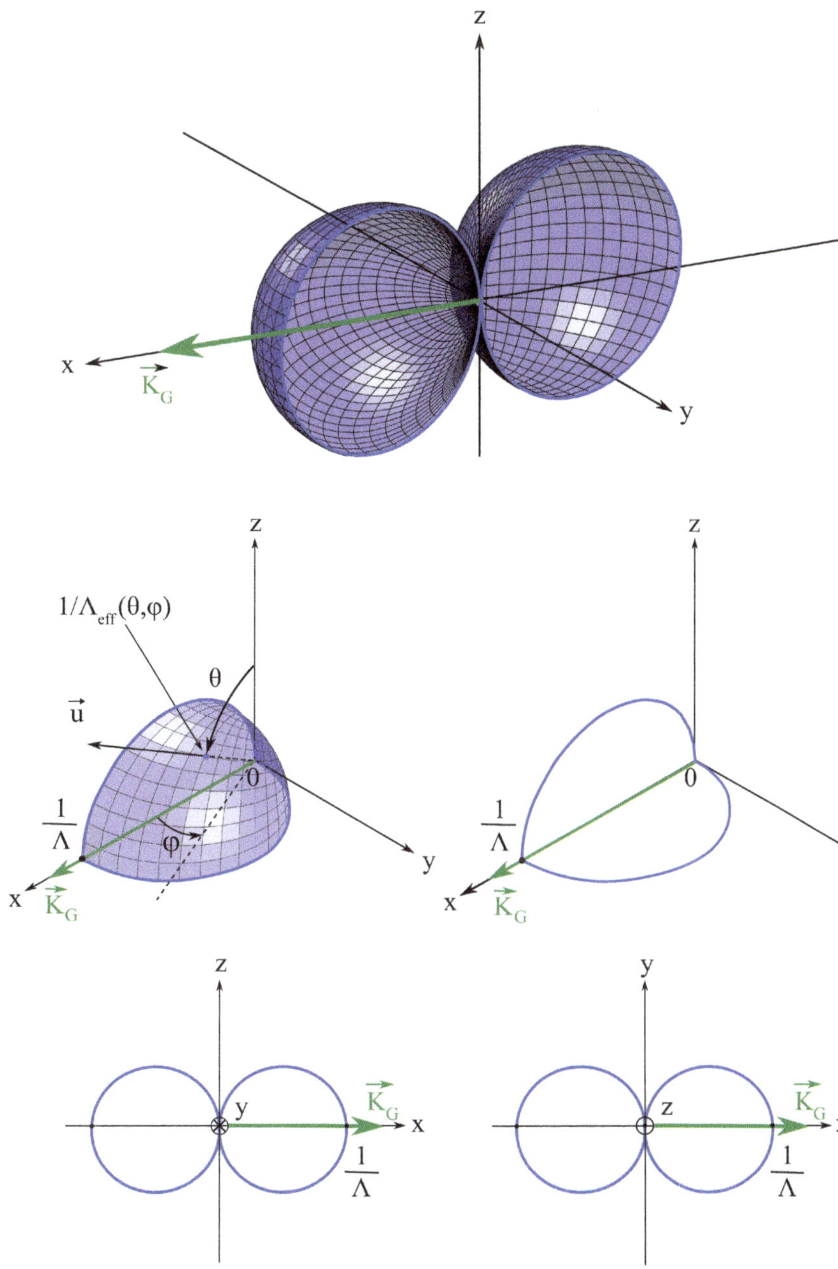

FIGURE 7.27 Grating indicatrix corresponding to a grating vector $\vec{k}_G\ //\ \overrightarrow{O_x}$. The indices (x, y, z) refer to the dielectric frame $(\overrightarrow{Ox},\overrightarrow{Oy},\overrightarrow{Oz})$. Top: half-space. Middle: space octant in which (θ,φ) are the angles of spherical coordinates of the unit wave vector $\vec{u}$ and $\Lambda_{eff}(\theta,\varphi)$ is the effective grating. Bottom: section of the grating indicatrix in the $(\overrightarrow{Ox},\overrightarrow{Oz})$ plane (left) and in the $(\overrightarrow{Ox},\overrightarrow{Oy})$ plane (right).

In the case of $\vec{K}_G // \overrightarrow{Oz}$ for example, Eq. (7.107) becomes according to Eq. (7.104):

$$(n_{2\omega} - n_{\omega}) - |\cos\theta_{AQPM}| \frac{\lambda_{2\omega}}{\Lambda} = 0 \tag{7.108}$$

Then the AQPM angle can be immediately calculated, *i.e.*:

$$\theta_{AQPM} = \pm ar\cos\left[(n_{2\omega} - n_{\omega})\frac{\Lambda}{\lambda_{2\omega}}\right] \tag{7.109}$$

From the geometrical point of view and according to Eq. (7.107) and Eq. (7.104), this angle corresponds to the intersection of the circle of radius $n_{2\omega}$ and the curve of equation $n_{\omega} + \frac{\lambda_{2\omega}}{\Lambda_{eff}(\theta)}$ with $\Lambda_{eff}(\theta,\varphi) = \frac{\Lambda}{|\cos\theta|}$.

Figure 7.28 shows that the intersection between $n_{\omega} + \frac{\lambda_{2\omega}}{\Lambda_{eff}(\theta)}$ and $n_{2\omega}$ is only possible if the following relation of order is verified:

$$n_{\omega} + \frac{\lambda_{2\omega}}{\Lambda} \le n_{2\omega} \tag{7.110}$$

Note that $n_{\omega} + \frac{\lambda_{2\omega}}{\Lambda} = n_{2\omega}$ corresponds to a QPM direction along $\overrightarrow{O_z}$.

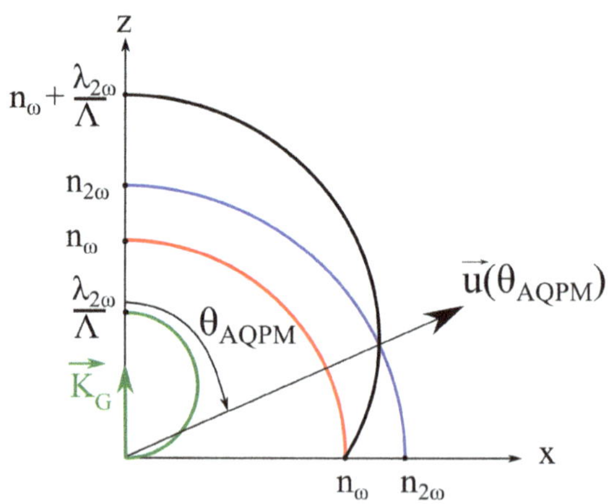

FIGURE 7.28 Geometrical determination of the AQPM angle θ_{AQPM} of SHG in a cubic crystal in which the grating vector $\vec{K}_G$ is along $\overrightarrow{O_z}$. The indices (x,y,z) refer to the dielectric frame $(\overrightarrow{Ox},\overrightarrow{Oy},\overrightarrow{Oz})$. $\vec{u}(\theta_{AQPM})$ is the unit vector of the corresponding AQPM direction. The red and blue circles correspond to n_{ω} and $n_{2\omega}$ respectively, *i.e.* to the section of the index surface at ω and 2ω in any plane containing $\overrightarrow{O_z}$. The green curve is proportional to the grating indicatrix, *i.e.* $\frac{\lambda_{2\omega}}{\Lambda_{eff}(\theta)}$ with $\Lambda_{eff}(\theta,\varphi) = \frac{\Lambda}{|\cos\theta|}$ where Λ is the grating period along $\overrightarrow{O_z}$. The black curve corresponds to $+\frac{\lambda_{2\omega}}{\Lambda_{eff}(\theta)}$.

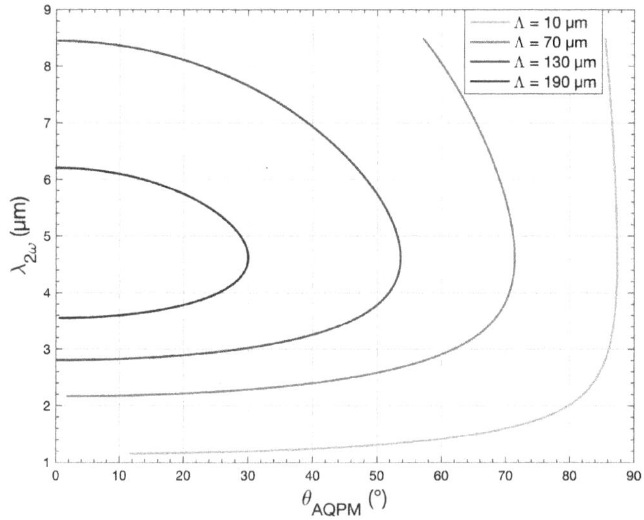

FIGURE 7.29 Second-harmonic wavelength $\lambda_{2\omega}$ as a function of the AQPM angle θ_{AQPM} in OP-GaAs for several values of grating period Λ.

Equation (7.108) shows that AQPM allows us to get wavelength tunability. Actually, Eq. (7.108) rewrites:

$$\lambda_{2\omega} = (n_{2\omega} - n_{\omega}) \frac{\Lambda}{|\cos\theta_{AQPM}|} \tag{7.111}$$

Equation (7.111) is plotted in the case of OP-GaAs in Figure 7.29 for different values of Λ.

Figure 7.29 shows for example that the SHG ($\lambda_{\omega} = 10\ \mu m \rightarrow \lambda_{2\omega} = 5\ \mu m$) can be quasi-phase-matched using angular quasi-phase-matching in orientation patterned GaAs using the following grating periods and AQPM angles (Λ, θ_{AQPM}): ($10\mu m, 87.36°$), ($70\mu m, 71.16°$), ($130\mu m, 53.16°$) and ($190\mu m, 28.80°$). It is reminded that birefringence phase-matching is not possible in a single crystal of GaAs according to Figure 7.7.

From the topological point of view, Eq. (7.109) corresponds to a cone with an aperture angle equal to θ_{AQPM} and with $\overrightarrow{Oz}$ as a revolution axis, as shown in Figure 7.30.

The symmetry group G_{AQPM} of the AQPM cone depicted in Figure 7.30 is $\left(\frac{\infty}{m}m\right)$, the same one as that of the grating indicatrix G_G as explained in the previous part. Actually, according to the AQPM relation given by Eq. (7.108), it comes:

$$G_{AQPM} = G_{\omega} \cap G_{2\omega} \cap G_G = G_G \tag{7.112}$$

with $G_{\omega} = G_{2\omega} = \left(\frac{\infty}{m}\frac{\infty}{m}\right)$ corresponding to the layers of the index surface at ω and 2ω, respectively.

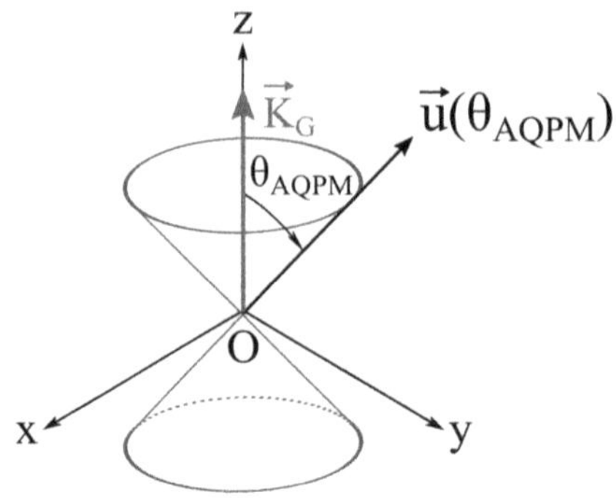

FIGURE 7.30 AQPM cone of a cubic crystal in which the grating vector $\vec{K}_G$ is along $\vec{Oz}$. The aperture angle is θ_{AQPM} and $\vec{u}(\theta_{AQPM})$ stands for the unit vector of any AQPM direction. (x, y, z) is the dielectric frame.

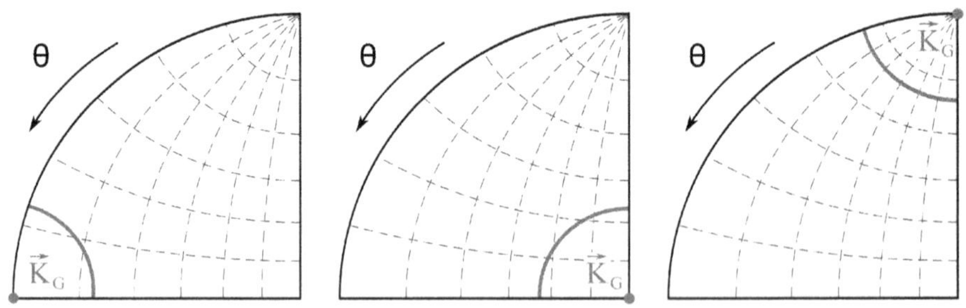

FIGURE 7.31 SFG or DFG AQPM cones in a cubic crystal in which the grating vector $\vec{K}_G$ is along $\vec{O_x}$ (left), $\vec{O_y}$ (middle), or $\vec{O_z}$ (right).

This result can be generalized to any SFG and DFG processes in isotropic crystals for which the AQPM relation writes:

$$\frac{n_3}{\lambda_3} - \frac{n_1}{\lambda_1} - \frac{n_2}{\lambda_2} - \frac{1}{\Lambda_{eff}\left(\theta_{AQPM}, \varphi_{AQPM}\right)} = 0 \tag{7.113}$$

Then, the corresponding group equation is:

$$G_{AQPM} = G_3 \cap G_1 \cap G_2 \cap G_G = G_G \tag{7.114}$$

with $G_1 = G_2 = G_3 = \left(\frac{\infty}{m}\frac{\infty}{m}\right)$ corresponding to the layers of the index surface at λ_1, λ_2, and λ_3 respectively, and $G_G = \left(\frac{\infty}{m}m\right)$ for the grating indicatrix.

The revolution axis of the AQPM cone, which corresponds to the infinite rotation symmetry axis, is along the grating vector $\vec{K}_G$, as shown in Figure 7.31 for three specific orientations.

7.4.4.3 Uniaxial Crystals

In any AQPM direction, with $\left(\theta_{AQPM},\varphi_{AQPM}\right)$ as angles of spherical coordinates, Eq. (105) can be written in terms of the Eigen refractive indices $n^{+}\left(\theta_{AQPM},\varphi_{AQPM}\right)$ and $n^{-}\left(\theta_{AQPM},\varphi_{AQPM}\right)$ at the wavelengths of the three interacting waves, which gives:

$$\frac{n_3^{\pm}\left(\theta_{AQPM},\varphi_{AQPM}\right)}{\lambda_3}-\frac{n_1^{\pm}\left(\theta_{AQPM},\varphi_{AQPM}\right)}{\lambda_1}-\frac{n_2^{\pm}\left(\theta_{AQPM},\varphi_{AQPM}\right)}{\lambda_2}\mp\frac{1}{\Lambda_{eff}\left(\theta_{AQPM},\varphi_{AQPM}\right)}=0 \tag{7.115}$$

As mentioned in Part 7.4.3, the eight configurations of polarization given in Table 7.3, *i.e.* types I to VIII, are compatible with the sign "–" in Eq. (7.115), which corresponds to the first-order positive QPM. But only types I, II, and III are compatible with the sign "+", which corresponds to the first-order negative QPM.

The following calculation deals with the example of type I SHG, *i.e.* $(\vec{e}_{2\omega}^{\,-},\vec{e}_{\omega}^{\,+},\vec{e}_{\omega}^{\,+})$, in a negative uniaxial crystal, *i.e.* $n_o\equiv n^{+}>n_e\equiv n^{-}$ which is the case of $LiNbO_3$. Then Eq. (7.114) written in terms of ordinary and extraordinary waves is expressed as:

$$\frac{1}{\lambda_{2\omega}}\left[\frac{1}{\sqrt{n_o^{-2}(2\omega)\cos^2(\theta_{AQPM})+n_e^{-2}(2\omega)\sin^2(\theta_{AQPM})}}-n_o(\omega)\right]\mp\frac{1}{\Lambda_{eff}\left(\theta_{AQPM},\varphi_{AQPM}\right)}=0 \tag{7.116}$$

In the case of $\vec{K}_G\,//\,\overrightarrow{Ox}$, which is the case of periodically poled $LiNbO_3$ (PPLN), and for a positive QPM, Eq. (7.116) writes:

$$\frac{1}{\sqrt{n_o^{-2}(2\omega)\cos^2(\theta_{AQPM})+n_e^{-2}(2\omega)\sin^2(\theta_{AQPM})}}-n_o(\omega)-\left|\sin\theta_{AQPM}\cos\varphi_{AQPM}\right|\frac{\lambda_{2\omega}}{\Lambda}=0 \tag{7.117}$$

The corresponding geometrical situation is depicted in Figure 7.32.

Equation (7.117) has no simple analytical solution so that a numerical resolution is required. It can lead to θ_{AQPM} as a function of φ_{AQPM}, as shown hereafter in Figure 7.33.

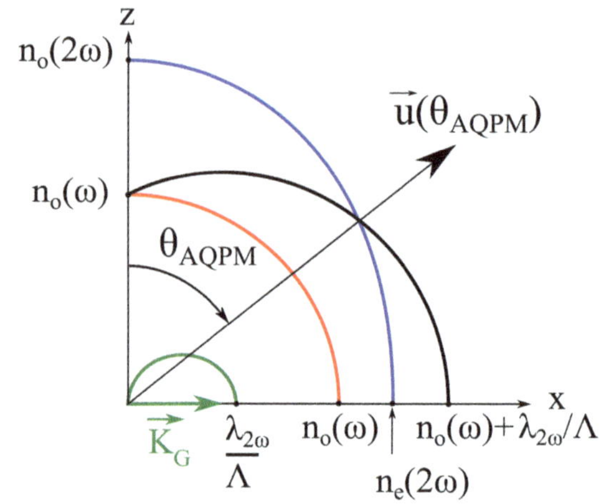

FIGURE 7.32 Geometrical determination of the AQPM angle θ_{AQPM} of type I SHG in a negative uniaxial crystal in which the grating vector $\vec{K_G}$ is along $\vec{O_x}$. The indices (x, y, z) refer to the dielectric frame $(\vec{Ox},\vec{Oy},\vec{Oz})$. $\vec{u}(\theta_{AQPM})$ is the unit vector of the corresponding AQPM direction. The red and blue circles correspond to $n_\omega^+(\theta)$ and $n_{2\omega}^-(\theta)$ respectively, *i.e.* to the section of the index surface at ω and 2ω in any plane containing $\vec{O_z}$, here in the xz-plane. n_o and n_e are the ordinary and extraordinary principle refractive indices, respectively. The green curve is proportional to the grating indicatrix, *i.e.* $\frac{\lambda_{2\omega}}{\Lambda_{eff}(\theta)}$ with $\Lambda_{eff}(\theta,\varphi)=\frac{\Lambda}{|\cos\theta\sin\theta|}$ where Λ is the grating period along $\vec{O_x}$. The black curve corresponds to $n_o(\omega)+\frac{\lambda_{2\omega}}{\Lambda_{eff}(\theta)}$.

In Figure 7.33, the AQPM directions form a non-circular cone joining the xz-plane and the yz-plane, as in the case of Figure (7.15) in the case of BPM in a biaxial crystal. The corresponding symmetry group is then $G_{AQPM}=(mmm)$, coming from $G_{AQPM}=G_{2\omega}^-\cap G_\omega^+\cap G_\omega^+\cap G_G$ with $G_{2\omega}^-=\left(\frac{\infty}{m}m\right)$ with a revolution axis along the z-axis, $G_\omega^+=\left(\frac{\infty}{m}\frac{\infty}{m}\right)$ and $G_G=\left(\frac{\infty}{m}m\right)$ with a revolution axis along the x-axis. Thus, only the three symmetry mirrors with respect to each of the three principal planes of the dielectric frame are maintained in biaxial crystals.

This cone of type I SHG around $\vec{O_z}$ in a negative uniaxial crystal exhibiting a grating along $\vec{O_x}$ can be also found in a positive uniaxial crystal and for types II and III SHG, but also for any SFG or DFG process. This result can be intuitively recovered by continuity, considering that in the yz-plane AQPM is no longer possible but only BPM, because $\Lambda_{eff}\left(\theta_{AQPM},\varphi_{AQPM}=90°\right)\to\infty$ which means infinite parametric coherence length, knowing that BPM is only possible with types I, II, III, and a cone around $\vec{O_z}$. Then AQPM tends towards BPM in the yz-plane.

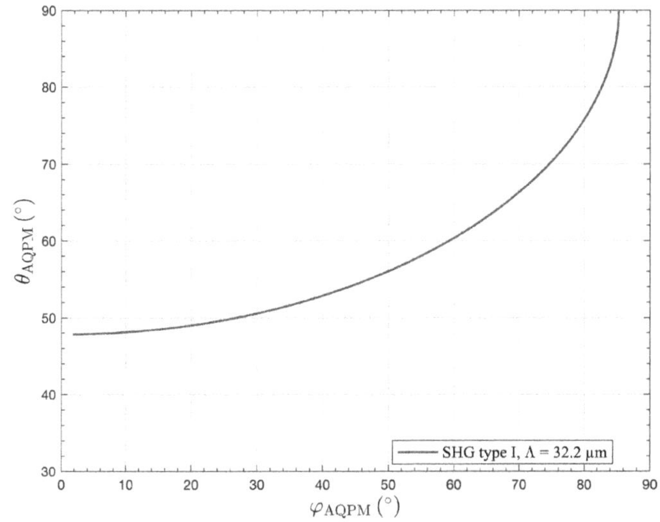

FIGURE 7.33 Positive first-order AQPM angles $(\theta_{AQPM}, \varphi_{AQPM})$ calculated for type I SHG in 5%MgO:PPLN exhibiting a grating period of $\Lambda = 32.2\ \mu m$ along $\overrightarrow{O_x}$ and pumped at the fundamental wavelength $\lambda_\omega = 4\ \mu m$. $\varphi_{AQPM} = 0°$ corresponds to the xz-plane and $\varphi_{AQPM} = 90°$ to the yz-plane.

This positive-order AQPM cone around $\overrightarrow{O_z}$ in a negative or positive uniaxial crystal exhibiting a grating along $\overrightarrow{O_x}$ is one of the possible topologies, the other one being a cone joining two AQPM directions located in the xz-plane. These AQPM loci cannot be of type I, II, or III since there is no intersection with the yz-plane and so no possibility of continuity with BPM. Then these two configurations concern types IV, V, VI, VII, and VIII.

The three possible topologies corresponding to positive-order AQPM are summarized in Figure 7.34.

Keeping the case of a grating vector $\overrightarrow{K_G}$ along $\overrightarrow{O_x}$ but now for a negative-order AQPM, it comes two possible topologies depicted in Figure 7.35: a cone around $\overrightarrow{O_z}$, as in the case of the positive-order AQPM, and a cone around $\overrightarrow{O_y}$, which is a new situation.

As seen in Part 7.4.3, a negative-order AQPM can be of types I, II, and III only, which is in accordance with the two possible topologies since the two cones intersect the yz-plane, which ensures the continuity with BPM.

The same approach for other orientations of grating vector $\overrightarrow{K_G}$ can be used for finding the possible AQPM loci.

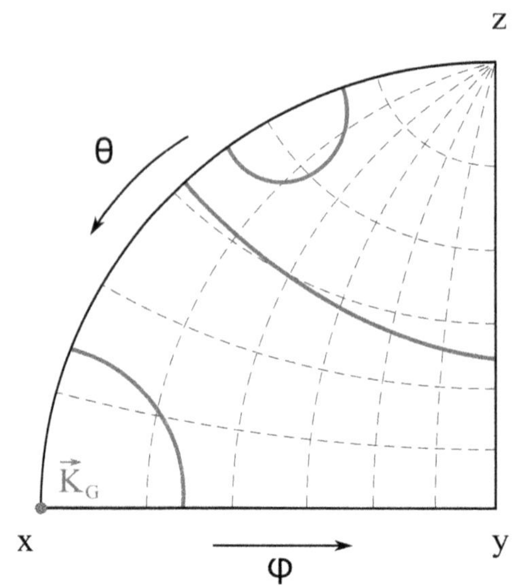

FIGURE 7.34 Possible SFG or DFG cones for a positive-order AQPM process in a positive or negative uniaxial crystal in which the grating vector $\vec{K}_G$ is along $\overrightarrow{O_x}$.

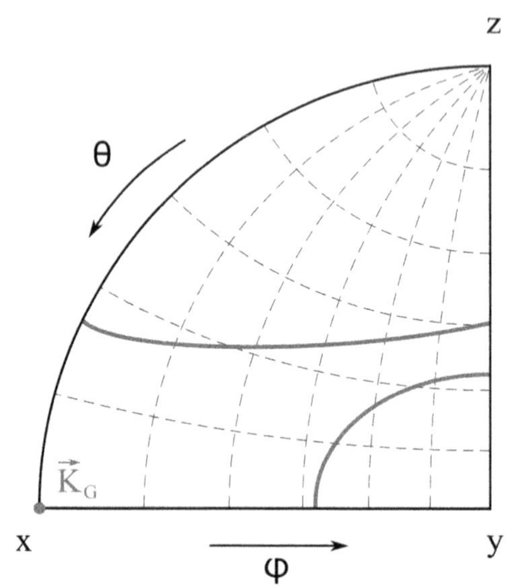

FIGURE 7.35 Possible SFG or DFG cones for a negative-order AQPM process in a positive or negative uniaxial crystal in which the grating vector $\vec{K}_G$ is along $\overrightarrow{O_x}$.

7.4.4.4 Biaxial Crystals

The AQPM relation is the same as for uniaxial crystals, *i.e.* Eq. (7.115). However, the Eigen refractive indices $n^{\pm}(\theta_{AQPM},\varphi_{AQPM})$ are expressed here as a function of the principal refractive indices n_x, n_y, and n_z. As example, Eq. (7.115) is solved in the case of PPKTP that is a positive biaxial crystal in which the grating vector $\overrightarrow{K_G}$ is along $\overrightarrow{O_x}$. The corresponding curves are shown Figure 7.36. (Petit 2020) [14].

By continuously increasing or reducing the grating period, Figure 7.36 shows that types I and II AQPM loci can form cones around $\overrightarrow{O_z}$, the AQPM order being positive or negative, while types IV and V, only possible for the positive AQPM order, surround $\overrightarrow{O_x}$. Moreover, for large enough grating periods, types I and II AQPM loci can also form cones around $\overrightarrow{O_y}$ for the negative AQPM order only.

A general approach of same type as that used for isotropic and uniaxial crystals allows us to identify five possible topologies for biaxial crystals, the AQPM order being positive or negative, as shown in Figure 7.37.

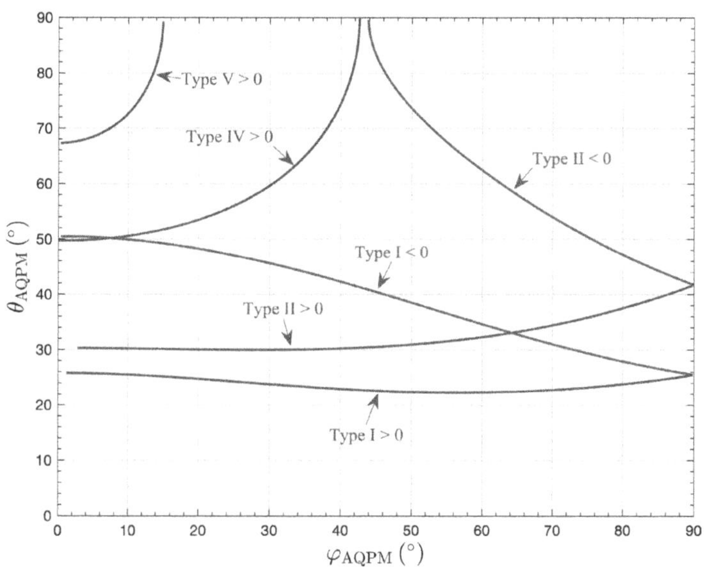

FIGURE 7.36 First-order AQPM angles $(\theta_{AQPM},\varphi_{AQPM})$ calculated for SHG in PPKTP exhibiting a grating period of $\Lambda = 38.52\ \mu m$ along $\overrightarrow{O_x}$ and pumped at the fundamental wavelength $\lambda_\omega = 2.15\ \mu m$: types I, II, IV, and V positive first-order AQPM, types I and II negative first-order AQPM. $\varphi_{AQPM} = 0°$ corresponds to the xz-plane, $\varphi_{AQPM} = 90°$ to the yz-plane, and $\theta_{AQPM} = 90°$ to the xy-plane.

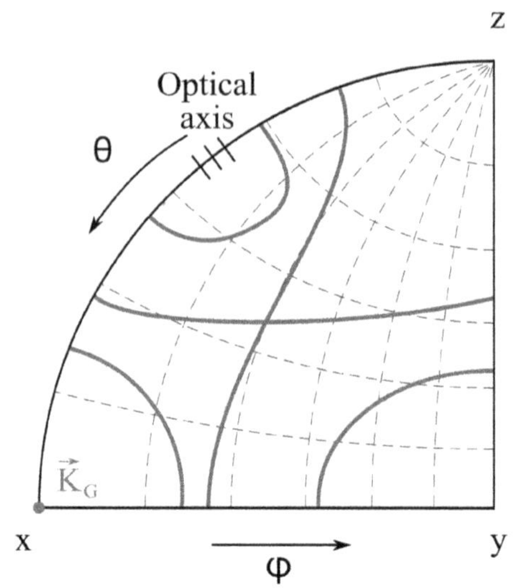

FIGURE 7.37 Possible SFG or DFG AQPM cones in a positive or negative biaxial crystal in which the grating vector $\vec{K}_G$ is along $\overrightarrow{O_x}$.

Figure 7.37 shows that AQPM includes the four possible BPM topologies depicted Figure 7.14 plus a new one surrounding $\overrightarrow{O_x}$. Note that the two cones joining the yz-plane can be of type I, II, or III only by continuity with BPM in this plane. The three other cones can be of any type.

7.5 EFFECTIVE COEFFICIENT

7.5.1 Symmetry of the Field Tensor

It is interesting for symmetry purposes to rewrite the effective coefficient given by Eq. (7.36) as the contracted product of two rank-3 tensors, *i.e.* in the direction of propagation $\vec{u}$ (Boulanger 1991, Duboisset 2024) [4,15]:

$$\begin{aligned}\chi_{eff}^{(2)}(\vec{u}) &= F^{(2)}\left(\omega_1 = \omega_3 - \omega_2, \vec{u}\right) \therefore \chi^{(2)}\left(\omega_1 = \omega_3 - \omega_2\right) \\ &= F^{(2)}\left(\omega_2 = \omega_3 - \omega_1, \vec{u}\right) \therefore \chi^{(2)}\left(\omega_2 = \omega_3 - \omega_1\right) \\ &= F^{(2)}\left(\omega_3 = \omega_1 + \omega_2, \vec{u}\right) \therefore \chi^{(2)}\left(\omega_3 = \omega_1 + \omega_2\right)\end{aligned} \tag{7.118}$$

with

$$\begin{aligned}F^{(2)}\left(\omega_1 = \omega_3 - \omega_2, \vec{u}\right) &= \vec{e}_1(\vec{u}) \otimes \vec{e}_3(\vec{u}) \otimes \vec{e}_2(\vec{u}) \\ F^{(2)}\left(\omega_2 = \omega_3 - \omega_1, \vec{u}\right) &= \vec{e}_2(\vec{u}) \otimes \vec{e}_3(\vec{u}) \otimes \vec{e}_1(\vec{u}) \\ F^{(2)}\left(\omega_3 = \omega_1 + \omega_2, \vec{u}\right) &= \vec{e}_3(\vec{u}) \otimes \vec{e}_1(\vec{u}) \otimes \vec{e}_2(\vec{u})\end{aligned} \tag{7.119}$$

$F^{(2)}$ is called the field tensor: it is a polar tensor built from the tensor product of the unit electric fields of the three interacting waves. $F^{(2)}$ depends on the direction of propagation $\vec{u}$ contrarily to the electric susceptibility tensor $\chi^{(2)}$. According to Eq. (2.26), it comes for Eq. (7.118) and Eq. (7.119) expressed in the dielectric frame:

$$\begin{aligned}\chi_{eff}^{(2)}(\vec{u}) &= \sum_{ijk} F_{ijk}^{(2)}(\omega_1 = \omega_3 - \omega_2, \vec{u})\chi_{ijk}^{(2)}(\omega_1 = \omega_3 - \omega_2)\\ &= \sum_{ijk} F_{ijk}^{(2)}(\omega_2 = \omega_3 - \omega_1, \vec{u})\chi_{ijk}^{(2)}(\omega_2 = \omega_3 - \omega_1)\\ &= \sum_{ijk} F_{ijk}^{(2)}(\omega_3 = \omega_1 + \omega_2, \vec{u})\chi_{ijk}^{(2)}(\omega_3 = \omega_1 + \omega_2)\end{aligned} \tag{7.120}$$

with

$$\begin{aligned}F_{ijk}(\omega_1 = \omega_3 - \omega_2, \vec{u}) &= e_i(\omega_1, \vec{u})e_j(\omega_3, \vec{u})e_k(\omega_2, \vec{u})\\ F_{ijk}(\omega_2 = \omega_3 - \omega_1, \vec{u}) &= e_i(\omega_2, \vec{u})e_j(\omega_3, \vec{u})e_k(\omega_1, \vec{u})\\ F_{ijk}(\omega_3 = \omega_1 + \omega_2, \vec{u}) &= e_i(\omega_3, \vec{u})e_j(\omega_1, \vec{u})e_k(\omega_2, \vec{u})\end{aligned} \tag{7.121}$$

The Cartesian indices i, j, and k are relative to the coordinates along the axes of the dielectric frame (x, y, z).

This rewriting of $\chi_{eff}^{(2)}$ allows us to access the symmetry effects due to the configuration of polarization of light. There are several symmetries relative to $F^{(2)}$ described hereafter.

A first symmetry leads to relations between field tensor coefficients at different circular frequencies due to the commutativity of the algebra product of the unit electric field vectors that compose the field tensors, *i.e.* according to Eq. (7.121) (Boulanger 1991) [4]:

$$\begin{aligned}&F_{ijk}(\omega_3 = \omega_1 + \omega_2, \vec{u}) = e_i(\omega_3, \vec{u})e_j(\omega_1, \vec{u})e_k(\omega_2, \vec{u})\\ &= F_{jik}(\omega_1 = \omega_3 - \omega_2, \vec{u}) = e_j(\omega_1, \vec{u})e_i(\omega_3, \vec{u})e_k(\omega_2, \vec{u})\\ &= F_{kij}(\omega_2 = \omega_3 - \omega_1, \vec{u}) = e_k(\omega_2, \vec{u})e_i(\omega_3, \vec{u})e_j(\omega_1, \vec{u})\end{aligned} \tag{7.122}$$

Equation (7.122) shows that the field tensor coefficients remain unchanged by concomitant permutations of the unit electric field vectors at the different frequencies and the corresponding Cartesian indices. These relations are the equivalents of Eq. (3.33), *i.e.* ABDP symmetry (Armstrong 1962) [1] regarding the electric susceptibility tensors in the case of a lossless medium. Then from Eq. (3.33) and Eq. (7.122), it comes for the effective coefficients of sum-frequency generation (SFG) and difference-frequency generation (DFG):

$$\begin{aligned}&\chi_{eff}^{SFG}(\omega_3 = \omega_1 + \omega_2, \vec{u}) = \chi^{(2)}(\omega_3 = \omega_1 + \omega_2) \therefore F^{(2)}(\omega_3 = \omega_1 + \omega_2, \vec{u})\\ &= \chi_{eff}^{DFG}(\omega_1 = \omega_3 - \omega_2, \vec{u}) = \chi^{(2)}(\omega_1 = \omega_3 - \omega_2) \therefore F^{(2)}(\omega_1 = \omega_3 - \omega_2, \vec{u})\\ &= \chi_{eff}^{DFG}(\omega_2 = \omega_3 - \omega_1, \vec{u}) = \chi^{(2)}(\omega_2 = \omega_3 - \omega_1) \therefore F^{(2)}(\omega_2 = \omega_3 - \omega_1, \vec{u})\end{aligned} \tag{7.123}$$

A second set of symmetries arise from ordinary (o) and extraordinary (e) polarizations. Then it concerns uniaxial crystals, or biaxial crystals in the principal planes only, as seen in Chapter 5. Then the $2^3 = 8$ possible configurations of polarization in these situations are (eee), (ooo), (eoo), (oee), (eoe), (eeo), (ooe), and (oeo). There are three kinds of symmetry in this framework: (i) annihilation of Cartesian coordinates of the unit electric field, (ii) orthogonality between ordinary and extraordinary polarizations, (iii) independence of polarization direction from circular frequency (Boulanger 2014) [7].

(i) As seen in Part 5.3, the ordinary polarization lies in the xy-plane for any direction of polarization, so that for example $F_{ijz}^{eoo} = F_{izk}^{eoo} = F_{izz}^{eoo} = 0$, which leads to the annihilation of fifteen field tensor components in this case; by keeping the same example but for a propagation in the xy-plane specifically, the extraordinary wave is polarized along the z-axis, so that $F_{xjk}^{eoo} = F_{yjk}^{eoo} = 0$ leading to the annihilation of ten more field tensor components. With the other example of (ooo), the annihilation concerns nineteen coefficients for any direction of propagation, *i.e.*: $F_{ijz}^{ooo} = F_{izk}^{ooo} = F_{zjk}^{ooo} = F_{izz}^{ooo} = F_{zjz}^{ooo} = F_{zzk}^{ooo} = F_{zzz}^{ooo} = 0$. That kind of symmetry never applies to the case of biaxial crystals out of the principal planes because all the Cartesian coordinates of the unit electric fields are non-zero.

(ii) Ordinary and extraordinary polarizations are always perpendicular so that for example $F_{xxk}^{eoo} + F_{yyk}^{eoo} + F_{zzk}^{eoo} = F_{xjx}^{eoo} + F_{yjy}^{eoo} + F_{zjz}^{eoo} = 0$, which leads to the following relations in addition to those of symmetry (i): $F_{xxx}^{eoo} = -F_{yxy}^{eoo} = -F_{yyx}^{eoo}$ and $F_{yyy}^{eoo} = -F_{xxy}^{eoo} = -F_{yyx}^{eoo}$. Note that this symmetry does not apply out of the principal planes of biaxial crystals either due to the non-orthogonality between the two Eigenmodes of polarization. Clearly, this symmetry is not relevant in cases (ooo) and (eee) either.

(iii) The ordinary polarization of a uniaxial crystal, or in the xz- and yz-planes of a biaxial crystal, does not depend on the circular frequency, whereas it is the case only for a direction of propagation at $\theta = \pi/2$, *i.e.* in the xy-plane. Again using the example of the configuration (eoo), there is a symmetry over the two last Cartesian indices, *i.e.* $F_{ijk}^{eoo} = F_{ikj}^{eoo}$, which leads to the equality $F_{zxy}^{eoo} = F_{zyx}^{eoo}$ in addition to the relationships provided by symmetries (i) and (ii). Note also that symmetry (iii) leads to the full symmetry for the (ooo) field tensor, *i.e.* $F_{ijk}^{ooo} = F_{ikj}^{ooo} = F_{jki}^{ooo} = F_{jik}^{ooo} = F_{kij}^{ooo} = F_{kji}^{ooo}$, and that for any direction of propagation.

A third kind of symmetry can come from equalities between the frequencies of interacting waves of same polarizations, which happens in the case of SHG ($2\omega = \omega + \omega$). This symmetry does not provide new equalities in the cases of the previous examples of (eoo) and (ooo) field tensors due to symmetry (iii). But it brings specific equalities when at least two extraordinary waves are involved; for example,

$F_{ijk}^{oee}(2\omega = \omega + \omega) = F_{ikj}^{oee}(2\omega = \omega + \omega)$. It is also the case in biaxial crystals for a propagation out of the principal planes, *i.e.* $F_{ijk}^{-++}(2\omega = \omega + \omega) = F_{ikj}^{-++}(2\omega = \omega + \omega)$.

7.5.2 Matrix Representation

Figure 7.38 to Figure 7.45 hereafter show the matrix representations of the field tensor based on symmetries (i), (ii), and (iii) described earlier for the eight possible configurations of polarization (ooo), (eee), (eoo), (oeo), (ooe), (oee), (eoe), and (eeo) and for any direction of propagation out of the principal axes $\overrightarrow{O_x}$, $\overrightarrow{O_y}$, and $\overrightarrow{O_z}$. The corresponding with the types, *i.e.* I, II, III, IV, V, VI, VII, and VIII, and the kinds of process, *i.e.* $SFG(\omega_3 = \omega_1 + \omega_2)$, $DFG(\omega_1 = \omega_3 - \omega_2)$, and $DFG(\omega_2 = \omega_3 - \omega_1)$, can be made using Table 7.3 knowing that the polarizations (o) and (e) correspond to (–) and (+) respectively in the case of a positive optical sign, and the reverse, *i.e.* (o) and (e) equivalent to (+) and (–) respectively for a negative sign. Note also that matrices (eoo), (oeo), and (eoo) can be deduced from each other by the proper permutation of the Cartesian indices. It is the same for the matrices (oee), (eoe), and (eeo).

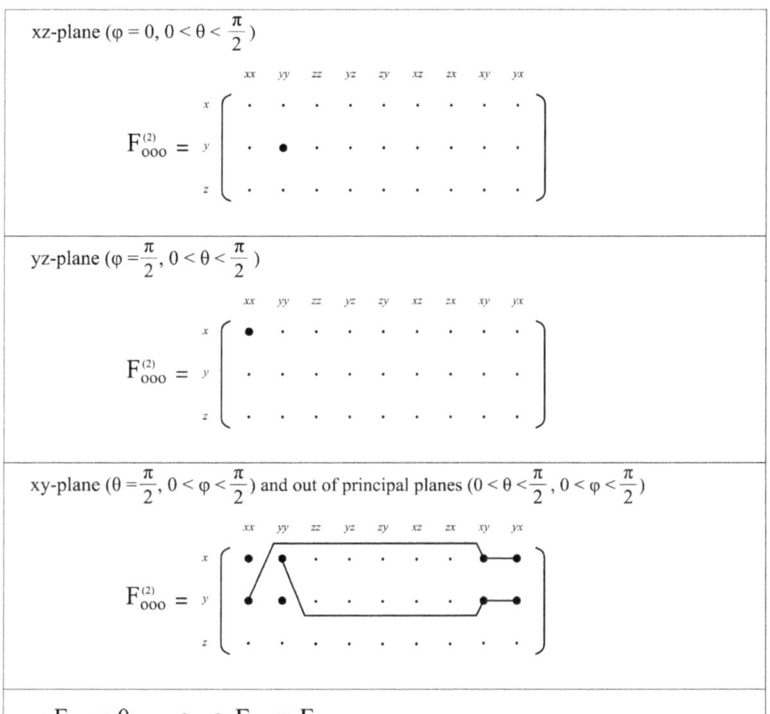

FIGURE 7.38 Field tensor matrices corresponding to the configuration of polarization (ooo) and for a propagation in the principal planes of uniaxial and biaxial crystals and out of the principal planes in uniaxial crystals only. θ and φ are the angles of spherical coordinates of the direction of propagation of the three interacting waves.

xz-plane ($\varphi = 0, 0 < \theta < \frac{\pi}{2}$)

$F^{(2)}_{eee} =$

	xx	yy	zz	yz	zy	xz	zx	xy	yx
x	●	·	●	·	·	●	●	·	·
y	·	·	·	·	·	·	·	·	·
z	●	·	●	·	·	●	●	·	·

yz-plane ($\varphi = \frac{\pi}{2}, 0 < \theta < \frac{\pi}{2}$)

$F^{(2)}_{eee} =$

	xx	yy	zz	yz	zy	xz	zx	xy	yx
x	·	·	·	·	·	·	·	·	·
y	·	●	●	●	●	·	·	·	·
z	·	●	●	●	●	·	·	·	·

xy-plane ($\theta = \frac{\pi}{2}, 0 < \varphi < \frac{\pi}{2}$)

$F^{(2)}_{eee} =$

	xx	yy	zz	yz	zy	xz	zx	xy	yx
x	·	·	·	·	·	·	·	·	·
y	·	·	·	·	·	·	·	·	·
z	·	·	●	·	·	·	·	·	·

Out of principal planes ($0 < \theta < \frac{\pi}{2}, 0 < \varphi < \frac{\pi}{2}$)

$F^{(2)}_{eee} =$

	xx	yy	zz	yz	zy	xz	zx	xy	yx
x	●	●	●	●	●	●	●	●	●
y	●	●	●	●	●	●	●	●	●
z	●	●	●	●	●	●	●	●	●

$\cdot\ F_{ijk} = 0$

FIGURE 7.39 Field tensor matrices corresponding to the configuration of polarization (eee) and for a propagation in the principal planes of uniaxial and biaxial crystals and out of the principal planes in uniaxial crystals only. θ and φ are the angles of spherical coordinates of the direction of propagation of the three interacting waves.

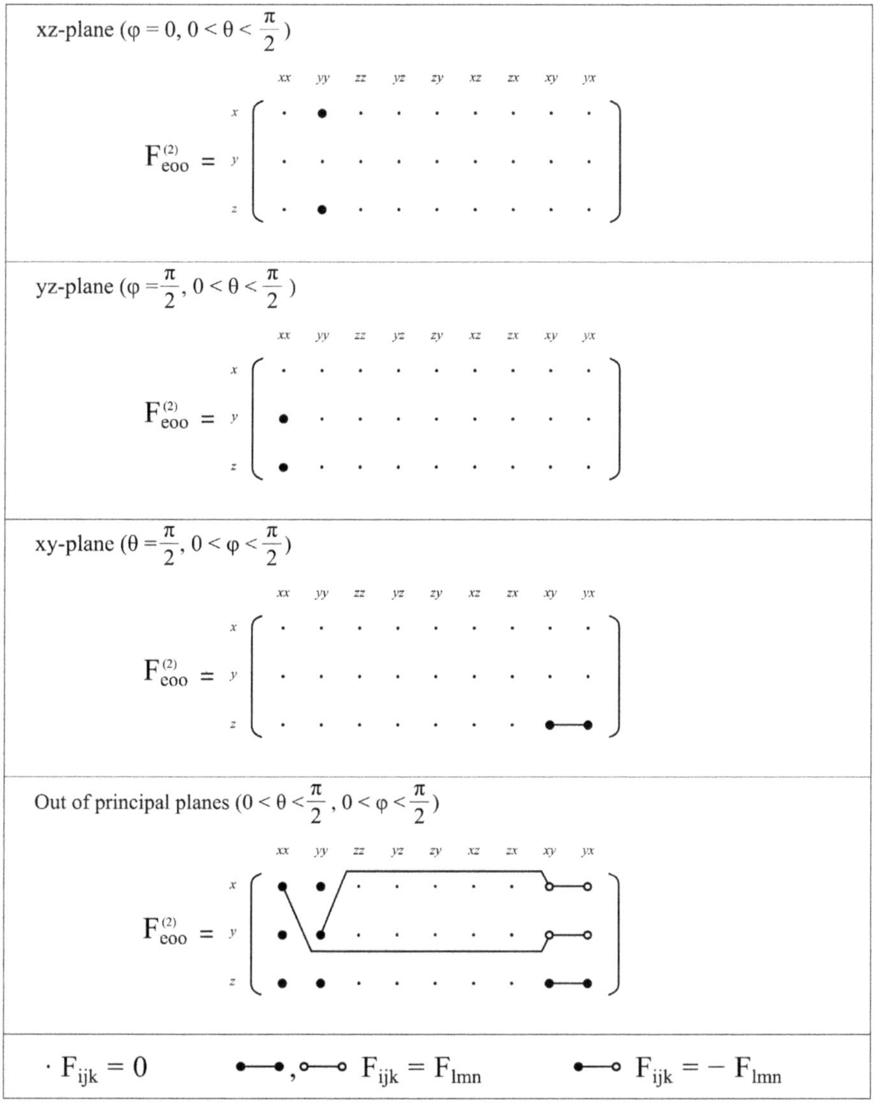

FIGURE 7.40 Field tensor matrices corresponding to the configuration of polarization (eoo) and for a propagation in the principal planes of uniaxial and biaxial crystals and out of the principal planes in uniaxial crystals only. θ and φ are the angles of spherical coordinates of the direction of propagation of the three interacting waves.

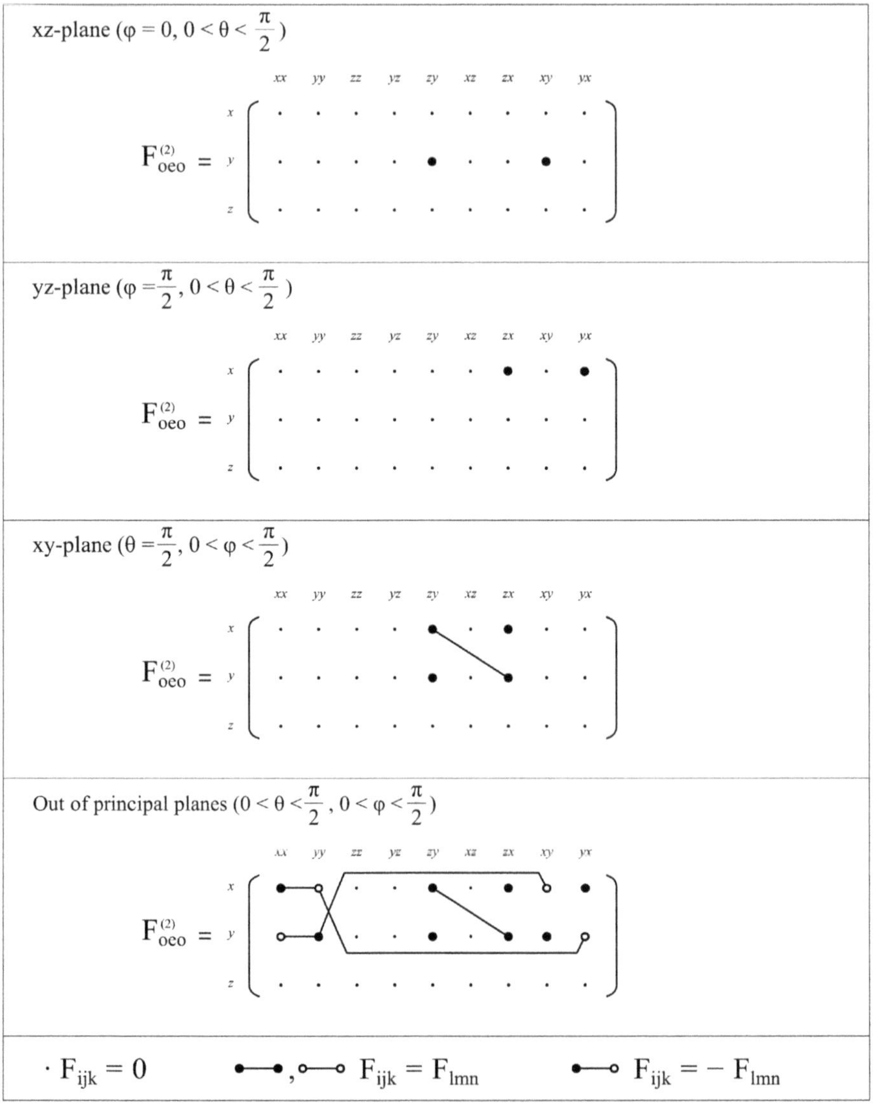

FIGURE 7.41 Field tensor matrices corresponding to the configuration of polarization (oeo) and for a propagation in the principal planes of uniaxial and biaxial crystals and out of the principal planes in uniaxial crystals only. θ and φ are the angles of spherical coordinates of the direction of propagation of the three interacting waves.

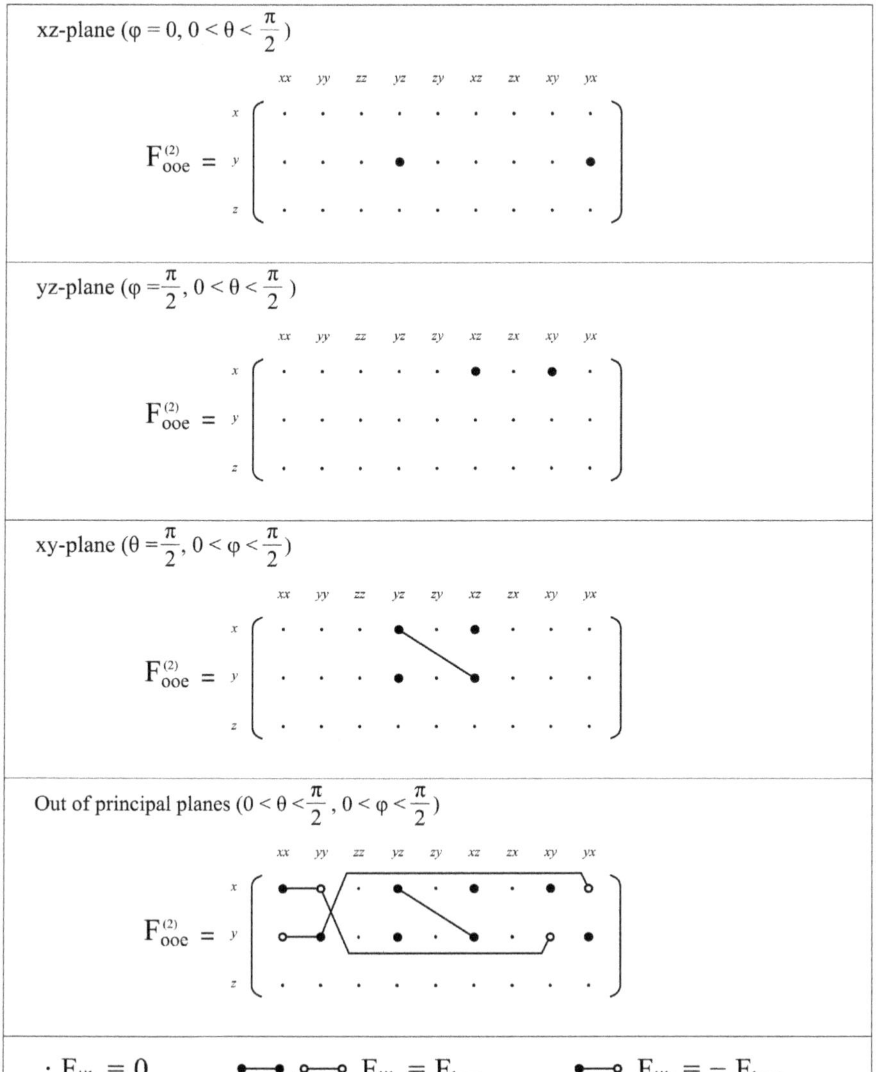

FIGURE 7.42 Field tensor matrices corresponding to the configuration of polarization (ooe) and for a propagation in the principal planes of uniaxial and biaxial crystals and out of the principal planes in uniaxial crystals only. θ and φ are the angles of spherical coordinates of the direction of propagation of the three interacting waves.

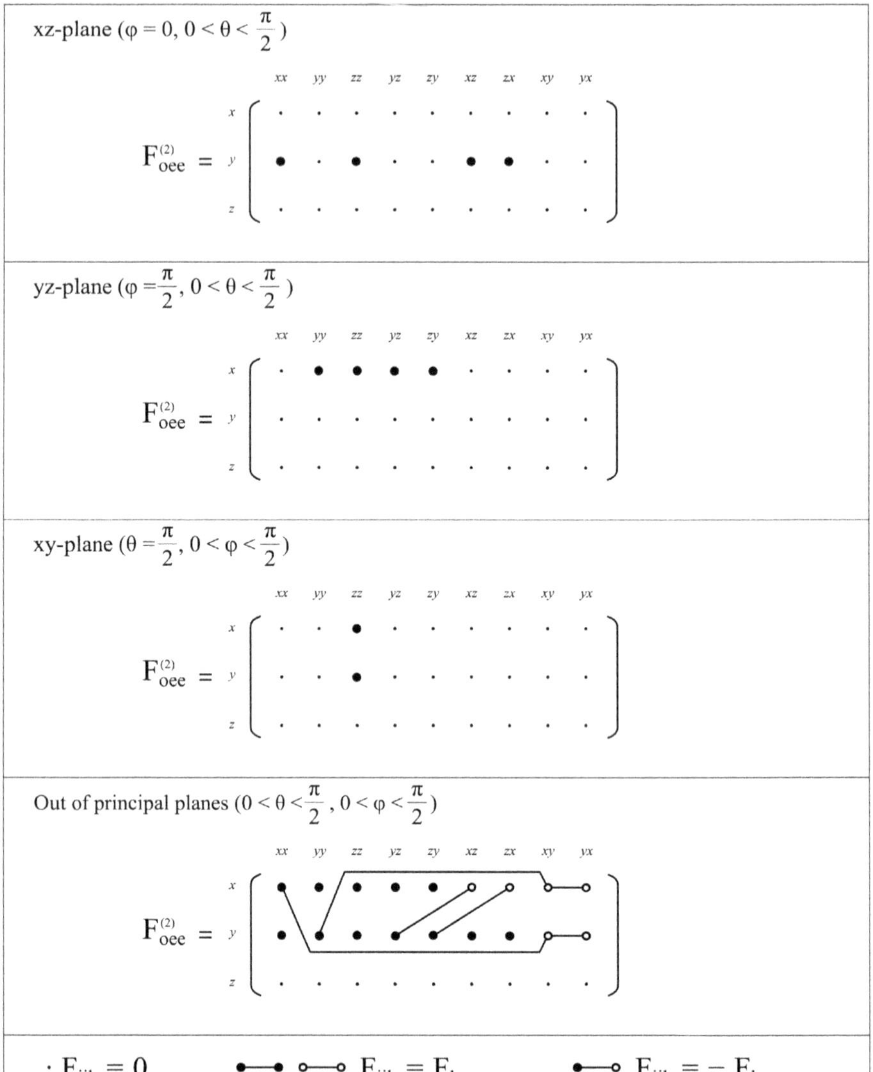

FIGURE 7.43 Field tensor matrices corresponding to the configuration of polarization (oee) and for a propagation in the principal planes of uniaxial and biaxial crystals and out of the principal planes in uniaxial crystals only. θ and φ are the angles of spherical coordinates of the direction of propagation of the three interacting waves.

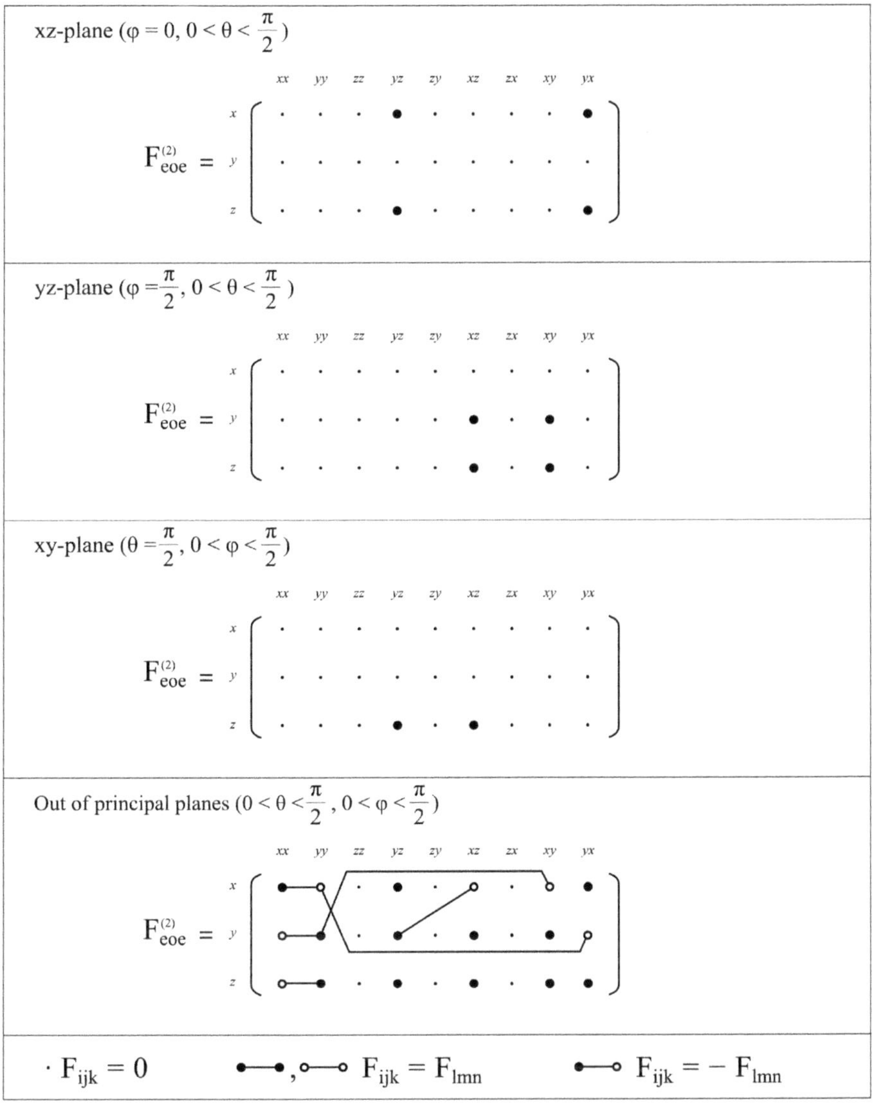

FIGURE 7.44 Field tensor matrices corresponding to the configuration of polarization (eoe) and for a propagation in the principal planes of uniaxial and biaxial crystals and out of the principal planes in uniaxial crystals only. θ and φ are the angles of spherical coordinates of the direction of propagation of the three interacting waves.

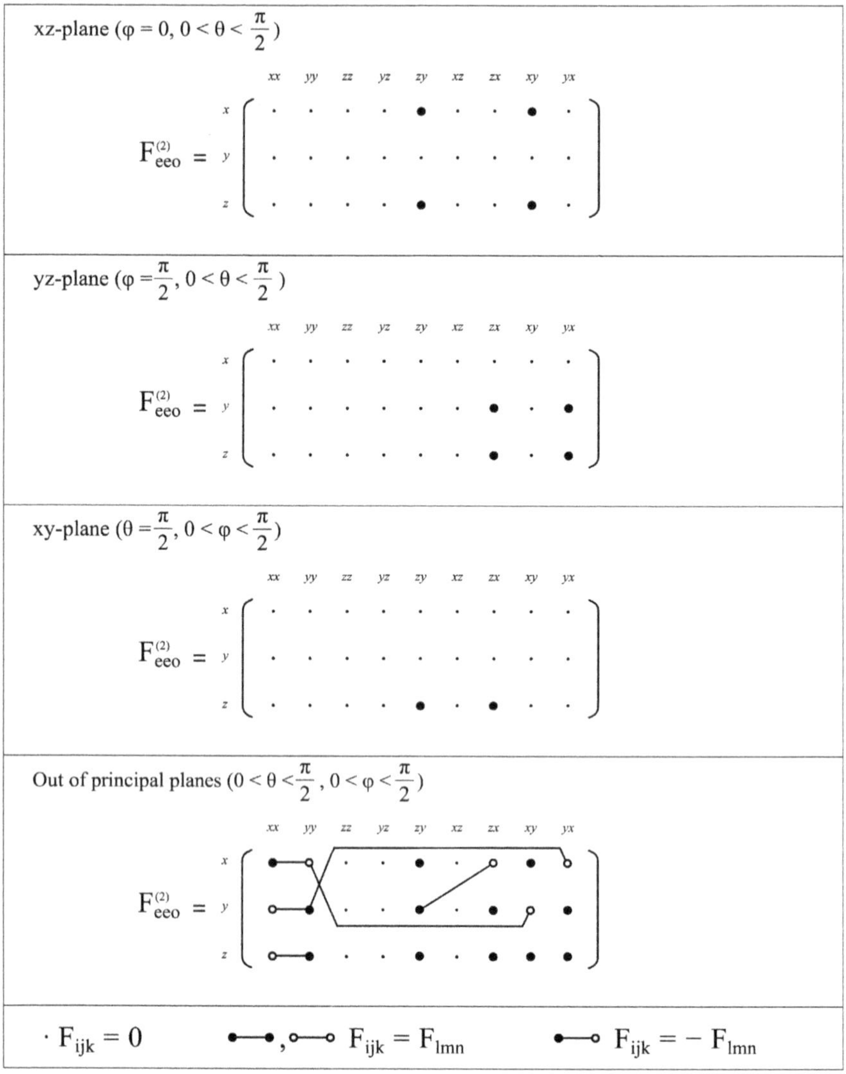

FIGURE 7.45 Field tensor matrices corresponding to the configuration of polarization (eeo) and for a propagation in the principal planes of uniaxial and biaxial crystals and out of the principal planes in uniaxial crystals only. θ and φ are the angles of spherical coordinates of the direction of propagation of the three interacting waves.

In the case of a propagation along the principle axes $\overrightarrow{O_x}$ and $\overrightarrow{O_y}$ in uniaxial and biaxial crystals, or along $\overrightarrow{O_z}$ in biaxial crystals, there is only one field tensor coefficient for each configuration of polarization, as shown in Table 7.4.

TABLE 7.4
Field Tensor Coefficients Corresponding to a Propagation along a Principal Axis in Uniaxial and Biaxial Crystals for the Eight Possible Configurations of Polarization

Optical class	Principal axis	ooo	eee	eoo	oeo	ooe	oee	eoe	eeo
Uniaxial and Biaxial	$\overrightarrow{O_x}$	$F^{(2)}_{yyy}$	$F^{(2)}_{zzz}$	$F^{(2)}_{zyy}$	$F^{(2)}_{yzy}$	$F^{(2)}_{yyz}$	$F^{(2)}_{yzz}$	$F^{(2)}_{zyz}$	$F^{(2)}_{zzy}$
	$\overrightarrow{O_y}$	$F^{(2)}_{xxx}$	$F^{(2)}_{zzz}$	$F^{(2)}_{zxx}$	$F^{(2)}_{xzx}$	$F^{(2)}_{xxz}$	$F^{(2)}_{xzz}$	$F^{(2)}_{zxz}$	$F^{(2)}_{zzx}$
Biaxial	$\overrightarrow{O_z}$	$F^{(2)}_{yyy}$	$F^{(2)}_{xxx}$	$F^{(2)}_{xyy}$	$F^{(2)}_{yxy}$	$F^{(2)}_{yyx}$	$F^{(2)}_{yxx}$	$F^{(2)}_{xyx}$	$F^{(2)}_{xxy}$

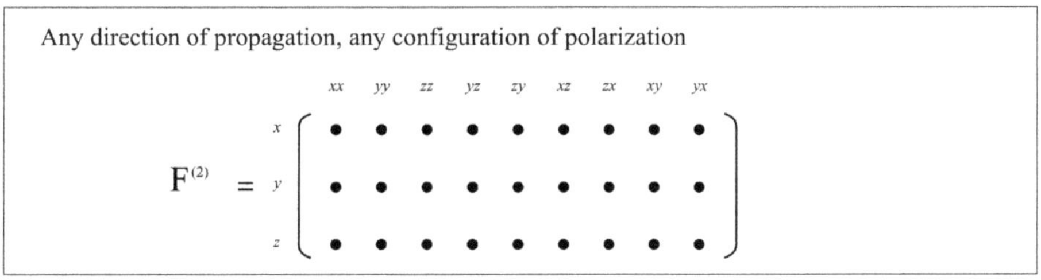

FIGURE 7.46 Field tensor matrix corresponding to the eight possible configurations of polarization (−−−), (+++), (−++), (+−+), (++−), (+−), (−+−), and (−−+) for a propagation out of the principal planes of biaxial crystals.

For a propagation along $\overrightarrow{O_z}$ in uniaxial crystals, any direction of polarization is allowed, as seen in Chapter 5. Thus there is an infinity of configurations of polarization, each of them leading to x and y Cartesian coordinates. Then in this case, the field tensor matrix is that of the (ooo) configuration depicted in Figure 7.38.

Finally, there is no symmetry, (i), (ii), or (iii), of the field tensor when the propagation occurs out of the principal planes in biaxial crystals since there are no ordinary and extraordinary polarizations. Then the field tensor exhibits twenty-seven independent coefficients whatever the configuration of polarization, as shown in Figure 7.46.

Then it is the same matrix as that of (eee) out of the principal planes of uniaxial crystals shown in Figure 7.39 (bottom).

Note also that all the matrices established in this section are possible for the isotropic optical class since any configuration is possible due to the lack of imposed Eigenmodes as seen in Chapter 5. But in this case, what is called an ordinary polarization corresponds to an electric field having x and y Cartesian coordinates only, the extraordinary polarization being orthogonal to the ordinary one so that it exhibits also a z-component.

7.5.3 Forbidden Crystal Classes

The symmetries of both the field tensor and electric susceptibility tensor can lead to situations in which their contraction is nil for a given configuration of polarization and a given direction of propagation $\vec{u}$, *i.e.*:

$$\begin{aligned}&\chi^{(2)}(\omega_3=\omega_1+\omega_2)\therefore F^{(2)}(\omega_3=\omega_1+\omega_2,\vec{u})\\&=\chi^{(2)}(\omega_1=\omega_3-\omega_2)\therefore F^{(2)}(\omega_1=\omega_3-\omega_2,\vec{u})\\&=\chi^{(2)}(\omega_2=\omega_3-\omega_1)\therefore F^{(2)}(\omega_2=\omega_3-\omega_1,\vec{u})\\&=0\end{aligned} \tag{7.124}$$

Thus in that case, there is orthogonality between the field tensor and the electric susceptibility tensor, which can be written:

$$\begin{cases}\chi^{(2)}(\omega_3=\omega_1+\omega_2)\perp F^{(2)}(\omega_3=\omega_1+\omega_2,\vec{u})\\\chi^{(2)}(\omega_1=\omega_3-\omega_2)\perp F^{(2)}(\omega_1=\omega_3-\omega_2,\vec{u})\\\chi^{(2)}(\omega_2=\omega_3-\omega_1)\perp F^{(2)}(\omega_2=\omega_3-\omega_1,\vec{u})\end{cases} \tag{7.125}$$

According to Eq. (7.34) and Eq. (7.36), Eq. (7.123) and Eq. (7.124) come from the orthogonality between the second-order induced polarization and the radiated field, *i.e.* $\vec{P}_i^{(2)} \perp \vec{e_i}$ with $i=(1,2,3)$.

This means that the corresponding conversion efficiency will be nil, even though phase-matching or quasi-phase-matching is achieved.

This orthogonality can occur in the full space of the directions of propagation $\vec{u}$, *i.e.* along the principal axes as well as in and out of the principal planes, for specific crystal classes and configurations of polarization. It happens in the following situations that concern the uniaxial optical class only: crystal classes 4*mm*, 422, $\bar{4}$, 4, 6*mm*, 622, 6 for the configuration of polarization (ooo); 422 and 622 for (eoo), (oeo), and (ooe); 4*mm* and 6*mm* for (oee), (eoe), and (eeo).

Figures 7.47 to 7.49 are given three examples of the corresponding matrices wherein the nullity of the contracted product is directly apparent, since it is given by the sum of the forward products.

In many situations, an orthogonality exists in the principal planes while it is not the case out of the latter ones, as shown in Figure 7.50.

This orthogonality criterion is of prime importance for choosing the best crystal and the most suitable configuration of polarization.

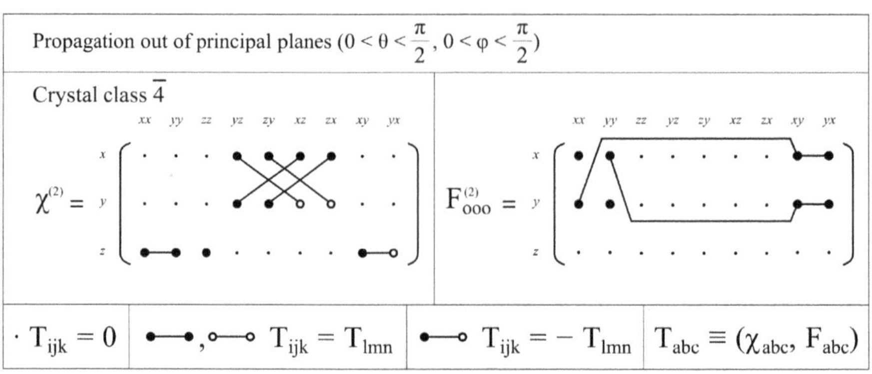

FIGURE 7.47 Second-order electric susceptibility tensor of crystal class $\bar{4}$ and second-order field tensor of configuration of polarization (ooo) for a propagation out of the principle planes whose contracted product is zero.

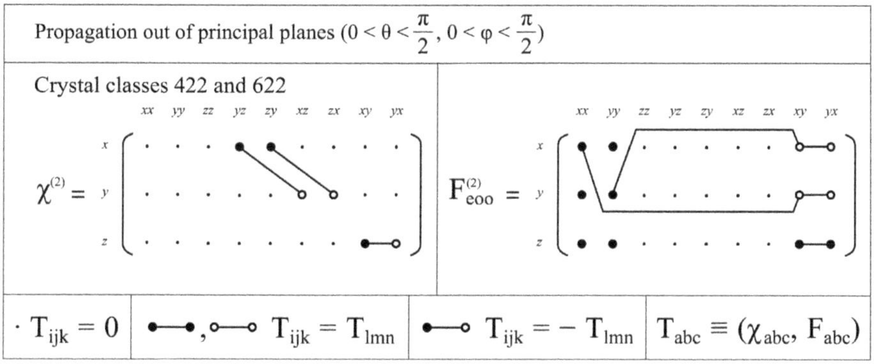

FIGURE 7.48 Second-order electric susceptibility tensor of crystal classes 422 and 622 and second-order field tensor of configuration of polarization (eoo) for a propagation out of the principle planes whose contracted product is zero.

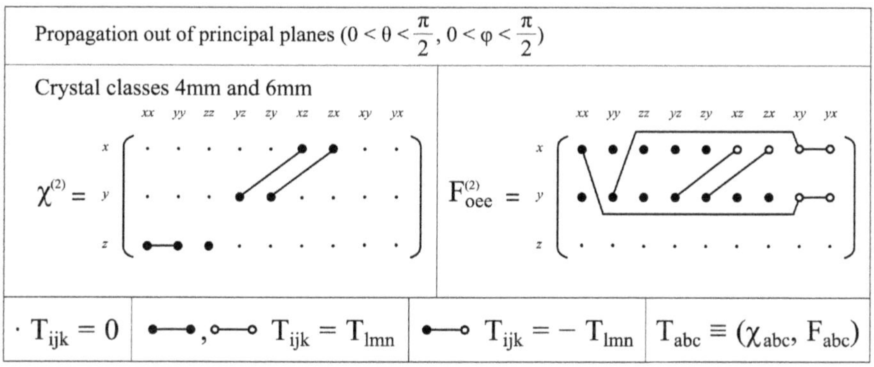

FIGURE 7.49 Second-order electric susceptibility tensor of crystal classes 4*mm* and 6*mm* and second-order field tensor of configuration of polarization (oee) for a propagation out of the principle planes whose contracted product is zero.

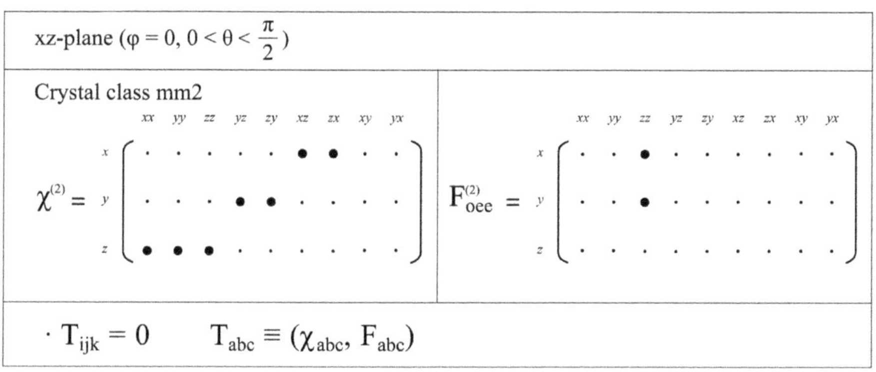

FIGURE 7.50 Second-order electric susceptibility tensor of crystal classes *mm*2 and second-order field tensor of configuration of polarization (oee) for a propagation in the *xz*-plane whose contracted product is zero.

7.5.4 Angular Distribution

The angular distribution of the field tensors coefficients can be properly established from the Cartesian coordinates of the components of the unit electric field vectors at the interacting circular frequencies, *i.e.* $e_i(\omega_j,\vec{u})$ with $i = (x, y, z)$ and $j = 1, 2, 3$, given in Part 5.3. As done in Chapter 5 and Chapter 6, it is convenient to plot the angular indicatrices as a function of the spherical angles coordinates (θ,φ) of the unit vector $\vec{u}$ of the direction of propagation.

It is useful here to begin by summarizing in a synthetic way what has been established in details in Part 5.3 for the ordinary and extraordinary Eigenmodes of polarization $\vec{e}^o(\omega,\vec{u})$ and $\vec{e}^e(\omega,\vec{u})$ defined for a propagation in uniaxial crystals. They are expressed as:

$$\vec{e}^o(\omega,\vec{u}) = \begin{pmatrix} e_x^o(\omega,\vec{u}) \\ e_y^o(\omega,\vec{u}) \\ e_z^o(\omega,\vec{u}) \end{pmatrix} = \begin{pmatrix} -\sin(\varphi) \\ \cos(\varphi) \\ 0 \end{pmatrix} \tag{7.126}$$

and

$$\vec{e}^e(\omega,\vec{u}) = \begin{pmatrix} e_x^e(\omega,\vec{u}) \\ e_y^e(\omega,\vec{u}) \\ e_z^e(\omega,\vec{u}) \end{pmatrix} = \begin{pmatrix} -\cos(\varphi)\cos\left(\theta \pm \rho^e(\omega,\vec{u})\right) \\ -\sin(\varphi)\cos\left(\theta \pm \rho^e(\omega,\vec{u})\right) \\ \sin\left(\theta \pm \rho^e(\omega,\vec{u})\right) \end{pmatrix} \tag{7.127}$$

with

$$\rho^e(\omega,\vec{u}) = Ar\cos\left[\frac{n_o^{-2}(\omega)\cos^2(\theta)+n_e^{-2}(\omega)\sin^2(\theta)}{\sqrt{n_o^{-4}(\omega)\cos^2(\theta)+n_e^{-4}(\omega)\sin^2(\theta)}}\right] \tag{7.128}$$

The sign in Eq. (7.126) is (+) for a negative uniaxial crystal, and (–) for a positive one. The quantity $\rho^e(\omega,\vec{u})$ is the spatial walk-off angle, whatever the positive or negative optical sign of the uniaxial crystal.

The description by ordinary and extraordinary Eigenmodes of polarization is also relevant in the case of a propagation in the principal planes of the dielectric frame of biaxial crystals. But it is necessary to replace (n_o, n_e) with (n_y, n_z) in the *xz*-plane $(\varphi = 0)$, with (n_x, n_z) in the *yz*-plane $(\varphi = \pi/2)$, and with (n_y, n_x) in the *xy*-plane $(\theta = \pi/2)$ in Eq. (7.127) for the extraordinary polarization. Equation (7.125), that describes the ordinary polarization, can also be used in the *xz*- and *yz*-planes, but not in the *xy*-plane where it has to be replaced by:

$$\begin{cases} \vec{e}^{\,o}(\omega,\vec{u}) = \begin{pmatrix} -\sin\left(\varphi \mp \rho^o(\omega,\vec{u})\right) \\ \cos\left(\varphi \mp \rho^o(\omega,\vec{u})\right) \\ 0 \end{pmatrix} \\ \rho^o(\omega,\vec{u}) = Ar\cos\left[\dfrac{n_y^{-2}(\omega)\cos^2(\varphi) + n_x^{-2}(\omega)\sin^2(\varphi)}{\sqrt{n_y^{-4}(\omega)\cos^2(\varphi) + n_x^{-4}(\omega)\sin^2(\varphi)}}\right] \end{cases} \quad (7.129)$$

The sign in Eq. (7.128) is (–) for a negative biaxial crystal, and (+) for a positive one.

For a propagation out of the principal planes of biaxial crystals, it is necessary to use the following equation:

$$e_i^{\pm}(\omega,\vec{u}) = \frac{[n^{\pm}(\omega,\vec{u})]^2}{[n^{\pm}(\omega,\vec{u})]^2 - [n_i]^2} u_i[\vec{u}.\overrightarrow{e^{\pm}}(\omega,\vec{u})] \quad (7.130)$$

where i = x, y, z, and $n^{\pm}(\omega,\vec{u})$ refers to the two possible values that the refractive index can take in the considered direction of propagation $\vec{u}$.

Thanks to these equations and those established in Part 7.5.1, it is easy to calculate the angular distribution of the relevant field tensor coefficients of any crystal class at first, and then the corresponding effective coefficient if the second-order electric susceptibility tensor coefficients are known. Next are given the examples of two uniaxial crystals belonging to different crystal classes, *i.e.* $CdSiP_2$ ($\bar{4}2m$) (Kemlin 2011) [5] and $La_3Ga_{5.5}Ta_{0.5}O_{14}$ (32) (Boursier 2014) [16], and two biaxial crystals: $KTiOPO_4$ (*mm*2) (Boulanger 1994) [8] and $BaGa_4Se_7$ (*m*) (Boursier 2016) [17].

$CdSiP_2$ (CSP) is a negative uniaxial crystal belonging to the tetragonal crystal class $\bar{4}2m$. According to Neumann's principle, its second-order electric susceptibility tensor is described by Figure 3.29 showing that there are six coefficients: $\chi^{(2)}_{xyz} = \chi^{(2)}_{yxz}$, $\chi^{(2)}_{xzy} = \chi^{(2)}_{yzx}$, and $\chi^{(2)}_{zxy} = \chi^{(2)}_{zyx}$. Then, the six corresponding field tensor coefficients are $F^{(2)}_{xyz}, F^{(2)}_{yxz}, F^{(2)}_{xzy}, F^{(2)}_{yzx}, F^{(2)}_{zxy}$, and $F^{(2)}_{zyx}$. Taking the example of type I SHG, whose phase-matching curve is shown in Figure 7.12, the configuration of polarization is (eoo). According to Eq. (7.125) and Eq. (7.126), the field tensor coefficients write:

$$\begin{cases} F^{(2)}_{xyz} = F^{(2)}_{yxz} = F^{(2)}_{xzy} = F^{(2)}_{yzx} = 0 \\ F^{(2)}_{zxy} = F^{(2)}_{zyx} = -\sin(\varphi)\cos(\varphi)\sin\left(\theta + \rho^e(2\omega,\theta)\right) \\ = -\dfrac{1}{2}\sin(2\varphi)\sin\left(\theta + \rho^e(2\omega,\theta)\right) \end{cases} \quad (7.131)$$

It is obvious that the relations between the field tensor coefficients in Eq. (7.30) correspond to those in the matrix of Figure 7.40 corresponding to the general case of a propagation out of the principal planes.

From Eq. (7.130) and according to Eq. (7.120), it comes for the effective coefficient of type I SHG in CSP:

$$\chi_{eff}^{(2)} = -\chi_{zxy}^{(2)}(2\omega)\sin(2\varphi)\sin\left(\theta + \rho^{e}(2\omega,\theta)\right) \tag{7.132}$$

Then the extremal value of the effective coefficient is reached at $\varphi = 45°$, which corresponds to the maximum of conversion efficiency since it is proportional to $\left[\chi_{eff}^{(2)}\right]^2$.

The curves of $F_{zxy}^{(2)} = F_{zyx}^{(2)}$ and $\left[\chi_{eff}^{(2)}\right]^2$ as a function of φ_{PM} are depicted Figure 7.51 in the case in which $\lambda_\omega = 5\ \mu m$, which gives a phase-matching angle $\theta_{PM} = 42.54°$ according to Figure 7.12. The calculations are performed with $\chi_{zxy}^{(2)} = \chi_{zyx}^{(2)} = 169\ pm/V$ and the ordinary and extraordinary principal refractive indices given in Kemlin (2011) [5].

The curves of Figure 7.51 are calculated along the upper half-space part of the phase-matching cone, *i.e.* at $\theta_{PM} = 42.54°$. In the lower half-space part, *i.e.* $90° \le \theta \le 180°$, there is a change only on the sign of the field tensor leading to a sign reversal of the lobes, the square of the effective coefficient remaining the same. Considering the full space, *i.e.* $0° \le \theta \le 180°$, the angular distribution of the square of the effective coefficient clearly exhibits a 4-fold axis along the z-axis and three symmetry mirrors corresponding to the three principal planes, *i.e.* the xz-, yz-, and xy-planes. The orientation symmetry is then $\frac{4}{m}mm$ (see the stereographic projection of Figure 2.13) which is not the same as that of the angular distribution of the phase-matching directions depicted in Figures 7.10 and 7.11 that belongs to the infinite point group $\left(\frac{\infty}{m}m\right)$. Still note that the property "square of the effective coefficient" is a supergroup of the crystal's orientation symmetry group, *i.e.* $\frac{4}{m}mm \supset \bar{4}2m$, which is in line with Curie's laws, in which the orientation symmetry elements of a physical response contains at least the symmetry elements of the cause of such a response.

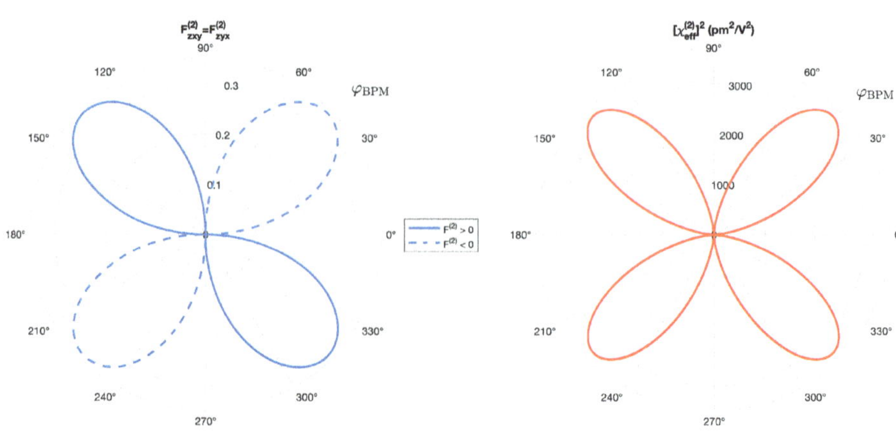

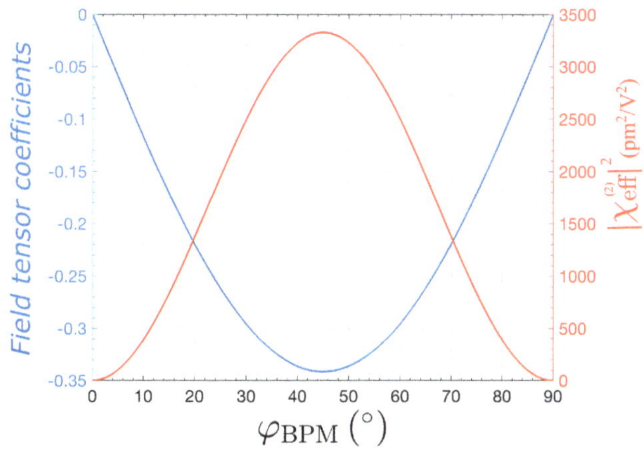

FIGURE 7.51 Second-order field tensor coefficients $F^{(2)}_{zxy} = F^{(2)}_{zyx}$ and square of the second-order effective coefficient $\left[\chi^{(2)}_{eff}\right]^2$ as a function of the phase-matching angle of spherical coordinate φ_{PM} in the $\bar{4}2m$ negative uniaxial crystal $CdSiP_2$ in the case of type I SHG ($\lambda_\omega = 5\ \mu m \rightarrow \lambda_{2\omega} = 2.5\ \mu m$), with the configuration of polarization (eoo). The phase-matching angle of spherical coordinate θ_{PM} is equal to 42.54°.

$La_3Ga_{5.5}Ta_{0.5}O_{14}$ (LGT) is a positive uniaxial crystal belonging to the trigonal crystal class 32. Its second-order electric susceptibility tensor is described by Figure 3.26 (left) showing that there are ten coefficients from Neumann's principle verifying: $\chi^{(2)}_{xxx} = -\chi^{(2)}_{xyy} = -\chi^{(2)}_{yxy} = -\chi^{(2)}_{yyx}$, $\chi^{(2)}_{xyz} = -\chi^{(2)}_{yxz}$, $\chi^{(2)}_{xzy} = -\chi^{(2)}_{yzx}$, and $\chi^{(2)}_{zxy} = -\chi^{(2)}_{zyx}$. Then, there are ten field tensor coefficients to consider: $F^{(2)}_{xxx}, F^{(2)}_{xyy}, F^{(2)}_{yxy}, F^{(2)}_{yyx}, F^{(2)}_{xyz}, F^{(2)}_{yxz}, F^{(2)}_{xzy}, F^{(2)}_{yzx}, F^{(2)}_{zxy}$, and $F^{(2)}_{zyx}$. The example of type I SHG is also considered here, but the corresponding configuration of polarization here is (oee) because the optical sign is positive. Then it comes:

$$\begin{cases} F^{(2)}_{xxx} = -F^{(2)}_{yxy} = -F^{(2)}_{yyx} = -\sin(\varphi)\left[\cos(\varphi)\right]^2\left[\cos\left(\theta - \rho^e(\omega,\theta)\right)\right]^2 \\ F^{(2)}_{xyy} = -\left[\sin(\varphi)\right]^3\left[\cos\left(\theta - \rho^e(\omega,\theta)\right)\right]^2 \\ F^{(2)}_{xyz} = F^{(2)}_{xzy} = \left[\sin(\varphi)\right]^2\cos\left(\theta - \rho^e(\omega,\theta)\right)\sin\left(\theta - \rho^e(2\omega,\theta)\right) \\ F^{(2)}_{yxz} = F^{(2)}_{yzx} = -\left[\cos(\varphi)\right]^2\cos\left(\theta - \rho^e(\omega,\theta)\right)\sin\left(\theta - \rho^e(\omega,\theta)\right) \\ F^{(2)}_{zxy} = F^{(2)}_{zyx} = 0 \end{cases} \tag{7.133}$$

Here also, the relations between field tensor coefficients are those appearing in the matrix of Figure 7.43 in the case of a propagation out of the principal planes, adding the symmetry over the two last indices thanks to SHG.

Then from Eq. (7.132) and the second-order electric susceptibility tensor for the crystal class 32 under Kleinman symmetry shown in Figure 3.26 in which the only non-zero coefficients are $\chi_{xxx}^{(2)} = -\chi_{xyy}^{(2)} = -\chi_{yxy}^{(2)} = -\chi_{yyx}^{(2)}$, the effective coefficient of type I SHG in LGT writes:

$$\chi_{eff}^{(2)} = -\chi_{xxx}^{(2)}(2\omega)\sin(3\varphi)\left[\cos\left(\theta - \rho^{e}(\omega,\theta)\right)\right]^{2} \tag{7.134}$$

Then the extremal value of the effective coefficient over an octant of the space is reached at $\varphi = 30°$ and $\varphi = 90°$.

The curves of $F_{xxx}^{(2)} = -F_{yxy}^{(2)} = -F_{yyx}^{(2)}, F_{xyy}^{(2)}, F_{xyz}^{(2)} = F_{xzy}^{(2)}, F_{yxz}^{(2)} = F_{yzx}^{(2)}$, and $\left[\chi_{eff}^{(2)}\right]^{2}$ as a function of φ_{PM} are shown Figure 7.52 in the case in which $\lambda_{\omega} = 1.34\ \mu m$, which gives a phase-matching angle $\theta_{PM} = 58.5°$ according to Boursier (2014) [16]. The calculations are performed with $\chi_{xxx}^{(2)} = 4.8\ pm/V$ and the ordinary and extraordinary principal refractive indices given in Boursier (2014) [16].

The curves of Figure 7.52 are calculated along the upper half-space part of the phase-matching cone, *i.e.* at $\theta_{PM} = 58.5°$. In the lower half-space part, there is a change only on the sign of the field tensor leading to a sign reversal of the lobes, as is the case for CSP described previously. Considering the full space, the angular distribution of the square of the effective coefficient clearly exhibits a 6-fold axis along the z-axis and three symmetry mirrors corresponding to the three principal planes, *i.e.* the xz-, yz-, and xy-planes. The orientation symmetry is then $\frac{6}{m}mm$ (see the stereographic representation of Figure 2.14), which is a supergroup of the point group of the crystal, *i.e.* $\frac{6}{m}mm \supset 32$, similar to what was discussed previously for $CdSiP_2$.

$KTiOPO_4$ (KTP) is a positive biaxial crystal that belongs to the orthorhombic crystal class *mm*2. The second-order electric susceptibility tensor is given Figure 3.23: there are seven coefficients according to Neumann's principle, *i.e.* $\chi_{xxz}^{(2)}, \chi_{xzx}^{(2)}, \chi_{yyz}^{(2)}, \chi_{yzy}^{(2)}, \chi_{zxx}^{(2)}, \chi_{zyy}^{(2)}$, and $\chi_{zzz}^{(2)}$, and so seven field tensor coefficients to consider, *i.e.* $F_{xxz}^{(2)}, F_{xzx}^{(2)}, F_{yyz}^{(2)}, F_{yzy}^{(2)}, F_{zxx}^{(2)}, F_{zyy}^{(2)}$, and $F_{zzz}^{(2)}$. The example of type II SHG is considered, the suited configuration of polarization being (−+−) or (−−+). Actually, these two configurations are equivalent because it is a SHG process. As said previously, it is possible to define ordinary and extraordinary waves in the principal planes, which gives (oeo) or (ooe) in the present case; (ooe) is taken for the following calculations. The example of the phase-matching curve given in Figure 7.15 is considered here: the phase-matching cone joins the xy-plane and the yz-plane.

For the xy-plane, it comes according to Eq. (7.126), Eq. (7.127), and Eq. (7.128):

$$\begin{cases} F_{xxz}^{(2)} = \sin\left(\varphi - \rho^{o}(2\omega,\vec{u})\right)\sin\left(\varphi - \rho^{o}(\omega,\vec{u})\right) \\ F_{yyz}^{(2)} = \cos\left(\varphi - \rho^{o}(2\omega,\vec{u})\right)\cos\left(\varphi - \rho^{o}(\omega,\vec{u})\right) \\ F_{xzx}^{(2)} = F_{yzy}^{(2)} = F_{zxx}^{(2)} = F_{zyy}^{(2)} = F_{zzz}^{(2)} = 0 \end{cases} \tag{7.135}$$

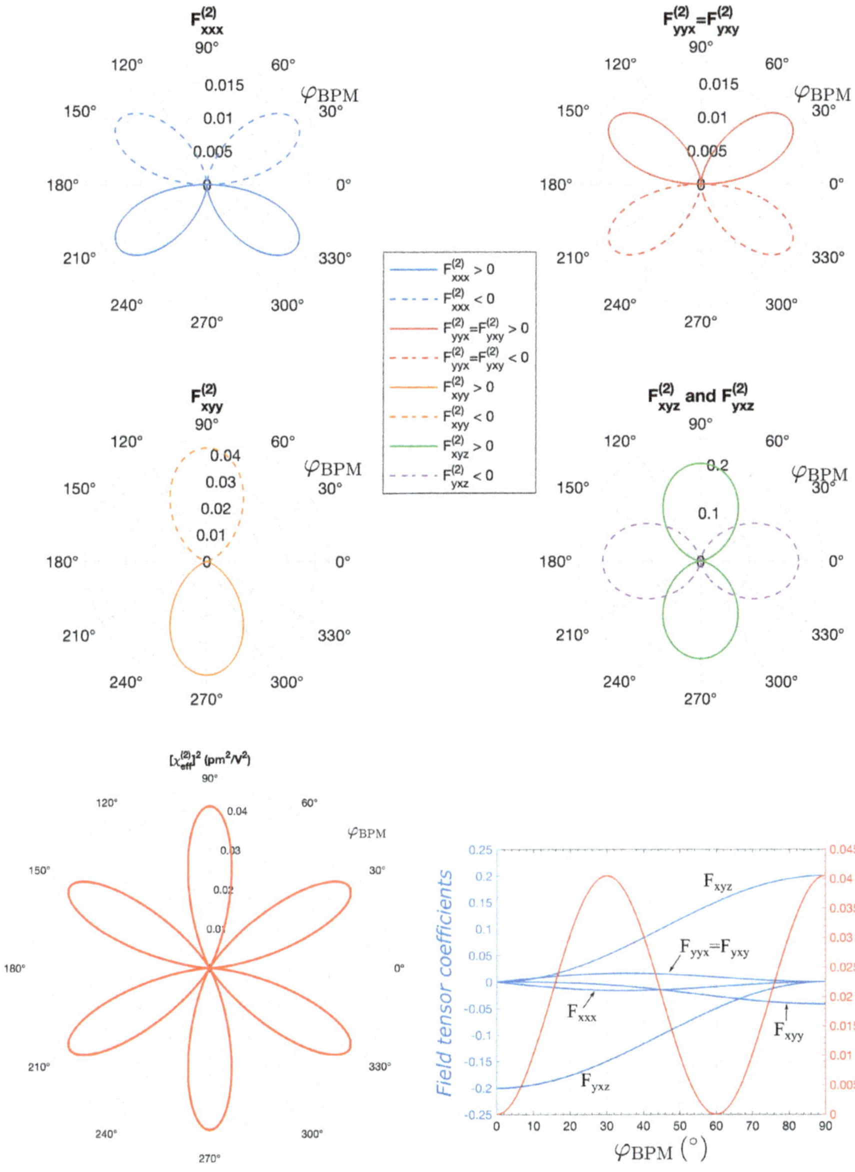

FIGURE 7.52 Second-order field tensor coefficients and square of the second-order effective coefficient as a function of the phase-matching angle of spherical coordinate φ_{PM} in the 32 positive uniaxial crystal $La_3Ga_{5.5}Ta_{0.5}O_{14}$ for type I SHG ($\lambda_\omega = 1.34\ \mu m \rightarrow \lambda_{2\omega} = 0.67\ \mu m$), with the configuration of polarization (eoo). The phase-matching angle of spherical coordinate θ_{PM} is equal to 58.5°.

Then the effective coefficient writes:

$$\chi_{eff}^{(2)} = \chi_{xxz}^{(2)}(2\omega)\left[\sin\left(\varphi - \rho^{o}(2\omega,\vec{u})\right)\sin\left(\varphi - \rho^{o}(\omega,\vec{u})\right)\right] + \chi_{yyz}^{(2)}(2\omega)\left[\cos\left(\varphi - \rho^{o}(2\omega,\vec{u})\right)\cos\left(\varphi - \rho^{o}(\omega,\vec{u})\right)\right] \quad (7.136)$$

In the yz-plane, the calculations from Eq. (7.125), Eq. (7.126), and Eq. (7.127) give:

$$\begin{cases} F_{xxz}^{(2)} = \sin\left(\theta - \rho^{e}(\omega,\vec{u})\right) \\ F_{yyz}^{(2)} = F_{xzx}^{(2)} = F_{yzy}^{(2)} = F_{zxx}^{(2)} = F_{zyy}^{(2)} = F_{zzz}^{(2)} = 0 \\ \chi_{eff}^{(2)} = \chi_{xxz}^{(2)}(2\omega)\sin\left(\theta - \rho^{e}(\omega,\vec{u})\right) \end{cases} \quad (7.137)$$

Out of the xy-plane and yz-plane, it is necessary to use Eq. (7.129).

The relations between the field tensor coefficients in Eq. (7.134) and Eq. (7.136) are well compatible with the matrices of Figure 7.42 for the xy-plane and yz-plane, respectively.

As explained in Part 3.2.3.1 about symmetries due to equalities between circular frequencies, it comes for KTP: $\chi_{xxz}^{(2)} = \chi_{xzx}^{(2)} \neq \chi_{yyz}^{(2)} = \chi_{yzy}^{(2)} \neq \chi_{zxx}^{(2)} \neq \chi_{zyy}^{(2)} \neq \chi_{zzz}^{(2)}$, which leads to five independent coefficients, *i.e.* $\chi_{15}^{(2)} \neq \chi_{24}^{(2)} \neq \chi_{31}^{(2)} \neq \chi_{32}^{(2)} \neq \chi_{33}^{(2)}$ using the contracted notation. The effective coefficient can then be also written:

$$\chi_{eff}^{(2)} = \chi_{15}^{(2)}(2\omega)\left[F_{xxz}^{(2)} + F_{xzx}^{(2)}\right] + \chi_{24}^{(2)}(2\omega)\left[F_{yyz}^{(2)} + F_{yzy}^{(2)}\right] + \chi_{31}^{(2)}(2\omega)\left[F_{zxx}^{(2)}\right] + \chi_{32}^{(2)}(2\omega)\left[F_{zyy}^{(2)}\right] + \chi_{33}^{(2)}(2\omega)\left[F_{zzz}^{(2)}\right] \quad (7.138)$$

$F_{xxz}^{(2)} + F_{xzx}^{(2)}, F_{yyz}^{(2)} + F_{yzy}^{(2)}, F_{zxx}^{(2)}, F_{zyy}^{(2)}, F_{zzz}^{(2)}$, and $\left[\chi_{eff}^{(2)}\right]^2$ of type II SHG ($\lambda_\omega = 1.064\ \mu m \rightarrow \lambda_{2\omega} = 0.532\ \mu m$) in KTP are plotted as a function of φ_{PM} in Figure 7.53. The corresponding phase-matching curve is shown in Figure 7.15. The calculation of the effective coefficient is performed assuming Kleinman's symmetry, *i.e.* $\chi_{15}^{(2)} = \chi_{31}^{(2)} = 2.80\ pm/V, \chi_{24}^{(2)} = \chi_{32}^{(2)} = 5.30\ pm/V$ and $\chi_{33}^{(2)} = 21.40\ pm/V$ (Boulanger 1994) [8], the principal refractive indices being given in Boulanger (1994) [8].

The curves of Figure 7.53 are calculated along the upper half-space part of the phase-matching cone. In the lower half-space part, there is a change only on the sign of the field tensor leading to a sign reversal of the lobes, as is the case for CSP and LGT. Considering the full space, the angular distribution of the square of the effective coefficient exhibits three mirror planes in the xy-, xz-, and yz-planes; then it belongs to the orthorhombic orientation symmetry class *mmm* (see the stereographic projection of Figure 2.11), the same as those of the phase-matching loci and of the index surface, and which verifies: $mmm \supset mm2$.

$BaGa_4Se_7$ (BGSe) is a positive biaxial crystal that belongs to the monoclinic crystal class m with the special axis along $\overrightarrow{Ox}$; it means that this axis is perpendicular to

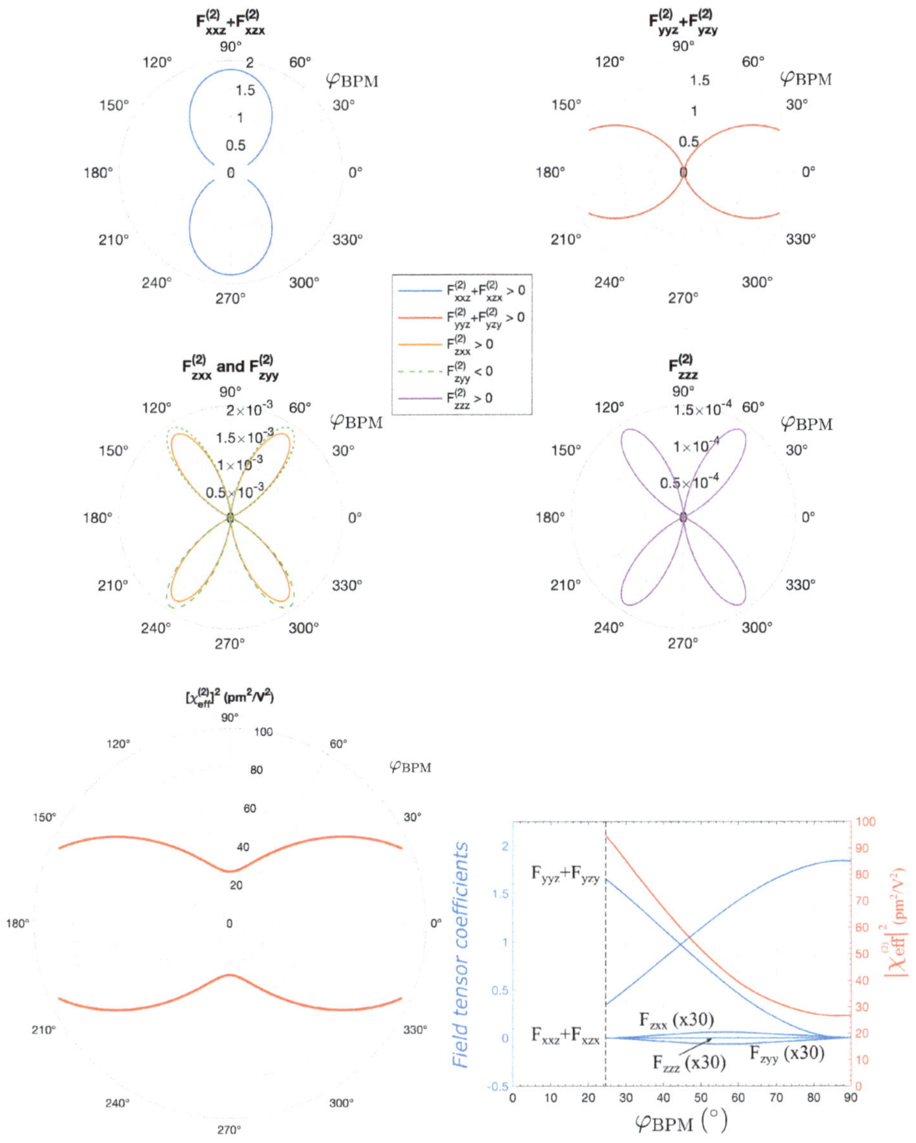

FIGURE 7.53 Second-order field tensor coefficients and square of the second-order effective coefficient as a function of the phase-matching angle of spherical coordinate φ_{PM} in the *mm*2 positive biaxial crystal KTiOPO$_4$ for type II SHG ($\lambda_\omega = 1.064\ \mu m \rightarrow \lambda_{2\omega} = 0.532\ \mu m$), with the configuration of polarization (−−+) equivalent to (−+−). The corresponding phase-matching angles of spherical coordinate θ_{PM} can be found using the relevant curve of Figure 7.15.

the mirror m. The second-order electric susceptibility tensor is not given in Chapter 3 but it can be obtained by that of Figure 3.21 by simply replacing the Cartesian indices (x,y,z) with (z,x,y). There are fourteen coefficients according to Neumann's principle, *i.e.* $\chi_{xxz}^{(2)},\chi_{xzx}^{(2)},\chi_{xxy}^{(2)},\chi_{xyx}^{(2)},\chi_{yxx}^{(2)},\chi_{yyy}^{(2)},\chi_{yzz}^{(2)},\chi_{yyz}^{(2)},\chi_{yzy}^{(2)},\chi_{zxx}^{(2)},\chi_{zyy}^{(2)},\chi_{zzz}^{(2)},\chi_{zyz}^{(2)},\chi_{zzy}^{(2)}$, which gives fourteen field tensor coefficients *i.e.* $F_{xxz}^{(2)},F_{xzx}^{(2)},F_{xxy}^{(2)},F_{xyx}^{(2)},F_{yxx}^{(2)},F_{yyy}^{(2)},F_{yzz}^{(2)},F_{yyz}^{(2)},F_{yzy}^{(2)}$, $F_{zxx}^{(2)},F_{zyy}^{(2)},F_{zzz}^{(2)},F_{zyz}^{(2)},F_{zzy}^{(2)}$. The example of type II SHG is taken as for KTP from earlier, and we consider a phase-matching cone joining the xz-plane and the yz-plane.

For the xz-plane, it comes according to Eq. (7.125), Eq. (7.126), and Eq. (7.127):

$$\begin{cases} F_{yyz}^{(2)} = \sin\left(\theta - \rho^e(\omega,\vec{u})\right) \\ F_{xxz}^{(2)}, F_{xzx}^{(2)} = F_{xxy}^{(2)} = F_{xyx}^{(2)} = F_{yxx}^{(2)} = F_{yyy}^{(2)} = F_{yzz}^{(2)} \\ = F_{yzy}^{(2)} = F_{zxx}^{(2)} = F_{zyy}^{(2)} = F_{zzz}^{(2)} = F_{zyz}^{(2)} = F_{zzy}^{(2)} = 0 \\ \\ \chi_{eff}^{(2)} = \chi_{xxz}^{(2)}(2\omega)\sin\left(\theta - \rho^e(\omega,\vec{u})\right) \end{cases} \tag{7.139}$$

- In the yz-plane, Eq. (7.125), Eq. (7.126), and Eq. (7.127) lead to:

$$\begin{cases} F_{xxz}^{(2)} = \sin\left(\theta - \rho^e(\omega,\vec{u})\right) \\ F_{xxy}^{(2)} = -\cos\left(\theta - \rho^e(\omega,\vec{u})\right) \\ F_{xzx}^{(2)} = F_{xyx}^{(2)} = F_{yxx}^{(2)} = F_{yyy}^{(2)} = F_{yzz}^{(2)} = F_{yyz}^{(2)} \\ = F_{yzy}^{(2)} = F_{zxx}^{(2)} = F_{zyy}^{(2)} = F_{zzz}^{(2)} = F_{zyz}^{(2)} = F_{zzy}^{(2)} = 0 \\ \\ \chi_{eff}^{(2)} = \chi_{xxz}^{(2)}(2\omega)\sin\left(\theta - \rho^e(\omega,\vec{u})\right) - \chi_{xxy}^{(2)}(2\omega)\cos\left(\theta - \rho^e(\omega,\vec{u})\right) \end{cases} \tag{7.140}$$

- Out of the xz-plane and yz-plane, it is necessary to use Eq. (7.129), as done for KTP.

The expression of the effective coefficient is heavy but it can be lightened by taking into account the symmetry of the second-order electric susceptibility tensor over the two last Cartesian indices since SHG is considered. Actually, it comes using the contracted notation:

$$\begin{cases} \chi_{xxz}^{(2)} = \chi_{xzx}^{(2)} \equiv \chi_{15}^{(2)} \neq \chi_{xxy}^{(2)} = \chi_{xyx}^{(2)} \equiv \chi_{16}^{(2)} \\ \neq \chi_{yxx}^{(2)} \equiv \chi_{21}^{(2)} \neq \chi_{yyy}^{(2)} \equiv \chi_{22}^{(2)}, \chi_{yzz}^{(2)} \equiv \chi_{23}^{(2)} \neq \chi_{yyz}^{(2)} = \chi_{yzy}^{(2)} \equiv \chi_{24}^{(2)} \\ \neq \chi_{zxx}^{(2)} \equiv \chi_{31}^{(2)} \neq \chi_{zyy}^{(2)} \equiv \chi_{32}^{(2)} \neq \chi_{zzz}^{(2)} \equiv \chi_{33}^{(2)} \neq \chi_{zyz}^{(2)} = \chi_{zzy}^{(2)} \equiv \chi_{34}^{(2)} \end{cases} \tag{7.141}$$

Then using Eq. (7.140), the effective coefficient writes:

$$\begin{cases} \chi_{eff}^{(2)} = \chi_{15}^{(2)}(2\omega)\left[F_{xxz}^{(2)} + F_{xzx}^{(2)}\right] + \chi_{16}^{(2)}(2\omega)\left[F_{xxy}^{(2)} + F_{xyx}^{(2)}\right] \\ \quad + \chi_{21}^{(2)}(2\omega)\left[F_{yxx}^{(2)}\right] + \chi_{22}^{(2)}(2\omega)\left[F_{yyy}^{(2)}\right] + \chi_{23}^{(2)}(2\omega)\left[F_{yzz}^{(2)}\right] \\ \quad + \chi_{24}^{(2)}(2\omega)\left[F_{yyz}^{(2)} + F_{yzy}^{(2)}\right] \\ \quad + \chi_{31}^{(2)}(2\omega)\left[F_{zxx}^{(2)}\right] + \chi_{32}^{(2)}(2\omega)\left[F_{zyy}^{(2)}\right] + \chi_{33}^{(2)}(2\omega)\left[F_{zzz}^{(2)}\right] \\ \quad + \chi_{34}^{(2)}(2\omega)\left[F_{zyz}^{(2)} + F_{zzy}^{(2)}\right] \end{cases} \tag{7.142}$$

$F_{xxz}^{(2)} + F_{xzx}^{(2)}, F_{xxy}^{(2)} + F_{xyx}^{(2)}, F_{yxx}^{(2)}, F_{yyy}^{(2)}, F_{yzz}^{(2)}, F_{yyz}^{(2)} + F_{yzy}^{(2)}, F_{zxx}^{(2)}, F_{zyy}^{(2)}, F_{zzz}^{(2)}, F_{zyz}^{(2)} + F_{zzy}^{(2)}$, and $\left[\chi_{eff}^{(2)}\right]^2$ of type II SHG ($\lambda_\omega = 5.0\ \mu m \rightarrow \lambda_{2\omega} = 2.5\ \mu m$) in BGSe are plotted as a function of φ_{PM} in Figure 7.54. The phase-matching cone joins the xz-plane at $\theta_{PM} = 50°$ and the yz-plane at $\theta_{PM} = 22.4°$ (Boursier 2016) [17]. Kleinman's symmetry is taken for the calculation of the effective coefficient, *i.e.* according to Miller's rule in order to be at $\lambda_{2\omega} = 2.5\ \mu m$:

$\chi_{xxz}^{(2)} = \chi_{xzx}^{(2)} = \chi_{zxx}^{(2)} = 14.3 pm/V$, $\chi_{xxy}^{(2)} = \chi_{xyx}^{(2)} = \chi_{yxx}^{(2)} = 5.2 pm/V$, $\chi_{yyy}^{(2)} = 18.2 pm/V$, $\chi_{yzz}^{(2)} = \chi_{zyz}^{(2)} = \chi_{zzy}^{(2)} = -20.6\ pm/V$, $\chi_{yyz}^{(2)} = \chi_{yzy}^{(2)} = \chi_{zyy}^{(2)} = -15.2\ pm/V$, and $\chi_{zzz}^{(2)} = -2.2 pm\ /V$ (Guo 2022) [18].

Figure 7.54 (bottom left and right) well shows that the full angular distribution of the square of the effective coefficient, *i.e.* by considering the upper and lower half-spaces, belongs to the monoclinic orientation symmetry class $\frac{2}{m}$ in which the mirror plane m lies in the y-z plane, *i.e.* $\varphi = 90°$, that also corresponds to the crystallographic mirror plane (Guo 2022) [18]. Then the binary axis 2 is along the x-axis, *i.e.* at $\theta = 90°, \varphi = 0°$. As a consequence, the corresponding stereographic projection is that of Figure 2.10 rotated by 90°. It also appears in Figure 7.54 that all the octants are not equivalent, which is of prime importance for the use of such a crystal belonging to the crystal class m. Furthermore, Curie's law is verified, *i.e.* $\frac{2}{m} \supset m$.

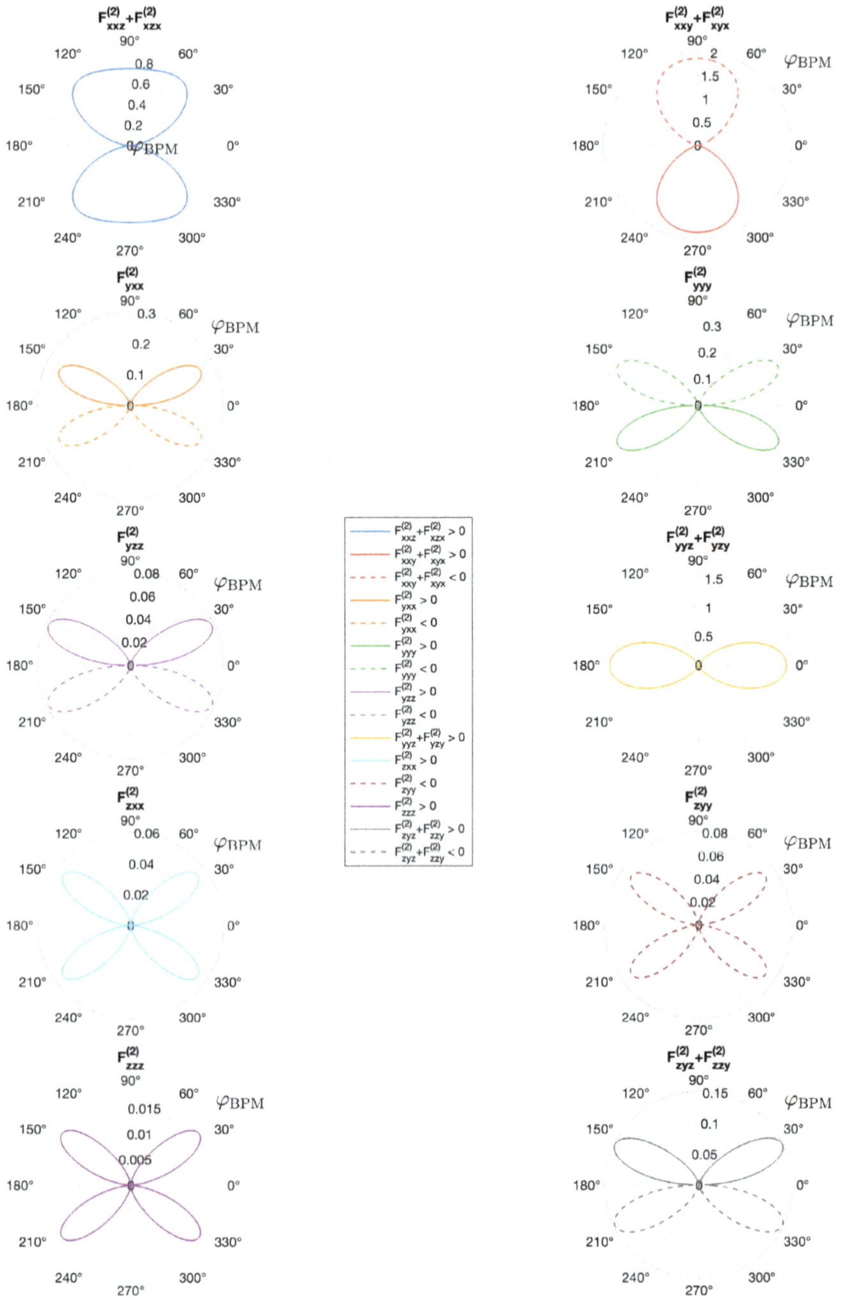

$F^{(2)}_{xxz}+F^{(2)}_{xzx}$
$F^{(2)}_{xxy}+F^{(2)}_{xyx}$
$F^{(2)}_{yxx}$
$F^{(2)}_{yyy}$
$F^{(2)}_{yzz}$
$F^{(2)}_{yyz}+F^{(2)}_{yzy}$
$F^{(2)}_{zxx}$
$F^{(2)}_{zyy}$
$F^{(2)}_{zzz}$
$F^{(2)}_{zyz}+F^{(2)}_{zzy}$
φ_{BPM}
$F^{(2)}_{xxz}+F^{(2)}_{xzx} > 0$
$F^{(2)}_{xxy}+F^{(2)}_{xyx} > 0$
$F^{(2)}_{xxy}+F^{(2)}_{xyx} < 0$
$F^{(2)}_{yxx} > 0$
$F^{(2)}_{yxx} < 0$
$F^{(2)}_{yyy} > 0$
$F^{(2)}_{yyy} < 0$
$F^{(2)}_{yzz} > 0$
$F^{(2)}_{yzz} < 0$
$F^{(2)}_{yyz}+F^{(2)}_{yzy} > 0$
$F^{(2)}_{zxx} > 0$
$F^{(2)}_{zyy} < 0$
$F^{(2)}_{zzz} > 0$
$F^{(2)}_{zyz}+F^{(2)}_{zzy} > 0$
$F^{(2)}_{zyz}+F^{(2)}_{zzy} < 0$

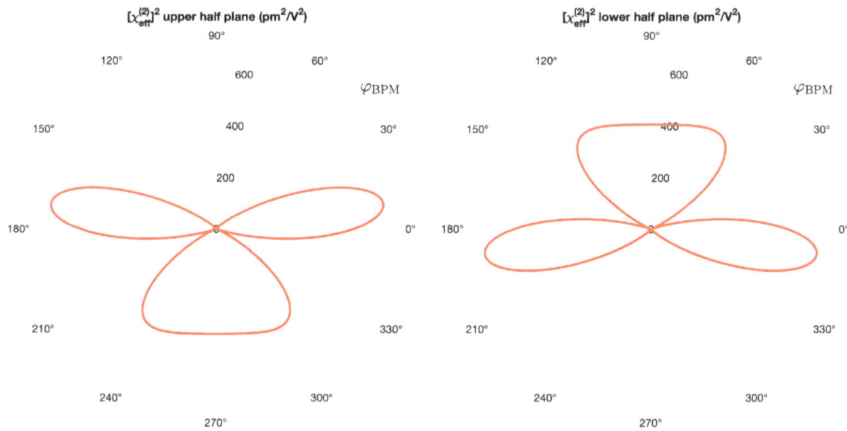

FIGURE 7.54 Second-order field tensor coefficients and square of the second-order effective coefficient as a function of the phase-matching angle of spherical coordinate φ_{PM} in the *m* positive biaxial crystal $BaGa_4Se_7$ for type II SHG ($\lambda_\omega = 5.0\ \mu m \rightarrow \lambda_{2\omega} = 2.5\ \mu m$), with the configuration of polarization (−−+) equivalent to (−+−). The phase-matching cone joins the xz-plane at ($\theta_{PM} = 50°, \varphi_{PM} = 0°$) and the yz-plane at ($\theta_{PM} = 22.4°, \varphi_{PM} = 90°$).

7.6 ACCEPTANCES

7.6.1 Definition

Equation (7.62) for birefringence phase-matching (BPM) and Eq. (7.89) in the case of quasi-phase-matching (QPM) show that the intensity of the generated wave is proportional to $\sin c^2[\frac{\Delta k Z}{2}]$ leading to a maximal value for $\Delta k = 0$. Note that here and in the following, the quantity Δk stands for both BPM and QPM, knowing that in Part 7.4, the quasi-phase-mismatch parameter has been defined as $\Delta k \mp K_G$.

The phase-mismatch parameter Δk depending on the refractive indices, it is interesting to consider the effect of a variation of any dispersive parameter ξ of the refractive indices, *e.g.* the direction of propagation ($\xi = \theta$) or ($\xi = \varphi$), the wavelength ($\xi = \lambda$), or the temperature ($\xi = T$), on the intensity of the generated wave. It is then relevant for that purpose to plot the intensity of the generated wave as a function of a given parameter ξ, all the others being fixed, as shown in Figure 7.55.

The quantity that is taken for the characterization of the effect of a variation of ξ is the acceptance bandwidth $\delta\xi$ defined as the deviation from the phase-matching or angular quasi-phase-matching value $\xi_{PM/AQPM}$ leading to a phase-mismatch variation Δk from 0 to $\frac{2\pi}{L}$, where L is the crystal length. As shown in Figure 7.55, $\delta\xi$ is also the full width of the intensity peak at 0.405. The calculation of the acceptances can be performed numerically or analytically by expanding Δk in the following Taylor series (Boulanger 2014) [7]:

$$\Delta k = \frac{2\pi}{L} = \left(\frac{\partial \Delta k}{\partial \xi}\right)_{\xi_{PM/AQPM}} \delta\xi + \frac{1}{2}\left(\frac{\partial^2 \Delta k}{\partial \xi^2}\right)_{\xi_{PM/AQPM}} \delta\xi^2 + \ldots \quad (7.143)$$

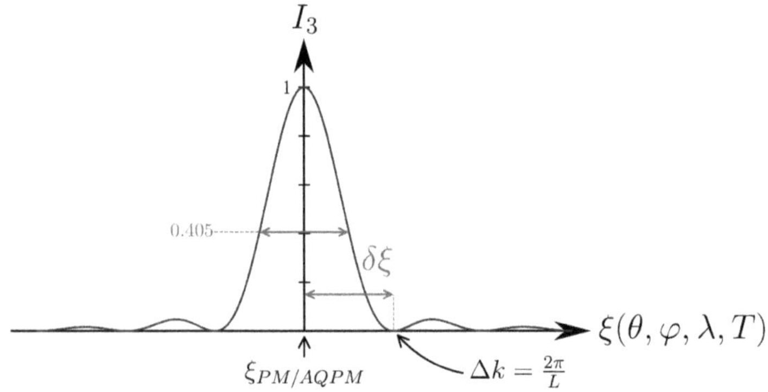

FIGURE 7.55 Intensity of the generated wave as function of a dispersive parameter ξ of the refractive indices: an angle of spherical coordinates (θ) or (φ), the wavelength (λ), or the temperature (T). The parameter $\xi_{PM/AQPM}$ corresponds to phase-matching (PM) or angular quasi-phase-matching (AQPM). The crystal length is L. The quantity $\delta\xi$ is the angle or wavelength or temperature acceptance bandwidth.

where Δk is given by Eq. (7.66) for phase-matching and Eq. (7.101) for angular quasi-phase-matching.

Two situations can be identified considering Eq. (7.142): non-critical phase-matching/angular quasi-phase-matching ($NC_{PM/AQPM}$) when $\left(\frac{\partial \Delta k}{\partial \xi}\right)_{\xi_{PM/AQPM}} = 0$; critical phase-matching/angular quasi-phase-matching ($C_{PM/AQPM}$) when $\left(\frac{\partial \Delta k}{\partial \xi}\right)_{\xi_{PM/AQPM}} \neq 0$. Then it appears according to Eq. (7.142) that these two regimes are characterized by a specific product $L\delta\xi$, *i.e.*:

for $NC_{PM/AQPM}$

$$L\delta\xi_{NCPM/AQPM} = \sqrt{\frac{4\pi L}{\left(\frac{\partial^2 \Delta k}{\partial \xi^2}\right)_{\xi_{PM/AQPM}}}} \tag{7.144}$$

for $C_{PM/AQPM}$

$$L\delta\xi_{CPM/AQPM} = \frac{2\pi}{\left(\frac{\partial \Delta k}{\partial \xi}\right)_{\xi_{PM/AQPM}}} \tag{7.145}$$

Equation (7.144) contains only the first-order term of the series, since it is the largest one. These two situations are illustrated by two examples depicted by Figure 7.56 calculated from Eq. (7.62), in which Δk is a parabolic function of ξ for the $NC_{PM/AQPM}$ case and a linear function for the $C_{PM/AQPM}$ case.

Figure 7.56 well shows that $\delta\xi_{NCPM/AQPM} > \delta\xi_{CPM/AQPM}$; then it comes for a given crystal length L:

$$L\delta\xi_{NCPM/AQPM} > L\delta\xi_{CPM/AQPM} \quad (7.146)$$

Then a small angular, wavelength, or thermal dispersion of the refractive indices, through $\left(\frac{\partial^2 \Delta k}{\partial \xi^2}\right)$ in Eq. (7.143) and $\left(\frac{\partial \Delta k}{\partial \xi}\right)$ in Eq. (7.144), leads to a high acceptance bandwidth regarding the angles, the wavelengths, or the temperature, respectively.

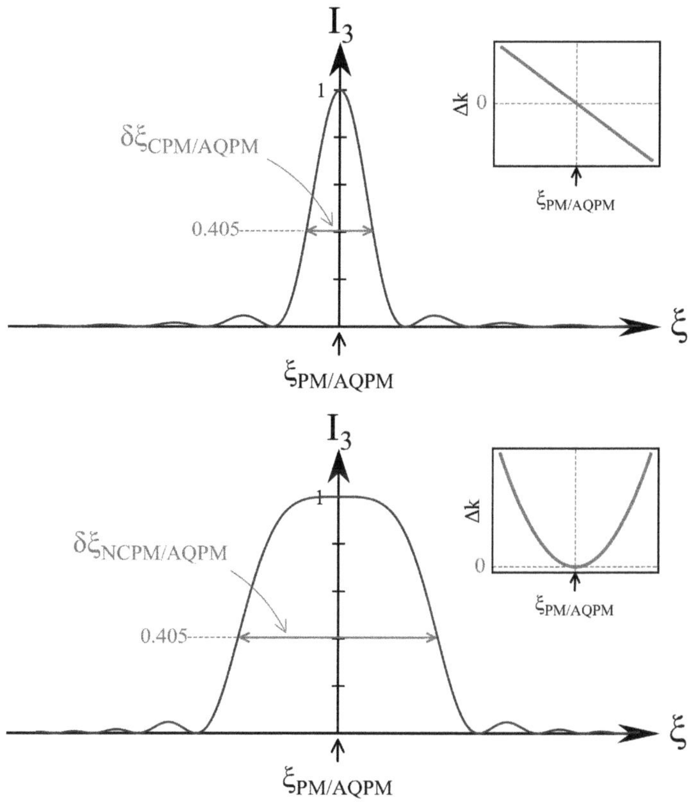

FIGURE 7.56 Normalized intensity of the generated wave calculated as function of a dispersive parameter ξ of the refractive indices for a given crystal length L. The parameter $\xi_{PM/AQPM}$ corresponds to phase-matching (PM) or angular quasi-phase-matching (AQPM): non-critical phase-matching/angular quasi-phase-matching ($NC_{PM/AQPM}$) in which the phase-mismatch Δk is a parabolic function of ξ (Top); critical phase-matching/angular quasi-phase-matching ($C_{PM/AQPM}$) in which the phase-mismatch Δk is a linear function of ξ (Bottom). The quantity $\delta\xi_{NCPM/AQPM}$ is the non-critical phase-matching/angular quasi-phase-matching acceptance bandwidth while $\delta\xi_{CPM/AQPM}$ is the non-critical phase-matching/angular quasi-phase-matching acceptance bandwidth.

Note also that the higher $L\delta\xi$, the lower the decrease of the intensity of the generated wave corresponding (i) to a given angular shift, due to a misorientation of cutting of the crystal or to the divergence of the pump beam, (ii) to the spectral bandwidth of the pump beam, especially in the femtosecond regime, or (iii) to the heating of the crystal due to absorption or external heating.

Hereafter are detailed the links between the angle acceptance as well as the wavelength acceptance and the behaviors of the PM or AQPM curves.

7.6.2 Angle Acceptance

There two angle acceptances: one relative to the φ angle, the other one to θ.

7.6.2.1 The Case of Birefringence Phase-Matching

Uniaxial crystals exhibit BPM cones of revolution around the z-axis as shown in Figure 7.11: it means that θ_{PM} remains the same at any value of φ_{PM} . This is due to an invariance in φ of the uniaxial index surface and thus of Δk . This geometrical consideration immediately leads to the following equalities:

$$\left.\frac{\partial \Delta k}{\partial \varphi}\right|_{(\theta_{PM},\varphi_{PM})} = \left.\frac{\partial^2 \Delta k}{\partial \varphi^2}\right|_{(\theta_{PM},\varphi_{PM})} = 0 \tag{7.147}$$

As a consequence, the acceptance in φ is non-critical, and according to Eq. (7.143), it comes: $L\delta\varphi_{NCPM} \to \infty$. Concerning the acceptance in θ , it is necessary to consider two situations for the phase-matching loci: out of the xy-plane, the acceptance is critical since $\frac{\partial \Delta k}{\partial \theta} \neq 0$, while $\frac{\partial \Delta k}{\partial \theta} = 0$ for a propagation in the xy-plane due to the fact that the latter one is a mirror of the index surface leading to a non-critical acceptance that has a finite value in that case contrary to the non-critical acceptance in φ. Figure 7.57 summarizes these features.

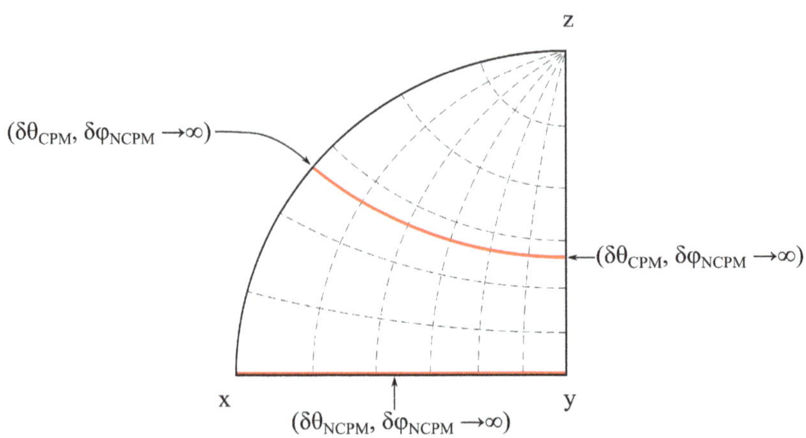

FIGURE 7.57 Angle acceptance distributions in the case of birefringence phase-matching in a positive or negative uniaxial crystal. The red lines correspond to the phase-matching loci. $\delta\varphi_{NCPM}$ and $\delta\theta_{NCPM}$ stand for non-critical acceptances in φ and θ respectively, while $\delta\theta_{CPM}$ corresponds to critical phase-matching in θ.

Figure 7.57 shows that angle non-critical phase-matching is associated with a horizontal tangent when φ is concerned, *i.e.* $\left.\frac{\partial \theta_{PM}}{\partial \varphi}\right|_{\varphi_{PM}} = 0$ while it is a vertical tangent when it deals with θ, *i.e.* $\left.\frac{\partial \varphi_{PM}}{\partial \theta}\right|_{\theta_{PM}=\frac{\pi}{2}} = 0.$

The following is given as an example expression of the angle critical phase-matching acceptance of type I SHG in the negative uniaxial crystal CSP. Using Eq. (7.144) and the uniaxial index surface given in Chapter 5, it comes for $0 < \theta_{PM} < \pi/2$:

$$L\delta\theta_{CPM} \simeq -\frac{2\pi c}{\omega n_o^3(\omega)\left[n_o^{-2}(2\omega) - n_e^{-2}(2\omega)\right]\sin 2\theta_{PM}} \tag{7.148}$$

In the case where $\lambda_\omega = 5\ \mu m \rightarrow \lambda_\omega = 2.5\ \mu m$ which corresponds to $\theta_{PM} = 42.54°$ according to Figure 7.12, it gives for L = 1 cm: $L\delta\theta_{CPM} \simeq 8.7\ cm.mrad$.

Then a fundamental beam with an isotropic divergency, *i.e.* $\delta\alpha$ along both θ and φ, can give birth to a second-harmonic (SH) beam with an anisotropy of divergency if $\delta\alpha > \delta\theta_{CPM}$: actually in that case, the SH beam exhibits a divergency equal to $\delta\theta_{CPM}$ along θ and $\delta\alpha$ along ϕ, since $L\delta\varphi_{NCPM} \rightarrow \infty$. It is also important to notice that the function $\sin c^2(\Delta kZ/2)$ plotted as a function of θ for $0 < \theta_{PM} < \pi/2$, is not symmetrical since $\left.\frac{\partial \Delta k}{\partial \theta}\right|_{(\theta_{PM},\varphi_{PM})}$ does not have the same behavior from either side of θ_{PM}, contrary to the case in which $\theta_{PM} = \pi/2$, *i.e.* in the xy-plane that is a mirror plane regarding θ.

In the case of biaxial crystals, the phase-matching loci that are given in Figure 7.14 never exhibit an infinite angle non-critical phase-matching acceptance as in the uniaxial case. And angle non-critical phase-matching exists only for a propagation in the principal planes, as shown in Figure 7.58, and only for the angular dependence perpendicularly to the considered principal plane.

It also appears that the "anisotropy" of the acceptance in term of criticality in the xz-plane and yz-plane, *i.e.* critical in θ and non-critical in φ, is the reverse of that of the xy-plane in which phase-matching is critical in φ and non-critical in θ.

Note that the "isotropy" of the acceptance exits along the x-axis, y-axis, and z-axis, in which it is non-critical in both φ and θ, and also at any phase-matching direction lying out of the principal planes, in which it is critical in both φ and θ.

As mentioned earlier for uniaxial crystals, the function $\sin c^2(\Delta kZ/2)$ plotted as a function of the angle is not always symmetrical. Actually, it is symmetrical only in the principal planes: as a function of φ in the xz-plane and yz-plane since these planes are mirror planes regarding φ; as a function of θ in the xy-plane since it is a mirror plane regarding θ. Since the principal axes belong to two principal planes, they exhibit a symmetrical function $\sin c^2(\Delta kZ/2)$ regarding both θ and φ along these specific directions.

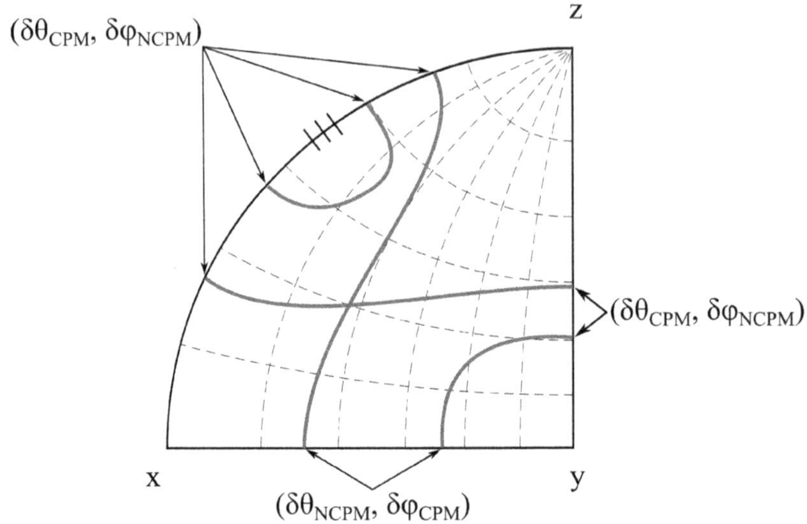

FIGURE 7.58 Angle acceptance distributions in the case of birefringence phase-matching in a positive or negative biaxial crystal. The lines correspond to the phase-matching loci. $\delta\varphi_{NCPM}$ and $\delta\theta_{NCPM}$ stand for non-critical acceptances in φ and θ respectively, while $\delta\varphi_{CPM}$ and $\delta\theta_{CPM}$ correspond to critical phase-matching.

7.6.2.2 The Case of Angular Quasi-Phase-Matching

The case of angular quasi-phase-matching can be deduced in part from the case of phase-matching. Actually, the phase-mismatch, which is the key parameter governing the criticality, writes here:

$$\Delta k_{AQPM} = \Delta k \mp K_G \tag{7.149}$$

Equation (7.147) correspond to the first-order AQPM, with "–" for the positive order and "+" for the negative order, as explained in Part 7.4.2. The quantity Δk is the one that is considered in the previous part relative to birefringence phase-matching and $K_G = \dfrac{2\pi}{\Lambda_{eff}}$ is the modulus of the grating vector where Λ_{eff} is the effective grating period: it is given by Eq. (7.102) when the grating vector is along the x-axis, which is the case of PPKTP and PPLN and the framework of the following calculations. The case of a positive order is taken. Then Eq. (7.147) becomes:

$$\Delta k_{AQPM} = \Delta k - \Lambda\left|\sin\theta\cos\varphi\right| \tag{7.150}$$

The angles of spherical coordinates being such as $\theta \in [0,\pi]$ and $\varphi \in [0,2\pi]$ by definition, Eq. (7.148) can be rewritten:

$$\Delta k_{AQPM} = \Delta k - \Lambda\sin\theta\left|\cos\varphi\right| \tag{7.151}$$

And the first derivatives of Δk_{AQPM} are expressed as:

$$\begin{cases} \left.\dfrac{\partial \Delta k_{AQPM}}{\partial \theta}\right|_{(\theta_{AQPM},\varphi_{AQPM})} = \left.\dfrac{\partial \Delta k}{\partial \theta}\right|_{(\theta_{AQPM},\varphi_{AQPM})} - \Lambda \cos\theta_{AQPM} \left|\cos\varphi_{AQPM}\right| \\ \left.\dfrac{\partial \Delta k_{AQPM}}{\partial \varphi}\right|_{(\theta_{AQPM},\varphi_{AQPM})} = \left.\dfrac{\partial \Delta k}{\partial \varphi}\right|_{(\theta_{AQPM},\varphi_{AQPM})} - \Lambda \sin\theta_{AQPM} \left[\pm\sin\varphi_{AQPM}\right] \end{cases} \tag{7.152}$$

with $-\sin\varphi$ for $\varphi \in \left\{\left[0,\dfrac{\pi}{2}\right],\left[\dfrac{3\pi}{2},2\pi\right]\right\}$ and $+\sin\varphi$ for $\varphi \in \left[\dfrac{\pi}{2},\dfrac{3\pi}{2}\right]$.

Equation (7.150) allows us to know the angle criticality, *i.e.* non-critical *versus* critical, according to the AQPM *loci*, in particular in the principal planes as shown in Table 7.5 and Table 7.6.

Table 7.6 shows that AQPM is different from BPM only in the yz-plane concerning the φ angle. Actually, it is non-critical here for AQPM, which is due to the discontinuity of the grating contribution since the periodicity tends toward infinity for a propagation in the yz-plane only, with a grating vector along the x-axis in the dielectric frame. This difference well appears in Figure 7.36 in the case of PPKTP in which it appears that the tangents to the AQPM curves are perpendicular to the xz-plane and the xy-plane while it is not the case for the yz-plane. These features are illustrated by the following Wulff representations according to the optical classes in Figure 7.59, Figure 7.60 and Figure 7.61.

TABLE 7.5
θ Angle Criticality or Non-Criticality of Angular Quasi-Phase-Matching according to the Principal Planes of the Index Surface and the Parameters of Eq. (7.150)

Principal planes	$\left.\dfrac{\partial \Delta k}{\partial \theta}\right\|_{(\theta_{AQPM},\varphi_{AQPM})}$	$-\Lambda \cos\theta_{AQPM} \left\|\cos\varphi_{AQPM}\right\|$	$\left.\dfrac{\partial \Delta k_{AQPM}}{\partial \theta}\right\|_{(\theta_{AQPM},\varphi_{AQPM})}$	Angle criticality in θ
xz-plane $(\varphi_{AQPM}=0)$	$\neq 0$	$\neq 0$	$\neq 0$	Critical
xy-plane $\left(\theta_{AQPM}=\dfrac{\pi}{2}\right)$	$=0$	$=0$	$=0$	Non-critical
yz-plane $\left(\varphi_{AQPM}=\dfrac{\pi}{2}\right)$	$\neq 0$	$\neq 0$	$\neq 0$	Critical

TABLE 7.6
φ Angle Criticality or Non-Criticality of Angular Quasi-Phase-Matching according to the Principal Planes of the Index Surface and the Parameters of Eq. (7.150)

Principal planes	$\left.\frac{\partial \Delta k}{\partial \varphi}\right\|_{(\theta_{AQPM},\varphi_{AQPM})}$	$-\Lambda \sin\theta_{AQPM}\left[\pm \sin\varphi_{AQPM}\right]$	$\left.\frac{\partial \Delta k_{AQPM}}{\partial \varphi}\right\|_{(\theta_{AQPM},\varphi_{AQPM})}$	Angle criticality in φ
xz-plane ($\varphi_{AQPM} = 0$)	$= 0$	$= 0$	$= 0$	Non-critical
xy-plane $\left(\theta_{AQPM} = \frac{\pi}{2}\right)$	$\neq 0$	$\neq 0$	$\neq 0$	Critical
yz-plane $\left(\varphi_{AQPM} = \frac{\pi}{2}\right)$	$= 0$	$\neq 0$	$\neq 0$	Critical

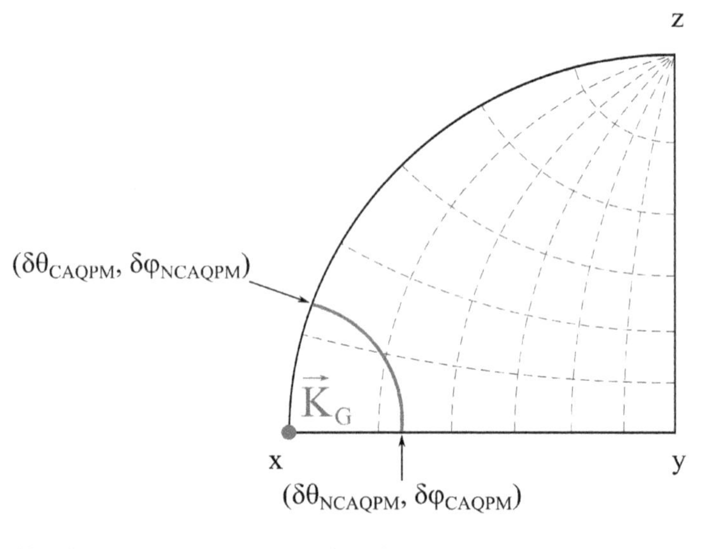

FIGURE 7.59 Angle acceptance distributions in the case of positive first-order angular quasi-phase-matching (AQPM) in an isotropic crystal in which the grating vector is along the x-axis. The line joining the xz-plane and xy-plane correspond to the AQPM loci. $\delta\varphi_{NCAQPM}$ and $\delta\theta_{NCAQPM}$ stand for non-critical acceptances in φ and θ respectively, while $\delta\varphi_{CAQPM}$ and $\delta\theta_{CAQPM}$ corresponds to critical phase-matching.

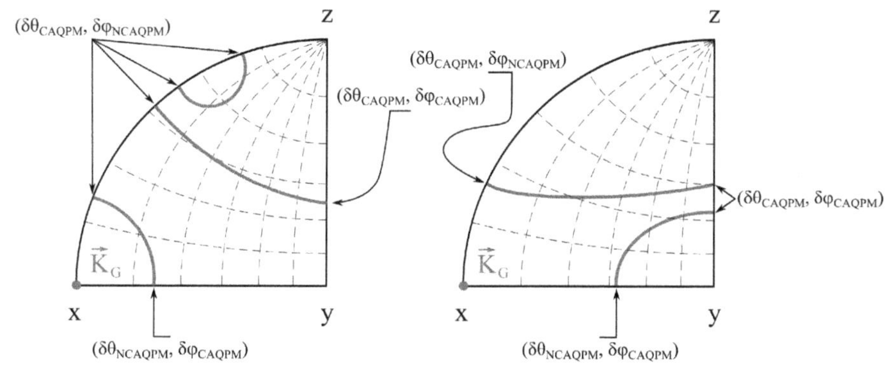

FIGURE 7.60 Angle acceptance distributions in the cases of positive first-order (left) and negative first-order (right) angular quasi-phase-matching (AQPM) in a positive or negative uniaxial crystal in which the grating vector is along the x-axis. The lines joining the principal planes correspond to the AQPM loci. $\delta\varphi_{NCAQPM}$ and $\delta\theta_{NCAQPM}$ stand for non-critical acceptances in φ and θ respectively, while $\delta\varphi_{CAQPM}$ and $\delta\theta_{CAQPM}$ corresponds to critical phase-matching.

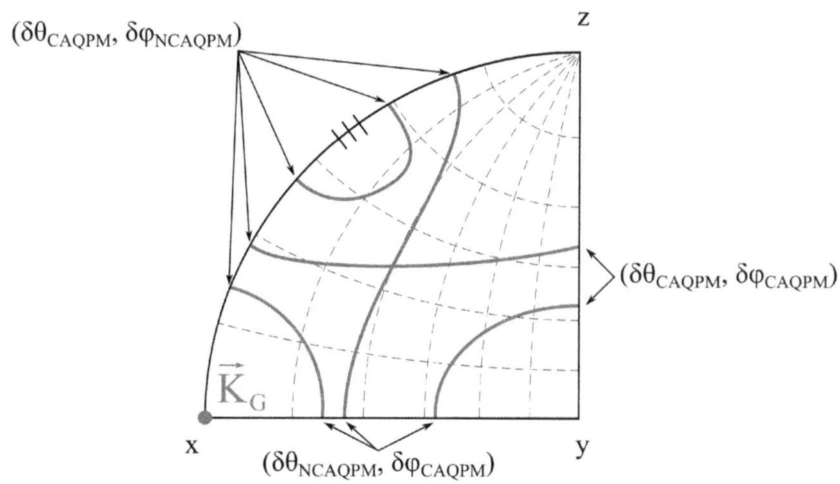

FIGURE 7.61 Angle acceptance distributions in the cases of positive or negative first-order angular quasi-phase-matching (AQPM) in a positive or negative biaxial crystal in which the grating vector is along the x-axis. The lines joining the principal planes correspond to the AQPM loci. $\delta\varphi_{NCAQPM}$ and $\delta\theta_{NCAQPM}$ stand for non-critical acceptances in φ and θ respectively, while $\delta\varphi_{CAQPM}$ and $\delta\theta_{CAQPM}$ corresponds to critical phase-matching.

7.6.3 Wavelength Acceptance

There is no need to distinguish BPM from AQPM since the grating vector of the latter one does not depend on wavelength. The wavelength dispersion of the refractive indices is then the only relevant parameter to consider for that purpose of wavelength acceptance.

As for angular acceptances, the consideration of the BPM or AQPM curves, here in wavelength, are of prime importance for identifying non-critical or critical situations. It is also important to notice that uniaxial and biaxial crystals exhibit the same kind of wavelength curves, as the one shown for example by Figure 7.12 relative to types I and II SHG in the uniaxial crystal CSP in which the fundamental wavelength λ is plotted as a function of θ_{PM}. Wavelength non-critical phase-matching, allowing the biggest wavelength acceptance, is associated with a vertical tangent, *i.e.* $\frac{\partial\theta}{\partial\lambda} = 0$ at $\theta_{PM} = 42.3°$ for type I and $\theta_{PM} = 72.5°$ for type II. It is critical at any other phase-matching angle, with the largest degree of criticality, *i.e.* the smallest wavelength acceptance, at $\theta_{PM} = 90°$. These two extreme situations are illustrated in Figure 7.62 in which the normalized intensity of type I SHG in CSP is calculated from Eq. (7.62) as a function of λ_ω around 5.3 μm at $\theta_{PM} = 42.3°$ and around 2.69 μm at $\theta_{PM} = 90°$.

Note that each of these two phase-matching peaks is not symmetrical with λ_ω, due to the Δk asymmetry in wavelength on either side of 0. This simply results from the fact that the refractive index dispersion is of course not symmetrical around any given wavelength, the Sellmeier dispersion curves being monotonic over the transparency range of crystals.

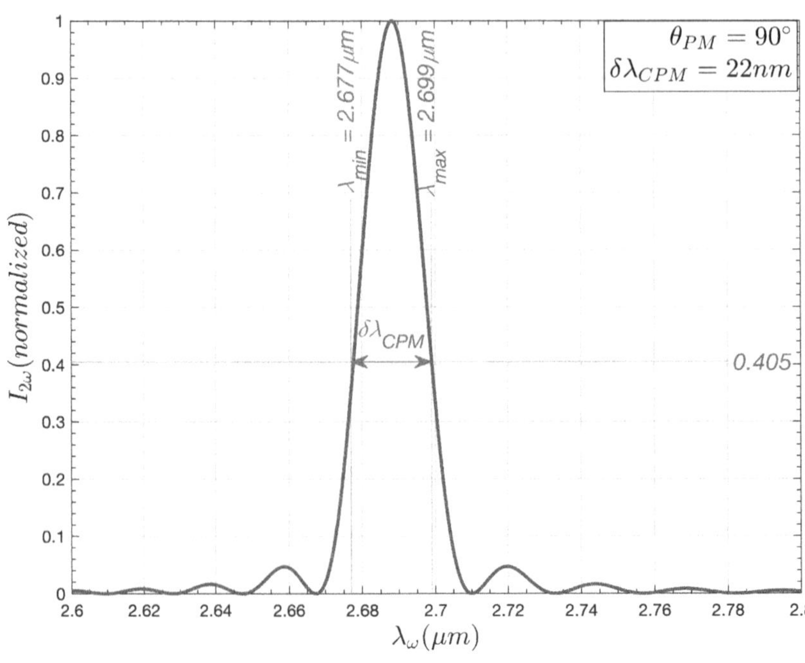

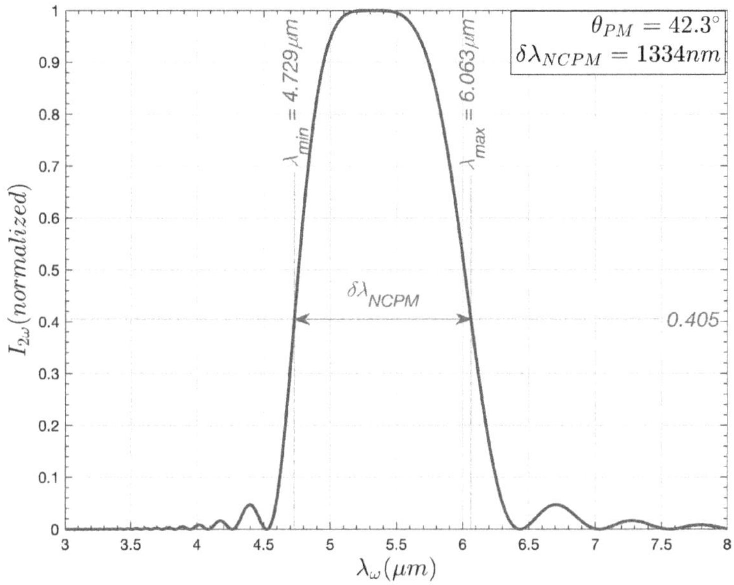

FIGURE 7.62 Wavelength acceptances of type I SHG in a 1-cm-long CSP crystal at $\theta_{PM} = 90°$ (Top) and $\theta_{PM} = 42.3°$ (Bottom). $I_{2\omega}$ is the normalized second-harmonic intensity calculated as a function of λ_ω that is the fundamental wavelength.

7.7 SPATIAL WALK-OFF ATTENUATION

The spatial walk-off effect described in Chapter 5, which is a linear optical phenomenon leading to double refraction as explained in Part 5.4.2, also has an influence on the conversion efficiency of nonlinear processes: this is due to the non-collinearity of the Umov-Poynting vectors of the interacting beams when different polarization modes are involved, despite a collinear wavevector propagation. This is systematically the case for birefringence phase-matching as shown in Part 7.3.2. Types I and II SHG are also used here to illustrate and explain the corresponding phenomenology.

7.7.1 Type I SHG

Figure 7.63 shows the example of a propagation out of the equatorial plane of the index surface, *i.e.* out of the xy-plane, in a negative uniaxial crystal.

Figure 7.63 shows a transverse spreading of the second-harmonic (SH) intensity in the walk-off plane over a distance equal to *L.tan(ρ)*. This feature can be intuited without any calculation. Actually, each "section" of the fundamental beam over the crystal length activates a nonlinear polarization source term that can locally emit a SH wave. And because $\vec{S}^{o}_{\omega}$ and $\vec{S}^{e}_{2\omega}$ are not collinear, then the SH wave radiated by the different sections intercept the exit surface at different transverse positions in the walk-off plane. Then the first effect of the spatial walk-off is the creation of an anisotropy of the transverse profile of intensity of the SH beam.

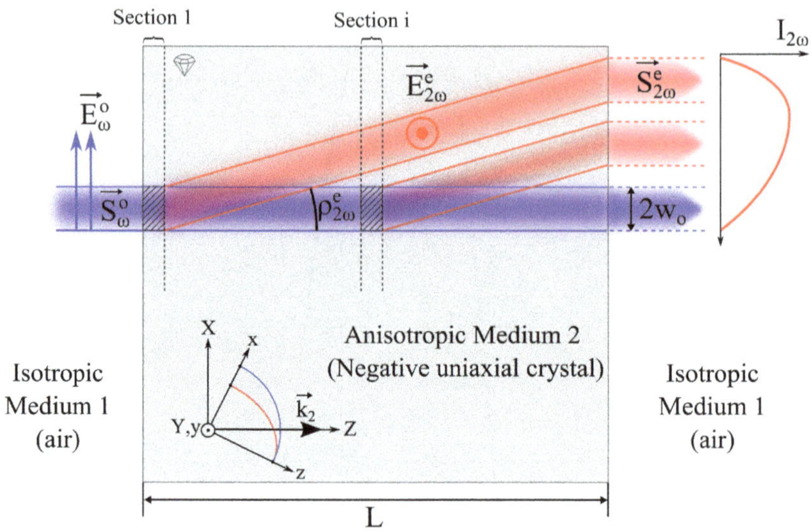

FIGURE 7.63 Spatial walk-off effects in the case of type I SHG phase-matched out of the equatorial plane of a negative uniaxial crystal. $\vec{E}^{o}_{\omega}$ is the ordinary fundamental electric field vector and $\vec{S}^{o}_{\omega}$ the associated Umov-Poynting vector. $\vec{E}^{e}_{2\omega}$ is the extraordinary second-harmonic electric field vector and $\vec{S}^{e}_{2\omega}$ the associated Umov-Poynting vector. $\rho = \rho^{e}_{2\omega}$ is the spatial walk-off angle due to the second-harmonic wave. w_0 is the radius of the fundamental beam taken as a parallel beam. $I_{2\omega}$ is the second-harmonic intensity at the exit of the nonlinear medium plotted as a function of the transverse spatial coordinate in the walk-off plane. Section 1 and section i are two sections of the fundamental beam.

A second effect is the possible attenuation of the SHG conversion efficiency because the SH and fundamental beams do not overlap each section, which can also be seen in Figure 7.63. An analytical modeling of this effect is possible in the parallel beam approximation, which can be applied in the case of a Gaussian beam when the crystal length is smaller than twice the Rayleigh length of the fundamental beam as shown in Figure 7.64.

The Rayleigh length is defined by the following equation (Siegman 1986) [19]:

$$Z_R = \frac{\pi n(\lambda)}{\lambda}\left(\frac{\mathrm{w}_0^2}{M^2}\right) \tag{7.153}$$

The parameter n is the refractive index at the wavelength λ, w_0 is the beam waist radius and $M2$ the beam quality factor.

If $L < 2Z_R$, then the transverse profile of the complex amplitude of the electric field is fixed over L and given by the following equation:

$$E(X,Y,Z) = E(0,0,Z)\exp\left(-\frac{X^2+Y^2}{\mathrm{w}_0^2}\right) \tag{7.154}$$

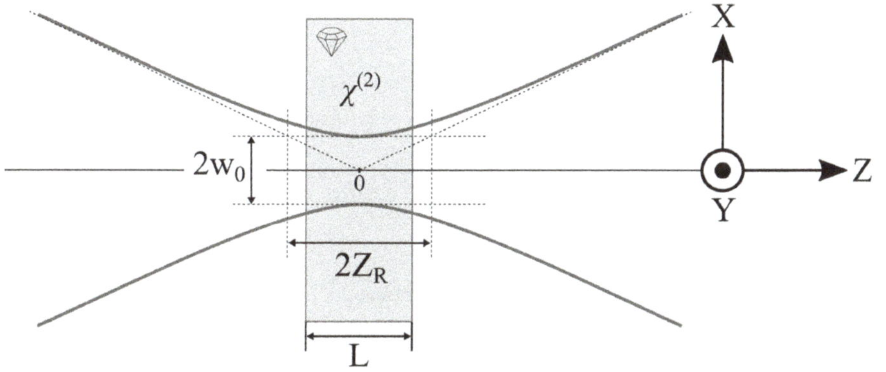

FIGURE 7.64 Longitudinal profile of a Gaussian beam focused on the center of a $\chi^{(2)}$ crystal for which the length L is smaller than twice the Rayleigh length Z_R. w_0 is the beam waist radius and (X,Y,Z) the laboratory frame in which Z is along the direction of propagation of light.

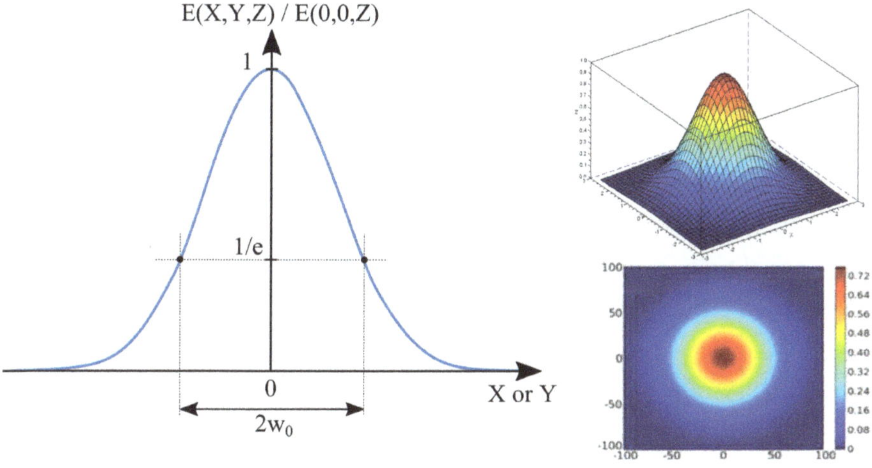

FIGURE 7.65 Transverse profile of the electric field amplitude of a Gaussian beam in the approximation of the parallel beam limit. X and Y are the transverse coordinates and w_0 the beam waist radius.

The corresponding behavior is depicted in Figure 7.65.

Taking into account the spatial walk-off and Eq. (7.152), the integration of Eq. (7.54) other the transverse coordinates (X,Y) leads to Eq. (7.62), being specified for type I BPM SHG that is consider here, multiplied by a spatial walk-off attenuation coefficient G^I (Boyd 1965) [20], which gives:

$$I_{2\omega}(Z) = 2\pi^2 \sqrt{\frac{\mu_0}{\varepsilon_0}} \frac{1}{\lambda_{2\omega}^2 (\cos\rho_{2\omega})^4} \frac{[\chi_{eff}^{(2)}]^2}{n_\omega^2 n_{2\omega}} I_\omega(0)^2 Z^2 G^I(\gamma) \qquad (7.155)$$

with

$$\begin{cases} G^I(\gamma) = \dfrac{\sqrt{\pi}}{\gamma} erf(\gamma) - \dfrac{1}{\gamma^2}[1-\exp(-\gamma^2)] \\ \gamma = \dfrac{\rho Z}{\mathrm{w}_0} \\ erf(\gamma) = \dfrac{2}{\sqrt{\pi}} \displaystyle\int_0^{\gamma} \exp(-x^2)dx \end{cases} \quad (7.156)$$

When $\gamma = 0$, *i.e.* $\rho = 0$, then $G^I(\gamma) = 1$, and for $\gamma \gg 1$, *i.e.* $\rho Z \gg \mathrm{w}_0$, it comes $G^I(\gamma) \simeq \sqrt{\pi}/\gamma$. Then when $\rho \neq 0$, the behavior of $I_{2\omega}(Z)$ is closed to Z^2 at small Z while it is a Z/ρ evolution at bigger Z as shown in Figure 7.66.

Note that $I_\omega(0)$ in Eq. (7.153) is the intensity relative to a single fundamental wave $E_\omega(0)$, so that $I_\omega(0) = I_\omega^{total}(0)/2$ where $I_\omega^{total}(0)$ represents the total incident fundamental intensity. Actually, type I SHG is an interaction between two identical fundamental waves and one second-harmonic wave.

Note that these calculations are also valid for QPM, but Eq. (7.153) has to be multiplied by the Fourier coefficient $(4/\pi^2)$ for the first-order QPM, Eq. (7.154) remaining the same.

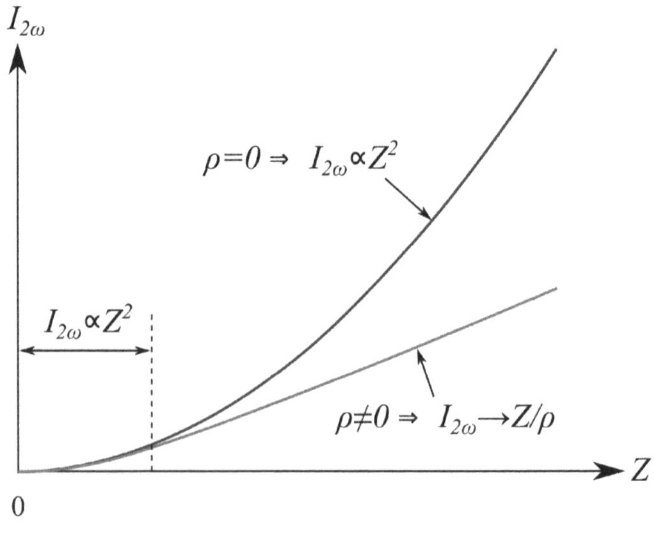

FIGURE 7.66 Behavior of type I SH intensity $I_{2\omega}$ in the undepleted pump approximation over the propagation distance Z when the spatial walk-off angle is $\rho = 0$ or $\rho \neq 0$.

The spatial walk-off angle ρ depends on the configuration of polarization: for the uniaxial optical class, $\rho = \rho^e(2\omega)$ in the case of an interaction between one extraordinary wave and two ordinary waves (eoo), which is the case of Figure 7.63, and $\rho = \rho^e(\omega)$ for a (oee) interaction, the calculation being done from Eq. (5.112) at the proper circular frequency; for the biaxial optical class, in the yz-plane it is $\rho = \rho^+(\omega)$ from Eq. (5.123) for a positive optical sign and $\rho = \rho^-(2\omega)$ from Eq. (5.126) for a negative optical sign; in the xz-plane for a positive optical sign, it is $\rho = \rho^+(\omega)$ from Eq. (5.129) at $V_z < \theta \leq \pi/2$ and $\rho = \rho^-(2\omega)$ from Eq. (5.128) at $0 \leq \theta < V_z$, the equations being the same for a negative optical sign by taken $\rho = \rho^-(2\omega)$ at $V_z < \theta \leq \pi/2$ and $\rho = \rho^+(\omega)$ at $0 \leq \theta < V_z$; in the xy-plane, it is $\rho = \rho^-(2\omega)$ from Eq. (5.120) for a positive optical sign and $\rho = \rho^+(\omega)$ from Eq. (5.121) for a negative optical sign; finally out of the principal planes, $\rho^-(2\omega)$ and $\rho^+(\omega)$ are non-zero so that the spatial walk-off angle is defined as $\rho = \left|\rho^+(\omega) - \rho^-(2\omega)\right|$.

Hereafter are given some numerical applications in the case of type I BPM SHG in the positive biaxial crystal KTP pumped at the fundamental wavelength $\lambda_\omega = 1.064\ \mu m$. The angles $\rho^+(\omega)$, $\rho^-(2\omega)$, ρ, and G^I are calculated as a function of the BPM angle φ_{PM} of Figure 7.15 using the equations mentioned earlier. The corresponding curves are plotted in Figure 7.67.

Figure 7.67 well shows that, at fixed values of the walk-off angle, the corresponding attenuation decreases, *i.e.* G^I increases, when the size of the beam radius increases.

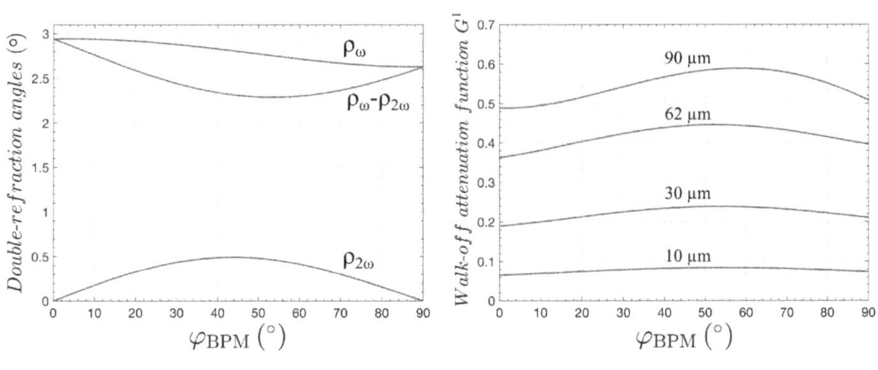

FIGURE 7.67 Spatial walk-off angles $\rho^+(\omega)$, $\rho^-(2\omega)$, $\rho = \left|\rho^+(\omega) - \rho^-(2\omega)\right|$ (left), and walk-off attenuation function G^I corresponding to different values of beam waist radius w_0 (right), calculated as a function of type I BPM SHG angle φ_{BPM} in a 5-mm-long KTP crystal pumped at the fundamental wavelength $\lambda_\omega = 1.064\ \mu m$.

It is also important to notice that the expression of $G^I(\gamma)$ given by Eq. (7.154) is valid for type I $SFG(\omega_3 = \omega_1 + \omega_2)$, type II $DFG(\omega_1 = \omega_3 - \omega_2)$, and type III $DFG(\omega_2 = \omega_3 - \omega_1)$ with $\gamma = \rho Z / w_0$ where ρ has to be taken as the maximum between $\rho = \left|\rho^-(\omega_3) - \rho^+(\omega_1)\right|$ and $\rho = \left|\rho^-(\omega_3) - \rho^+(\omega_2)\right|$. Then Eq. (7.153) writes:

$$I_a(Z) = 2\pi^2 \sqrt{\frac{\mu_0}{\varepsilon_0}} \frac{1}{\lambda_a^2 (\cos\rho_a)^4} \frac{[\chi_{eff}^{(2)}]^2}{n_a n_b n_c} I_b(0) I_c(0) Z^2 G^I(\gamma) \tag{7.157}$$

with $(\omega_a, \omega_b, \omega_c)$ stand for $(\omega_3, \omega_1, \omega_2)$, $(\omega_1, \omega_3, \omega_2)$, and $(\omega_2, \omega_3, \omega_1)$ in the cases of $SFG(\omega_3 = \omega_1 + \omega_2)$, $DFG(\omega_1 = \omega_3 - \omega_2)$, and $DFG(\omega_2 = \omega_3 - \omega_1)$ respectively.

7.7.2 Type II SHG

Figure 7.68 shows the example of a propagation out of the xy-plane of a negative uniaxial crystal.

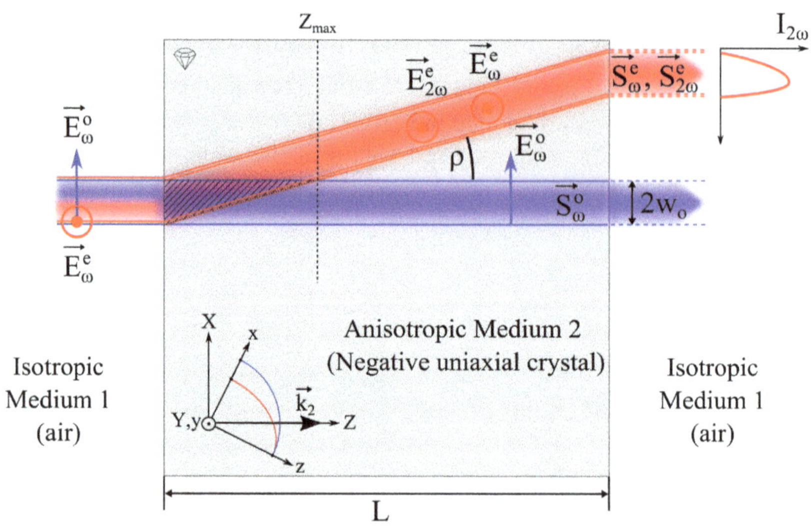

FIGURE 7.68 Spatial walk-off effects in the case of type II SHG phase-matched out of the equatorial plane of a negative uniaxial crystal. $\left(\vec{E}_\omega^o, \vec{E}_\omega^e\right)$ are the ordinary fundamental electric field vectors and $\left(\vec{S}_\omega^o, \vec{S}_\omega^e\right)$ the associated Umov-Poynting vectors. $\vec{E}_{2\omega}^e$ is the extraordinary second-harmonic electric field vector and $\vec{S}_{2\omega}^e$ the associated Umov-Poynting vector. $\rho = \rho_\omega^e$ is the spatial walk-off angle due to the fundamental wave. w_0 is the radius of the fundamental beam taken as a parallel beam. Z_{max} is the maximal interacting length. $I_{2\omega}$ is the second-harmonic intensity at the exit of the nonlinear medium plotted as a function of the transverse spatial coordinate in the walk-off plane.

Figure 7.68 shows that, contrary to type I, there is no spatial spreading of the SH intensity. The other difference with type I is the limitation of the interacting length, above $Z_{max} = 2w_0 / \tan(\rho)$, due to a spatial walk-off at the level of the fundamental beam, *i.e.* $\rho = \rho_{\omega}^{e}$. Thus the nonlinear interaction cannot occur beyond this distance, which leads to a saturation of the SH intensity.

An analytical modeling can be performed in the parallel beam approximation described in Part 7.7.1. Then it comes:

$$I_{2\omega}(Z) = 2\pi^2 \sqrt{\frac{\mu_0}{\varepsilon_0}} \frac{1}{\lambda_{2\omega}^2 (\cos\rho_\omega)^4} \frac{[\chi_{eff}^{(2)}]^2}{n_\omega^+ n_\omega^- n_{2\omega}^-} I_\omega^+(0) I_\omega^-(0) Z^2 G^{II}(\gamma) \qquad (7.158)$$

with (Mehendale and Gupta 1988) [21]:

$$\begin{cases} G^{II}(\gamma) = \dfrac{2}{\sqrt{\pi}} \displaystyle\int_{-\infty}^{+\infty} F^2(a,\gamma) da \\ F(a,\gamma) = \left(\dfrac{1}{\gamma}\right) \exp(-a^2) \displaystyle\int_0^{\tau} \exp[-(a+\tau)^2] d\tau \\ \gamma = \dfrac{\rho Z}{w_0}, a = \dfrac{v}{w_0}, \tau = \dfrac{\rho u}{w_0} \end{cases} \qquad (7.159)$$

u and v are the Cartesian coordinates of an orthonormal frame defined in the walk-off plane where u is along the wave vectors of the three interacting waves.

In this model, ρ_ω^- and $\rho_{2\omega}^-$, *i.e.* ρ_ω^e and $\rho_{2\omega}^e$ in the example of Figure 7.68, are supposed to be identical and in the same plane: this is a reasonable hypothesis since birefringence tends to show a more moderate dispersion than that of the principal refractive indices. Then ρ is classically taken as the maximum value between $\left|\rho^+(\omega) - \rho^-(2\omega)\right|$ and $\left|\rho^+(\omega) - \rho^-(\omega)\right|$. As for type I, ρ depends on the configuration of polarization and can be calculated using Eq. (5.112) to Eq. (5.129).

Equations (7.157) show that $G^{II}(\gamma) = 1$ when $\gamma = 0$, and $G^{II}(\gamma) \simeq 2\arctan(\sqrt{2}) / \gamma^2$ for $\gamma \gg 1$ which corresponds to the saturation of the SH intensity already mentioned. Then when $\rho \neq 0$, the behavior of $I_{2\omega}(Z)$ is closed to Z^2 at small Z while it saturates at a value proportional to $1/\rho^2$ as shown in Figure 7.69. This saturation is due to the full splitting of the two fundamental beams beyond Z_{max} as shown in Figure 7.68.

Note that in Eq. (7.155), $I_\omega^+(0) = I_\omega^-(0)$ when the incident fundamental beam is properly polarized, *i.e.* at 45° of the two Eigen vibration planes associated with the considered phase-matching direction. Then $I_\omega^{+,-}(0) = I_\omega^{total}(0)/2$ where $I_\omega^{total}(0)$ is the total incident fundamental intensity.

It is important to notice that type II is less suitable than type I in the case of strong focusing, *i.e.* small beam waist radius w_0, because of the complete splitting of the

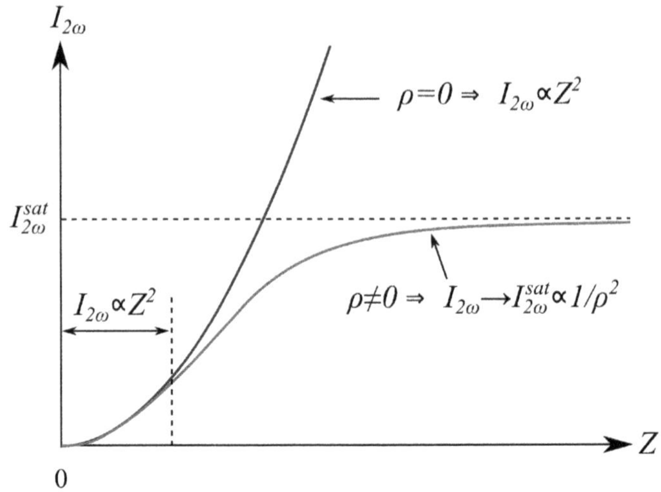

FIGURE 7.69 Behavior of type II SH intensity $I_{2\omega}$ in the undepleted pump approximation over the propagation distance Z when the spatial walk-off angle is $\rho = 0$ or $\rho \neq 0$.

two fundamental beams, which does not occur for type I. Note also that this splitting leads to an increase in all the acceptances since Z_{max} is smaller than the crystal length L (Fève 1995) [22].

As for type I, Eq. (7.156) has to be multiplied by the Fourier coefficient $(4/\pi^2)$ for the positive or negative first-order QPM, and Eq. (7.157) remains the same.

Numerical applications plotted in Figure 7.70 concern type II BPM SHG in the positive biaxial crystal KTP pumped at the fundamental wavelength $\lambda_\omega = 1.064\ \mu m$. The angles $\rho^+(\omega)$, $\rho^-(\omega)$, $\rho^-(2\omega)$, and $\rho = \left|\rho^+(\omega) - \rho^-(\omega)\right|$ that corresponds to the maximum difference between the walk-off angles are plotted as a function of the BPM angle φ_{PM} of Figure 7.15 using the equations mentioned earlier; G^{II} is also calculated for different beam sizes.

Equation (7.156) is also valid for types II and III $SFG(\omega_3 = \omega_1 + \omega_2)$, and the corresponding types of $DFG(\omega_1 = \omega_3 - \omega_2)$ and $DFG(\omega_2 = \omega_3 - \omega_1)$ according to Table 7.1.

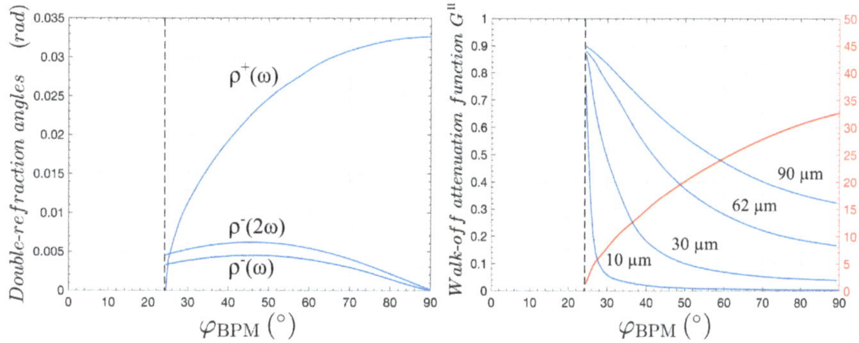

FIGURE 7.70 Spatial walk-off angles $\rho^+(\omega)$, $\rho^-(\omega)$, $\rho^-(2\omega)$, $\rho=\left|\rho^+(\omega)-\rho^-(\omega)\right|$ (left), and walk-off attenuation function G^{II} corresponding to different values of beam waist radius w_0 (right), calculated as a function of type II BPM SHG angle φ_{PM} in a 5-mm-long KTP crystal pumped at the fundamental wavelength $\lambda_\omega = 1.064\ \mu m$.

7.8 SECOND-HARMONIC GENERATION AND SECOND-ORDER SPONTANEOUS PARAMETRIC DOWN-CONVERSION

7.8.1 Second-Harmonic Generation

The case of a type I phase-matched second-harmonic generation (SHG) is considered here, so that Eq. (7.37) reduces to:

$$\begin{cases} \dfrac{\partial E_\omega(Z)}{\partial Z} = i\varepsilon_0\kappa_\omega\chi_{eff}^{(2)}E_{2\omega}(Z)E_\omega^*(Z) & (1) \\ \dfrac{\partial E_\omega(Z)}{\partial Z} = i\varepsilon_0\kappa_\omega\chi_{eff}^{(2)}E_{2\omega}(Z)E_\omega^*(Z) & (2) \\ \dfrac{\partial E_{2\omega}(Z)}{\partial Z} = i\varepsilon_0\kappa_{2\omega}\chi_{eff}^{(2)}E_\omega(Z)^2 & (3) \end{cases} \tag{7.160}$$

Equations (1) and (2) of this system are identical because type I SHG is considered here so that the two fundamental waves have the same complex amplitude $E_\omega(Z)$.

On the other hand, Manley-Rowe's relations, given by Eq. (7.48), can be written as follows:

$$2I_\omega(Z)+I_{2\omega}(Z)=2I_\omega(0) \tag{7.161}$$

where I_ω is the intensity relative to a single fundamental wave E_ω, so that $2I_\omega$ represents the total fundamental intensity I_ω^{total}.

From Eq. (7.43), the intensity writes:

$$\begin{cases} I_j(Z) = \dfrac{n_j}{2}\sqrt{\dfrac{\varepsilon_0}{\mu_0}}\left\|E_j(Z)\right\|^2 \\ j \equiv (\omega, 2\omega) \end{cases} \tag{7.162}$$

Then it comes from Eq. (7.159) and Eq. (7.160):

$$\left\|E_\omega(Z)\right\|^2 = \left\|E_\omega(0)\right\|^2 - \frac{1}{2}\left\|E_{2\omega}(Z)\right\|^2 \tag{7.163}$$

The following equation can be then obtained by combining Eq. (7.161) and equations (1) and (3), or equivalently (2) and (3), of the system given by Eq. (7.158):

$$\frac{\partial\left\|E_{2\omega}(Z)\right\|}{\partial Z} = \varepsilon_0\kappa_{2\omega}\chi_{eff}^{(2)}\left[\left\|E_\omega(0)\right\|^2 - \frac{1}{2}\left\|E_{2\omega}(Z)\right\|^2\right] \tag{7.164}$$

The integration of Eq. (7.162) over the crystal length $Z = L$ leads to:

$$\left\|E_{2\omega}(L)\right\|^2 = 2\left\|E_\omega(0)\right\|^2 \tanh^2\left[\frac{\sqrt{2}}{2}\varepsilon_0\kappa_{2\omega}\chi_{eff}^{(2)}\left\|E_\omega(0)\right\|L\right] \tag{7.165}$$

Then it comes by using Eq. (7.161) one more time:

$$\left\|E_\omega(L)\right\|^2 = \left\|E_\omega(0)\right\|^2 \operatorname{sech}^2\left[\frac{\sqrt{2}}{2}\varepsilon_0\kappa_{2\omega}\chi_{eff}^{(2)}\left\|E_\omega(0)\right\|L\right] \tag{7.166}$$

Finally, Eq. (7.163) and Eq. (7.164) allow us to calculate the fundamental and second-harmonic intensities at the exit of the crystal, *i.e.*:

$$\begin{cases} I_{2\omega}(L) = I_\omega^{total}(0)\tanh^2\left[\mu L\right] \\ I_\omega^{total}(L) = I_\omega^{total}(0)\operatorname{sech}^2\left[\mu L\right] \end{cases} \tag{7.167}$$

with

$$\mu = \frac{\omega}{c}\left(\frac{\mu_0}{\varepsilon_0}\right)^{1/4}\frac{\chi_{eff}^{(2)}}{(\cos\rho_{2w})^2 n_\omega^{3/2}}\sqrt{\frac{I_\omega^{total}(0)}{2}} \tag{7.168}$$

where $n_\omega = n_\omega^+$ according to Table 7.1.

These equations are also valid for type II phase-matched SHG by replacing $n_\omega^{3/2}$ in Eq. (7.166) by $\left(n_{2\omega}^- n_\omega^+ n_\omega^-\right)^{1/2}$, with $n_{2\omega}^- = \left(n_\omega^+ + n_\omega^-\right)/2$.

Equation (7.165) and Eq. (7.166) describe SHG in the depleted pump regime. An illustration is given by Figure 7.71.

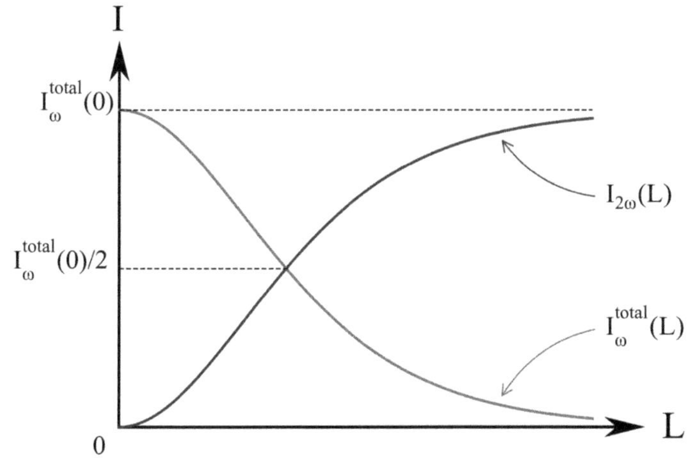

FIGURE 7.71 Schematic behavior of the fundamental intensity I_ω^{total} and second-harmonic intensity $I_{2\omega}$ as a function of the crystal length L in the pump depletion regime.

It is important to notice that $\left(\tanh^2[\mu L], \text{sech}^2[\mu L]\right) \to \left([\mu L]^2, 1\right)$ when $[\mu L] \to 0$, which corresponds to the undepleted pump approximation (UPA). In that case, $I_\omega^{total}(L) \to I_\omega^{total}(0)$ and $I_{2\omega}(L)$ given by Eq. (7.165) tends toward Eq. (7.62) by setting $\Delta k = 0$ and $I_1(0) = I_2(0) = \frac{I_\omega^{total}(0)}{2}$, and by making the suited replacements.

7.8.2 Second-Order Spontaneous Parametric Down-Conversion

Second-order spontaneous parametric down-conversion (SPDC) $(1/\lambda_3 \to 1/\lambda_1 + 1/\lambda_2)$ is the reverse of SHG, as explained in Chapter 1 and illustrated by Figure 1.16. In the case of phase-matching and in the UPA, Eq. (7.37) reduces to:

$$\begin{cases} \dfrac{\partial E_1(Z)}{\partial Z} = i\varepsilon_0 \kappa_1 \chi_{eff}^{(2)} E_3(0) E_2^*(Z) \\ \dfrac{\partial E_2(Z)}{\partial Z} = i\varepsilon_0 \kappa_2 \chi_{eff}^{(2)} E_3(0) E_1^*(Z) \\ \dfrac{\partial E_3(Z)}{\partial Z} = 0 \end{cases} \tag{7.169}$$

This system of equations leads to:

$$\begin{cases} \dfrac{\partial^2 E_2(Z)}{\partial Z^2} - \beta^2 E_2^*(Z) = 0 \\ E_1(Z) = i \dfrac{1}{\varepsilon_0 \kappa_2 \chi_{eff}^{(2)} E_3^*(0)} \dfrac{\partial E_2(Z)}{\partial Z} \end{cases} \tag{7.170}$$

where

$$\beta = \varepsilon_0 \chi_{eff}^{(2)} \left\| E_p(0) \right\| \sqrt{\kappa_1 \kappa_2} \tag{7.171}$$

Then it comes from the integration of Eq. (7.168) over the crystal length $Z = L$ (Yariv and Yeh 2002) [23]:

$$\begin{cases} E_1(L) = E_1(0)\cos h(\beta L) + i\sqrt{\dfrac{\omega_1 n_2}{\omega_2 n_1}} \dfrac{E_3(0)}{\left\|E_3(0)\right\|} E_2^*(0)\sin h(\beta L) \\ E_2(L) = E_2(0)\cos h(\beta L) + i\sqrt{\dfrac{\omega_2 n_1}{\omega_1 n_2}} \dfrac{E_3(0)}{\left\|E_3(0)\right\|} E_1^*(0)\sin h(\beta L) \end{cases} \tag{7.172}$$

The goal being to calculate $E_1(L)$ and $E_2(L)$, it then necessary to know the initial fields, *i.e.* $E_1(0)$ and $E_2(0)$, while no beam is sent at the entrance of the crystal. A piece of quantum optics is required at this step. Actually, a good description of SPDC can be done by keeping Eq. (7.170) that is relative to classical fields, and to take the quantum fluctuations of vacuum as initial fields, *i.e.* $E_1(0) \equiv \Delta E_1^{\text{vacuum}}$ and $E_2(0) \equiv \Delta E_2^{\text{vacuum}}$ (Bertrand 2025) [24]. That means that the quantum fluctuations of vacuum are considered as the stimulation of the scission of the pump photons giving birth to pairs of photons at ω_1 and ω_2. Then the $\sin h(\beta L)$ contributions in Eq. (7.170) describes the optical parametric amplification (OPA) of vacuum while the $\cos h(\beta L)$ contributions describe the generation over the mode not initially present at the entrance of the crystal.

The quantum fluctuations of vacuum are expressed as (Bertrand 2025, Riek 2015) [24,25]:

$$\begin{cases} \Delta E_j^{\text{vacuum}} = \sqrt{\dfrac{\hbar\omega_j}{4\pi\varepsilon_0 n_j S}\Delta\omega_j} \\ j = (1,2) \end{cases} \tag{7.173}$$

The spectral width $\Delta\omega_j$ can be estimated by calculating the SPDC spectral acceptance in the considered phase-matching direction, while S is the beam cross section.

Because SPDC is used in a quantum optics context, it is more useful to calculate the photons pairs flux $N_1 = N_2 \equiv N_{\text{pairs}}$, expressed in [Hz] (pairs/s), rather than the pairs intensity. The photons pairs photon flux can be deduced from the generated electric field $E_{1,2}^{\text{generated}}(L) = E_{1,2}(L) - \Delta E_{1,2}^{\text{vacuum}}$, and using $N_{1,2} = \dfrac{\varepsilon_0 n_{1,2} c S}{4\hbar\omega_{1,2}} \left\|E_{1,2}\right\|^2$ coming from Eq. (7.43) and Eq. (7.49). Then it comes:

$$N_{\text{pairs}}(L) = \frac{\varepsilon_0 n_{1,2} c S}{4\hbar\omega_1} \left[\left\|\Delta E_1^{\text{vacuum}}\right\| \left(\cos h(\beta L) - 1\right) + \sqrt{\frac{\omega_2 n_1}{\omega_1 n_2}} \left\|\Delta E_2^{\text{vacuum}}\right\| \sin h(\beta L) \right]^2 \tag{7.174}$$

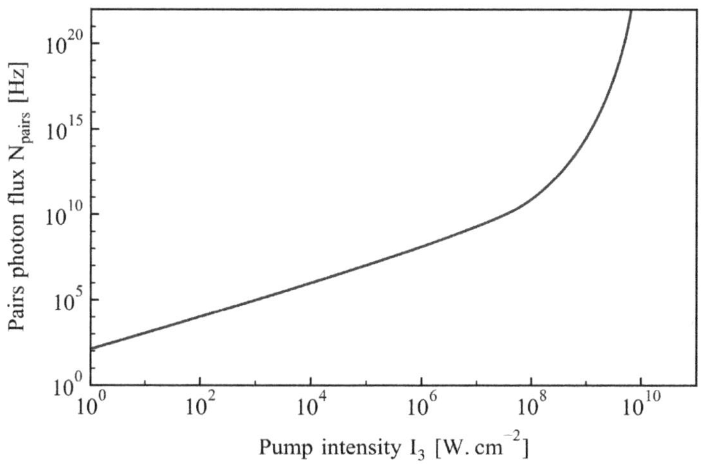

FIGURE 7.72 Calculated SPDC photon flux generated in a 1-cm-long x-cut KTP crystal as a function of the incident pump intensity.

Equation (7.172) is plotted, in Figure 7.72, in the case of a SPDC pumped at $\lambda_3 = 532nm$ in KTP crystal pumped along the x-axis and generating pairs of photons at $(\lambda_1 = 1037\ nm, \lambda_2 = 1092\ nm)$ (Bertrand 2025) [24].

This curve exhibits a linear part and an exponential part, which can be explained by the existence of two different regimes of SPDC according to the relative value of $E_{s,i}$ and $\Delta E_{s,i}^{\text{vacuum}}$: one below a limit pump intensity, for which $E_{s,i} < \Delta E_{s,i}^{\text{vacuum}}$, and the other one above this limit that corresponds to $E_{s,i} > \Delta E_{s,i}^{\text{vacuum}}$. From Eq. (7.172), it is possible to obtain the analytical expressions describing these two regimes (Bertrand 2025) [24].

7.9 SUMMARY

From the equation of propagation in the nonlinear regime given at the beginning of the chapter, and using the slowly varying envelop approximation, it has been possible to establish a system of three coupled differential equations relative to the complex amplitude of the electric field of the three interacting waves with respect to the space coordinate in the direction of propagation. Before performing their integration, these equations allowed us to write three systems of three coupled differential equations regarding (i) the real amplitudes, (ii) the intensity, and (iii) the photon flux density, which led respectively to (i) the demonstration of the quadrature between the initial phases of the incident and generated waves, (ii) the conservation and exchange of energy, and (iii) the non-conservation of the numbers of photons. All these equations are governed by two parameters of prime importance: the effective coefficient and the phase-mismatch. The integration of the complex amplitudes system in the undepleted pump approximation shows that the intensity of the generated wave depends on the square of the effective coefficient and on an interference function between the nonlinear polarization and the radiated field. This interference "corresponds" to momentum conservation in the quantum picture. Two techniques allow the interference to be constructive over the full crystal length: birefringence phase-matching, requiring uniaxial or biaxial crystals, and quasi-phase matching that is performed in an artificial medium, belonging to the isotropic, uniaxial, or biaxial optical classes, in which the sign of the effective coefficient is periodically reversed along the direction of propagation. The phase-matching or quasi-phase-matching directions form surfaces with specific orientation symmetries. The effective coefficient can be expressed as the scalar product of two rank-3 tensors: the second-order electric susceptibility tensor that characterizes the nonlinear response matter, and the field tensor built from the tensor product of the unit electric field vectors of the three interacting waves. The coefficients of the field tensor give the weight of solicitation of the second-order electric susceptibility tensor coefficients. It can happen that these two tensors are orthogonal, which leads to a nil conversion efficiency, and that even if phase-matching or quasi-phase-matching is fulfilled. The chapter ends with (i) the notion of angle and wavelength acceptances corresponding to the width of the interference function due to the angle and wavelength dispersion of the refractive index, and (ii) the attenuation of the conversion efficiency due to the spatial walk-off between the beams of the interacting waves.

REFERENCES

[1] Armstrong J. A., Bloembergen N., Ducuing J., and Pershan P. S. 1962. Interactions between light waves in a nonlinear dielectric. Physical Review 127: 1918–1939.

[2] Manley J. M. and Rowe H. E. 1959. General energy relations in nonlinear reactances. Proceedings of the IRE 47: 2115–2116.

[3] Skauli T., Kuo P. S., Vodopyanov K. L., Pinguet T. J., Levi O., Eyres L. A., Harris J. S., Fejer M. M., Gerard B., Becouarn L., and Lallier E. 2003. Improved dispersion relations for GaAs and applications to nonlinear optics. Journal of Applied Physics 94(10): 6447–6455.

[4] Boulanger B. and Marnier G. 1991. Field factor calculation for the study of the relationships between all the 3-wave non-linear optical interactions in uniaxial and biaxial crystals. Journal of Physics: Condensed Matter 3: 8327–8350.

[5] Kemlin V., Boulanger B., Petrov V., Segonds P., Ménaert B., Schunemann P. G., and Zawilski K. T. 2011. Nonlinear, dispersive, and phase-matching properties of the new chalcopyrite $CdSiP_2$. Optical Materials Express 1(7): 1292–1300.

[6] Fève J. P., Boulanger B., and Marnier G. 1993. Calculation and classification of the direction loci for collinear types I, II and III phase-matching of 3-wave non-linear optical parametric interactions in uniaxial and biaxial acentric crystals. Optics Communications 99: 284–302.

[7] Boulanger B. and Zyss J. 2013. Non-linear optical properties—chapter 1.7. In International Tables for Crystallography, Vol. D: Physical Properties of Crystals, edited by A. Authier, pp. 181–222. Dordrecht, Netherlands: International Union of Crystallography, Kluwer Academic Publisher.

[8] Boulanger B., Fève J. P., Marnier G., Ménaert B., Cabirol X., Villeval P., and Bonnin C. 1994. Relative sign and absolute magnitude of d(2) nonlinear coefficients of KTP from second-harmonic-generation measurements. Journal of the Optical Society of America B 11(5): 750–757.

[9] Trophème B., Boulanger B., and Mennerat G. 2012. Phase-matching loci and angular acceptance of non-collinear optical parametric amplification. Optics Express 20(24): 26176–26183.
[10] Fejer M. M., Magel G. A., Jundt D. H., and Byer R. L. 1992. Quasi-phase matched second harmonic generation: Tuning and tolerances. IEEE Journal of Quantum Electronics 28(11): 2631.
[11] Karlson H. and Laurell F. 1997. Electric field poling of flux grown $KTiOPO_4$. Applied Physics Letters 71: 3474.
[12] Schunemann P. G., Zawilski K. T., Pollak T. M., Zelmon D. E., Fernelius N. C., and Hopkins F. K. 2008. Inorganic nonlinear optical materials. Advanced Solid-State Photonics, OSA Technical Digest Series (CD), Paper MG6.
[13] Petit Y., Boulanger B., Segonds P., and Taira T. 2007. Angular quasi-phase matching. Physical Review A 76(6): 063817.
[14] Petit Y., Peña A., Segonds P., Debray J., Joly S., Zukauskas A., Laurell F., Pasiskevicius V., Canalias C., and Boulanger B. 2020. Development and experimental demonstration of negative first-order quasi-phase-matching in a periodically-poled Rb-doped $KTiOPO_4$ crystal. Optics Letters 45(21): 6026–6029.
[15] Duboisset J., Boulanger B., Brasselet S., Segonds P., and Zyss J. 2024. Nonlinear optics through the field tensor formalism. Laser and Photonics Reviews. https://doi.org/10.1002/lpor.202400411.
[16] Boursier E., Segonds P., Boulanger B., Félix C., Debray J., Jegouso D., Ménaert B., Roshchupkin D., and Shoji I. 2014. Phase-matching directions, refined Sellmeier equations and second order nonlinear coefficient of the infrared Langatate crystal $La_3Ga_{5.5}Ta_{0.5}O_{14}$. Optics Letters 39(13): 4033–4036.
[17] Boursier E., Segonds P., Ménaert B. J., Inácio P. L., Panyutin V., Badikov V., Badikov D., Petrov V., and Boulanger B. 2016. Phase-matching directions and refined Sellmeier equations of the monoclinic acentric crystal $BaGa_4Se_7$. Optics Letters 41(12): 2731–2734.
[18] Guo F., Boursier E., Segonds P., Peña A., Debray J., Badikov V., Panyutin V., Badikov D., Petrov V., and Boulanger B. 2022. Second-order nonlinear optical coefficients of the monoclinic crystal $BaGa_4Se_7$. Optics Letters 47(4): 842–845.
[19] Siegman A. E. 1986. Lasers. Mill Valley, CA: University Science Books.
[20] Boyd G. D., Ashkin A., Dziedzic J. M., and Kleinman D. A. 1965. Second-harmonic generation of light with double refraction. Physical Review 137: 1305–1320.
[21] Mehendale S. C. and Gupta P. K. 1988. Effect of double refraction on type II phase matched second harmonic generation. Optics Communications 68: 301–304.
[22] Fève J. P., Boulanger B., and Marnier G. 1995. Experimental study of walk-off attenuation for type II second harmonic generation in KTP. IEEE Journal of Quantum Electronics 31(8): 1569–1571.
[23] Yariv A. and Yeh P. 2002. Optical Waves in Crystals. New York: Wiley.
[24] Bertrand J., Boutou V., and Boulanger B. 2025. Semi-classical modelling of second-order spontaneous down-conversion measured over 10 decades of pump intensity in a type II phase-matched KTP crystal. https://doi.org/10.48550/arXiv.2502.04815.
[25] Riek C., Seletskiy D. V., Moskalenko A. S., Schmidt J. F., Krauspe P., Eckart S., Eggert S., Burkard G., and Leitenstorfer A. 2015. Direct sampling of electric-field vacuum fluctuations. Science 350(6259): 420–423. https://doi.org/10.1126/science.aac9788.

8 Third-Order Interactions

8.1 COUPLED EQUATIONS

8.1.1 COMPLEX AMPLITUDES

Starting from Eq. (7.1), with $\vec{P}^{NL}(\omega,\vec{r}) = \vec{P}^{(3)}(\omega,\vec{r})$ and using the procedure described by Eq. (7.2) to Eq. (7.31), it comes for the complex amplitude of a plane wave:

$$\frac{\partial E_0(\omega,\vec{r})}{\partial Z} = i\frac{\omega^2\mu_0}{2k(\omega)\cos^2(\rho(\omega))}\vec{e}(\omega).\vec{P}^{(3)}(\omega,\vec{r})\mathrm{e}^{-ik(\omega)Z} \tag{8.1}$$

As for the second-order case studied in Chapter 7, it is considered here that the four interacting waves propagate in the same direction, along the Z-axis, so that Eq. (8.1) can be applied to each wave, which gives:

$$\begin{cases} \dfrac{\partial E_1(Z)}{\partial Z} = i\dfrac{\omega_1^2\mu_0}{2k_1\cos^2\rho_1}\vec{e}_1.\vec{P}_1^{(2)}(Z)\exp[-ik_1Z] \\ \dfrac{\partial E_2(Z)}{\partial Z} = i\dfrac{\omega_2^2\mu_0}{2k_2\cos^2\rho_2}\vec{e}_2.\vec{P}_2^{(2)}(Z)\exp[-ik_2Z] \\ \dfrac{\partial E_3(Z)}{\partial Z} = i\dfrac{\omega_3^2\mu_0}{2k_3\cos^2\rho_3}\vec{e}_3.\vec{P}_3^{(2)}(Z)\exp[-ik_3Z] \\ \dfrac{\partial E_4(Z)}{\partial Z} = i\dfrac{\omega_4^2\mu_0}{2k_4\cos^2\rho_4}\vec{e}_4.\vec{P}_4^{(2)}(Z)\exp[-ik_4Z] \end{cases} \tag{8.2}$$

The same writing convention in Chapter 7 is used, *i.e.* $E_j(Z) \equiv E_0(\omega_j,\vec{r})$, $k_j \equiv k(\omega_j)$, $\rho_j \equiv \rho(\omega_j)$, $\vec{e}_j \equiv \vec{e}(\omega_j)$, and $\vec{P}_j^{(3)}(Z) \equiv \vec{P}^{(3)}(\omega_j,\vec{r})$.

The system given by Eq. (8.2) is coupled with the four third-order induced polarizations that are defined by Eq. (3.3) and Eq. (3.4), which gives:

In the case $\omega_4 = \omega_1 + \omega_2 + \omega_3$

DOI: 10.1201/9780429091735-8

$$\left\{\begin{array}{l}\vec{P}_1^{(3)}(Z)=\varepsilon_0[\chi^{(3)}(\omega_1=\omega_4-\omega_2-\omega_3).\therefore\vec{e}_4\otimes\vec{e}_2\otimes\vec{e}_3] \\ \qquad E_4(Z)E_2^*(Z)E_3^*(Z)\exp[i(k_4-k_2-k_3)Z \\ \vec{P}_2^{(3)}(Z)=\varepsilon_0[\chi^{(3)}(\omega_2=\omega_4-\omega_1-\omega_3).\therefore\vec{e}_4\otimes\vec{e}_1\otimes\vec{e}_3] \\ \qquad E_4(Z)E_1^*(Z)E_3^*(Z)\exp[i(k_4-k_1-k_3)Z \\ \vec{P}_3^{(3)}(Z)=\varepsilon_0[\chi^{(3)}(\omega_3=\omega_4-\omega_1-\omega_2).\therefore\vec{e}_4\otimes\vec{e}_1\otimes\vec{e}_2] \\ \qquad E_4(Z)E_1^*(Z)E_2^*(Z)\exp[i(k_4-k_1-k_2)Z \\ \vec{P}_4^{(3)}(Z)=\varepsilon_0[\chi^{(2)}(\omega_4=\omega_1+\omega_2+\omega_3).\therefore\vec{e}_1\otimes\vec{e}_2\otimes\vec{e}_3] \\ \qquad E_1(Z)E_2(Z)E_3(Z)\exp[i(k_1+k_2+k_3)Z\end{array}\right. \tag{8.3}$$

In the case that $\omega_4+\omega_1=\omega_2+\omega_3$

$$\left\{\begin{array}{l}\vec{P}_1^{(3)}(Z)=\varepsilon_0[\chi^{(3)}(\omega_1=-\omega_4+\omega_2+\omega_3).\therefore\vec{e}_4\otimes\vec{e}_2\otimes\vec{e}_3] \\ \qquad E_4^*(Z)E_2(Z)E_3(Z)\exp[i(-k_4+k_2+k_3)Z \\ \vec{P}_2^{(3)}(Z)=\varepsilon_0[\chi^{(3)}(\omega_2=\omega_4+\omega_1-\omega_3).\therefore\vec{e}_4\otimes\vec{e}_1\otimes\vec{e}_3] \\ \qquad E_4(Z)E_1(Z)E_3^*(Z)\exp[i(k_4+k_1-k_3)Z \\ \vec{P}_3^{(3)}(Z)=\varepsilon_0[\chi^{(3)}(\omega_3=\omega_4+\omega_1-\omega_2).\therefore\vec{e}_4\otimes\vec{e}_1\otimes\vec{e}_2] \\ \qquad E_4(Z)E_1(Z)E_2^*(Z)\exp[i(k_4+k_1-k_2)Z \\ \vec{P}_4^{(3)}(Z)=\varepsilon_0[\chi^{(2)}(\omega_4=-\omega_1+\omega_2+\omega_3).\therefore\vec{e}_1\otimes\vec{e}_2\otimes\vec{e}_3] \\ \qquad E_1^*(Z)E_2(Z)E_3(Z)\exp[i(-k_1+k_2+k_3)Z\end{array}\right. \tag{8.4}$$

By developing the four contracted products in Eq. (8.3) and Eq. (8.4) assuming the ABDP symmetry given by Eq. (3.33), and according to the commutativity of the products of unit electric field, a third-order effective coefficient $\chi_{eff}^{(3)}$ can be defined as follows:

In the case $\omega_4=\omega_1+\omega_2+\omega_3$

$$\begin{aligned}\chi_{eff}^{(3)}&=\vec{e}_1.[\chi^{(3)}(\omega_1=\omega_4-\omega_2-\omega_3).\therefore\vec{e}_4\otimes\vec{e}_2\otimes\vec{e}_3] \\ &=\vec{e}_2.[\chi^{(3)}(\omega_2=\omega_4-\omega_1-\omega_3).\therefore\vec{e}_4\otimes\vec{e}_1\otimes\vec{e}_3] \\ &=\vec{e}_3.[\chi^{(3)}(\omega_3=\omega_4-\omega_1-\omega_2).\therefore\vec{e}_4\otimes\vec{e}_1\otimes\vec{e}_2] \\ &=\vec{e}_4.[\chi^{(3)}(\omega_4=\omega_1+\omega_2+\omega_3).\therefore\vec{e}_1\otimes\vec{e}_2\otimes\vec{e}_3]\end{aligned} \tag{8.5}$$

In the case that $\omega_4 + \omega_1 = \omega_2 + \omega_3$

$$\begin{aligned}
\chi_{eff}^{(3)} &= \vec{e}_1.[\chi^{(3)}(\omega_1 = -\omega_4 + \omega_2 + \omega_3) \therefore \vec{e}_4 \otimes \vec{e}_2 \otimes \vec{e}_3] \\
&= \vec{e}_2.[\chi^{(3)}(\omega_2 = \omega_4 + \omega_1 - \omega_3) \therefore \vec{e}_4 \otimes \vec{e}_1 \otimes \vec{e}_3] \\
&= \vec{e}_3.[\chi^{(3)}(\omega_3 = \omega_4 + \omega_1 - \omega_2) \therefore \vec{e}_4 \otimes \vec{e}_1 \otimes \vec{e}_2] \\
&= \vec{e}_4.[\chi^{(3)}(\omega_4 = -\omega_1 + \omega_2 + \omega_3) \therefore \vec{e}_1 \otimes \vec{e}_2 \otimes \vec{e}_3]
\end{aligned} \tag{8.6}$$

Then Eq. (8.2) writes:

- In the case $\omega_4 = \omega_1 + \omega_2 + \omega_3$

$$\begin{cases}
\dfrac{\partial E_1(Z)}{\partial Z} = j\varepsilon_0\kappa_1\chi_{eff}^{(3)}E_4(Z)E_2^*(Z)E_3^*(Z)\exp[j\Delta kZ] \\
\dfrac{\partial E_2(Z)}{\partial Z} = j\varepsilon_0\kappa_2\chi_{eff}^{(3)}E_4(Z)E_1^*(Z)E_3^*(Z)\exp[j\Delta kZ] \\
\dfrac{\partial E_3(Z)}{\partial Z} = j\varepsilon_0\kappa_3\chi_{eff}^{(3)}E_4(Z)E_1^*(Z)E_2^*(Z)\exp[j\Delta kZ] \\
\dfrac{\partial E_4(Z)}{\partial Z} = j\varepsilon_0\kappa_4\chi_{eff}^{(3)}E_1(Z)E_2(Z)E_3(Z)\exp[-j\Delta kZ] \\
\kappa_i = \dfrac{\omega_i^2\mu_0}{2k_i\cos^2\rho_i} \qquad i = (1,2,3,4) \\
\Delta k = k_4 - k_1 - k_2 - k_3
\end{cases} \tag{8.7}$$

- In the case $\omega_4 + \omega_1 = \omega_2 + \omega_3$

$$\begin{cases}
\dfrac{\partial E_1(Z)}{\partial Z} = j\varepsilon_0\kappa_1\chi_{eff}^{(3)}E_2(Z)E_3(Z)E_4^*(Z)\exp[-j\Delta kZ] \\
\dfrac{\partial E_2(Z)}{\partial Z} = j\varepsilon_0\kappa_2\chi_{eff}^{(3)}E_4(Z)E_1(Z)E_3^*(Z)\exp[j\Delta kZ] \\
\dfrac{\partial E_3(Z)}{\partial Z} = j\varepsilon_0\kappa_3\chi_{eff}^{(3)}E_4(Z)E_1(Z)E_2^*(Z)\exp[j\Delta kZ] \\
\dfrac{\partial E_4(Z)}{\partial Z} = j\varepsilon_0\kappa_4\chi_{eff}^{(3)}E_2(Z)E_3(Z)E_1^*(Z)\exp[-j\Delta kZ] \\
\kappa_i = \dfrac{\omega_i^2\mu_0}{2k_i\cos^2\rho_i} \qquad i = (1,2,3,4) \\
\Delta k = k_4 + k_1 - k_2 - k_3
\end{cases} \tag{8.8}$$

The quantity ΔkZ is the spatial mismatch between the third-order induced polarization and the radiated field for each Fourier component. As in the case of a second-order process, the effective coefficient, here $\chi_{eff}^{(3)}$, and Δk are two key parameters: the nonlinear coupling being stronger if $\chi_{eff}^{(3)}$ is larger and Δk smaller, with an optimum at $\Delta k = 0$. The later condition, called "phase-matching", corresponds to the "photon momentum conservation" given by Eq. (1.106).

8.1.2 Real Amplitudes

By writing $E_j(Z) = A_j(Z)e^{i\Phi_j}$, where $A_j(Z)$ is the real amplitude and Φ_j the initial phase, and by using $e^{i\alpha} = \cos\alpha + i\sin\alpha$, it is possible to access a differential equations system relative to the real amplitudes of the electric fields of the four interacting waves.

In the case $\omega_4 = \omega_1 + \omega_2 + \omega_3$, it comes:

$$\begin{cases} \dfrac{\partial A_1(Z)}{\partial Z} = -\varepsilon_0\kappa_1\chi_{eff}^{(3)}A_4(Z)A_2(Z)A_3(Z)\sin[\Delta kZ + \Delta\Phi] \\ \dfrac{\partial A_2(Z)}{\partial Z} = -\varepsilon_0\kappa_2\chi_{eff}^{(3)}A_4(Z)A_1(Z)A_3(Z)\sin[\Delta kZ + \Delta\Phi] \\ \dfrac{\partial A_3(Z)}{\partial Z} = -\varepsilon_0\kappa_3\chi_{eff}^{(3)}A_4(Z)A_1(Z)A_2(Z)\sin[\Delta kZ + \Delta\Phi] \\ \dfrac{\partial A_4(Z)}{\partial Z} = \varepsilon_0\kappa_4\chi_{eff}^{(3)}A_1(Z)A_2(Z)A_3(Z)\sin[\Delta kZ + \Delta\Phi] \\ \\ \Delta\Phi = \Phi_4 - \Phi_1 - \Phi_2 - \Phi_3 \end{cases} \tag{8.9}$$

When $\Delta k = 0$, Eq. (8.9) writes:

$$\begin{cases} \dfrac{\partial A_1(Z)}{\partial Z} = -\varepsilon_0\kappa_1\chi_{eff}^{(3)}A_4(Z)A_2(Z)A_3(Z)\sin[\Delta\Phi] \\ \dfrac{\partial A_2(Z)}{\partial Z} = -\varepsilon_0\kappa_2\chi_{eff}^{(3)}A_4(Z)A_1(Z)A_3(Z)\sin[\Delta\Phi] \\ \dfrac{\partial A_3(Z)}{\partial Z} = -\varepsilon_0\kappa_3\chi_{eff}^{(3)}A_4(Z)A_1(Z)A_2(Z)\sin[\Delta\Phi] \\ \dfrac{\partial A_4(Z)}{\partial Z} = \varepsilon_0\kappa_4\chi_{eff}^{(3)}A_1(Z)A_2(Z)A_3(Z)\sin[\Delta\Phi] \\ \\ \Delta\Phi = \Phi_4 - \Phi_1 - \Phi_2 - \Phi_3 \end{cases} \tag{8.10}$$

Equation (8.10) shows that there is a maximal energy exchange for $\sin[\Delta\Phi] = \pm 1$, *i.e.* for $\Delta\Phi = \pm\pi/2$ respectively.

$\Delta\Phi = +\pi/2$ corresponds to an optimal photon fusion $\{\omega_1 + \omega_2 + \omega_3 \rightarrow \omega_4\}$ requiring $\left\{\frac{\partial A_1(Z)}{\partial Z} < 0, \frac{\partial A_2(Z)}{\partial Z} < 0, \frac{\partial A_3(Z)}{\partial Z} < 0, \frac{\partial A_4(Z)}{\partial Z} > 0\right\}$.

Then $\Phi_4 = \Phi_1 + \Phi_2 + \Phi_3 + \pi/2$ as shown in Figure 8.1.

$\Delta\Phi = -\pi/2$ corresponds to an optimal photon scission $\{\omega_4 \rightarrow \omega_1 + \omega_2 + \omega_3\}$ requiring $\left\{\frac{\partial A_1(Z)}{\partial Z} > 0, \frac{\partial A_2(Z)}{\partial Z} > 0, \frac{\partial A_3(Z)}{\partial Z} > 0, \frac{\partial A_4(Z)}{\partial Z} < 0\right\}$; then $\Phi_1 + \Phi_2 + \Phi_3 = \Phi_4 + \pi/2$ for a spontaneous scission as shown in Figure 8.2.

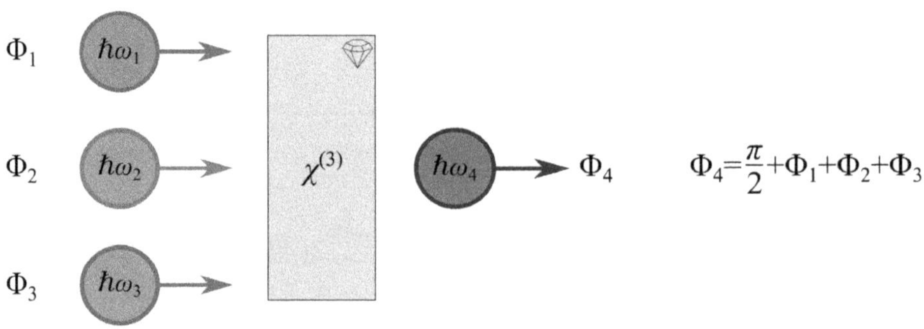

FIGURE 8.1 Initial phase configuration for the phase-matched third-order photon fusion process $\{\omega_1 + \omega_2 + \omega_3 \rightarrow \omega_4\}$.

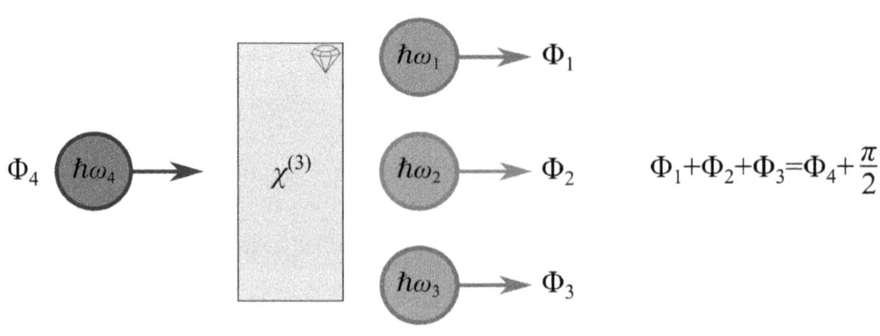

FIGURE 8.2 Initial phase configuration for the phase-matched third-order photon spontaneous scission process $\{\omega_4 \rightarrow \omega_1 + \omega_2 + \omega_3\}$.

In the case that $\omega_4+\omega_1=\omega_2+\omega_3$, it comes:

$$\begin{cases} \dfrac{\partial A_1(Z)}{\partial Z} = \varepsilon_0\kappa_1\chi_{eff}^{(3)}A_4(Z)A_2(Z)A_3(Z)\sin[\Delta kZ+\Delta\Phi] \\ \dfrac{\partial A_2(Z)}{\partial Z} = -\varepsilon_0\kappa_2\chi_{eff}^{(3)}A_4(Z)A_1(Z)A_3(Z)\sin[\Delta kZ+\Delta\Phi] \\ \dfrac{\partial A_3(Z)}{\partial Z} = -\varepsilon_0\kappa_3\chi_{eff}^{(3)}A_4(Z)A_1(Z)A_2(Z)\sin[\Delta kZ+\Delta\Phi] \\ \dfrac{\partial A_4(Z)}{\partial Z} = \varepsilon_0\kappa_4\chi_{eff}^{(3)}A_1(Z)A_2(Z)A_3(Z)\sin[\Delta kZ+\Delta\Phi] \\ \\ \Delta\Phi = \Phi_4+\Phi_1-\Phi_2-\Phi_3 \end{cases} \tag{8.11}$$

When $\Delta k=0$, Eq. (8.11) writes:

$$\begin{cases} \dfrac{\partial A_1(Z)}{\partial Z} = \varepsilon_0\kappa_1\chi_{eff}^{(3)}A_4(Z)A_2(Z)A_3(Z)\sin[\Delta\Phi] \\ \dfrac{\partial A_2(Z)}{\partial Z} = -\varepsilon_0\kappa_2\chi_{eff}^{(3)}A_4(Z)A_1(Z)A_3(Z)\sin[\Delta\Phi] \\ \dfrac{\partial A_3(Z)}{\partial Z} = -\varepsilon_0\kappa_3\chi_{eff}^{(3)}A_4(Z)A_1(Z)A_2(Z)\sin[\Delta\Phi] \\ \dfrac{\partial A_4(Z)}{\partial Z} = \varepsilon_0\kappa_4\chi_{eff}^{(3)}A_1(Z)A_2(Z)A_3(Z)\sin[\Delta\Phi] \\ \\ \Delta\Phi = \Phi_4+\Phi_1-\Phi_2-\Phi_3 \end{cases} \tag{8.12}$$

The energy exchange is maximal for $\Delta\Phi=\pm\pi/2$.

$\Delta\Phi=+\pi/2$ corresponds to the optimal process $\{\omega_4+\omega_1\to\omega_2+\omega_2\}$ requiring $\left\{\dfrac{\partial A_1(Z)}{\partial Z}<0, \dfrac{\partial A_4(Z)}{\partial Z}<0, \dfrac{\partial A_2(Z)}{\partial Z}>0, \dfrac{\partial A_3(Z)}{\partial Z}>0\right\}$.

Then $\Phi_2+\Phi_3=\Phi_4+\Phi_1+\pi/2$ as shown in Figure 8.3.

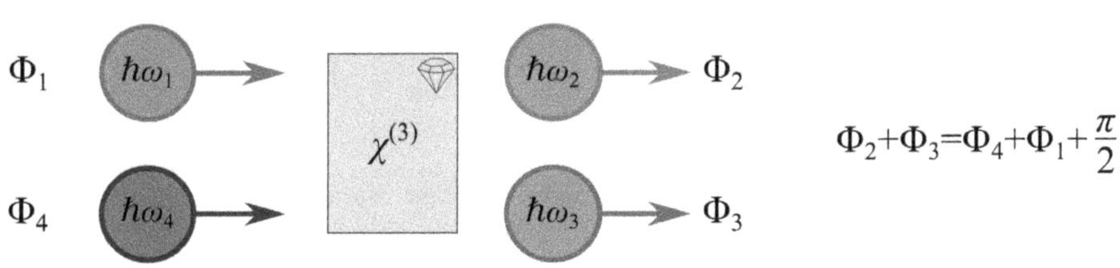

FIGURE 8.3 Initial phase configuration for the phase-matched third-order process $\{\omega_4+\omega_1\to\omega_2+\omega_2\}$.

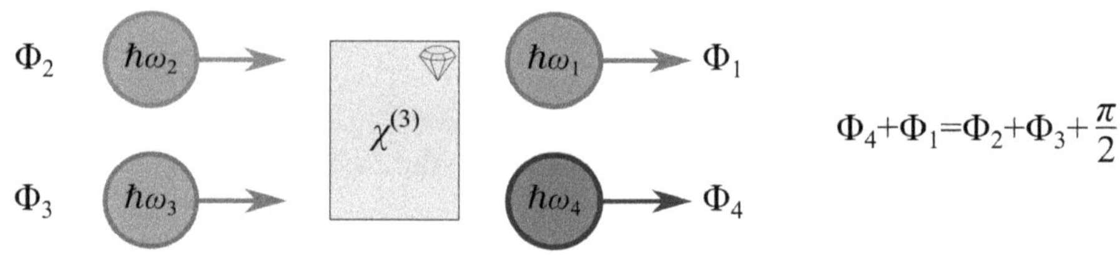

FIGURE 8.4 Initial phase configuration for the phase-matched third-order process $\{\omega_2+\omega_3\rightarrow\omega_4+\omega_1\}$.

$\Delta\Phi=-\pi/2$ corresponds to the reverse process $\{\omega_2+\omega_3\rightarrow\omega_4+\omega_1\}$ requiring $\left\{\frac{\partial A_2(Z)}{\partial Z}<0,\frac{\partial A_3(Z)}{\partial Z}<0,\frac{\partial A_4(Z)}{\partial Z}>0,\frac{\partial A_1(Z)}{\partial Z}>0\right\}$;

then $\Phi_4+\Phi_1=\Phi_2+\Phi_3+\pi/2$ as shown in Figure 8.4.

8.1.3 Intensities

Equation (8.7) and Eq. (8.8), using Eq. (7.43) and Eq. (7.44) and $[jC^*-jC]=2\,\mathrm{Im}\{C\}$, lead to the coupled intensities equations for the two families of four-wave interactions.

In the case that $\omega_4=\omega_1+\omega_2+\omega_3$, it comes:

$$\begin{cases}\dfrac{\partial I_1(Z)}{\partial Z}=-\dfrac{\omega_1}{2}\Psi(Z)\\ \dfrac{\partial I_2(Z)}{\partial Z}=-\dfrac{\omega_2}{2}\Psi(Z)\\ \dfrac{\partial I_3(Z)}{\partial Z}=-\dfrac{\omega_3}{2}\Psi(Z)\\ \dfrac{\partial I_4(Z)}{\partial Z}=+\dfrac{\omega_4}{2}\Psi(Z)\\ \\ \Psi(Z)=\varepsilon_0\chi_{eff}^{(3)}\,\mathrm{Im}\{E_4(Z)E_1^*(Z)E_2^*(Z)E_3^*(Z)\exp[j\Delta kZ]\}\end{cases}\tag{8.13}$$

Equation (8.3) allows us to establish the following relation:

$$\frac{1}{\omega_1}\frac{\partial I_1(Z)}{\partial Z}=\frac{1}{\omega_2}\frac{\partial I_2(Z)}{\partial Z}=\frac{1}{\omega_3}\frac{\partial I_3(Z)}{\partial Z}=-\frac{1}{\omega_4}\frac{\partial I_4(Z)}{\partial Z}\tag{8.14}$$

These equalities indicate the conservation of the total intensity, *i.e.*:

$$\frac{\partial}{\partial Z}[I_1(Z)+I_2(Z)+I_3(Z)+I_4(Z)]=\frac{\Psi(Z)}{2}[\omega_4-\omega_1-\omega_2-\omega_3]=0\tag{8.15}$$

In the case that $\omega_4 + \omega_1 = \omega_2 + \omega_3$, the system writes:

$$\begin{cases} \dfrac{\partial I_1(Z)}{\partial Z} = +\dfrac{\omega_1}{2}\Psi(Z) \\ \dfrac{\partial I_2(Z)}{\partial Z} = -\dfrac{\omega_2}{2}\Psi(Z) \\ \dfrac{\partial I_3(Z)}{\partial Z} = -\dfrac{\omega_3}{2}\Psi(Z) \\ \dfrac{\partial I_4(Z)}{\partial Z} = +\dfrac{\omega_4}{2}\Psi(Z) \\ \\ \Psi(Z) = \varepsilon_0 \chi_{eff}^{(3)} \operatorname{Im}\{E_4(Z)E_1(Z)E_2^*(Z)E_3^*(Z)\exp[j\Delta kZ]\} \end{cases} \tag{8.16}$$

which leads to

$$\frac{1}{\omega_2}\frac{\partial I_2(Z)}{\partial Z} = \frac{1}{\omega_3}\frac{\partial I_3(Z)}{\partial Z} = -\frac{1}{\omega_1}\frac{\partial I_1(Z)}{\partial Z} = -\frac{1}{\omega_4}\frac{\partial I_4(Z)}{\partial Z} \tag{8.17}$$

and so

$$\frac{\partial}{\partial Z}[I_1(Z)+I_2(Z)+I_3(Z)+I_4(Z)] = \frac{\Psi(Z)}{2}[\omega_4+\omega_1-\omega_2-\omega_3] = 0 \tag{8.18}$$

Equation (8.15) and Eq. (8.18) correspond to intensity conservation.

8.1.4 Photon Flux Densities

In the same way, it is possible to access the coupled photon flux densities equations starting from Eq. (8.7) and Eq. (8.8) and using Eq. (7.49).

In the case that $\omega_4 = \omega_1 + \omega_2 + \omega_3$, it comes:

$$\begin{cases} \dfrac{\partial N_1(Z)}{\partial Z} = -\dfrac{\Psi(Z)}{2\hbar} \\ \dfrac{\partial N_2(Z)}{\partial Z} = -\dfrac{\Psi(Z)}{2\hbar} \\ \dfrac{\partial N_3(Z)}{\partial Z} = -\dfrac{\Psi(Z)}{2\hbar} \\ \dfrac{\partial N_4(Z)}{\partial Z} = +\dfrac{\Psi(Z)}{2\hbar} \end{cases} \tag{8.19}$$

and so

$$\begin{cases} \dfrac{\partial N_1(Z)}{\partial Z} = \dfrac{\partial N_2(Z)}{\partial Z} = \dfrac{\partial N_3(Z)}{\partial Z} = -\dfrac{\partial N_4(Z)}{\partial Z} \\ \dfrac{\partial}{\partial Z}[N_1(Z)+N_2(Z)+N_3(Z)+N_4(Z)] = -\dfrac{\Psi(Z)}{\hbar} \neq 0 \end{cases} \tag{8.20}$$

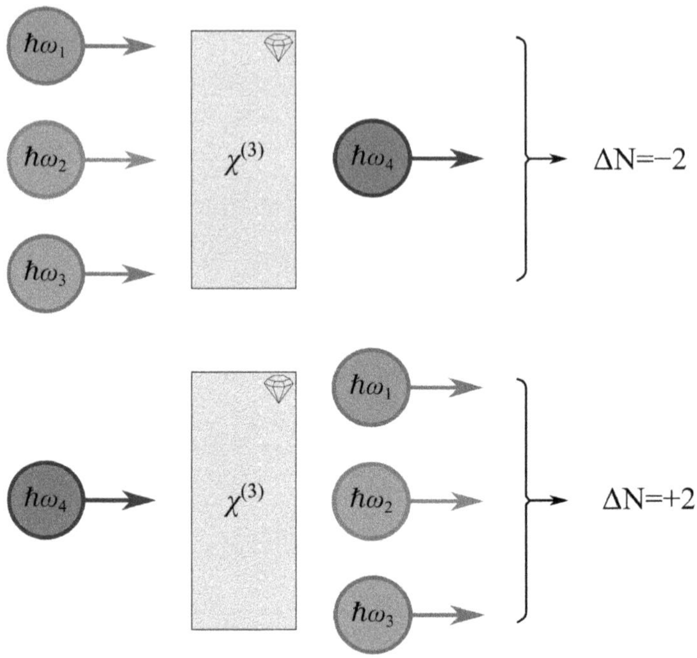

FIGURE 8.5 Schematic differential balance of the photon flux density ΔN due to photon fusion (top) and photon scission (down).

Equation (8.20) shows that the total number of photons is not conserved during the nonlinear process as illustrated by Figure 8.5.

In the case $\omega_4 + \omega_1 = \omega_2 + \omega_3$, it comes:

$$\begin{cases} \dfrac{\partial N_1(Z)}{\partial Z} = +\dfrac{\Psi(Z)}{2\hbar} \\ \dfrac{\partial N_2(Z)}{\partial Z} = -\dfrac{\Psi(Z)}{2\hbar} \\ \dfrac{\partial N_3(Z)}{\partial Z} = -\dfrac{\Psi(Z)}{2\hbar} \\ \dfrac{\partial N_4(Z)}{\partial Z} = +\dfrac{\Psi(Z)}{2\hbar} \end{cases} \tag{8.21}$$

and so

$$\begin{cases} \dfrac{\partial N_2(Z)}{\partial Z} = \dfrac{\partial N_3(Z)}{\partial Z} = -\dfrac{\partial N_1(Z)}{\partial Z} = -\dfrac{\partial N_4(Z)}{\partial Z} \\ \dfrac{\partial}{\partial Z}[N_1(Z) + N_2(Z) + N_3(Z) + N_4(Z)] = 0 \end{cases} \tag{8.22}$$

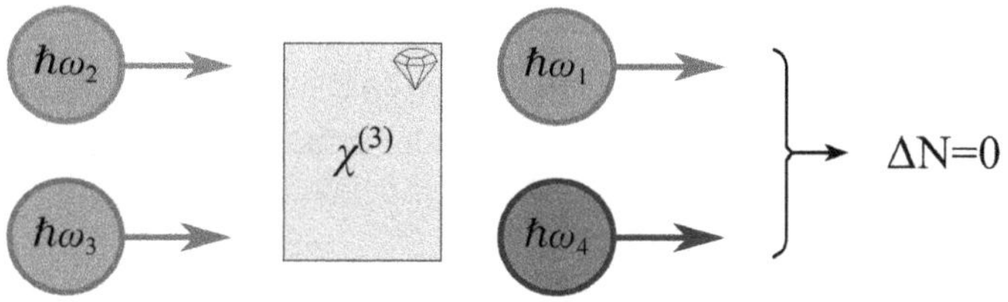

FIGURE 8.6 Schematic differential balance of the photon flux density ΔN due to a photon fusion-scission.

Thus, in that case, contrary to the previous one, the total number of photons is conserved during the nonlinear process, which is well shown by Figure 8.6.

8.2 OUT OF PHASE-MATCHING

8.2.1 Generated Intensity in the Undepleted Pump Approximation

In this case, since $\Delta k \neq 0$, the "undepleted pump approximation" can be applied, which corresponds to $E_1(Z) = E_1(0)$, $E_2(Z) = E_2(0)$, and $E_3(Z) = E_3(0)$ for the example of sum-frequency generation ($\omega_1 + \omega_2 + \omega_3 \rightarrow \omega_4$). Then Eq. (8.7) reduces to:

$$\begin{cases} \dfrac{\partial E_1(Z)}{\partial Z} = \dfrac{\partial E_2(Z)}{\partial Z} = \dfrac{\partial E_3(Z)}{\partial Z} = 0 \\ \dfrac{\partial E_4(Z)}{\partial Z} = i\varepsilon_0 \kappa_4 \chi_{eff}^{(3)} E_1(0) E_2(0) E_3(0) \exp[-i\Delta k Z] \end{cases} \tag{8.23}$$

Thanks to Eq. (7.54) and Eq. (7.55), the integration of $\dfrac{\partial E_4(Z)}{\partial Z}$ leads to:

$$\|E_4(Z)\|^2 = 4[\varepsilon_0 \kappa_4]^2 [\chi_{eff}^{(3)}]^2 \|E_1(0)\|^2 \|E_2(0)\|^2 \|E_3(0)\|^2 \frac{1}{[\Delta k]^2} \sin^2[\frac{\Delta k Z}{2}] \tag{8.24}$$

Using Eq. (7.38), Eq. (7.43), and $\mu_0 \varepsilon_0 c^2 = 1$, the generated intensity writes:

$$I_4(Z) = 16\pi^2 \frac{\mu_0}{\varepsilon_0} \frac{1}{\lambda_4^2 (\cos\rho_4)^4} \frac{[\chi_{eff}^{(3)}]^2}{n_1 n_2 n_3 n_4} I_1(0) I_2(0) I_3(0) \frac{1}{[\Delta k]^2} \sin^2\left[\frac{\Delta k Z}{2}\right] \tag{8.25}$$

where n_j stands for the refractive index $n(\omega_j)$.

Note that Eq. (8.25) is also valid for difference-frequency generation, *i.e.* ($\omega_4 - \omega_2 - \omega_3 \rightarrow \omega_1$), ($\omega_4 - \omega_1 - \omega_3 \rightarrow \omega_2$) or ($\omega_4 - \omega_1 - \omega_2 \rightarrow \omega_3$), by permuting the index 4 by 1, 2, or 3 respectively. This equation also describes the other category of third-order interaction, *i.e.* ($\omega_4 + \omega_1 \rightarrow \omega_2 + \omega_3$), if $E_2(0) \neq 0$ or $E_3(0) \neq 0$: in that case, index 4 has to be permuted with indices 3 or 2, respectively.

8.2.2 Parametric Coherence Length

As shown by Eq. (7.58), the intensity of the generated wave oscillates as a function of the space coordinate Z, as in the case of a second-order interaction illustrated by Figure 7.5. The periodicity is $2l_c$, where l_c is the parametric coherence length defined by:

$$\begin{cases} l_c = \dfrac{\pi}{\Delta k} = \dfrac{\pi c}{\omega_4 n_4 \pm \omega_1 n_1 - \omega_2 n_2 - \omega_3 n_3} \\ \\ \pm \left| \begin{array}{l} + : (\omega_4 + \omega_1 = \omega_2 + \omega_3) \\ - : (\omega_4 = \omega_1 + \omega_2 + \omega_3) \end{array} \right. \end{cases} \tag{8.26}$$

As for the second-order interactions, there is a progressive building of the generated wave over the first coherence length followed by its progressive destruction over the second one, and so on. Furthermore, the parametric coherence length is very small, typically ranging between 1 µm and 100 µm. These characteristics make this configuration of out of phase-matching unattractive.

8.2.3 Configurations of Polarization

As seen in Chapter 3, any kind of crystal, being non-centrosymmetric or not, can have non-zero coefficients of the third-order electric susceptibility tensor, and so a non-zero third-order effective coefficient $\chi_{eff}^{(3)}$. Thus, the three optical classes, *i.e.* isotropic, uniaxial, and biaxial, are concerned. In the case of the isotropic optical class, there is no restriction for the direction of the electric field as seen in Chapter 5, so that any configuration of polarizations is allowed. In the cases of uniaxial and biaxial crystals, any wave can have only two allowed directions of polarization, the so-called Eigenmodes of polarization or birefringence (+) and (–), in any direction of propagation except along an optical axis as explained in Chapter 5. Because four waves are considered here, it gives $2^4 = 16$ possible combinations of Eigenmodes that are given in Table 8.1; (+,–) corresponds to (e,o) in the case of positive uniaxial crystals while it is (o,e) for negative uniaxial crystals, with $\omega_4 > \omega_3 > \omega_2 > \omega_1$. Types 1 and 16 can also exist in isotropic crystals.

In the case of a sum-frequency generation with a configuration (+–+–) for example, it comes:

$$\begin{cases} \chi_{eff}^{(3)} = \vec{e}_4^{\,+} .[\chi^{(2)}(\omega_4 = \omega_1 + \omega_2 + \omega_3) \therefore \vec{e}_1^{\,-} \otimes \vec{e}_2^{\,+} \otimes \vec{e}_3^{\,-}] \\ I_4(Z) = 16\pi^2 \dfrac{\mu_0}{\varepsilon_0} \dfrac{1}{\lambda_4^2 (\cos\rho_4)^4} \dfrac{[\chi_{eff}^{(3)}]^2}{n_1^- n_2^+ n_3^- n_4^+} \\ \qquad I_1(0) I_2(0) I_3(0) \dfrac{1}{[\Delta k]^2} \sin^2 \left[\dfrac{\Delta k Z}{2}\right] \\ l_c = \dfrac{\pi}{\Delta k} = \dfrac{\pi c}{\omega_4 n_4^+ - \omega_1 n_1^- - \omega_2 n_2^+ - \omega_3 n_3^-} \end{cases} \tag{8.27}$$

TABLE 8.1
The Sixteen Possible Combinations of Eigenmodes of Polarization and Corresponding Types of a Third-Order Interaction in Uniaxial and Biaxial Crystals

Types	ω_4	ω_1	ω_2	ω_3
1	−	−	−	−
2	+	−	−	−
3	−	−	−	+
4	−	−	+	−
5	−	+	−	−
6	+	−	−	+
7	+	−	+	−
8	+	+	−	−
9	−	−	+	+
10	−	+	−	+
11	−	+	+	−
12	+	−	+	+
13	+	+	−	+
14	+	+	+	−
15	−	+	+	+
16	+	+	+	+

8.3 BIREFRINGENCE PHASE-MATCHING

As in the case of the three-wave interactions, the use of birefringence allows phase-matching to be achieved. But here, crystals having an inversion center are concerned, in addition to the acentric ones. Thus, any uniaxial or biaxial crystal is potentially interesting for that purpose.

8.3.1 Generated Intensity in the Undepleted Pump Approximation

A sum-frequency generation is also considered here. As did in Chapter 7, it is interesting when $\Delta k = 0$ to write Eq. (8.25) as follows:

$$I_4(Z) = 4\pi^2 \frac{\mu_0}{\varepsilon_0} \frac{1}{\lambda_4^2 (\cos\rho_4)^4} \frac{[\chi_{eff}^{(3)}]^2}{n_1 n_2 n_3 n_4} I_1(0) I_2(0) I_3(0) Z^2 \sin c^2 \left[\frac{\Delta k Z}{2}\right] \quad (8.28)$$

The plotting of Eq. (8.28) is the same as Figure 7.6, in which the maximum of generation is obtained for $\Delta k = 0$, which gives a Z^2 behavior of $I_4(Z)$ that corresponds to the constructive interference between the third-order induced polarization and the radiated field over the whole crystal length.

8.3.2 Configurations of Polarization

The determination of the possible configurations of polarization in terms of modes (+) and (–) allowing phase-matching requires accounting for both energy and momentum conservations as a starting point, *i.e.*:

$$\begin{cases} \omega_4 - \omega_1 - \omega_2 - \omega_3 = 0 \\ \omega_4 n_4 - \omega_1 n_1 - \omega_2 n_2 - \omega_3 n_3 = 0 \end{cases} \tag{8.29}$$

or

$$\begin{cases} \omega_4 + \omega_1 - \omega_2 - \omega_3 = 0 \\ \omega_4 n_4 + \omega_1 n_1 - \omega_2 n_2 - \omega_3 n_3 = 0 \end{cases} \tag{8.30}$$

It is relevant at this step to define the following parameter N coming from Eq. (8.29) as well as from Eq. (8.30) (Fève 2002) [1]:

$$N = \frac{\omega_4 n_4 + \omega_1 n_1}{\omega_4 + \omega_1} = \frac{\omega_2 n_2 + \omega_3 n_3}{\omega_2 + \omega_3} \tag{8.31}$$

In the usual case of normal dispersion, *i.e.* $n_1 < n_2 < n_3 < n_4$ when $(0 <)\omega_1 < \omega_2 < \omega_3 < \omega_4$, Eq. (8.31) leads to the following inequalities:

$$\begin{cases} 0 < \dfrac{n_3 - N}{N - n_2} < 1 \\ 0 < \dfrac{n_4 - N}{N - n_1} < 1 \end{cases} \tag{8.32}$$

From Eq. (8.32), it is possible to show that only seven combinations of modes (+/–) are compatible with the case $\omega_4 - \omega_1 - \omega_2 - \omega_3 = 0$ corresponding to Eq. (8.29), *i.e.* types 3, 4, 5, 9, 10, 11, and 15 given in Table 8.1 (Boulanger 1993) [2]. By using the same way, it can be shown that the sixteen types are allowed for the case $\omega_4 + \omega_1 - \omega_2 - \omega_3 = 0$ corresponding to Eq. (8.30) (Fève 2002) [1]. Type 1 (–,–,–,–) and type 16 (+,+,+,+) are exotic situations because they do not involve birefringence and they can exist in anisotropic crystals belonging to the uniaxial or biaxial optical classes. But because there is no need of birefringence, types 1 and 16 can also exist in an isotropic crystal.

8.3.3 Angular Distribution

The approach used in Part 7.3.4 can be fully followed in this part. The example of third-harmonic generation (THG) can be considered. For type 15 (–,+,+,+) in a positive uniaxial crystal, *i.e.* $n_o < n_e$, the phase-matching relation writes:

$$n_{3\omega}^{-}(\theta_{PM}, \varphi_{PM}) = n_{\omega}^{+}(\theta_{PM}, \varphi_{PM}) \tag{8.33}$$

Then according to Eq. (5.61), it comes:

$$n_o(3\omega) = \frac{1}{\sqrt{n_o^{-2}(\omega)\cos^2(\theta_{PM}) + n_e^{-2}(\omega)\sin^2(\theta_{PM})}} \tag{8.34}$$

Then the phase-matching angle θ_{PM} is expressed as:

$$\theta_{PM} = \arcsin\sqrt{\frac{n_o^{-2}(3\omega) - n_o^{-2}(\omega)}{n_e^{-2}(\omega) - n_o^{-2}(\omega)}} \tag{8.35}$$

The corresponding geometrical representation is given in Figure 8.7.

Then all the phase-matching directions form a revolution cone of aperture angle θ_{PM} around the z-axis, as is the case for type I SHG in uniaxial crystals. It can be shown that this result dealing with uniaxial crystals applies for fourteen types, *i.e.* from type 2 to type 15 corresponding to the situations for which birefringence is required. It can also be shown that for biaxial crystals, the possible loci of phase-matching for these fourteen types are also similar to those of the second-order interactions. These results are summarized in the Wulff abaci shown Figure 8.8.

Figure 8.8 shows that the symmetry group of the phase-matching angular distribution is the infinite group of the cone $G = \left(\frac{\infty}{m}m\right)$ for the uniaxial case, while it is the orthorhombic finite group $G = (mmm)$.

As seen earlier, type 1 (–,–,–,–) and type 16 (+,+,+,+) are particular cases because they can be implemented in both isotropic and anisotropic crystals, and it is reminded that they are valid only for the case $\omega_4 + \omega_1 - \omega_2 - \omega_3 = 0$ (Fève 2002) [1].

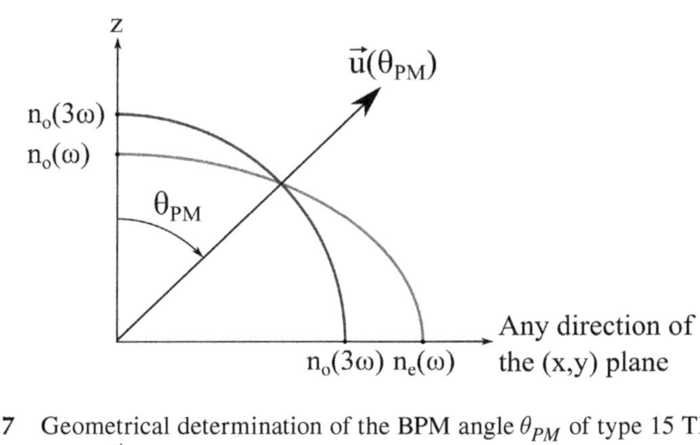

FIGURE 8.7 Geometrical determination of the BPM angle θ_{PM} of type 15 THG in a positive uniaxial crystal. $\vec{u}(\theta_{PM})$ is the unit vector of the corresponding phase-matching direction. The circle (blue line) and ellipse (red line) are the section of the index surface in any plane containing the z-axis. (x, y, z) is the dielectric frame.

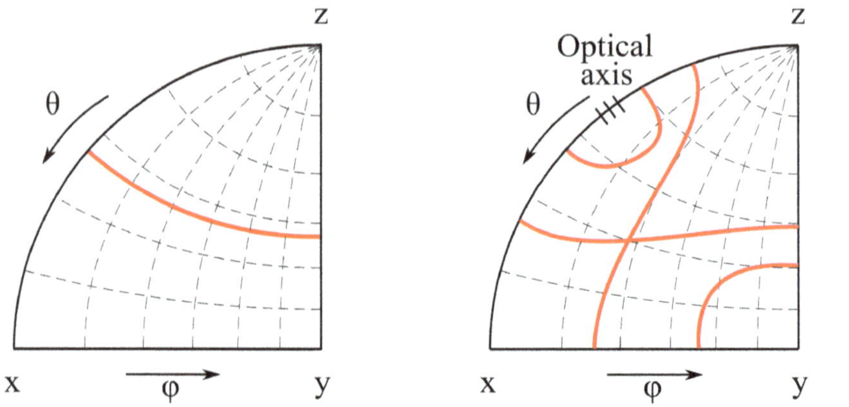

FIGURE 8.8 Possible birefringence phase-matching (BPM) cones (red lines) of a uniaxial crystal (left) and of a biaxial crystal (right) for fourteen types, from type 2 to type 15 given in Table 8.1. These fourteen types are compatible with $\omega_4 + \omega_1 - \omega_2 - \omega_3 = 0$ while types 3, 4, 5, 9, 10, 11, and 15 are only allowed for the case $\omega_4 - \omega_1 - \omega_2 - \omega_3 = 0$. (x, y, z) is the dielectric frame.

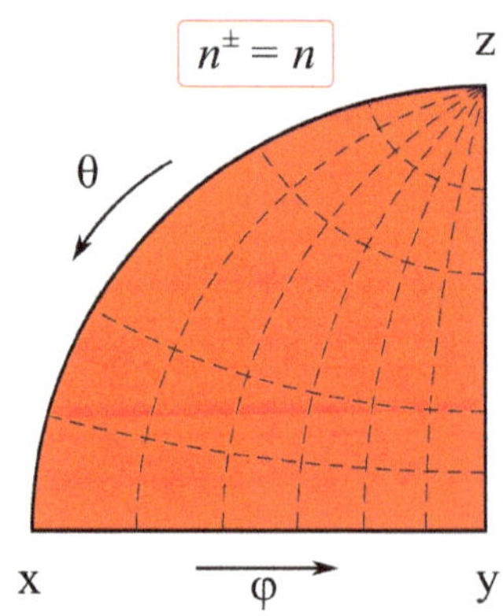

FIGURE 8.9 Possible phase-matching directions (red) for type 1 $(\omega_4^- + \omega_1^- - \omega_2^- - \omega_3^- = 0)$ and type 16 $(\omega_4^+ + \omega_1^+ - \omega_2^+ - \omega_3^+ = 0)$ in an isotropic crystal, in which (+) and (–) refer to the Eigen propagation modes. (x, y, z) is the dielectric frame and $n^{\pm} = n$ stands for the refractive index.

Concerning isotropic crystals, if Eq. (8.30) is fulfilled for a given set of pulsations, then phase-matching is allowed in any direction of propagation for this set, as shown in Figure 8.9. As a consequence, the corresponding angular distribution belongs to the infinite point group of the sphere, *i.e.* $G = \left(\frac{\infty}{m}\frac{\infty}{m}\right)$.

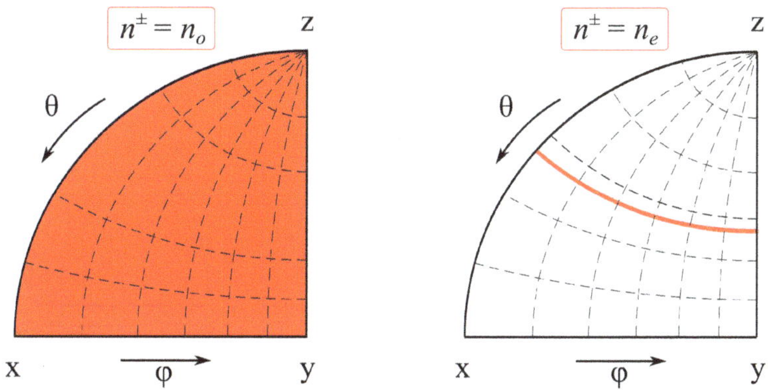

FIGURE 8.10 Phase-matching directions (red) for type 1 $(\omega_4^- + \omega_1^- - \omega_2^- - \omega_3^- = 0)$ and type 16 $(\omega_4^+ + \omega_1^+ - \omega_2^+ - \omega_3^+ = 0)$ in an uniaxial crystal, in which (+) and (–) refer to the Eigen propagation modes. (x, y, z) is the dielectric frame and $n^{\pm}$ stands for the refractive index relative to modes (+) and (–), while n_o and n_e correspond to the ordinary and extraordinary refractive indices.

In the case of a uniaxial crystal, the situation is comparable to the isotropic one if the four waves are ordinary polarized, *i.e.* (o,o,o,o). Besides, the phase-matching directions form a cone of revolution around the z-axis when extraordinary polarizations are concerned, *i.e.* (e,e,e,e) leading to the infinite group of the cone $G = \left(\frac{\infty}{m} m\right)$. The corresponding Wulff *abaci* are given in Figure 8.10.

For a biaxial crystal, there are also two situations to distinguish while considering type 1 and type 16: if the four waves are relative to the circular section of the index surface, *i.e.* n_x in the yz-plane, n_y in the xz-plane and n_z in the xy-plane, then the whole phase-matching directions lie only in the corresponding portion of the circle, as shown by Figure 8.11 (top and bottom, left and middle), while no phase-matching solution exists out of the related dielectric plane: the symmetry group is then $G = (mmm)$. For the other cases, that is to say when the refractive index is relative to the elliptical section in a principal plane, the phase-matching directions join the relevant portions of the principal planes as depicted in Figure 8.11 (top and bottom, right). The corresponding group is also $G = (mmm)$.

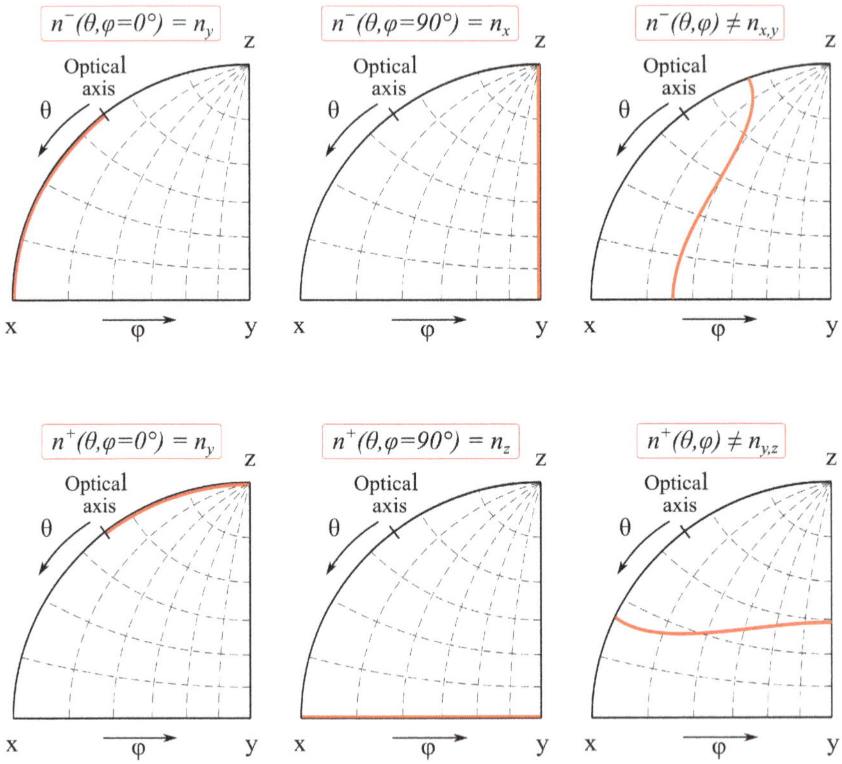

FIGURE 8.11 Possible phase-matching directions (red) for type 1 $(\omega_4^- + \omega_1^- - \omega_2^- - \omega_3^- = 0)$ and type 16 $(\omega_4^+ + \omega_1^+ - \omega_2^+ - \omega_3^+ = 0)$ in a positive biaxial crystal, in which (+) and (–) refer to the Eigen propagation modes. (x,y,z) is the dielectric frame and $n^\pm$ stands for the refractive index relative to modes (+) and (–), while n_y and n_z are principal refractive indices. The indices n^+ and n^- have to be permuted in the case of a negative biaxial crystal.

Non-collinear phase-matching can also be implemented, and the corresponding phase-matching relations take into account the angles between the wave vectors of the interacting waves as described in Part 7.3.5.

8.4 QUASI-PHASE-MATCHING

8.4.1 Reversal of the Sign of $\chi^{(3)}$

The concepts and approaches developed in Part 7.4 can be fully applied here. In the following are only detailed some configurations of sign reversal of $\chi^{(3)}$ based on sequence of matter alternately rotated by an angle π. It is then easy to identify the coefficients for which the sign is going to be reversed thanks to such a rotation of π, or possibly $\pi/2$, around a given axis. Some examples are given in Figure 8.12, Figure 8.13, Figure 8.14, Figure 8.15, and Figure 8.16. The matrices that are shown correspond to that of the triclinic crystal class 1 and $\bar{1}$ for which all the coefficients are not nil, which gives the full picture of the third-order case.

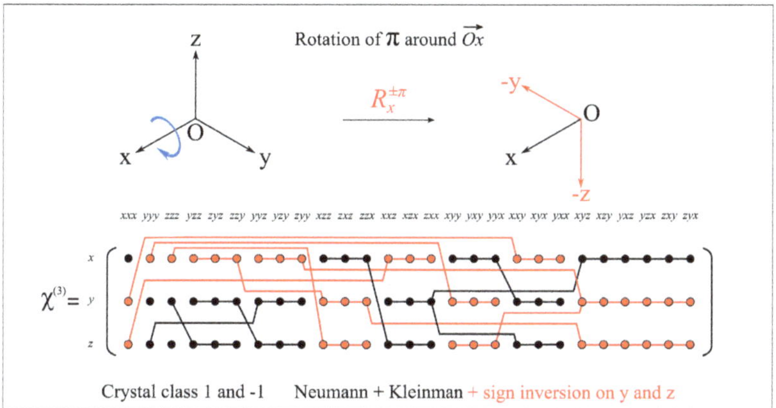

FIGURE 8.12 Third-order electric susceptibility tensor coefficients (red dots) for which the sign is reversed due to a rotation of π around the x-axis of the dielectric frame.

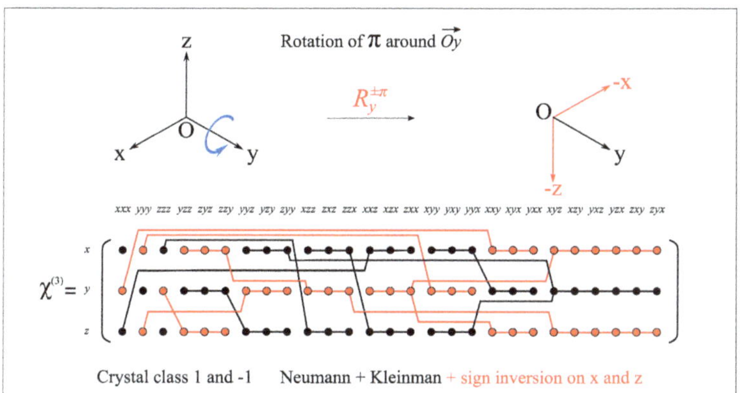

FIGURE 8.13 Third-order electric susceptibility tensor coefficients (red dots) for which the sign is reversed due to a rotation of π around the y-axis of the dielectric frame.

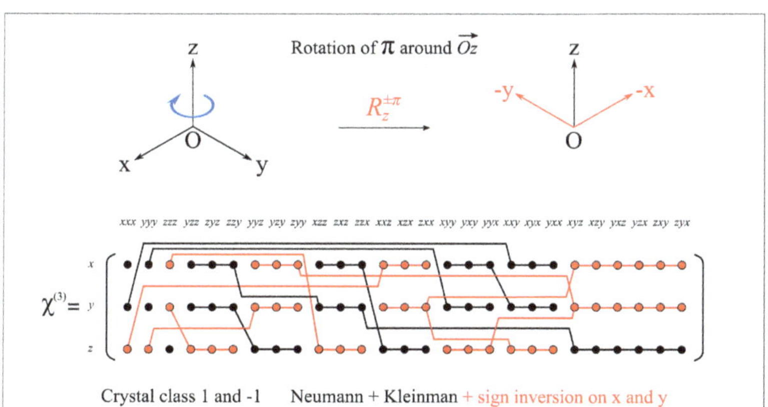

FIGURE 8.14 Third-order electric susceptibility tensor coefficients (red dots) for which the sign is reversed due to a rotation of π around the z-axis of the dielectric frame.

Note that the rotation of π around the z-axis (Figure 8.14) and the rotation of π around an axis oriented at $\pi / 4$ from the x- and y-axis in the xy-plane (Figure 8.16) lead to the same tensor configuration.

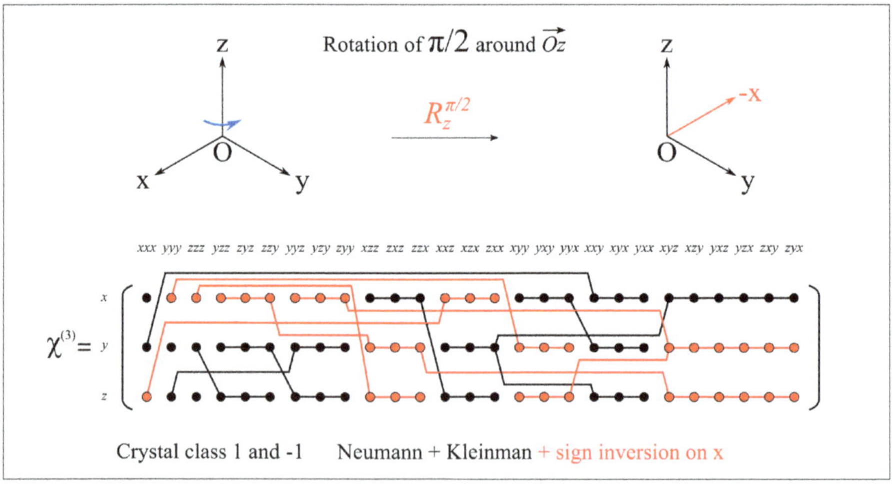

FIGURE 8.15 Third-order electric susceptibility tensor coefficients (red dots) for which the sign is reversed due to a rotation of $\pi / 2$ around the z-axis of the dielectric frame.

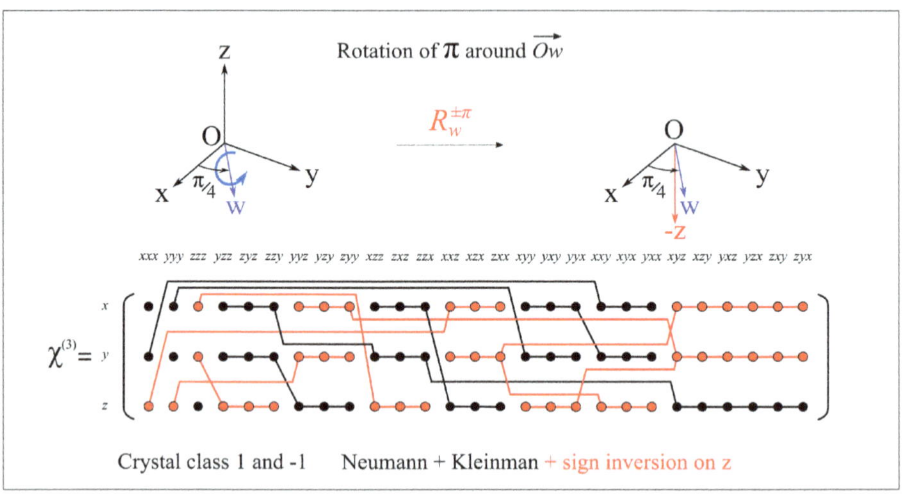

FIGURE 8.16 Third-order electric susceptibility tensor coefficients (red dots) for which the sign is reversed due to a rotation of π around an axis oriented at $\pi / 4$ from the x- and y-axis in the xy-plane.

8.4.2 Positive and Negative m^th^-Order Quasi-Phase-Matching

As explained in Part 7.4 for a second-order process, a square periodic modulation of $\chi^{(3)}(Z)$ is also well described by a Fourier series of odd terms only, *i.e.*:

$$\begin{cases} \chi^{(3)}(Z) = \frac{2}{\pi}\chi_{eff}^{(3)} \exp[i\left|\vec{K}_{Gm}\right| Z] \\ \left|\vec{K}_{Gm}\right| \equiv K_{Gm} = \frac{2\pi m}{\Lambda} \\ m = \pm 1, \pm 3, \pm 5, ... \end{cases} \tag{8.36}$$

The effective coefficient $\chi_{eff}^{(3)}$ is defined by Eq. (8.5) and Eq. (8.6), and K_{Gm} is the modulus of the grating vector $\vec{K}_{Gm}$ along Z associated with the grating periodicity Λ such as $\Lambda = 2l_c$ where l_c is the parametric coherent length expressed by Eq. (8.26).

Then the QPM relation writes:

$$\begin{cases} \frac{n_4}{\lambda_4} \pm \frac{n_1}{\lambda_1} - \frac{n_2}{\lambda_2} - \frac{n_3}{\lambda_3} - \frac{m}{\Lambda} = 0 \\ \pm \begin{vmatrix} + : (\omega_4 + \omega_1 = \omega_2 + \omega_3) \\ - : (\omega_4 = \omega_1 + \omega_2 + \omega_3) \end{vmatrix} \end{cases} \tag{8.37}$$

The first-order QPM, *i.e.* $m = \pm 1$, is the most efficient. It is illustrated in Figure 8.17 and Figure 8.18.

Equation (8.37) in the case $m > 0$ can be verified for any of the 2^4 possible combinations of Eigen refractive indices $n_i^{\pm}$ given in Table 8.1. The approach described in Part 7.3.3 allows us to determine the possible combinations for $m < 0$, which requires at least the existence of phase-matching conditions for such polarization combinations.

As for the second-order processes, QPM can also be realized for the third-order processes at any angle of spherical coordinates $\left(\theta_{AQPM}, \varphi_{AQPM}\right)$ with respect to the grating vector $\vec{K}_G$, which corresponds to the so-called angular quasi-phase-matching (AQPM). Thus Eq. (8.37) becomes:

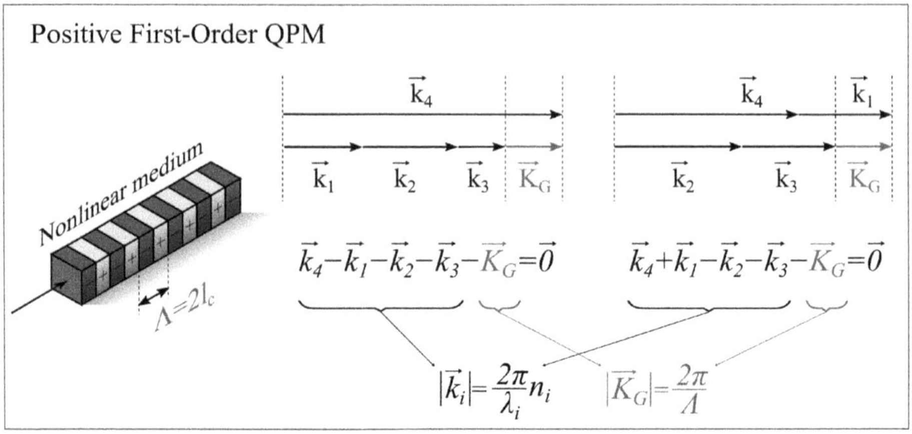

FIGURE 8.17 Conditions for achieving positive first-order (m = 1) quasi-phase-matching (QPM) in the two possible configurations of a third-order process. The nonlinear medium exhibits a periodical spatial modulation of the sign (+/–) of the third-order electric susceptibility. $(\vec{k}_1,\vec{k}_2,\vec{k}_3,\vec{k}_4)$ are the wave vectors of the three interacting waves at the wavelengths $(\lambda_1,\lambda_2,\lambda_3,\lambda_4)$. $\vec{K}_G$ is the grating vector associated with the grating period Λ equal to twice the parametric coherence length l_c. The arrow at the entrance of the crystal corresponds to the direction of propagation of the incident light.

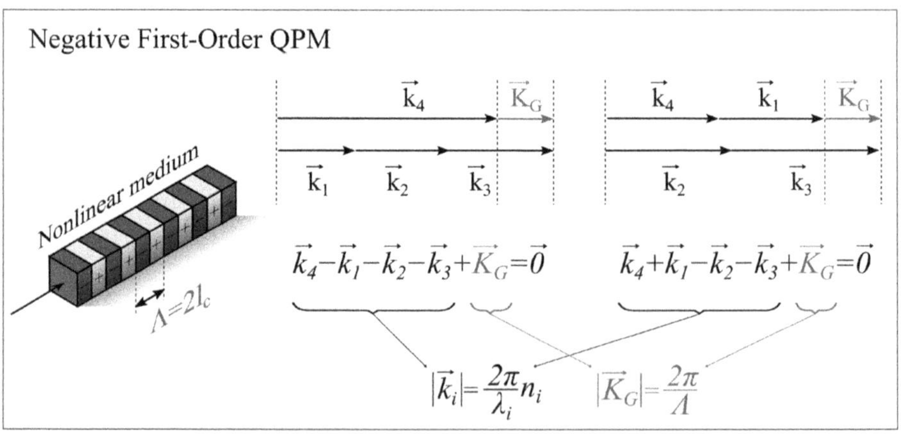

FIGURE 8.18 Conditions for achieving negative first-order (m = −1) quasi-phase-matching (QPM) in the two possible configurations of a third-order process. The nonlinear medium exhibits a periodical spatial modulation of the sign (+/–) of the third-order electric susceptibility. $(\vec{k}_1,\vec{k}_2,\vec{k}_3,\vec{k}_4)$ are the wave vectors of the three interacting waves at the wavelengths $(\lambda_1,\lambda_2,\lambda_3,\lambda_4)$. $\vec{K}_G$ is the grating vector associated with the grating period Λ equal to twice the parametric coherence length l_c. The arrow at the entrance of the crystal corresponds to the direction of propagation of the incident light.

$$\begin{cases} \dfrac{n_3\left(\theta_{AQPM},\varphi_{AQPM}\right)}{\lambda_4} \pm \dfrac{n_1\left(\theta_{AQPM},\varphi_{AQPM}\right)}{\lambda_1} - \dfrac{n_2\left(\theta_{AQPM},\varphi_{AQPM}\right)}{\lambda_2} \\ -\dfrac{n_3\left(\theta_{AQPM},\varphi_{AQPM}\right)}{\lambda_3} - \dfrac{m}{\Lambda_{eff}\left(\theta_{AQPM},\varphi_{AQPM}\right)} = 0 \\ \pm \left| \begin{array}{l} + : (\omega_4 + \omega_1 = \omega_2 + \omega_3) \\ - : (\omega_4 = \omega_1 + \omega_2 + \omega_3) \end{array} \right. \end{cases} \tag{8.38}$$

The quantity $\Lambda_{eff}\left(\theta_{AQPM},\varphi_{AQPM}\right)$ denotes the effective grating periodicity in the direction of propagation $\left(\theta_{AQPM},\varphi_{AQPM}\right)$: it is given by Eq. (7.102), Eq. (7.103), and Eq. (7.104) when $\vec{K}_G$ lies along $\overrightarrow{Ox}$, $\overrightarrow{Oy}$, and $\overrightarrow{Oz}$, respectively. Equation (8.38) can be solved numerically in any case.

The intensity of the generated beam can be calculated using Eq. (8.28) with a modified phase-mismatched parameter and the suited Fourier factor, which gives:

$$\begin{cases} I_4(Z) = \dfrac{1}{m^2} \dfrac{\mu_0}{\varepsilon_0} \dfrac{16}{\lambda_4^2 (\cos\rho_4)^4} \dfrac{[\chi_{eff}^{(3)}]^2}{n_1 n_2 n_3 n_4} I_1(0) I_2(0) I_3(0) Z^2 \\ \qquad \sin c^2 \left[\dfrac{(\Delta k - K_{Gm}) Z}{2} \right] \\ \Delta k = k_4 \pm k_1 - k_2 - k_3 \\ m = \pm 1, \pm 3, \pm 5, \ldots \end{cases} \tag{8.39}$$

The intensity is maximal when $\Delta k - K_{Gm} = 0$, that is to say when QPM is achieved, which corresponds to Eq. (8.37).

8.5 EFFECTIVE COEFFICIENT

8.5.1 Definition of the Field Tensor

The effective coefficient $\chi_{eff}^{(3)}$ given by Eq. (8.5) and Eq. (8.6) can be written as the contracted product of two rank-4 tensors, *i.e.* in the direction of propagation $\vec{u}$ (Boulanger 1993, Boulanger 2013, Duboisset 2024) [2,3,4]:

In the case $\omega_4 = \omega_1 + \omega_2 + \omega_3$

$$\begin{aligned} \chi_{eff}^{(3)}(\vec{u}) &= F^{(3)}\left(\omega_1 = \omega_4 - \omega_2 - \omega_3, \vec{u}\right) :: \chi^{(3)}\left(\omega_1 = \omega_4 - \omega_2 - \omega_3\right) \\ &= F^{(3)}\left(\omega_2 = \omega_4 - \omega_1 - \omega_3, \vec{u}\right) :: \chi^{(3)}\left(\omega_2 = \omega_4 - \omega_1 - \omega_3\right) \\ &= F^{(3)}\left(\omega_3 = \omega_4 - \omega_1 - \omega_2, \vec{u}\right) :: \chi^{(3)}\left(\omega_3 = \omega_4 - \omega_1 - \omega_2\right) \\ &= F^{(3)}\left(\omega_4 = \omega_1 + \omega_2 + \omega_3, \vec{u}\right) :: \chi^{(3)}\left(\omega_4 = \omega_1 + \omega_2 + \omega_3\right) \end{aligned} \tag{8.40}$$

with

$$\begin{aligned}
F^{(3)}\left(\omega_1=\omega_4-\omega_2-\omega_3,\vec{u}\right)&=\vec{e}_1(\vec{u})\otimes\vec{e}_4(\vec{u})\otimes\vec{e}_2(\vec{u})\otimes\vec{e}_3(\vec{u})\\
F^{(3)}\left(\omega_2=\omega_4-\omega_1-\omega_3,\vec{u}\right)&=\vec{e}_2(\vec{u})\otimes\vec{e}_4(\vec{u})\otimes\vec{e}_1(\vec{u})\otimes\vec{e}_3(\vec{u})\\
F^{(3)}\left(\omega_3=\omega_4-\omega_1-\omega_2,\vec{u}\right)&=\vec{e}_3(\vec{u})\otimes\vec{e}_4(\vec{u})\otimes\vec{e}_1(\vec{u})\otimes\vec{e}_2(\vec{u})\\
F^{(3)}\left(\omega_4=\omega_1+\omega_2+\omega_3,\vec{u}\right)&=\vec{e}_4(\vec{u})\otimes\vec{e}_1(\vec{u})\otimes\vec{e}_2(\vec{u})\otimes\vec{e}_3(\vec{u})
\end{aligned} \tag{8.41}$$

In the case $\omega_4+\omega_1=\omega_2+\omega_3$

$$\begin{aligned}
\chi_{eff}^{(3)}(\vec{u})&=F^{(3)}\left(\omega_1=-\omega_4+\omega_2+\omega_3,\vec{u}\right)::\chi^{(3)}\left(\omega_1=-\omega_4+\omega_2-\omega_3\right)\\
&=F^{(3)}\left(\omega_2=\omega_4+\omega_1-\omega_3,\vec{u}\right)::\chi^{(3)}\left(\omega_2=\omega_4+\omega_1-\omega_3\right)\\
&=F^{(3)}\left(\omega_3=\omega_4+\omega_1-\omega_2,\vec{u}\right)::\chi^{(3)}\left(\omega_3=\omega_4+\omega_1-\omega_2\right)\\
&=F^{(3)}\left(\omega_4=-\omega_1+\omega_2+\omega_3,\vec{u}\right)::\chi^{(3)}\left(\omega_4=-\omega_1+\omega_2+\omega_3\right)
\end{aligned} \tag{8.42}$$

with

$$\begin{aligned}
F^{(3)}\left(\omega_1=-\omega_4+\omega_2+\omega_3,\vec{u}\right)&=\vec{e}_1(\vec{u})\otimes\vec{e}_4(\vec{u})\otimes\vec{e}_2(\vec{u})\otimes\vec{e}_3(\vec{u})\\
F^{(3)}\left(\omega_2=\omega_4+\omega_1-\omega_3,\vec{u}\right)&=\vec{e}_2(\vec{u})\otimes\vec{e}_4(\vec{u})\otimes\vec{e}_1(\vec{u})\otimes\vec{e}_3(\vec{u})\\
F^{(3)}\left(\omega_3=\omega_4+\omega_1-\omega_2,\vec{u}\right)&=\vec{e}_3(\vec{u})\otimes\vec{e}_4(\vec{u})\otimes\vec{e}_1(\vec{u})\otimes\vec{e}_2(\vec{u})\\
F^{(3)}\left(\omega_4=-\omega_1+\omega_2+\omega_3,\vec{u}\right)&=\vec{e}_4(\vec{u})\otimes\vec{e}_1(\vec{u})\otimes\vec{e}_2(\vec{u})\otimes\vec{e}_3(\vec{u})
\end{aligned} \tag{8.43}$$

The field tensor $F^{(3)}$ is a polar tensor built from the tensor product of the unit electric fields of the four interacting waves. Thus $F^{(3)}$, similarly to the case of second-order interactions detailed in Chapter 7, depends on the direction of propagation $\vec{u}$ contrarily to the electric susceptibility tensor $\chi^{(3)}$.

These tensor equations projected on the dielectric frame write as follows:

In the case $\omega_4=\omega_1+\omega_2+\omega_3$

$$\begin{aligned}
\chi_{eff}^{(3)}(\vec{u})&=\sum_{ijkl}F_{ijkl}^{(3)}\left(\omega_1=\omega_4-\omega_2-\omega_3,\vec{u}\right)\chi_{ijkl}^{(3)}\left(\omega_1=\omega_4-\omega_2-\omega_3\right)\\
&=\sum_{ijkl}F_{ijkl}^{(3)}\left(\omega_2=\omega_4-\omega_1-\omega_3,\vec{u}\right)\chi_{ijkl}^{(3)}\left(\omega_2=\omega_4-\omega_1-\omega_3\right)\\
&=\sum_{ijkl}F_{ijkl}^{(3)}\left(\omega_3=\omega_4-\omega_1-\omega_2,\vec{u}\right)\chi_{ijkl}^{(3)}\left(\omega_3=\omega_4-\omega_1-\omega_2\right)\\
&=\sum_{ijkl}F_{ijkl}^{(3)}\left(\omega_4=\omega_1+\omega_2+\omega_3,\vec{u}\right)\chi_{ijkl}^{(3)}\left(\omega_4=\omega_1+\omega_2+\omega_3\right)
\end{aligned} \tag{8.44}$$

with

$$\begin{aligned} &F^{(3)}_{ijkl}\left(\omega_1=\omega_4-\omega_2-\omega_3,\vec{u}\right)=e_i(\omega_1,\vec{u})\otimes e_j(\omega_4,\vec{u})\otimes e_k(\omega_2,\vec{u})\otimes e_l(\omega_3,\vec{u}) \\ &F^{(3)}_{ijkl}\left(\omega_2=\omega_4-\omega_1-\omega_3,\vec{u}\right)=e_i(\omega_2,\vec{u})\otimes e_j(\omega_4,\vec{u})\otimes e_k(\omega_1,\vec{u})\otimes e_l(\omega_3,\vec{u}) \\ &F^{(3)}_{ijkl}\left(\omega_3=\omega_4-\omega_1-\omega_2,\vec{u}\right)=e_i(\omega_3,\vec{u})\otimes e_j(\omega_4,\vec{u})\otimes e_k(\omega_1,\vec{u})\otimes e_l(\omega_2,\vec{u}) \\ &F^{(3)}_{ijkl}\left(\omega_4=\omega_1+\omega_2+\omega_3,\vec{u}\right)=e_i(\omega_4,\vec{u})\otimes e_j(\omega_1,\vec{u})\otimes e_k(\omega_2,\vec{u})\otimes e_l(\omega_3,\vec{u}) \end{aligned} \quad (8.45)$$

In the case $\omega_4+\omega_1=\omega_2+\omega_3$

$$\begin{aligned} \chi^{(3)}_{eff}(\vec{u}) &= \sum_{ijkl} F^{(3)}_{ijkl}\left(\omega_1=-\omega_4+\omega_2+\omega_3,\vec{u}\right)\chi^{(3)}_{ijkl}\left(\omega_1=-\omega_4+\omega_2-\omega_3\right) \\ &= \sum_{ijkl} F^{(3)}_{ijkl}\left(\omega_2=\omega_4+\omega_1-\omega_3,\vec{u}\right)\chi^{(3)}_{ijkl}\left(\omega_2=\omega_4+\omega_1-\omega_3\right) \\ &= \sum_{ijkl} F^{(3)}_{ijkl}\left(\omega_3=\omega_4+\omega_1-\omega_2,\vec{u}\right)\chi^{(3)}_{ijkl}\left(\omega_3=\omega_4+\omega_1-\omega_2\right) \\ &= \sum_{ijkl} F^{(3)}_{ijkl}\left(\omega_4=-\omega_1+\omega_2+\omega_3,\vec{u}\right)\chi^{(3)}_{ijkl}\left(\omega_4=-\omega_1+\omega_2+\omega_3\right) \end{aligned} \quad (8.46)$$

with

$$\begin{aligned} &F^{(3)}_{ijkl}\left(\omega_1=-\omega_4+\omega_2+\omega_3,\vec{u}\right)=e_i(\omega_1,\vec{u})\otimes e_j(\omega_4,\vec{u})\otimes e_k(\omega_2,\vec{u})\otimes e_l(\omega_3,\vec{u}) \\ &F^{(3)}_{ijkl}\left(\omega_2=\omega_4+\omega_1-\omega_3,\vec{u}\right)=e_i(\omega_2,\vec{u})\otimes e_j(\omega_4,\vec{u})\otimes e_k(\omega_1,\vec{u})\otimes e_l(\omega_3,\vec{u}) \\ &F^{(3)}_{ijkl}\left(\omega_3=\omega_4+\omega_1-\omega_2,\vec{u}\right)=e_i(\omega_3,\vec{u})\otimes e_j(\omega_4,\vec{u})\otimes e_k(\omega_1,\vec{u})\otimes e_l(\omega_2,\vec{u}) \\ &F^{(3)}_{ijkl}\left(\omega_4=-\omega_1+\omega_2+\omega_3,\vec{u}\right)=e_i(\omega_4,\vec{u})\otimes e_j(\omega_1,\vec{u})\otimes e_k(\omega_2,\vec{u})\otimes e_l(\omega_3,\vec{u}) \end{aligned} \quad (8.47)$$

8.5.2 Symmetries of the Field Tensor

The commutativity of the algebra product of the unit electric field vectors in Eq. (8.45) and Eq. (8.47) leads to relations between field tensor coefficients at different circular frequencies due to, *i.e.*:

$$\begin{cases} F^{(3)}_{ijkl}\left(\omega_1=\omega_4-\omega_2-\omega_3,\vec{u}\right)=F^{(3)}_{kjil}\left(\omega_2=\omega_4-\omega_1-\omega_3,\vec{u}\right) \\ =F^{(3)}_{ljik}\left(\omega_3=\omega_4-\omega_1-\omega_2,\vec{u}\right)=F^{(3)}_{jikl}\left(\omega_4=\omega_1+\omega_2+\omega_3,\vec{u}\right) \\ \\ F^{(3)}_{ijkl}\left(\omega_1=-\omega_4+\omega_2+\omega_3,\vec{u}\right)=F^{(3)}_{kjil}\left(\omega_2=\omega_4+\omega_1-\omega_3,\vec{u}\right) \\ =F^{(3)}_{ljik}\left(\omega_3=\omega_4+\omega_1-\omega_2,\vec{u}\right)=F^{(3)}_{jikl}\left(\omega_4=-\omega_1+\omega_2+\omega_3,\vec{u}\right) \end{cases} \quad (8.48)$$

Equation (8.48) shows that the field tensor coefficients remain unchanged by concomitant permutations of the unit electric field vectors at the different frequencies and the corresponding Cartesian indices.

Following the approach described in Part 7.5, it is possible to determine the symmetries arising from ordinary (o) and extraordinary (e) polarizations which then concern uniaxial crystals for any direction of propagation, as well as biaxial crystals but in the principal planes only. Among the $2^4 = 16$ possible configurations of polarization, as given in Table 8.1, are detailed in Figure 8.19 the cases of 5 configurations: (oooo), (eeee), (eooo), (oeee), and (ooee) for any direction of propagation out of the principal planes of uniaxial crystals. Note that the other matrices, *i.e.* (oeoo), (ooeo), etc. can be deduced from (eooo), (oeee), and (ooee) by the proper permutations of the Cartesian indices.

Note that the field tensor matrix corresponding to the sixteen possible configurations of polarization for a propagation out of the principal planes of biaxial crystals is the same matrix as that of the (eeee) configuration of uniaxial crystals while properly considering the involved unit electric fields out of the biaxial dielectric planes, shown in Figure 8.19.

The symmetries of both the field tensor and electric susceptibility tensor can lead to situations in which their contraction is nil for a given configuration of polarization and a given direction of propagation $\vec{u}$. This corresponds to the orthogonality between the field tensor and the electric susceptibility tensor, *i.e.* $F^{(3)}(\vec{u}) \perp \chi^{(3)}$, leading to a nil conversion efficiency even if phase-matching or quasi-phase-matching is achieved. In the case of the uniaxial optical class related to the hexagonal crystal classes $6mm$, $\bar{6}2m$, 622, and $\dfrac{6}{m}mm$, this orthogonality exists for the configurations of polarization (eooo), (oeoo), (ooeo), (oooe), (oeee), (eoee), (eeoe), and (eeeo) for all the directions of propagation.

As for the second-order case, the angular distributions of the field tensors coefficients $F_{ijkl}^{(3)}(\vec{u})$ and of the square of the effective coefficient $\chi_{eff}^{(3)}(\vec{u})$ can be established from the Cartesian coordinates of the components of the unit electric field vector at the interacting circular frequencies, using Eq. (7.125) to Eq. (7.129).

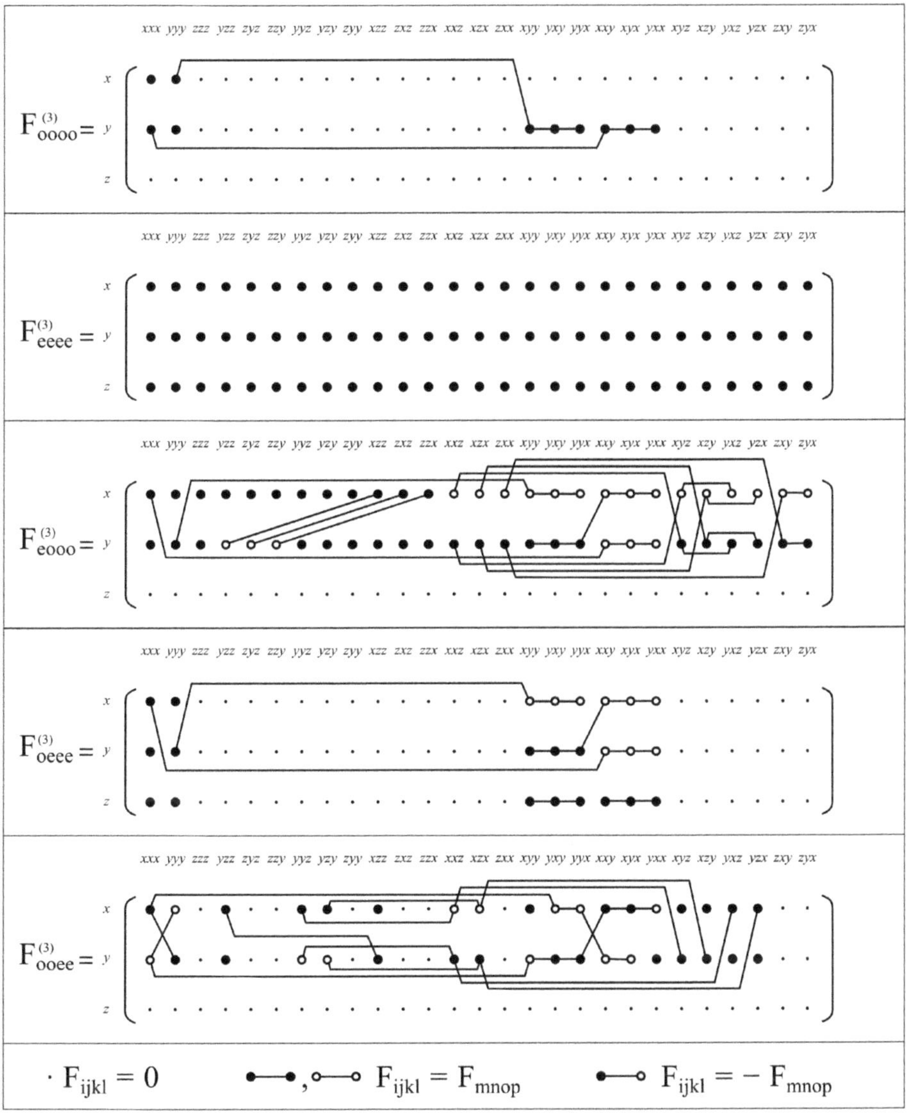

FIGURE 8.19 Field tensor matrices corresponding to the configurations of polarization (oooo), (eeee), (eooo), (oeee), and (ooee) for any direction of propagation out of the principal planes of uniaxial crystals. θ and φ are the angles of spherical coordinates of the direction of propagation of the four interacting waves.

8.6 THIRD-HARMONIC GENERATION AND THIRD-ORDER PARAMETRIC DOWN-CONVERSION

8.6.1 Third-Harmonic Generation

Phase-matched third-harmonic generation (THG) is a photon fusion process, so that the starting point for calculation is Eq. (8.10) with $\Delta\Phi = \pi / 2$, which gives:

$$\begin{cases} \dfrac{\partial A_1(Z)}{\partial Z} = -\varepsilon_0 \kappa_1 \chi_{eff}^{(3)} A_4(Z) A_2(Z) A_3(Z) \\ \dfrac{\partial A_2(Z)}{\partial Z} = -\varepsilon_0 \kappa_2 \chi_{eff}^{(3)} A_4(Z) A_1(Z) A_3(Z) \\ \dfrac{\partial A_3(Z)}{\partial Z} = -\varepsilon_0 \kappa_3 \chi_{eff}^{(3)} A_4(Z) A_1(Z) A_2(Z) \\ \dfrac{\partial A_4(Z)}{\partial Z} = \varepsilon_0 \kappa_4 \chi_{eff}^{(3)} A_1(Z) A_2(Z) A_3(Z) \end{cases} \tag{8.49}$$

From Eq. (7.43), it comes at each circular frequency ω_i (i = 1, 2, 3, 4):

$$\frac{\partial I_i(Z)}{\partial Z} = n_i \sqrt{\frac{\varepsilon_0}{\mu_0}} A_i(Z) \frac{\partial A_i(Z)}{\partial Z} \tag{8.50}$$

Then according to Eq. (8.49), the differential of the generated intensity writes:

$$\frac{\partial I_4(Z)}{\partial Z} = 4\pi \left(\frac{\mu_0}{\varepsilon_0}\right)^{3/2} \frac{\chi_{eff}^{(3)}}{(n_4)^2 \cos^2(\rho_4) \lambda_4} \sqrt{\frac{I_4(Z) I_1(Z) I_2(Z) I_3(Z)}{n_4 n_1 n_2 n_3}} \tag{8.51}$$

Then from Manley-Rowe's relations given by Eq. (8.17), it comes:

$$\frac{\partial I_4(Z)}{\partial Z} = \Lambda \sqrt{I_4(Z)[\gamma_1 - I_4(Z)][\gamma_2 - I_4(Z)][\gamma_3 - I_4(Z)]} \tag{8.52}$$

with

$$\begin{cases} \gamma_1 = \dfrac{\lambda_1}{\lambda_4} I_1(0) + I_4(0) \\ \gamma_2 = \dfrac{\lambda_2}{\lambda_4} I_2(0) + I_4(0) \\ \gamma_3 = \dfrac{\lambda_3}{\lambda_4} I_3(0) + I_4(0) \\ \Lambda = 4\pi \left(\dfrac{\mu_0}{\varepsilon_0}\right)^{3/2} \dfrac{\chi_{eff}^{(3)}}{(n_4)^2 \cos^2(\rho_4)} \dfrac{\lambda_4}{\sqrt{n_4 n_1 n_2 n_3} \sqrt{\lambda_4 \lambda_1 \lambda_2 \lambda_3}} \end{cases} \tag{8.53}$$

In the case of THG, it comes: $\lambda_1 = \lambda_2 = \lambda_3 (\equiv \lambda_\omega)$ and $\lambda_4 (\equiv \lambda_{3\omega}) = \lambda_\omega / 3$. Thus, Eq. (8.52) writes:

$$\frac{\partial I_{3\omega}(Z)}{\partial Z} = \Lambda \sqrt{I_{3\omega}(Z)} [I_\omega(0) - I_{3\omega}(Z)]^{3/2} \tag{8.54}$$

where $I_\omega(0) = \dfrac{I_\omega^{total}(0)}{3}$ is the intensity of each fundamental mode at the entrance of the crystal, $I_\omega^{total}(0)$ being the total fundamental intensity.

The integration of Eq. (8.54) over the crystal length L leads to the intensity of the generated third-harmonic wave $I_{3\omega}(L)$ knowing that $I_{3\omega}(0)=0$, *i.e.*:

$$\begin{cases} \Lambda L = \displaystyle\int_0^{I_{3\omega}(L)} \frac{dI_{3\omega}}{\sqrt{I_{3\omega}}\,[I_\omega(0)-I_{3\omega}]^{3/2}} \\ I_{3\omega}(L) = I_\omega(0)\dfrac{(\Lambda I_\omega(0)L)^2}{4+(\Lambda I_\omega(0)L)^2} \end{cases} \quad (8.55)$$

Then, from Eq. (8.55) and Manley-Rowe's relations from Eq. (8.18), it comes for the total fundamental intensity at the exit of the crystal:

$$I_\omega^{total}(L) = I_\omega^{total}(0) - I_{3\omega}(L) = 3I_\omega(0) - I_{3\omega}(L) \quad (8.56)$$

where $I_{3\omega}(L)$ is given by Eq. (8.55).

The curves corresponding to Eq. (8.55) and Eq. (8.56) are plotted Figure 8.20 in the case of TiO_2 rutile for a THG phase-matched with the configuration $[n_4 = n_o(3\omega), n_1 = n_2 = n_e(\omega), n_3 = n_o(\omega)]$ along the x-axis of the crystal at $\lambda_\omega = 1839.6\ nm$. The calculation is performed with the suited values of refractive indices and effective coefficient (Gravier 2006) [5].

Figure 8.20 shows that the pump is almost depleted at L = 20 mm since the conversion efficiency $I_{3\omega}(L)/I_{3\omega}^{total}(0)$ is equal to 91%.

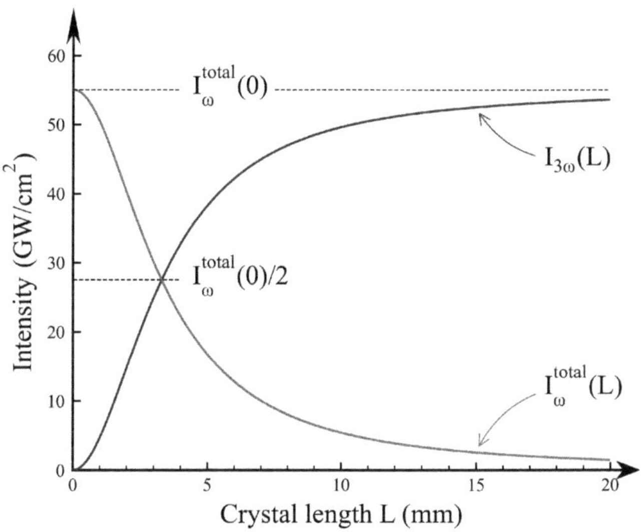

FIGURE 8.20 Fundamental intensity I_ω, with $I_\omega^{total}(0) = 55\ GW/cm^2$, and third-harmonic intensity $I_{3\omega}$ as a function of the crystal length L in a TiO_2 rutile crystal phase-matched along the x-axis of the crystal at $\lambda_\omega = 1839.6\ nm$.

8.6.2 Third-Order Parametric Down-Conversion

A phase-matched triple-photon generation (TPG) $\left[1/\lambda_4 \rightarrow 1/\lambda_1 + 1/\lambda_2 + 1/\lambda_3\right]$ is a photon scission process described by Eq. (8.10) in which $\Delta\Phi = -\pi/2$, which gives:

$$\begin{cases} \dfrac{\partial A_1(Z)}{\partial Z} = \varepsilon_0 \kappa_1 \chi_{eff}^{(3)} A_4(Z) A_2(Z) A_3(Z) \\ \dfrac{\partial A_2(Z)}{\partial Z} = \varepsilon_0 \kappa_2 \chi_{eff}^{(3)} A_4(Z) A_1(Z) A_3(Z) \\ \dfrac{\partial A_3(Z)}{\partial Z} = \varepsilon_0 \kappa_3 \chi_{eff}^{(3)} A_4(Z) A_1(Z) A_2(Z) \\ \dfrac{\partial A_4(Z)}{\partial Z} = -\varepsilon_0 \kappa_4 \chi_{eff}^{(3)} A_1(Z) A_2(Z) A_3(Z) \end{cases} \tag{8.57}$$

Using the same approach as that followed in Part 6.5.1, it comes for $I_1(Z)$:

$$\frac{\partial I_1(Z)}{\partial Z} = \Lambda \sqrt{I_1(Z)[\gamma_4 - I_1(Z)][\gamma_2 + I_1(Z)][\gamma_3 + I_1(Z)]} \tag{8.58}$$

with

$$\begin{cases} \gamma_4 = \dfrac{\lambda_4}{\lambda_1} I_4(0) + I_1(0) \\ \gamma_2 = \dfrac{\lambda_2}{\lambda_1} I_2(0) - I_1(0) \\ \gamma_3 = \dfrac{\lambda_3}{\lambda_1} I_3(0) - I_1(0) \\ \Lambda = 4\pi \left(\dfrac{\mu_0}{\varepsilon_0}\right)^{3/2} \dfrac{\chi_{eff}^{(3)}}{(n_1)^2 \cos^2(\rho_1)\sqrt{n_4 n_1 n_2 n_3}} \dfrac{\lambda_1}{\sqrt{\lambda_4 \lambda_1 \lambda_2 \lambda_3}} \end{cases} \tag{8.59}$$

Then with the limit conditions $I_1(0) = 0, I_2(0) \neq 0, I_3(0) \neq 0, I_4(0) \neq 0$, which corresponds to the photonic diagram of Figure 1.24 by permuting ω_1 and ω_2, the intensities of the four interacting waves can be expressed on the basis of the Jacobi elliptic functions $cn(u\,|\,1-m)$ and $sn(u\,|\,1-m)$ (Fève 2002) [1]:

$$
\begin{cases}
I_1(L) = \dfrac{\gamma_3\gamma_4 sn^2(aL\,|\,1-m)}{D} \\
I_2(L) = \dfrac{I_2(0)(\gamma_3+\gamma_4)[sn^2(aL\,|\,1-m)+cn^2(aL\,|\,1-m)]}{D} \\
I_3(L) = \dfrac{I_3(0)(\gamma_3+\gamma_4)[sn^2(maL\,|\,1-m)+cn^2(aL\,|\,1-m)]}{D} \\
I_4(L) = \dfrac{I_4(0)(\gamma_3+\gamma_4)cn^2(aL\,|\,1-m)}{D} \\
\\
D = \gamma_3 m sn^2(aL\,|\,1-m)+(\gamma_3+\gamma_4)cn^2(aL\,|\,1-m)
\end{cases}
\tag{8.60}
$$

with

$$
\begin{cases}
m = \dfrac{\gamma_2(\gamma_3+\gamma_4)}{\gamma_3(\gamma_2+\gamma_4)} \\
a = \dfrac{\Lambda}{2}\sqrt{\gamma_3(\gamma_2+\gamma_4)}
\end{cases}
\tag{8.61}
$$

The case of the TPG $\left[1/\lambda_4 \rightarrow 1/\lambda_1 + 1/\lambda_2 + 1/\lambda_3\right]$ stimulated over the two modes at λ_2 and λ_3 in KTP is plotted in Figure 8.21 (Bencheikh 2022) [6].

The pump can be depleted at L around 1.3 cm according to Figure 8.21.

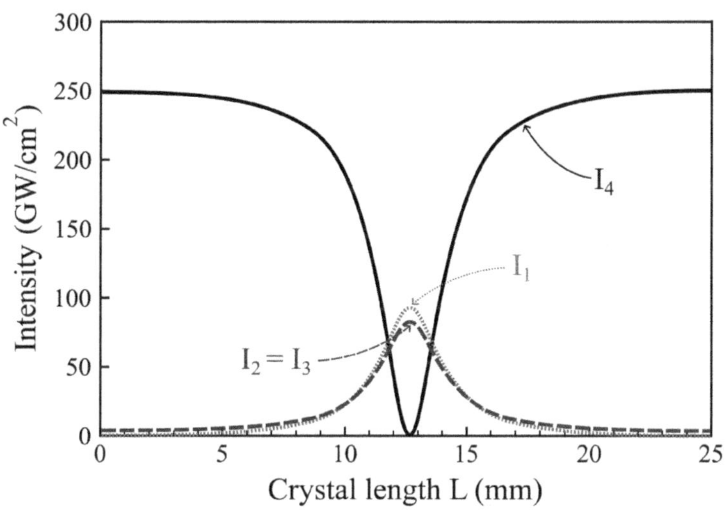

FIGURE 8.21 Triple-photon generation $\left[1/\lambda_4 \rightarrow 1/\lambda_1 + 1/\lambda_2 + 1/\lambda_3\right]$ with $\lambda_4 = 532\ nm$, $\lambda_1 = 1449\ nm$, $\lambda_2 = \lambda_3 = 1681\ nm$ in a KTP crystal phase-matched along the x-axis. I_4, I_1, I_2, I_3 are the intensities of the four interacting waves, with $I_4(0) = 250\ GW/cm^2$, $I_1(0) = 0\ GW/cm^2, I_2(0) = I_3(0) = 3.25\ GW/cm^2$.

8.7 SUMMARY

The tensor calculations are heavier in the case of third-order nonlinear processes than those of second-order nonlinear interactions, but there is no conceptual novelty, at any level of the modeling. On the other hand, third-order nonlinear processes do open up a wider field of possibilities: firstly, because they are compatible with the existence of a center of inversion so that the thirty-two crystalline classes are allowed to be active; secondly, because of a higher tensor rank, the number of possible polarization configurations is brought up to sixteen. Quasi-phase-matching (QPM) is rarely used today, due to the impossibility of inverting the sign of $\chi^{(3)}_{zzzz}$ that is the highest nonlinear coefficient in the case of ferroelectric crystals for which the electric poling is possible. But there are many other coefficients for which it is possible, which allows us to realistically envisage the realization of QPM for performing third-order nonlinear optics. Regarding conversion efficiency, the examples of third-harmonic generation and triple-photons generation show that it is possible to reach significant values.

REFERENCES

[1] Fève J. P., Boulanger B., and Douady J. 2002. Specific properties of cubic optical parametric interactions compared to quadratic interactions. Physical Review A 66: 063817-1-11.

[2] Boulanger B., Fève J. P., and Marnier G. 1993. Field factor formalism for the study of the tensorial symmetry of the 4-wave non-linear optical parametric interactions in uniaxial and biaxial crystal classes. Physical Review E 48(6): 4730–4751.

[3] Boulanger B. and Zyss J. 2013. Non-linear optical properties—chapter 1.7. In International Tables for Crystallography, Vol. D: Physical Properties of Crystals, edited by A. Authier., pp. 181–222. Dordrecht, Netherlands: International Union of Crystallography, Kluwer Academic Publisher.

[4] Duboisset J., Boulanger B., Brasselet S., Segonds P., and Zyss J. 2024. Nonlinear optics through the field tensor formalism. Laser and Photonics Reviews. https://doi.org/10.1002/lpor.202400411.

[5] Gravier F. and Boulanger B. 2006. Cubic parametric frequency generation in rutile single crystal. Optics Express 14(24): 11715–11720.

[6] Bencheikh K., Cenni M. F. B., Oudot E., Boutou V., Félix C., Prades J. C., Vernay A., Bertrand J., Bassignot F., Chauvet M., Bussières F., Zbinden H., Levenson A., and Boulanger B. 2022. Demonstrating quantum properties of triple photons generated by χ(3) processes. European Physical Journal D—Topical Issue on "Quantum Optics of Light and Matter: Honoring Alain Aspect" 76: 186. https://doi.org/10.1140/epjd/s10053-022-00514-3.

Index

Note: Page numbers in *italics* indicate a figure and page numbers in **bold** indicate a table on the corresponding page.

D

E

N

O

P

T

U

For Product Safety Concerns and Information please contact our EU representative GPSR@taylorandfrancis.com
Taylor & Francis Verlag GmbH, Kaufingerstraße 24, 80331 München, Germany

www.ingramcontent.com/pod-product-compliance
Ingram Content Group UK Ltd.
Pitfield, Milton Keynes, MK11 3LW, UK
UKHW022013111225
465993UK00001B/18

* 9 7 8 1 4 9 8 7 7 4 5 6 7 *